BOTANICHESKII INSTITUT IM. V.L. KOMAROVA AKADEMII NAUK SSSR

Botanical Institute of the Academy of Sciences of the USSR

FLORA of the U.S.S.R.

(Flora SSSR)

(Series initiated by V.L. Komarov)

Volume XXIV

Dipsacaceae, Cucurbitaceae, Campanulaceae

Chief Editor B. K. Shishkin
Volume Editors B.K. Shishkin and E.G. Bobrov

Compiled by
E.G. Bobrov, An.A. Fedorov, S.G. Gorshkova, and I.T. Vasil'chenko

Izdatel'stvo Akademii Nauk SSSR
Moskva-Leningrad
1957

Translated from Russian

Israel Program for Scientific Translations
Jerusalem 1972

TT 72-50003

Published Pursuant to an Agreement with
THE SMITHSONIAN INSTITUTION
and
THE NATIONAL SCIENCE FOUNDATION, WASHINGTON, D.C.

IPST Cat. No. 600674

ISBN 0 7065 1254 5

Translated by R. Lavoott
Edited by Prof. J. Lorch and U. Plitman

Printed in Jerusalem by Keter Press
Binding: Wiener Bindery Ltd., Jerusalem

Available from the
U.S. DEPARTMENT OF COMMERCE
National Technical Information Service
Springfield, Va. 22151

VI/7/3

TABLE OF CONTENTS

SUBJECTS AND CONTRIBUTORS

Indexes	Editorial Staff
Families Morinaceae, Dipsacaceae	Arranged by E.G. Bobrov
Family Cucurbitaceae	Arranged by I.T. Vasil'chenko
Family Campanulaceae	Arranged by A.A. Fedorov
Family Lobeliceae	Arranged by S.G. Gorshkova

Addenda – Descriptiones plantarum novarum in tomo XXIV Florae URSS commemoratarum

The plates were drawn by the following artists: N.A. Moiseeva – I–VIII; E.S. Gaskevich – IX–XXVII.

SYSTEMATIC INDEX OF SPECIES IN VOLUME XXIV*

Family CLV. **Morinaceae** (Dum.) Van Tiegh.

Genus 1409. **Morina** L.

Section I. *Cryptothladia* Bge.

Family CLVI. **Dipsacaceae** Lindl.

Tribe I. **Knautieae** Van Tiegh.

Genus 1410. **Knautia** L.

Subgenus I. *Trichera* (Schrad.) Rouy

Section I. *Eutricherae* Szabo

Subsection I. Sylvaticae Krasan

Tribe 2. **Dipsaceae** Van Tiegh.

Genus 1411. **Dipsacus** L.

* [This index has been reproduced photographically from the Russian original.]

** [Russian page numbers appear in the left-hand margin of the text.]

Tribe 5. **Phyteumateae** Fed.

Genus 1440. **Phyteuma** L.

Section 1. *Spicata* R. Schultz

Section 2. *Tetramera* (R. Schulz) Fed.

Genus 1441. **Asyneuma** Griseb. et Schenk

Section 1. *Eupodanthum* (Boiss.) Fed.

Section 2. *Sceptrum* Fed.

Genus 1442. **Sergia** Fed.

Genus 1443. **Cryptocodon** Fed.

Genus 1444. **Cylindrocarpa** Rgl.

Genus 1445. **Legouzia** Durand

PREFACE

Volume XXIV of the Flora of the USSR is being published before Volume XXIII. The two volumes were prepared simultaneously, but as Volume XXIV was completed first the editorial board decided to go ahead with publication. Volume XXIII, published slightly later, does not upset the order of things, nor does it interfere with the general enumeration of the genera and species.

There are now data on the taxonomy of Campanulaceae (A. A. Fedorov). In addition to six new tribes, five new genera are described. There are eight new species of Campanula and twelve new subsections. Also subsections have been established and new species described for other genera.

The hitherto neglected Dipsacaceae have been re-examined. The separation of the genus Morina in the family Morinaceae has resulted in a more precise determination of the family boundaries. A new section was introduced for Scabiosa; some new species have been described and previously described species of Scabiosa have been considerably reduced. Treatment of Cucurbitaceae, with its wealth of species of cucurbit, vegetable and ornamental plants, is by I. T. Vasil'chenko. The author has purposely refrained from dwelling in detail on Curcurbita, Melo, and Citrullus, which are presented in conformity with the Flora series.

1 Family CLV. **MORINACEAE** (DUM.) VAN TIEGH.*

Flowers irregular, bilaterally symmetrical, monoecious, forming spurious whorl in axils of whorled terminal bracts, comprising opposite cymose inflorescences, short-pediceled, each flower with campanulate-tubular involucel; outer calyx cupular (similar to Dipsacaceae but without grooves, ribs or pits), apparently produced by the concrescence of 4 bracts, i. e., 4-merous in origin. Ovary inferior, laterally compressed, planoconvex, with 1 midvein and 2 lateral veins on plane side, the same on the convex side, making it 6-lobed (and not 8-lobed as in Dipsacaceae); the bilateral transverse symmetry of this sphere becomes more pronounced in the fruits. Ovary crowned with calyx, campanulate below, divided into 2 leaf-shaped lobes above, more or less 2-lipped. The calyx is interpreted as a reduced 5-merous posterior sepal. Corolla zygomorphic, 2-lipped, the distinct 5-petal lobes alternating with the five 'sepals' of the zygomorphic calyx. Androecium typically 5-merous, with 2 fertile lateral posterior stamens, and a reduced median and 2 lateroventral stamens, the latter sometimes enclosed in tube, rudimentary, the median stamen obsolete. Anthers 4-locular, dehiscing longitudinally inside flower. Pollen grains (Table 1, Figure 3) tricolpate or triporate, ovoid or elongate-ellipsoid, in polar view rounded, rarely rounded-triangular; polar diameter 120–155 (200)μ equatorial 80–140μ; pores of complex structure, clavate or dome-shaped, with equatorial girdle; exine not sculptured or irregularly tuberculate in equatorial zone, fine structure indistinct. The pollen grains of Morinaceae cannot be related to those of Dipsacaceae or Valerianaceae.

2 Pistils of 3 carpels, 1 ventral and 2 posterior, connate to tip of style, terminating in disklike stigma. Of the 3, there normally develops only 1 of the lateral posterior pair. Hence ripe ovary 1-locular with single pendulous campylotropous ovule. Pistils zygomorphic, ovary spherical, transversely symmetrical, calyx, corolla and androecium also zygomorphic, but longitudinally symmetrical. Thus, summing up, the structure of the flowers is as follows: 4-merous involucre, 5-merous calyx with 1 reduced sepal, 5-merous corolla, 5-merous androecium (3 stamens reduced), 3-merous pistils with 2 sterile carpels. Fruit an achene, freely enclosed in broad involucre, crowned with calyx, tangentially symmetrical, laterally flattened, the flat internal side with deep groove, covered by fold of pericarp; the thin brown seed coat has vascular bundles at its tangential side; the viscous, mealy, compact, starchless, endosperm encloses the embryo with upper radicle and 2 broad thin inward turning cotyledons occupying the sides with vascular bundles.

* Treatment by E.G. Bobrov.

Morina should be removed from the family Dipsacaceae on the basis of numerous features in the structure of flowers: the 4-merous involucel, 5-merous calyx, corolla and androecium (the first and last of which are reduced), 4-merous gynoecium reduced to 1-locular, and also on the basis of the structure of the inflorescences and structure of the pollen.

Morina is distinguished from the Valerianaceae most notably by the absence of gamophyllous involucels around each flower, and by the albuminous seeds.

In a comparison of Morina with Triplostegia Wall. ex DC and Hoeckia Engl. et Gr., van Tiegh. (in Ann. Sc. Nat, ser. 9, X (1909) 200) proposed that because of the character of the involucel the latter genera should be separated into a specific tribe, Triplostegieae, within the Morinaceae. However, the differences between Morina and these genera in the structure of the flower, the simple arrangement of the calyx, the character of the inflorescence and the pollen grains are so striking that such an approach would disrupt the natural character of the Morinaceae. We submit that the genera Triplostegia Wall. ex DC. and Hoeckia Engl. et Gr. should be included in the family Triplostegiaceae, which is closer to the Valerianaceae than to the Morinaceae.

Morinaceae are a monotypic family.

Genus 1409. **MORINA** * L.

L. Sp. pl. (1753) 28.— Asaphes Spreng. Syst. IV, 2 Curae post. (1827) 222.

Flowers short-peduncled, loosely surrounded by tubular-campanulate involucels with prickly spinose-toothed margins, the two opposite prickles
3 the longest. Inferior ovary planoconvex, truncate, crowned with calyx. Calyx broadly campanulate below, 2-lobed above, lobes leaf-shaped, emarginate or 2-toothed, exerted from involucel. Corolla 3.5–5 cm or ca. 1 cm long, with long tube and 2-lipped limb, the upper lip 2-lobed, the lower 3-lobed. Fertile stamens 2, adnate to corolla at throat, the 2 rudimentary ones sometimes sterile, enclosed in tube. Style long, sometimes longer than stamens, with disklike stigma. Achene planoconvex, obliquely truncate, light brown, tuberculate-rugose. Perennials with simple stem and whorled leaves, usually pinnatifid or spinose-toothed, rarely entire. Flowers numerous, forming spurious whorl in axils of whorled terminal bracts, whorls scattered, together forming an interrupted spicate inflorescence.

Type species: M. persica L.

The 17 species of this genus are distributed in the mountains of southern Asia, from Greece and European Turkey to the provinces of Sichwan and Yunnan in S. China. In the South, the genus is represented in Syria and S. Iran, in the North, in W. Tien Shan and Dzungarian Ala-Tau.

1. Radical and cauline leaves cut-lobed and spinous-toothed or pinnatilobate with spinous-toothed lobes . . . . 2.
+ Radical and cauline leaves entire, acuminate, very rarely small-toothed, terminating in prickle . . . . 4. **M. kokanica** Rgl.

* Named by Tournefort in 1703 in honor of Louis Morin (1635–1715), member of the French Academy, who was associated with the Paris Botanical Garden.

2. Corolla ca. 1 cm long, barely exerted from involucel; calyx ca. 8 mm long in fruit, exerted from involucel for 3 mm; stems 20–50 cm . . . . 1. **M. parviflora** Kar. et Kir.
\+ Corolla 3.5 or ca. 5 cm long, 3 to 5 times as long as involucel; calyx 10–14 or ca. 20 mm long, exerted from involucel for 5–7 mm; stems 50–70 cm . . . . 3.
3. Corolla pale yellow, ca. 3.5 cm long; involucel 12–14 mm long, its largest prickle as long as tube, and the one opposite nearly one-third as long; calyx ca. 2 cm long, lobes unequal in upper part and short-cleft, or 2-toothed . . . . 2. **M. lehmanniana** Bge.
\+ Corolla pale purple, ca. 5 cm long; involucel 10–13 mm long, the two largest opposite prickles as long as tube or even longer, 13–15 mm long; calyx 10–14 mm long, lobes obtuse, slightly cleft, or barely 2-toothed . . . . 3. **M. persica** L.

4 Section 1. CRYPTOTHLADIA Bge. in Mém. Ac. Sc. Pétersb. par div. sav. VII (1854) 321. – Corolla ca. 1 cm long, barely exerted from involucel; fertile stamens 2.

1. M. parviflora Kar. et Kir. in Bull. Soc. Nat. Mosc. XV (1842) 373. – Perennial; root 15 cm thick in upper part; stems 20–50 cm, solitary or 2, covered at base with remnants of numerous leaves from previous years, 0.6–1 cm across below, lower internodes glabrous, upper spreadingly pubescent; radical leaves few, usually 5–6, coriaceous, glabrous, 10–15 cm long, 1–1.5 cm wide, linear-lanceolate, with cut-spinous-toothed margins; cauline leaves usually 2-whorled, 5–6 leaves per whorl, like lower leaves 6–8 cm long, 1 cm wide, with pubescent margins; peduncles 3–4-whorled in upper part of stem, spreading in fruit; bracts usually in whorls of 5, like cauline leaves 3–3.5 cm long, narrowly lanceolate, with spinous-toothed margins, more densely pubescent beneath, leaves free in lower whorls, proximally connate in upper whorls. Involucel cupular, tubular-infundibular, 8–10 mm long, sessile, netted-veined, with 6–8 obscure longitudinal veins, with scattered crisp hairs, with 6–8 marginal prickles of which 1 ca. 8 mm long, the opposite one slightly shorter or half as long, the intermediate ones much shorter, irregular; fruit crowned with calyx, deciduous at end, ca. 8 mm long, exerted from outer calyx for 3 mm, more than half connate in lower part, campanulate, 2-lobed for more than half, each lobe cleft-spinous-toothed distally, lobes in upper part more densely pubescent above; corolla dull red, ca. 1 cm long, barely exerted from involucel (calyx weakly developed at anthesis), limb 2-lobed, upper lip slightly larger, nearly 3-toothed, the lower barely 2-lobed; 2 fertile stamens in corolla tube, 2 rudimentary ones at its base; achene obscurely tuberculate. August-September. (Plate I, Figure 1).

In alpine zone, 3,000–4,000 m. – Centr. Asia: Dzu-Tarb. (Dzungarian
7 Ala-Tau), T. Sh. (E. Kirghizia: Atbash, Kokdzhar, Naryn Range). **Gen. distr.:** Dzu.-Kash. Described from the upper reaches of the Sarkand River in the alpine zone of Dzungarian-Ala-Tau. Type in Leningrad.

Note. The authors of this species mention its unpleasant odor, reminiscent of some of Labiatae.

(5)

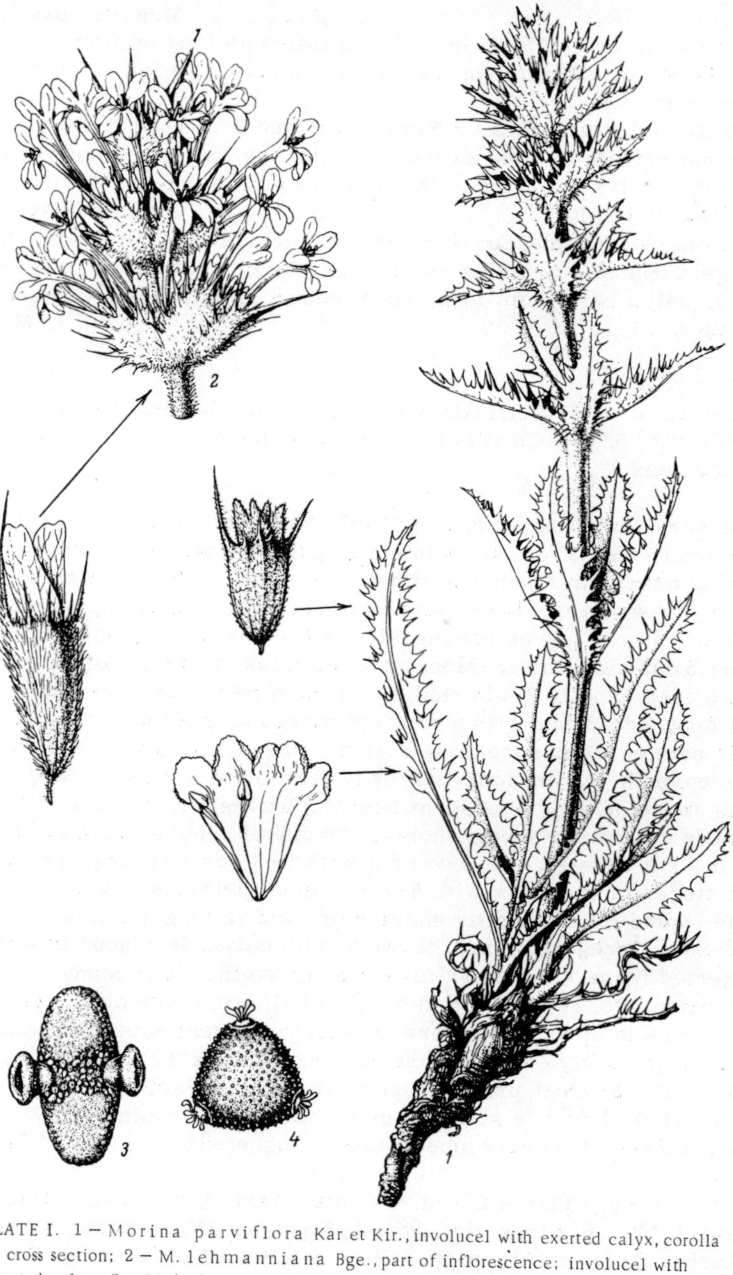

PLATE I. 1 — Morina parviflora Kar et Kir., involucel with exerted calyx, corolla in cross section; 2 — M. lehmanniana Bge., part of inflorescence; involucel with exerted calyx; 3 — M. kokanica Rgl., typical pollen grains of Morinaceae; 4 — Dipsacus pilosus L., typical pollen grains of Dipsacaceae. (These drawings of the pollen by L. Vinokurova, from her thesis; Leningrad University, 1954.)

Section 2. DIOTOCALYX DC. Prodr. IV (1830) 644; Bge. in Mém. Ac. Sc. Pétersb. par div. sav. VII (1854) 321. — Corolla 3.5—5 cm long, nearly four times as long as involucel; fertile stamens 2. Diploid number of chromosomes in this section: 2n = 34 (Kachidze in Planta 7, 4, 1929, 500).

2. M. lehmanniana Bge. in Mém. Ac. Sc. Pétersb. par div. sav. VII (1854) 321; Boiss. Fl. or. III, 114.

Perennial; root 2 cm thick in upper part; stems 50—70 cm, simple, straight, 0.6—1 cm thick in lower part, in upper part densely covered with spreading white hairs, and more intensely colored; radical leaves 10—25, glabrous, 10—20 (25) cm long, 2 cm wide, stiffly coriaceous, pinnatilobate, 10—15 lobes at each side, each with 2—5 acute stiff spines, limb tapering to rather wide petiole with softer marginal spines; cauline leaves 2-whorled, 5—6 whorls per leaf, similar to radical leaves but much shorter, sessile and glabrous. Peduncles 4—5-whorled in upper part of stem; lower part, sometimes also the terminal part, with fewer flowers, the median part with ca. 8—15 flowers; bracts 5-whorled, 5 cm long, narrowly lanceolate, tapering to acute spine, with shorter marginal simple stiff spines, densely covered beneath with long soft hairs and above with short scattered appressed hairs; involucel cupular, tubular-infundibular, 12—14 mm long, on stipe 4 mm long, netted-veined, with 10 obscurely longitudinal veins, rather densely covered with white hairs, margin with 10 prickles, one of which as long as tube, the opposite one nearly one-third as long, the others unequal and much shorter; calyx crowned with fruit, ca. 2 cm long, campanulate for one-third, covered with long white hairs at base, divided for two-thirds into 2 glabrous liguliform lobes becoming short-cleft or inflated; corolla pale yellow, tube ca. 1 mm across, ca. 3.5 cm long, soft-haired outside, limb 2-lipped, upper lip 2-lobed, the lower 3-lobed, lobes obovate, 1 cm long, 0.5 cm wide; stamens 2, fertile, attached between lobes to upper and lower lips, 2 rudimentary sterile ones below throat, between lateral and median lobes of lower lip; achene subtrihedral, slightly rugose-tuberculate. August. (Plate I, Figure 2.)

8 Subalpine zone, from 2,500 to 3,500 m. — Centr. Asia: Pam.-Al. (from Zaravshan to Shugnan, inclusively). Endemic. Described from upper Zaravshan, collected near the village of Fon. Type in Paris.

3. M. persica L. Sp. pl. (1753) 28; Boiss. Fl. or III, 114. — M. orientalis Mill. Gard. Dict. ed. VIII (1768). — M. virticillata Moench, Meth. pl. Suppl. (1794) 186. — M. Tournefortii Jaub. et Sp. Ill. pl. or. V (1853—1857) 31. — M. graeca Jaub. et Sp. l. c. 32. — M. Aucheri Jaub. et Sp. l. c. — M. persica ssp. turcica Hal. in Oester. bot. Zeitschr. XLI (1891) 409. — M. turcica Deg. et Hal. ex Čelak. in Bot. Jahrb. XVII (1893) 396. — M. spectabilis Gontsch. in herb. — Ic.: Jaub. et Sp. l. c. tab. 429; Bot. Jahrb. XVII, tab. IX.

Perennial; root straight, ca. 1.5 cm thick in upper part; stems 50—70 cm, solitary, with rosette of leaves of the next year, grooved-faceted, 7 mm thick at base, colored, subglabrous in lower part, white-hairy and glandular in upper part; radical leaves 15—25 cm long, 1—2.5 cm wide, oblong-lanceolate, coriaceous, glabrous, cut-lobed, spinous-toothed, tapering to broad-winged petiole with spinous margin; cauline leaves 3—4-whorled, usually quadrifoliate, similar in shape, smaller and sessile. Peduncles 4–5-whorled, whorls 5–8 cm apart at end of flowering; bracts 4-whorled, 4–5 cm long, ovate-lanceolate, tapering to spine, with 3–5 marginal spines, netted-veined, covered outside with predominantly short hairs; involucel short-stiped, 10–13 mm long, campanulate-tubular, with 12–15 unequal prickles of which 2 opposite, the largest nearly as long as tube or much longer, 10–15 mm long; calyx 10–17 mm long, 5–6 mm longer than spines but usually shorter, white-haired at base, connate for not more than lower one-third, campanulate, with 2 obtuse lobes, or slightly cleft or barely 2-toothed, netted-veined, very short-haired outside; corolla pale purple, lower lip more intensively colored, 5 cm long, tube long, thin, 3.5–4 cm long, limb 2-lipped 12–15 mm long, upper lip 2-lobed, the lower 3-lobed; stamens 2, fertile, with long filaments, alternating with lips of limb attached to rictus; fruit planoconvex, tuberculate-rugose, 6–7 mm long, ca. 3 mm wide. Fl. August, Fr. September.

Stony slopes and outcrops in subalpine zone at 2,400 to 3,200 m. – Centr. Asia: Pam.-Al. Dasht-i-Zhum District: Kugi-Pur and Kugi-Frush mountains and the Sufai-Mir-Atau mountains (to the north of Bal'dzhuan). – **Gen. distr.:** Bal.-As. Min., Arm.-Kurd., Iran (to Gindukush in the north). Described from Isfagan District. Type in London.

9 Note. The distribution area of this species extends from Greece and S. Bulgaria in the west to Afghanistan in the east; the habitat closest to the southern borders of the USSR is E. Anatolia. It is not encountered in Transcaucasia. S. Tadzhikistan represents the extreme northeast limit of the distribution area. Here it is readily distinguished from M. lehmanniana Bge., endemic to Pam.-Al., by its purple flowers, as against the yellow flowers of the latter.

4. M. kokanica Rgl. in Bull. Soc. Nat. Mosc. XL, 1 (1867) 14; Vved. in Sched. ad Herb. Fl. As. Med. XIV–XX, 83. – M. Sewerzowii Rgl. l. c. tab. I (IV). – Ic.: Rgl. l. c. – Exs.: HFAM, No. 477 a, b.

Perennial; root straight, 1.5 cm thick in upper part; stems 60–80 cm, solitary, with sterile rosette of leaves of the coming year, grooved-faceted, completely glabrous below, profusely covered with simple hairs, mixed with glandular ones in inflorescence; radical leaves 25–35 cm long, 2–3 cm wide, coriaceous, oblong-lanceolate, entire, acuminate, tapering to long narrowly winged proximally broadened petiole, with single prominent nerve, terminating in prickle, very rarely margin small-toothed; cauline leaves similar to radical, in 2–3 whorls of 4, sessile in upper whorls, in lower with winged petioles, proximally connate; upper whorled leaves 7–12 cm long, 1–1.5 cm wide, the lower 20–25 cm long, 1.5–2.5 cm wide. Peduncles in upper part of stem 4–5-whorled, 5–7 cm apart; bracts 4-whorled, ovate, acuminate, tapering to spine, with 2–3 spines at each side, more densely white-hairy beneath; involucel 10–12 mm long, short-stipitate, tubular, truncate, with 11–13 unequal prickles of which 2 opposite, the largest nearly as long as but not longer than half the tube, i. e., 5–6 mm long; calyx 12–14 mm long, exerted for 7–8 mm from the outer calyx, white-hairy at base, lower one-third adnate, campanulate, upper two thirds 2-lobed, the lobes cleft and unequally 2-toothed, covered on both sides with scattered white hairs, rather distinctly netted-veined inside; corolla pale purple or pink, 4 cm long, tube long, thin, limb 11–13 mm long, 2-lipped, upper lip 2-lobed, the lower 3-lobed, more intensively colored; stamens 2, filaments long, attached to rictus of corolla; style slightly longer than stamens; achene planoconvex, ca. 7 mm long, 3 mm wide. wide. August–September. (Plate I, Figure 3.)

10 Savin juniper and subalpine zone, 1,500–3,000 m. – Centr. Asia: T. Sh. (SE T. Sh.), Pam.-Al. (only W. Kashka-Dar'ya, Dzhaus – reported by A. I. Vvedenskii). Endemic. Probably described from the Severtsov collections made in the vicinity of the village of Khodzhakent, below Chirchik. Type in Leningrad.

Family CLVI. **DIPSACACEAE** Lindl.*

Flowers monoecious, with organs arranged in 5 whorls, connate in lower part, 4 whorls usually 4-merous and gynoecium 2-merous, partially reduced in central flowers of capitulum, actinomorphic, peripheral ones usually zygomorphic (pistils always zygomorphic). Involucel (outer calyx) unicolored, nearly always sessile, sacciform or infundibular-tubular, more or less ribbed or plicate, either 4- or 8-ribbed, without corona, or 8-ribbed with corona, longitudinally ribbed-plicate or pitted-grooved, rarely tetrahedrally prismatic; involucel presumably formed by the concrescence of 4 bracts arranged crosswise. Calyx on top of ovary, 4- or 5-merous, cupular or saucer-shaped, according to genus, with spinous-toothed or setaceous margin. Corolla with long tube and 4- or 5-lobed limb, usually enlarged in marginal flowers. Androecium of 4 equal filiform stamens, attached to corolla tube, alternating with the lobes; where corolla 5-lobed, the stamen between the two posterior lobes is absent; filaments dorsifix, 3-locular, anthers introrse, dehiscing longitudinally. Pollen grains 3-colpate, 3-colporate, or rarely 4-colpate, nearly spherical, orbicular-triangular or 3-lobed, the greatest diameter 70–115 μ; pores covered with lid having digitate processes; exine overlapping, with prickly ornamentation and a distinctly columnar inner structure. The pollen grains of Dipsacaceae resemble those of Valerianaceae. Ovary inferior, with 2 carpels, anatropous pendulous ovule of the anterior carpel usually fertile, the posterior carpel ovuleless, reduced to the style. In some cases the anterior carpel has a style of its own, in others it contains only the ovule. In the former the style of the anterior carpel concresces for its whole length with that of the sterile posterior carpel. As a result of unequal development, the surface of the broadened stigma tapers somewhat. In the latter cases the
11 anterior carpel lacks a style, which is then formed only by the sterile posterior one. Fruit an achene topped by the calyx, in a compactly enclosing involucel, seeds starchless with viscous endosperm; embryo erect, with upper radicle and 2 equal plicate epigeal cotyledons. Inflorescence heads surrounded with bracts of the choriphyllous involucre; these more or less equal, numerous, saucer-shaped, or the outer bracts larger, green and leaf-shaped, or bracts bearing spinous processes. Receptacle nearly flat, hemispherical, narrowly conical or fusiform, bearing scarious bracts (paleae); these rarely absent, and then receptacle covered with bristles. Perennial (sometimes woody at base), biennial or annual herbs, with opposite exstipulate leaves, simple or truncate proximally, connate in pairs.

* Treatment by E.G. Bobrov.

The distribution of the members of this family is associated with the countries of the Ancient Mediterranean; some extend to China and Japan in the east, barely reaching the taiga zone. Others grow in the mountains of East Africa and beyond, to the southern part of that continent. Dipsacaceae do not grow in the wild state in the Western hemisphere.

Nine genera embracing about 250 species. The type genus is Dipsacus L.

Key to Genera

1. Receptacle covered with bristles, without paleae; involucel 4-ribbed . . . . 1410. **Knautia** L.
\+ Receptacle with paleae, each flower in axil of scale; involucel 8-ribbed . . . . 2.
2. Stems and peduncles spinous; receptacle conical; calyx saucer-shaped, 4-toothed-lobed, densely pubescent . . . . 1411. **Dipsacus** L.
\+ Stems and peduncles not spinous; receptacle flat, vaulted or even fusiform . . . . 3.
3. Involucral bracts 1–2 seriate . . . . 4.
\+ Involucral bracts multiseriate, imbricate; calyx saucer-shaped, densely pubescent, margin with numerous small teeth . . . . 1412. **Cephalaria** Schrad.
4. Calyx with 10 bristles; involucel 8-grooved . . . . 1413. **Pterocephalus** Vaill.
\+ Calyx with 5 teeth and bristles or without . . . . 5.
5. Corolla 4-lobed . . . . 6.
12 + Corolla 5-lobed . . . . 1416. **Scabiosa** L.
6. Calyx without teeth and bristles; involucel urceolate or terete . . . . 1415. **Succisella** Beck.
\+ Calyx with 5 teeth and bristles; involucel prismatic . . . . 1414. **Succisa** Neck.

Tribe 1. **KNAUTIEAE** Van Tiegh. in Ann. Sc. Nat. sér. 9, X (1909) 169. – Receptacle covered with bristles, without paleae; involucel short-stiped, 4-ribbed, without corona; calyx 4-merous, corolla 3-lobed, style double.

Genus 1414. **KNAUTIA*** L.

L. Gen. pl. ed. 5 (1754) 47, s. l.; Coult. Mém. Dips. (1824) 73; Szabo, Monogr. gen. Knautia (1911) 173.– Scabiosa L. Gen. pl. ed. 5 (1754) 43, pro parte.– Trichera Schrad. Cat. sem. hort. Götting. (1814).

Flowers in very short-peduncled heads, covered with bristles, without scales; involucral bracts numerous, herbaceous, rarely few at first erect, becoming recurved; heads with numerous, rarely few flowers, the pistillate small, barely radial symmetrical, the male larger, more radially symmetrical; corolla tetrahedral, with unequal lobes; stamens 4, filaments exserted

* Named after Christian Knaut, naturalist and physician (Halle, 1654–1716).

from corolla; style filiform, also exserted from corolla, stigma short-2-lobed; ovary 1-ovuled; achene short-stiped, tetrahedral, truncate, crowned with barely noticeable teeth with pits between ribs above; calyx saucer-shaped, 8–16-toothed, bristly. Perennial, biennial and annual herbs with erect, often hollow stem, and bristly or pubescent axillary branches; peduncles axillary and terminal, bristly, sometimes also glandular; leaves decussate, elliptic or ovate, the cauline usually sessile, the lower tapering to petiole, entire, dentate, lyrate or pinnatifid, bristly, pubescent or subglabrous.

A genus of about 50 species, distributed in Europe, outside the Arctic region and in the Mediterranean countries – including southwest Asia and northern Africa – mainly in the mountainous parts.

1. Annual heads, 5–10 flowers per head, involucre cylindrical at first, bracts becoming spreading later, few, narrowly lanceolate or linear; corolla blood-red . . . . 6. **K. orientalis** L.
\+ Perennials and biennials; heads with numerous flowers; involucre
13 multifoliate, bracts spreading, lanceolate or ovate, at onset of flowering not straight . . . . 2.
2. Perennials; flowers bluish-lilac or purple-lilac . . . . 3.
\+ Biennials; flowers white, pale yellow or yellow . . . . 4.
3. Leaves lyrate or pinnatifid, sometimes entire, and then widest in upper third . . . . 2. **K. arvensis** (L.) Coult.
\+ Leaves entire, widest at middle (Carpathia) . . . . 1. **K. dipsacifolia** (Host.) Gren et Godr .
4. Flowers yellow; outer involucral bracts ovate or broadly lanceolate, 14 mm long, 7–9 mm wide; plant 0.5–0.8 m high . . . . 5. **K. involucrata** Somm. et Lev.
\+ Flowers white or yellowish-white; involucral bracts lanceolate; plant 1–2 m high . . . . 5.
5. Achenes 6–8 mm long; eastern regions of the European part of the USSR and the Urals . . . . 3. **K. tatarica** (L.) Litw.
\+ Achenes 5–6 mm long; Caucasus . . . . 4. **K. montana** (M. B.) DC.

Subgenus I. **Trichera** (Schrad.) Rouy, Fl. France, VIII (1903) 105; Szabo, Monogr. 196. – Trichera Schrad. Cat. sem. hort. Götting. (1814) p. p. pro genere. – Perennials and biennials (Soviet annuals not represented); involucre multifoliate, bracts lanceolate, elliptic or ovate, herbaceous; heads with numerous flowers; calyx saucer-shaped, with 8–12 bristles; achenes barely dentate above.

Section 1. EUTRICHERAE Szabo, Monogr. 106. – Perennials.

Subsection 1. SYLVATICAE Krašan in Mitth. naturw. Ver. Steiermark (1899) 94; Szabo, Monogr. 308. – Rootstock without terminal sterile rosette; leaves entire, elliptic, widest at middle.

1. **K. dipsacifolia** (Host) Gren. et Godr. Fl. de France, II (1850) 72; Szaf., Kulcz., Pavl. Rosl. Polskie, 586. – K. silvatica Duby, Bot. gallicum, I (1828) 256; DC. Prodr. IV, 651; Ldb. Fl. Ross. II, 451; Shmalh., Fl. II, 27; Szabo, Monogr. 331. – Scabiosa dipsacifolia Host, Fl. Austr. I (1827) 191. – Ic.: Rchb. Ic. Fl. Germ. XII, 1350, 1352; Szabo, l. c. tab. XXXIX; Hegi, Ill. Fl. VI, 1 Taf. 254, 3. – Exs.: Schultz, Herb. norm. No. 681; Fl. exs. austro-hung. No. 2276.

Perennial; stems 40–100 cm, erect, branching, usually leafy, especially below, spreading, bristly; leaves chartaceous, oblong-lanceolate, tapering to petiole, entire, dentate, sometimes only the upper leaves incised-dentate, pubescent, rarely subglabrous, bright green. Heads 4 cm across; involucral
14 bracts ovate-lanceolate, slightly pubescent, with margin ciliate; flowers radial, lilac or purple-lilac, obtusely lobed, oblong; achenes ovoid-oblong, pubescent, 5–6 mm long; calyx pubescent with 8 setae. June–September.

Forests and shrubs in forest belt and high-mountain meadows. – European part: U. Dnp. (Carpathia), Balt. (also early records for Lithuania and Tartu, where probably introduced). **Gen. distr.:** W. Eur. Described from Austria. Type in Vienna.

Subsection 2. ARVENSES Krasan in Mitth. naturw. Ver. Steiermark (1899) 101; Szabo, Monogr. 214. – Rootstock multicipital, producing wintering rosette in the fall; leaves lyrate or pinnatifid, if entire then widest in upper third.

2. **K. arvensis** (L.) Coult. Mém. dips. (1824) 41; DC. Prodr. IV, 651; Ldb. Fl. Ross. II, 450; Boiss. Fl. or III, 126; Shmal'g., Fl. II, 27; Szabo, Monogr. 224; Grossg., Fl. Kavk. IV, 53; Kryl., Fl. Zap. Sib. X, 2622. – Scabiosa arvensis L. Sp. pl. (1753) 99. – S. campestris Andrz. ex Bess. Enum. pl. Volhyn. (1822) 7. – S. crenata Luce, Topogr. Nachr. Oesel (1823) 28, non Cyr. – S. laciniata Luce, l. c. non Licht. – Trichera arvensis Schrad. Cat. sem. hort. Götting. (1814). – Ic.: Rchb. Ic. Fl. Germ. XII, tab. 980, f. 1353; Szabo, l. c. tab. XIV; Hegi, III. Fl. VI, 1, Taf. 254, 2; Maevskii, Fl. Vol. 7-e, p. 682. – Exs.: GRF, No. 318; Pl. pol. exs. No. 639; Pl. Finl. exs. No. 964.

Perennial; rootstock multicipital, decurrent; stems 30–80 cm, erect, simple or often branching, grooved, covered with downturned hairs beneath, short-haired or subglabrous above; peduncles pubescent, sometimes glandular; leaves lanceolate, usually with stiff hairs, lyrate or pinnatifid, sometimes entire (var. integrifolia Coult.). Heads 2–3 cm long, involucral bracts lanceolate, acuminate, with ciliate margin; flowers bluish-lilac, rarely reddish or yellowish, outer flowers larger, sometimes all flowers of equal length, not radial (var. campestris Andrz. pro sp.); achene oblong-ovoid, 5–6 mm long, 2 mm wide, small-toothed above, with stiff hairs; calyx pubescent, of 8 prickles, Fl. July, Fr. August.

Meadows, mostly dry valleys, edges of light forests. – European part: all regions with the exception of far north and Arctic; Caucasus: Cisc.; W. Siberia: Ob (S.), U. Tob.; Centr. Asia: Ar.-Casp. (Turgai); Far East: Sakh. (introduced). **Gen. distr.:** W. Eur., Iran. Described from Europe. Type in London.

Note. West-European authors have established dozens of epithets and numerous varieties and forms for this exceptionally polymorphic species. Of Soviet authors, Andrzheevskii has described S. campestris Andrz. as
15 an independent species, distinguished by the equal length of the flowers. This is common in the USSR, apparently corresponding to the discoid forms widely known among Compositae. One such plant is described by Litvinov ("Gerbarii russkoi flory," GRF, No. 318, Herbarium of Russian Flora) as a variety bearing the name proposed by Andrzheevskii. K. arvensis, with discoid flowers, is frequent in the steppe regions of the USSR, particularly on the right bank of the Dnieper, in the Ukraine. This form of Scabiosa has, in addition to the name indicated, ten different synonyms proposed by West-European authors.

Subsection 3. ALBESCENTES Szabo, Monogr. 206. — Biennials; flowers white, pale yellow or yellow.

3. K. tatarica (L.) Litw. in Tr. Bot. muz. AH, XV (1916) 141. — K. montana auct. fl. Ross. or non DC.; Szabo, Monogr. 206, pl. cauc. excel. — Scabiosa tatarica L. Sp. pl. (1753) 99. — S. altissima Mill. Gard. dict. ed. 8 (1768) p. p. — S. ciliata Spreng. in Schrad. Journ. 1800, 2 (1800) 199; Rchb. Ic. Pl. rar. III, 64, tab. 273. — S. tatarica Hornem. Hort. Hafn. I (1812) 126. — Ic.: L. in Acta Reg. Soc. Upsal, a. 1744—1750, tab. I; Rchb. l. c.; Fl. yugo-vost. VI, 279. — Exs.: GRF, No. 2554.

Biennial; stems 2 m, cylindrical, hollow inside at lower part, ca. 2 cm across at base, green and for much of its length rather densely covered with downturned bristles, strongly branching; radical leaves lanceolate, ovate, usually entire, 20—25 cm long, 12—15 cm wide, with long winged petiole; cauline leaves sessile, opposite, semiamplexicaul, connate, acuminate, bristly, large-toothed. Peduncles elongate with spreading hairs, glandular or eglandulose; involucral bracts lanceolate; capitula loose, 2—4 cm across; corolla white or somewhat yellowish, 12—15 mm long, with long lobes; achene with stiff hairs, 6—8 mm long, 2—2.5 mm wide, elliptic, truncate, short-fibriate, slightly dentate; calyx with 8—12 bristly teeth, nearly half the length of the outer calyx. Fl. July, Fr. August.

Grass plots and edges of broad-leaved and mixed forests and shrubs, chiefly in the Urals. — European part: V.-Don (Zheguli), V.-Kama (SE), Transv. (E.). Endemic. Described from the Urals. Type in the Linnean herbarium.

Note. D. I. Litvinov's detailed studies of nomenclature (op. cit.) leave no doubt that a specimen from the Urals served as type for this Linnean species, so well known to investigators of the flora of the Urals. We are convinced that the habitat of K. tatarica is in Zheguli (E. V.-Don) although
16 from 1912 onward the literature (Litvinov, op. cit., Maevskii, Fl. 4 and 5 ed., etc.) does not refer to any later investigations. The first report from near Elgushi Lake was made by V. I. Smirnova (Prilozh. k protok. zased. Obshch. estestvoisp. pri Kaz. univ. No. 231, 1904).

4. K. montana (M. B.) DC. Prodr. IV (1830) 651; Ldb. Fl. Ross. II, 451; Boiss. Fl. or III, 127; Shmal'g., Fl. II, 27, p. p.; Szabo, Monogr. 206, excl.

pl. ural.; Grossg., Fl. Kavk. IV, 54. – K. heterotricha Koch in Linnaea, XXIV (1851) 444; Grossg., op. cit. IV, 54; Kolak., Fl. Abkh. IV, 158. – K. montana β heterotricha Boiss. Fl. or. III (1875) 128; Szabo, l. c. 208. – Scabiosa montana M. B. Fl. taur.-cauc. I (1808) 95. – Ic.: Jacq. f. Eclogae, I (1811–1816) tab. 60 et access. I.

Biennial; plant extraordinarily like the preceding in all its characters; the only difference noticeable in herbarium samples is the shorter achene, 5–6 mm long, not fimbriate above, shortly and irregularly dentate. Fl. June, Fr. July.

Forests, forest edges, grass plots, shrubs, upper timberline to subalpine zone. – Caucasus: all regions outside plains and high mountains, with the exception of Tal. **Gen. distr.**: As.-Min. (E. Anatolia). Described from Kislovodsk. Type in Leningrad.

5. **K. involucrata** Somm. et Lev. in Tr. Peterb. bot. sada, XIII, 1 (1893) 46, XVI (1900) 216; Lipskii, Fl. Kavk. 339; Szabo, Monogr. 212; Grossg., Fl. Kavk. IV, 53. – K. integrifolia Koch in Linnaea, XXIV (1851) 443, non Bertol. 1835; Boiss. Fl. or. III, 129. – K. flaviflora Borbas in Magyar Orvosok Munkal. (1894) 276 et in Beih. z. Bot. Zentralbl. II (1895) 98; Szabo, l. c. 310.

Biennial; stems erect, usually simple, 0.5–0.8 m, with 3, rarely 1 or few heads, with downturned bristles below, appressed-hairy above, with small bristles or glands, rarely subglabrous, always bristly under heads; leaves entire, rarely the lowermost pinnatifid at base, lower leaves elliptic, gradually tapering into petiole, acuminate, entire, appressed-hairy on both sides; cauline leaves ovate or lanceolate, broad-cordate, sessile, amplexicaul, pubescent or puberulent, rarely subglabrous. Capitula 3–4 cm across; outer involucral bracts ovate or lanceolate, 14 mm long, 7–9 mm wide, canescent or pubescent, inner bracts narrowly lanceolate, acuminate; flowers radial, yellow, with exserted anthers and 2-lobed elongate stigma; achene
19 hairy, 7–8 mm long, short-marginate above, slightly dentate; calyx with 6–9 bristles, pubescent. Fl. July, Fr. August (Plate II, Figure 1.)

Alpine and subalpine meadows. – Caucasus: all high-mountain regions except for Tal. **Gen. distr.**: As.-Min. (Turkish Armenia). Described from Svanetia. Type in Leningrad.

Subgenus II. **Lychnoidea** Rouy, Fl. France, VIII (1903) 105; Szabo, Monogr. 176. – Annuals; involuere straight cylindrical, the bracts few, linear, yellowish, becoming recurved; flowers 5–10 per capitulum.

6. **K. orientalis** L. Sp. pl. (1753) 101; Ldb. Fl. Ross. II, 450; Boiss. Fl. or III, 126; Szabo, Monogr. 177; Grossg., Fl, Kavk. IV, 53. – Ic.: Szabo, l. c. tab. IV. – Exs.: Heldr. Herb. Graec. No. 848.

Annual; stems 30–60 cm, leafy, stiff-haired below, barely hairy above, pubescent and glandular under heads; leaves sessile, more or less amplexicaul, the radicle nearly rosettelike, recurved, lanceolate, upper leaves narrowly lanceolate or linear, acuminate, entire, dentate even pinnatipartite, subglabrous or pubescent, tapering into petiole. Flowers 5–10 per head, radial; involucral bracts straight-cylindrical, 1–2-seriate, 8–10, entire, the

PLATE II. 1 – Knautia involucrata Somm. et Lev.; 2 – Dipsacus Gmelini M.B., upper part of plant; 3 – D. azureus Schrenk, radical leaves, capitulum in fruit.

outer 10–12 mm long, 1–3 mm wide, acuminate, the inner linear, multiveined, ciliate and glandular, recurved after flowering; corolla blood red, outer lobe broadly ovate, larger than the others; achene 5–6 mm long, nearly tetrahedral, cylindrical, pubescent, with small cartilaginous teeth above; calyx cupular, with 12–16 bristly teeth. Fl. June–July.

Dry slopes. – Caucasus: W. Transc. (W. Georgia). **Gen. distr.**: E. Med., Bal.-As. Min. Described from the East. Type in London.

Note. This plant is rare in the USSR. Lebedour's report for W. Georgia was based on Nordmann's collections. No further collections seem to have been made there.

Tribe 2. **DIPSACEAE** Van Tiegh. in Ann. Sc. Nat. sér. 9, X (1909) 169. – Receptacle with epicalyx scales; involucre sessile, 8-ribbed, ribs visible from edge, without corona; calyx 4-merous; corolla 4-lobed; style simple.

Genus 1411. **DIPSACUS** * L.

L. Sp. pl. (1753) 97.

Heads subtended below by few ascending or recurved involucral bracts, usually markedly different in shape and size from the paleae; receptacle terete or globose, densely covered with cuneate, lanceolate or lanceolate-
20 linear paleae, tapering to acuminate spines; marginal flowers not radial; corolla tube long and limb 4-lobed; stigma oblong, undivided; stamens 4; outer calyx octahedral, grooved, truncate, with small plicate margin or 4 small teeth, continuing the ribs; calyx flattened, tetrahedral, bristly. Biennial, rarely perennial herbs, with high furcately branching, usually prickly stem; cauline leaves opposite, sometimes basally connate.

Type species D. fullonum L.

About 15 species distributed in W. Europe, the Mediterranean countries, North Africa (Ethiopia), S. and E. Asia (to Ceylon and the Himalayas and E. China and Japan).

1. Cauline leaves opposite, basally connate into cup-shaped conceptacle. . . . . 2.
+ Cauline leaves opposite, sessile, not connate . . . . 4.
2. Paleae longer than flowers; tapering to flexible erect mucro . . . . . 3.
+ Paleae shorter than or nearly as long as flowers, tapering into hard elastic recurved mucro; cultivated plant . . . . 3. **D. sativus** (L.) Honck.
3. Involucral bracts arcuate, recurved; heads with long spinous paleae; cauline leaves pinnatipartite, with bristly ciliate margin . . . . 1. **D. laciniatus** L.
+ Involucral bracts arcuately ascending; heads nearly without pappus above; cauline leaves entire or serrately dentate, smooth or with scattered prickles along margin . . . . 2. **D. fullonum** L.

* From the Greek dipsaein – thirst, because the connate leaves at base form a depression in which rainwater is collected.

4. Heads at end of flowering and in fruit ovoid or elliptic; paleae lanceolate, pale, tapering to an awn as long as paleae; flowers ca. 5 mm long . . . . 4. **D. gmelini** M. B.

\+ Heads globose or subglobose; paleae broadly obovate or broadly cuneate; flowers larger . . . . 5.

5. Flowers reddish-yellowish or blue, 13–15 mm long; paleae pale, with
21 uncolored awn one-third to one-half the length of the paleae; stems not
prickly . . . . 5. **D. azureus** Schrenk.

\+ Flowers yellowish-white; paleae with dark-colored awn nearly as long as or longer than paleae; facets of stems bristly, covered with short prickles . . . . 6.

6. Heads 20–30 mm across; paleae broadly obovate, tapering to awn longer than paleae, total length 15–20 mm . . . . . 7. **D. strigosus** Willd.

\+ Heads 12–17 mm across; paleae cuneate, tapering to awn not longer than paleae, total length 10–12 mm . . . . 6. **D. pilosus** L.

1. D. laciniatus L. Sp. pl. (1753) 97; M. B. Fl. taur.-cauc. I, 91; Ldb. Fl. Ross. II, 444; Boiss. Fl. or. III, 116; Grossg., Fl. Kavk. IV, 48. – D. fullonum β. laciniatus Schmalh. Fl. II (1897) 24. – Ic.: Hegi, Ill. Fl. VI, 1, Fig. 157.

Biennial; stems 0.5–1.5 m, erect, ribbed, prickly, branching above; radical leaves in rosette short-petioled, oblong-obovate, dentate or pinnatilobate, bristly on both sides; cauline leaves opposite, broadly connate into cuplike conceptacle, oblong, acuminate, lyrate-pinnatifid to pinnatipartite, irregularly dentate or incised-lobed, midrib prickly beneath, margin bristly. Heads oblong-ovoid, 5–8 cm long; involucral bracts subulate-lanceolate, usually shorter than head; appressed or at first recurved, margins and midrib prickly, paleae ciliate, tapering to erect long mucro, longer than flowers; flowers pale lilac or white, ca. 13 mm long. Fl. June–July, Fr. August.

Shrubs in central mountain zone and river valleys. – European part: M. D., Bes., Bl., L. Don, Crim.; Caucasus: all regions; Centr. Asia: T. Sh., Pam.-Al., Mtn. Turkm. **Gen. distr.**: Centr. Eur. and E. Med., Bal.-As. Min., Iran. Described from Alsace and Priazovskoe. Type in London.

2. D. fullonum L. Sp. pl. (1753) 97, pro max. parte, var. β. excl. – D. silvestris Huds. Fl. angl. (1762) 49; Mill. Gard. dict. ed. 8; Willd. Sl. pl. 1, 544; M. B. Fl. taur.-cauc. III, 98; Ldb. Fl. Ross. II, 444; Boiss. Fl. or. IV, 115; Hegi, Ill. Fl. VI, 1, 281. – D. fullonum α. silvestris Schmalh. Fl. II (1897) 24. – Ic.: Hegi, l. c. Taf. 253, 5. – Exs.: Fl. exs. reip. Boh.-Slov. No. 1076.

22 Biennial; stems 0.5–2 m, erect, branching above, ribbed, prickly along ribs; radical leaves in rosettes short-petioled, oblong-obovate, 20–30 cm long, usually glabrous, on both sides prickly, margin ciliate; cauline leaves opposite, basally connate, oblong-lanceolate, entire or serrate, prickly midrib and margins, margin not ciliate. Heads oblong-ovoid, 4–8 cm long; involucral bracts linear-lanceolate, glabrous, prickly, arcuate, unequal in length, the longest exceeding head; paleae ciliate, produced into erect spinous mucro, longer than flowers; flowers violet, rarely white, tube ca. 10 mm long; outer calyx acuminate above; calyx tetrahedral; achene ca. 5 mm long. Fl. June, Fr. August.

River valleys, shrubby formations and weedy habitats. – European part: U. Dnp., Bes., Bl., L. Don, Crim.; Caucasus: Cisc. (rarely). **Gen. distr.**: Atl. and Centr. Eur., Bal.-As. Min., Iran. Described from France. Type in London.

Economic importance. When dry, this plant contains a blue coloring matter (dipsacotine) similar to indigo but differing from it by its solubility in water (Hegi, l. c.). Because of the flexibility of the paleae, this plant is not used for napping in the woolen industry.

Note. By combining D. sylvestris and D. sativus under D. fullonum, Linnaeus complicated their nomenclature. D. sativus, noted by earlier authors as a cultivated plant, was made into a variety under the letter β. Subsequently, Hudson (l. c.) was the first to distinguish the wild and cultivated species, naming them D. sylvestris and D. fullonum, respectively. Later, Honkeny (1782) correctly named the cultivated species D. sativus (L.) Honck. Since then many authors have followed Hudson by calling the wild plant D. sylvestris. Proceeding from the premise that the Linnean name (D. fullonum) cannot be abolished and that he applied it to the wild plant, we decided to follow him. It should also be noted that D. fullonum is expressly noted as the type genus. The only result of this solution is that the wild plant, which is of no use to the textile industry, retains the name Fuller's teasel (D. fullonum).

3. **D. sativus** (L.) Honck. Vollst. syst. Verz. (1782) 16; Honck. Synops. pl. Germ. II, 4; Gmel. Fl. Bad. I, 314; Hegi, III, Fl. VI, 1, 283; Grossg., Fl.
23 Kavk. IV, 48. – D. fullonum β. L. Sp. pl. (1753) 97. – D. fullonum Huds. Fl. angl. (1762) 49. – D. fullonum Mill. Gard. dict. ed. 8 (1768); Ldb. Fl. Ross. II, 445; Boiss. Fl. or. III, 116. – D. fullonum γ. sativus Gmel. ex Shmal'g., Fl. II (1897) 24. – Ic.: Hegi, l. c. VI, 1, Fig. 157; Gun'ko, Vorsyanka, 1932.

Biennial; very similar to the preceding, from which it is distinguished by erect – not arcuately curved– involucral bracts, not exceeding the head, and becoming recurved and by paleae which are slightly exceeded by the flowers, and terminate in compact reflexed mucros, and by longer, 12–13 mm long flowers, Fl. July, Fr. August.

Cultivated for commercial purposes in the Crimea, Caucasus, and Central Asia. Described from France. Type in London.

Economic importance. Teasel is cultivated for its heads, which are used for raising a nap on cloth, including cloth from wool, flannel and velvet. In the past few years textile manufactures have been trying to replace teasel with metallic cards, but in spite of good success, particularly for costly woolens, the natural product cannot quite be replaced. For studies on the cultivation of teasel, see M. Altukhov. Vorsoval'naya shishka, Sel'khozgiz. 1931; G. K. Gun'kov. Vorsyanka, Izd. Vsesoyuzn. Inst. rastenievodstva. 1932.

4. **D. gmelini** M. B. Fl. taur.-cauc. I (1808) 92; Ldb. Fl. Ross. II, 444; Shmal'g., Fl. II, 24; Bobrov in Fl. yugo-vost. VI, 275; Kryl., Fl. Zap. Sib. X, 2620.

Biennial; stems 0.5–1 m, deeply grooved, prickly for entire length, bristly above; lower leaves lanceolate, acuminate, incised-dentate, tapering to winged petiole; cauline leaves sessile, pinnatifid, with bristly margins and midrib, uppermost leaves deeply pinnatisect, with lanceolate-linear bristly lobes. Heads subglobose at first, becoming ovoid, 3–5 cm long; involucral bracts linear, much shorter than heads, and recurved; paleae

10–12 mm long, narrow, oblanceolate or nearly cuneate, ciliate, tapering into slender, bristly awn as long as paleae; flowers bluish, ca. 5 mm long. Fl. July, Fr. August. (Plate II, Figure 2.)

Moist banks of rivers and lakes, moist steppe depressions in the southern steppe and semidesert zones. – European part: L. Don (S.), L. V.; Caucasus: Cisc. (along the Kuma River); W. Siberia: U. Tob. (S.), Irt. (S.); Centr. Asia: Ar.-Casp., Balkh., Dzu-Tarb. Endemic. Described from the banks of the Kuma River. Type in Leningrad.

5. D. azureus Schrenk, Enum. pl. nov. I (1841) 53; Ldb. Fl. Ross. II, 446; Kryl., Fl. Zap. Sib. X, 2621. – *Cephalaria dipsacoides* Kar.
24 et Kir. in Bull. Soc. Nat. Mosc. XIV (1841) No. 417, 434. – Exs.: HFAM, No. 295.

Biennial; stems 0.6–1.3 m, smooth, prickleless, with downturned bristles, confined to lower part, branching above; radical leaves entire, lanceolate, 40 cm long; lower leaves sessile, 15–30 cm long, 4–8 cm wide, proximally pinnatifid, with lanceolate decurrent lobes and terminal larger, rhombic lobe; upper leaves narrower and short, the uppermost lanceolate, linear or pinnatisect or sometimes dentate-incised at base. Heads subglobose, 2.5–4 cm long; involucral bracts stiff, spinous, subulate, nearly as long as head; paleae oblong, 12–14 mm long, whitish, cartilaginous, cuneate, slightly convex, dorsally subcarinate, pubescent, tapering into spinous acuminate awn one-third to one-half as long as paleae; flowers 13–15 mm, pink-violet or blue, hairy outside. Fl. July, Fr. August. (Plate II, Figure 3.)

Shrubby formations in pebbly slopes and mountain river valleys. – Centr. Asia: Balkh. (E.), Dzu-Tarb. (Dzhungarian Ala-Tau), T. Sh., Syr D. (E.), Pam.-Al. Endemic. Described from the Tentek River valley, flowing into the Alakul. Type in Leningrad.

Economic importance. The roots of this species contain 0.3% alkoloids; similar in composition with those of *Gentiana krylowii* and *Erythraea centaurium* (Zhurn. obshch. khim., XVIII, 8, 1948, 1510).

6. D. pilosus L. Sl. pl. (1753) 97; M. B. Fl. taur.-cauc. I, 92; Ldb. Fl. Ross. II, 445, p. p.; Boiss. Fl. or. III, 116; Shmal'g., Fl. II, 25, p. p.; Grossg., Fl. Kavk. IV, 48. – *Cephalaria pilosa* Gren. et Godr. Fl. Fr. II (1850) 69. – Ic.: Hegi, III. Fl. VI, 1, Fig. 158. – Exs.: Fl. exs. reip. Boh.-Slov. No. 630.

Biennial; stems 0.5–1.5 m, erect, grooved-faceted, facets with short prickles and bristles, sometimes subglabrous; leaves thin, coriaceous, petiolate, not connate, elliptic or ovate, lower leaves with longer petioles, sometimes lyrate-pinnatifid, attenuate into petioles, the uppermost leaves stiff-haired or ciliate, often with 2 small lateral lobes and a large terminal one. Head drooping at first, globose, on bristly peduncle, 12–17 cm across; involucral bracts coriaceous, lanceolate, recurved, bristly-ciliate, slightly longer than paleae, the latter as long as flowers, bristly-ciliate, cuneate, tapering into awn; outer calyx not grooved, dentate above; flowers yellowish-
25 white, infundibuliform, 8–10 mm long; anthers dark violet; achene pubescent, ca. 5 mm long. Fl. July, Fr. August. (Plate I, Figure 4.)

Shady places, forest edges and shrubby formations. – European part: U. Dns., M. D., V.-Don (SW), L. Don (S.), Bes., Bl. (?), Crim.; Caucasus:

all regions with the exception of S. Transc. and Tal. except for the high mountain areas. W. Siberia: Irt. (Mikhailovskoe near Barnaul, introduced). **Gen. distr.:** Atl. and Centr. Eur., N. Italy and the Balkans, Iran. Described from England. Type in London.

7. D. strigosus Willd. ex Roem. et Schult. Syst. III (1818) 520; Boiss. Fl. or. III, 116; Bobrov in Fl. yugo-vost. VI, 275; Grossg., Fl. Kavk. III, 48. – D. pilosus β. strigosus Ldb. Fl. Ross. II (1844) 445; Shmal'g., Fl. II, 25. – Ic.: Fl. yugo-vost. VI, Fig. 657. – Exs.: Kotschy, Pl. Pers. bor. No. 597.

Biennials; stems 0.5–1.5 m, erect, faceted, bristly and subulate; leaves dentate, oblong, acuminate, the lower entire, petiolate, the cauline proximally pinnatifid, with 2–5 lateral lobes, ciliate. Heads globose, ca. 3 cm across; involucral bracts lanceolate, acuminate, shorter than head; paleae similar to involucral bracts, tapering into long greenish-blackish bristly awn, together with paleae 15–20 mm long. Fl. June, Fr. July–August.

Shrub thickets in shady places, and as a weed. – European part: U. V. (Torzhok, introduced), V.-Don, Bl., L. Don, L. V., Crim.; Caucasus: throughout the submontane and montane zone; Centr. Asia: Mtn. Turkm. **Gen. distr.:** As. Min. (E. Anatolia), Arm.-Kurd., Iran., introduced into Sweden. Described from Gilyan. Type in Berlin.

Note. This species and the preceding species as well as D. laciniatus yielded some pharmacologically active alkaloids. (Dokl. Ak. n. Arm. SSR, V, 1946, 2.)

Genus 1412. **CEPHALARIA** * Schrad.

Schrad. Car. sem. hort. Götting. (1814) ex Roem. et Schult. Syst. III (1818) 1, nomen conservandum; Szabo in Math. Termtud. Ertes. XXXIX (1922) 267 et in Mag. Bot. Lapok, 1/12 (1926) 49; Ej. Monogr. gen. Cephalaria, 1940 (non vidi). – Lepicephalus Lag. Gen. et sp. nov. (1816) 7.

Inflorescence capitate, globose or ovoid; involucral bracts imbricate, squamiform, shorter than paleae, the latter lanceolate, acuminate, sometimes tapering into awn or broadly triangular or obtuse above, herbaceous or cartilaginous, pale or lilac above and at back, brown, reddish or blackish, glabrous or pubescent; calyx saucer-shaped or cupular, bristly, margin with numerous bristly prickles; flowers usually radial; corolla 4-lobed,
26 yellow or pale blue; style filiform; achene fusiform, tetrahedral, outer calyx pubescent, 8-toothed above, sometimes with small barely discernible teeth; teeth rarely absent. Perennial or annual herbs, often to 1 m high or higher; leaves lyrate-pinnatifid; radical leaves sometimes entire or barely pinnatifid proximally.

Of the 60 species of this genus, 23 are represented in the USSR. The 15 species of the subgenus Lobatocarpus Szabo grow in S. Africa; the 5 species of subgenus Fimbriatocarpus Szabo are distributed in the western part of the Mediterranean area. All the Soviet species may be referred to subgenus Denticarpus Szabo distributed in the eastern part of the Mediterranean area, taken in the broad sense, and in southern Europe.

* From the Greek kephale – head, alluding to the shape of the inflorescence.

1. Perennial plants . . . . 2.

\+ Annual plants . . . . 19.

2. Ripe outer calyx distinctly 8-toothed, the corner teeth reaching 1–3 mm; involucral bracts and palaeae light rufous, brown or dark brown and therefore heads darkish beneath; leaf lobes dentate, acuminate (Atrocephalae) . . . . 3.

\+ Calyx crowned with tubercles or with 8 small ribs corresponding to the cartilaginous teeth; involucral bracts and paleae pale, lilac-brownish on the outside; leaf lobes entire, narrow or ovate (Leucocephalae) . . . . 15.

3. All teeth nearly equal in length, one-sixth to one-third as long as outer calyx; paleae gradually acuminate . . . . 4.

\+ Intermediate teeth much shorter; involucral bracts abruptly or shortly acuminate . . . . 5.

4. Involucral bracts and paleae appressed-hairy, with long bristles; paleae with sparse short cilia along margins; outer flowers 2.5 cm long; stems 1–1.5 m (Caucasus) . . . . . . 1. **C. gigantea** (Ldb.) Bobr.

\+ Involucral bracts and paleae appressed-hairy, without long bristles; paleae with dense long cilia along margin; outer flowers 2 cm long; stems 1–2 m. Plants of the Middle Don . . . . . 2. **C. litvinovii** Bobr.

5. Angular teeth of outer calyx 1 mm long; heads 1–1.5 cm across; stems simple; leaves mostly radicle (Dagestan). . . 13. **C. dagestanica** Bobr.

\+ Angular teeth of outer calyx 2–3 mm long; heads 2–3 cm across;
27 stems branching, very rarely nearly simple; leaves mostly borne on lower part of stem, not radicle . . . . 6.

6. Plant 20–40 cm high, with nearly simple stems and solitary heads; lower cauline leaves entire or lyrate, 8 cm long (limestones of Mingrelia) . . . . 7. **C. calcarea** Alb.

\+ Plant 0.5–1 m high and more; stems branching; heads few; lower cauline leaves larger . . . . 7.

7. Limbs of all leaves lyrate-pinnatifid . . . . 8.

\+ Lower cauline leaves entire or nearly so . . . . 12.

8. Stems glabrous; lower leaves 30–40 cm long, nearly lyrate-pinnatifid . . . . 3. **C. procera** Fisch. et Lallem.

\+ Stems often covered with retrorse hairs in lower part, often densely bristly under heads; leaves smaller, distinctly pinnatifid . . . . . . 9.

9. Paleae ca. 7 mm long, broadly triangular above, slightly acuminate; stems with soft hairs . . . . . 6. **C. brevipalea** (Somm. et Lev.) Litv.

\+ Paleae 10–12 mm long, ovate, oblong-ovate or cuneate, acuminate above; stems covered below with retrorse bristles or hairs . . . . 10.

10. Stems under heads subglabrous; lower leaves 11–13-lobed; achene 6–8 mm long . . . . 8. **C. tchihatchewii** Boiss.

\+ Stems under heads densely covered with long bristles; lower leaves 3–5-lobed; achene 9–11 mm long . . . . 11.

11. Lower leaves not crowded in rosette; petioles 10–15 cm long; paleae with stiff bristles on the outside; teeth of outer calyx 2.5–3 mm long . . . . 4. **C. kotschyi** Boiss. et Hohen.

\+ Lower leaves crowded in rosette; petioles 5–7 cm long; paleae with soft bristles on the outside; angled teeth of peridium 1.5–2 mm long . . . . 5. **C. grossheimii** Bobr.

12. Lower part of stem densely covered with retrorse bristles; paleae coriaceous, lanceolate, 10 mm long; plant of subalpine zone of Nakhichevan ASSR . . . . 9. **C. nachiczevanica** Bobr.

\+ Lower part of stem covered with spreading or short hairs, without retrorse bristles; palea cartilaginous, oblanceolate or oblong, ca. 8 mm long . . . . 13.

13. Leaves with spreading bristles on both sides and along margin; stems glabrous above . . . . 11. **C. pilosa** Boiss. et Huet.

\+ Leaves short-pubescent on both sides and along margin; stems with scattered short hairs or short-lanate above . . . . 14.

14. Flowers radial, the outer ca. 14 mm long, the inner ca. 9 mm; stems with scattered short hairs; plant of S. and E. Transcaucasia . . . . 10. **C. armeniaca** Bordz.

28 + Flowers barely radial, 10–11 mm long; stems short-lanate. (Mtn. Turkm.) . . . . 12. **C. sublanata** (Bornm.) Szabo.

15. Stems glabrous, lustrous; outer calyx ribbed-tuberculate . . . . . . 16.

\+ Stems hairy or bristly below; outer calyx crowned with small teeth . . . . 17.

16. Leaves usually entire, coriaceous, shiny, rarely lyrate or pinnatifid; paleae oblanceolate, short, acuminate . . 15. **C. coriacea** (Willd.) Steud.

\+ Cauline leaves pinnatifid, coriaceous, glabrous, with oblong lanceolate lobes; paleae lanceolate, acuminate, nearly spinous . . . . 14. **C. demetrii** Bobr.

17. Outer calyx with 4 small angular teeth (0.5–1 cm) and 4 obsolete intermediate teeth . . . . 16. **C. uralensis** (Murr.) Schrad.

\+ Outer calyx with 8 equal obsolete teeth . . . . 18.

18. Stems pubescent with sparse retrorse hairs below; limbs of leaves with short scattered hairs, on both sides and along margins with appressed bristles; involucral bracts short-haired . . . . 17. **C. media** Litw.

\+ Stems densely covered with retrorse bristles below; limbs of leaves with soft and long sericeous hairs on both sides and along margins . . . . 18. **C. velutina** Bobr.

19. Outer calyx edenticulate or with obsolete teeth, and therefore achenes fusiform . . . . 20.

\+ Outer calyx distinctly 8-toothed . . . . 22.

20. Outer calyx edenticulate . . . . 21.

\+ Outer calyx tuberculate-dentate above with 4 distal lateral teeth corresponding to angles and 4 intermediate ones above grooves; involucral bracts ovate, obtuse; paleae oblong, acuminate without awn . . . . 19. **C. microdonta** Bobr.

21. Flowers pale yellow, the outer twice as long as the inner and 3–4 times as wide; paleae with dense seriaceous hairs on the outside, awn one-half to three quarters as long as paleae . . . . 21. **C. aristata** C. Koch.

\+ Flowers pale blue, the outer 13 mm, the inner 8 mm long; paleae white tomentose, pubescent, with ciliate margins, awn one quarter as long as paleae . . . . 22. C. **armena** Grossh.

22. Outer calyx with 8 nearly equal small teeth, shorter than calyx; paleae lanceolate, awned-acuminate, reddish brown above, shorter than flowers . . . . 20. **C. transsylvanica** (L.) Schrad.

29 + Outer calyx 8-toothed, the 4 angular teeth much longer than the intermediate teeth, and longer than calyx; paleae oblong-lanceolate, tapering into awn, awn as long as paleae, exceeding adjacent flowers . . . . 23. **C. syriaca** (L.) Schrad.

Subgenus I. **Denticarpus** Szabo in Math. Termtud. Ertes. XXXIX (1922) 273. – Outer calyx 8-toothed above, sometimes teeth poorly developed.

Section 1. ATROCEPHALAE Szabo, l. c. 274. – Paleae triangular, dark brown above; angular teeth of outer calyx longer, intermediate teeth usually smaller, sometimes all teeth nearly equal, or obsolete; leaves lyrate-pinnatifid, the radicle sometimes entire or barely lyrate.

In addition to the Soviet species this section includes another 6–7 species native to the mountains of S . Europe (Alps, Balkans) and Anatolia.

Series 1. **Tataricae** Bobr. in Bot. zhurn. SSSR, XVII (1932) 490. – Angular and intermediate teeth of outer calyx nearly equal, one-fifth to one-sixth as long as outer calyx; paleae gradually acuminate; stems 1–2 m, subglabrous or with sparse hairs; leaves lyrate-pinnatifid, 20–40 cm long, bristly.

Forest-steppe and mountain meadow rivers of the Caucasus and forest-steppe plains (M. D.).

1. C. gigantea (Ldb.) Bobr. in Bot. zhurn. SSSR. XVII (1932) 490. – C. elata Schrad. Cat. sem. hort. Götting. (1814) nomen. – C. tatarica Roem. et Schult. Syst. III (1818) 51; DC. Prodr. IV, 647; Fisch. et Lallem. in Ind. sem. hort. Petrop. VII, 46; Ldb. Fl. Ross. II, 447, ex p.; Shmal'g., Fl. II, 26, ex p. – C. caucasica Litv. in Spiske rast. GRF, VI (1908) 157. – Scabiosa gigantea Ldb. Suppl. I Enum. pl. hort. Dorpat. (1811) 4. – S. atrata Lag. Fl. hort. reg. Madr. (1805) nomen. – S. tatarica M. B. Fl. taur.-cauc. I (1808) 92, III, 98, non L. – Lepicephalus atratus Lag. Gen. et sp. nov. (1916) 7. – L. caucasicus Litv. in Tr. Bot. muzeya AN, XV (1916) 153. – Exs.: GRF, No. 1981 (sub Ceph. cauc.).

Perennial; stems 1–1.5 m, subglabrous or with scattered hairs, sulcate, short-haired above, with retrorse hairs below; leaves lyrate – pinnatifid, with 2–6 pairs of lobes, acuminate, dentate, nearly decurrent, bristly on both sides, terminal leaf larger. Heads globose, to 2.5 cm across in fruit; involucral bracts blackish, ovate-triangular, the inner oblong, acuminate, with short appressed hairs and long bristles on back and along margin; paleae nigrescent, lanceolate, acuminate, short-haired, with long-bristles on back and along margin, 13 mm long, 4 mm wide; achene ca. 7 mm long, pubescent,
30 tetrahedral, grooved; angular teeth of outer calyx ca. 1 mm long, one-sixth to one-fifth as long as achene, intermediate teeth nearly as long, teeth not exceeding calyx in fruit; flowers greenish yellow, pubescent outside, radial, the outer 2.5 cm long; calyx saucerlike, bristly-aristate with 2 mm long bristles. Fl, June, Fr. July.

Foothills to alpine meadows, grass plots in mountain forest-steppe belt and high mountain meadows. – Caucasus: all regions of the Greater and Lesser Caucasus with the exception of Tal. Endemic. Described from cultivated specimens. Type in Leningrad.

Economic importance. Chemical analysis of the roots has yielded 0.25–0.17% alkaloids and of the flowers 0.2–0.13% (Dokl. Arm. AN, IV (1946) 3). Earlier reports refer to the use of the inflorescences by local populations to dye wool bright yellow.

Note. The distribution of this species from forest plots of mountain forest-steppes to subalpine meadows hints at its morphological variability. It is most variable in size of plant, pubescence of stem and leaves, size, number and consistency of leaf lobes, size of the marginal flowers, degree of pubescence of the involucral bracts, etc. A comparison of plants in remote localities reveals differences which, however, fade on comparison with plants from nearer localities. The ecological variability of this species is reflected in the fact that some of its ecological types have been described as separate species. Thus, C. caucasica Litw., has been described from Teberda and C. balkharica E. Busch from the high mountain meadows of Balkariya (Bot. zhurn. SSSR, XXI (1936) 426). Busch, describing her species, maintained its independence, as well as that of C. caucasica Litw. According to her C. gigantea (Ldb.) Bobr. is distributed in tall herbaceous vegetation and the species established by her populate the densely turfy alpine meadows. Also according to Busch C. caucasica and C. balkharica are distributed in the high mountains of the Central Caucasus (Balkariya). The unusual closeness of this and the other form leave no room for doubt; the fact that their habitats are in close proximity proves the possibility of a common habitat for closely related species. All things considered, C. balkharica and C. caucasica are surely the high mountain forms of Ledebour's species. This also seems to be the case with C. sosnowskyi Kolak. (Zam. po sistem. i geogr. rast. Tbil. Bot. inst. 10 (1941) 111; Fl, Abkhazii, IV (1949) 157), described from Gagry Range. The type specimen was not seen but a study of the plants collected on Gagry
31 Range by P. S. Panyutin indicate that they should be referred to C. gigantea.

2. **C. litvinovii** Bobr. in Bot. zhurn. SSSR. XVI (1932) 495; Fl. yugo-vost VI, 276. – C. tatarica Ldb. Fl. Ross. II (1841) 447, p. p. non Roem. et Schult.; Shmal'g., Fl. II, 26, p. p. – Scabiosa tatarica Gmel. Reise, I (1774) 159, non L.; Pall. Reise, III, 686; Falk, Beitr. II, 111. – Lepicephalus tataricus Litv. in Tr. Bot. muz. AN, XV (1916) 148, p. p. non Lagasca; Maevskii, Fl., Vol. 5-e, 270.

Perennial; stems 2 m, glabrous, sulcate, short-haired above; leaves lyrate-pinnatifid, lobes 4–6-paired, acuminate, dentate, nearly decurrent, with scattered bristles on both sides, the terminal lobe larger. Heads globose, 2.5 cm across in fruit; involucral bracts blackish, ovate-triangular, the inner oblong, acuminate, dorsally with short apressed hairs, and ciliate margins; paleae nigrescent, lanceolate, acuminate, dorsally short-haired, margins ciliate, 13 mm long, 4 mm wide; achene ca. 7 mm long, tetrahedral, pubescent, grooved; angular teeth of outer calyx 1–1.5 mm, the intermediate nearly as long, not exceeding calyx in fruit; flowers sulphur-yellow, pubescent outside, markedly radial, outer flowers 2 cm long; calyx saucer-like, with bristly-aristate margin. Fl. July, Fr. August.

Shrubby formations in valleys and ravines. – European part: L. Don (N.). Endemic. Described from Borisoglebsk District in Tambov Region. Type in Moscow.

Note. This species is very close to the preceding one, from which it differs by appressed pubescence, without long hairs of the involucral bracts and paleae and the more densely long-ciliate margin of the paleae; the outer flowers are less radiant, the stems higher and more glabrous, and the distribution area is different. There is no question now about the distribution of this species in the Khoper basin, as the earlier questionable reports are confirmed. The plants reported by the eighteenth century authors Gmelin, Pallas, and Falk, which were lost, certainly belonged to this species. Bieberstein collected this species between Tambov and Novokhopersk (1820), Litvinov (1880 and later Antonov (1885) near Burnak in the Borisoglebsk District, Tambov Region. The most recently discovered locality is near Dmitrovka in the Kozlov District (V. B. Il'inskaya, 1940, who expressly searched for the plant at our request).

Series 2. **Procerae** Bobr. in Bot. zhurn, SSSR, XVII (1932) 497. – Angular teeth of outer calyx 1.5–2.5 mm, intermediate teeth much smaller;
32 paleae abruptly acuminate; stems 1 m, subglabrous or with scattered bristles; leaves lyrate-pinnatifid, 20–40 cm long, more or less bristly.

Mountain forest-steppes of Anatolia, Transcaucasia, and NW Iran.

3. C. procera Fisch. et Lallem. in Ind. sem. hort. Petrop. VII (1840) 46; Bobrov in Bot. zhurn. SSSR, XVII, 497; Grossg., Fl. Kavk. IV, 52. – C. decipiens Gandog. Dec. pl. nov. I (1875) 43.

Perennial; stems 1 m and more, glabrous, sulcate; leaves lyrate-pinnatisect, with acuminate lobes, dentate, decurrent, more or less bristly on both sides, lower leaves 30–40 cm long, bristly at base, long-petioled, with terminal lobe 12 cm long, 6 cm wide, lateral lobes decurrent, 7 cm long, 3 cm wide. Heads globose, ca. 3 cm across; involucral bracts ovate-rhombic, nigrescent, back and margins bristly; paleae 12 mm long, 4–5 mm wide, sublanceolate (the outer broad at base), acuminate, nigrescent, bristly backs and margins, recurved distally in fruit; achene tetrahedral, grooved, pubescent, 8 mm long; angular teeth of outer calyx 1.5–2.5 mm, one-quarter to one-third as long as tube, intermediate teeth much shorter, sometimes cleft, in fruit angular teeth longer than or nearly as long as calyx; flowers sulphur-yellow, pubescent outside, strongly radiating, outer flowers 15 mm, inner 10 mm long; calyx saucerlike, with bristly margin, irregularly aristate. Fl. July, Fr. August.

Forest and subalpine zones in forest plots and edges. – Caucasus: W. Transc., S. Transc. (SW). **Gen. distr.**: As.-Min. Described from specimens cultivated from seeds collected in Anatolia. Type in Leningrad.

4. C. kotschyi Boiss. et Hohen. in Boiss. Diagn. I ser. 10 (1849) 76; Bobrov in Bot. zhurn. SSSR, XVII, 498; Grossg., Fl. Kavk. IV, 52. – C. procera var. Kotschyi Boiss, Fl. or. III (1875) 126.

Perennial; stems 50–80 cm, branching, sulcate, with short scattered bristles above, more dense long bristles under heads, bristles at base of

stems retrorse; lower leaves long-petioled, lyrate, petioles 10–15 cm long, terminal lobe 12–18 cm long, 3–5 cm wide, lateral lobes 2–3-paired, 3 cm long, 1.5 cm wide, terminal lobe dentate, pubescent on both sides, median and cauline leaves lyrate, the terminal lobe 8 cm long, 3 cm wide, lateral lobes 3–4, 3–4 cm long, 1.5–2 cm wide, decurrent, pubescent, dentate; upper leaves lanceolate or linear, pinnatifid, pubescent. Heads globose, 2–3 cm across; involucral bracts triangular-ovate, blackish, dorsally bristly; paleae oblong-
33 ovate, 12 mm long, 5 mm wide, acuminate, brownish, pubescent above, with ciliate margin; achene 11 mm long, tetrahedral, grooved, pubescent, angular teeth 2.5–3 mm long, one-quarter to one-third as long as calyx; flowers sulphur-yellow, pubescent outside, the outer 14 mm long, the inner 8 mm; calyx saucerlike, bristly, with irregularly aristate margin, bristles 1 mm long. Fl. June, Fr. July.

Steppe slopes in the montane belt. – Caucasus: Tal. **Gen. distr.**: NW Iran. Described from Duder Pass, Elburz Range. Cotype in Leningrad.

5. **C. grossheimii** Bobr. in Bot. zhurn. SSSR, XVII (1932) 499; Grossg., Fl. Kavk. IV, 52.

Perennial; stems 30–50 cm, with few branches, sulcate, with soft scattered bristles, becoming densely long bristled under heads, bristles in lower part of stem retrorse; lower cauline leaves approaching rosette, lyrate, pubescent on both sides, dentate, petioles 5–7 cm long, terminal lobe 10–16 cm long, 5–7 cm wide, lateral lobew 2–4, 4 cm long, 1.5 cm wide; median cauline leaves lyrate, pubescent, with deeply dentate terminal lobe 6 cm long, 2–2.5 cm wide, lateral lobes 4–6, 2.5 cm long, 1 cm wide; upper leaves pinnatifid, with deeply dentate lobes, pubescent mainly along margins. Heads globose, 1.5–2 cm across; involucral bracts orbicular or rhombically orbicular, dark brown, dorsally covered with soft bristles; paleae ovate, 10 mm long, 5 mm wide, acuminate, dark brown above, margin with 1 mm long cilia, dorsally short-haired, especially short-haired above, with longer and softer hairs along keel; achene 9–11 mm long, tetrahedral, pubescent, angular teeth 1.5–2 mm, one-sixth to one-fifth as long as outer calyx, intermediate teeth 1 mm long, sometimes cleft, in fruit nearly equal to calyx; flowers sulphur-yellow, pubescent outside, radial, outer ca. 12 mm long, inner ca. 10 mm; calyx saucerlike, bristly, with irregularly aristate margin, awns 1.5 mm long, bristly. Fl. July.

Montane belt, among crops. – Caucasus: Tal. Endemic. Described from Razgov? Type in Tbilisi.

Note. Close to the preceding species from which it differs by its soft-bristly or hairy involucral bracts and paleae, shorter teeth of outer calyx, and shorter petioled lower leaves forming a rosette. It was collected only once (Grossgeim, 1917).

6. **C. brevipalea** (Somm. et Lev.) Litv. in Spiske rast. GRF, VI (1908) 156; Bobrov in Bot. zhurn. SSSR, XVII, 499; Grossg., Fl. Kavk. IV, 52; Kolak.,
34 Fl. Abkh. IV, 154. – C. tatarica var. brevipalea Somm. et Lev. in Tr. Peterb. bot sada, XIII (1893) 45. – Exs.: GRF, No. 1980.

Perennial; stems 1 m, sulcate, soft-haired; leaves lyrate-pinnate, with 2–4 opposite decurrent lateral lobes and larger terminal one, appressed soft-haired on both sides, irregularly dentate; upper leaves linear-lanceolate. Heads globose, ca. 2 cm across; involucral bracts broadly ovate, blackish, short-haired; paleae ca. 7 mm long, above broadly triangular, barely acuminate, blackish, short haired, dorsally pubescent, with sparingly ciliate margin; achene tetrahedral, pubescent, ca. 6 mm long, angular teeth ca. 1 mm, one-sixth to one-fifth as long as outer calyx, not exceeding calyx in fruit, intermediate teeth half as long as outer calyx; flowers sulphur-yellow, pubescent outside, slightly radial, the outer 12 mm long, the inner 10 mm; calyx saucerlike, margin with numerous irregular prickles. Fl. July, Fr. August.

Light forests in high mountain belt, limestones. – Caucasus: Cisc. (W.), W. Transc. Endemic. Described from Teberda. Type published.

Note. C. svanetica Kolak. (Zam. po sist. i geogr. rast. Tbil. bot. inst. 10 (1941) 113), described from Svanetia, should tentatively be referred to C. brevipalea Litv.

7. **C. calcarea** Alb. in Tr. Tifl. bot. sada. I (Prodr. Fl. colch.) (1895) 126; Bobrov in Bot. zhurn. SSSR, XVII, 499; Grossg., Fl. Kavk., IV, 52.

Perennial; low plant with obliquely branching root; stems 20–40 cm, glabrous or with retrorse hairs below, glabrous above, with solitary heads; radical leaves 3.5–8 cm long, 1.3–5 cm wide, entire, ovate, dentate, tapering to long petiole or lyrate, with 1–2 lateral lobes, limbs appressed-hairy or glabrous on both sides with ciliate margin; cauline leaves 2–6, similar to radicle, lyrate; upper leaves very narrow. Heads medium-sized, slightly radial; involucral bracts with stiff appressed hairs, blackish, the outer rounded-ovate, obtuse, the intermediate ovate-triangular, the inner oblong-rhombic; paleae lanceolate, distally lacerated; achene tetrahedral, pubescent, with 8 irregular teeth, one-fifth to one-third as long as outer calyx; barely as long or nearly equal to calyx.

Limestones of high mountain belt. – Caucasus: W. Transc. (Mingreliya). Endemic. Described from alpine pastures of Migariya (Dzhvari). Type in Geneva.

35 Note. This species, once collected by Albov (1893), is not represented in our collections. Judging by the description it presumably comes somewhere between the series Procerae and Hirsutae.

Series 3. **Hirsutae** Bobr. in Bot. zhurn. SSSR, XVII (1932) 500. – Angular teeth of outer calyx 2–3 mm long, intermediate teeth half as long; paleae short-acuminate; stems covered with retrorse bristles below; leaves lyrate-pinnatifid, with large terminal lobe and smaller lateral one.

Mountain steppe species of E. Anatolia, Arm.-Kurd., NW Iran., S. and E. Transc.

8. **C. tchihatchewii** Boiss. in Tchihat. Bot. As. Min. (1860) 220; Bobrov in Bot. zhurn. SSSR, XVII, 500; Grossg., Fl. Kavk. IV, 51. – Ic.: Tchihat. l. c. Atlas, tab. 26.

Perennial; stems 1 m, sulcate, subglabrous above, with retrorse hairs below; leaves pinnatifid, with 11–13 decurrent bristly lobes, with irregularly dentate margin, terminal lobes much larger, upper leaves subglabrous, the lower with stiff hairs beneath. Heads globose, ca. 2.5 cm across; involucral bracts ovate, blackish, short-haired, dorsally pubescent; paleae 11 mm long, 4 mm wide, cuneate, triangularly broadening above and short-acuminate, blackish in upper quarter, short haired, dorsally pubescent, with slightly ciliate margin; acheme 6–8 mm long, tetrahedral, pubescent; angular teeth of outer calyx ca. 2 mm long, almost exceeding calyx in fruit, intermediate teeth ca. 1.5 mm long; flowers sulphur-yellow, pubescent outside, radial, inner flowers 15–18 mm long; calyx saucerlike, with numerous irregular marginal prickles. July.

Open slopes of montane belt. – Caucasus: Tal. **Gen. distr.**: Arm.-Kurd., NW Iran. Described from sources of the Araks River.

9. **C. nachiczevanica** Bobr. sp. nov. in Addenda XXIII, p. 325.

Perennial; stems ca. 0.8 m; sulcate, subglabrous above, below more densely pubescent, with retrorse bristles, slightly branching below heads; lower leaves subentire, long-petioled, the median and upper leaves with 1–2 pairs of lateral lobes, these lanceolate, 10–15 cm long, 2–3 cm wide, long-decurrent; limbs pubescent on both sides, with slightly dentate or entire margin, lobes of terminal leaves sublinear. Heads globose, ca. 2 cm across; involucral bracts nearly coriaceous, ovate-lanceolate, brownish only above, with long appressed hairs; paleae coriaceous, lanceolate, 10 mm long, almost acumi-
36 nate, dorsally acuminate, with pubescent margin, turning brownish; achene 6–8 mm (?), tetrahedral, pubescent, angular teeth ca. 3 mm long, one-third to one-half as long as tube, the latter exceeding calyx in fruit; intermediate teeth nearly half as long as the angular; calyx with numerous unequal hairy marginal prickles; flowers sulphur-yellow, hairy outside, inner flowers ca. 10 mm, the outer larger. June.

Subalpine belt. – Caucasus: S. Transc. (Nakhichevan ASSR: Lyakatakh, Bichenakh mountains). **Gen. distr.**: probably distributed in Arm.-Kurd. Described from Lyakatakh Mountain. Type in Leningrad.

Note. Distinguished from C. hirsuta Stapf by less densely pubescent leaves, 1–2 (as against 3–4) pairs of small lateral lobes, ovate-lanceolate involucral bracts (not orbicular), and by lanceolate, not oblong-cuneate paleae.

Series 4. **Microcephalae** Bobr. in Bot. zhurn. SSSR, XVII (1932) 502. – Angular teeth of outer calyx 3 mm long, intermediate teeth half as long; stems 80 cm, slightly branching; radical leaves entire, rarely pinnatifid at base.

Mountain steppe species of E. Anatolia, Arm.-Kurd., N. Iran., S. Transc. and Mtn. Turkm.

10. C. armeniaca Bordz. in Tr. Yur'evsk. bot sada, XIII (1912) 22; Bobrov in Bot. zhurn. SSSR, XVII, 502; Grossg., Fl. Kavk. IV, 51. – C. melanolepis Fisch. et Mey. nom. in herb. Leninopol. – Lepicephalus armeniacus Bordz. in Izv. Kievsk. bot. sada, VII-VIII (1928) 22.

Perennial; stems 80 cm, branching, sulcate, with very short scattered hairs, to below heads; lower and radical leaves oblong-lanceolate, acuminate, dentate, subglabrous or with very short scattered hairs with ciliate margin; median leaves pinnatifid, with 2–4 linear lateral lobes and large oblong-lanceolate terminal one; upper leaves pinnatifid, with linear-lanceolate entire lobes. Heads globose, ca. 2 cm across; involucral bracts ovate, blackish, short-haired; paleae ca. 8 mm long, oblanceolate, acuminate, pubescent, blackish in upper third, dorsally with long appressed hairs, with sparsely short-ciliate margin; achene tetrahedral, grooved, pubescent, 6–8 mm long; angular teeth 2.5 mm long, half as long as outer calyx, exceeding calyx in fruit, intermediate teeth 1.5 mm long; flowers pale yellow; pubescent outside, outer flowers 14 mm, inner flowers ca. 9 mm; calyx saucerlike, with unequal marginal prickles. Fl. July, Fr. August.

Subalpine meadows. – Caucasus: S. Transc. (Alagez, Lake Mada-Tapa, Akhalkalaki). Endemic. Described from Alagez. Type in Leningrad.

37 **11. C. pilosa** Boiss. et Huet in Boiss. Diagn. ser. 2, II (1856) 122; Bobrov in Bot. zhurn. SSSR, XVII, 503; Grossg., Fl. Kavk. IV, 51.

Perennial; stems 50 cm, with stiff hairs in lower part, glabrous above; leaves coriaceous, with prominent veins, bristly on both sides and at margin; radical leaves apparently entire, the lower pinnatifid, with narrow lateral lobes, the upper with very narrow lobes. Heads globose, ca. 2 cm across; involucral bracts triangular-ovate, blackish, dorsally bristly; paleae 8 mm long, oblong, triangularly acuminate, becoming blackish, bristly dorsally and along margin; achene 6–7 mm long, tetrahedral, grooved, pubescent; angular teeth 2–2.5 mm, one-third as long as outer calyx, exceeding calyx in fruit, intermediate teeth half as long as outer calyx; flowers pale yellow, pubescent outside, inner and outer flowers nearly equal, ca. 10 mm long; calyx saucerlike, bristly, with bristly marginal awns. Fl. July, Fr. August.

High mountain steppe. – Caucasus: S. Transc. (Novyi Bayazet). **Gen. distr.:** Arm.-Kurd., NW Iran. Described from Erzerum District. Type in Geneva.

12. C. sublanata (Bornm.) Szabo in Ung. Bot. Bl. Heft 1/12 (1925) 11; Bobrov in Bot. zhurn. SSSR, XVII, 504. – C. microcephala var. sublanata Bornm. in Beih. z. Bot. Zentralbl. XIX, 2 (1906) 269.

Perennial; stems 50–70 cm, sulcate, lanate, rather more densely lanate in lower part; radical leaves subentire, 10 cm long, 1–1.5 cm wide, tapering to long petiole, dentate, pubescent on both sides and along margin, the lower proximally pinnatifid, the terminal lobe 10 cm long, 1.5 cm wide, lateral lobes 1–1.5 cm long, upper leaves linear, pinnatifid. Heads subglobose, loose, 1.5–2 cm across; involucral bracts ovate, greenish, brownish above, with crowded bristles; paleae 8–10 mm long, lanceolate, triangularly acuminate, dorsally covered with appressed hairs. At margin with scattered bristles, becoming brownish; achene 6–7 mm long, tetrahedral, pubescent, grooved; angular teeth 3 mm long, one-third to one-half as long as outer calyx, intermediate teeth as long as calyx in fruit; flowers pale yellow, pubescent outside, outer and inner flowers nearly equal, 10–11 mm long; calyx saucerlike, bristly, with unequal marginal bristles. Fl. June, Fr. July.

Mountain steppes. – Centr. Asia.: Mtn. Turkm. (Kopet Dagh). **Gen. distr.:** N. Iran. Described from Ashkhabad District. Cotype in Leningrad.

Series 5. **Simplices** Bobr. in Bot. zhurn. SSSR, XVII (1932) 502. – Angular teeth of outer calyx 1 mm long, intermediate teeth much shorter;
38 heads 1–1.7 mm across; stems simple; leaves radical, lyrate.

This series comprises a single genus, with one species.

13. C. dagestanica Bobr. In Bot. zhurn. SSSR, XVII (1932) 502; Grossg., Fl. Kavk. IV, 51. – Ic.: Bobrov, op. cit. 501.

Perennial; stems 2–3, 30–40 cm, simple, sulcate, covered with white retrorse bristles below, with scattered bristles above, bristles dense under heads, leafless, sometimes with lyrate leaves below; leaves usually radical, 5–10 cm long, lyrate, with 5–8 decurrent lobes on each side, the lower 2–3 mm long, 1 mm wide, the terminal lobe ovate, 4 cm long, 2 cm wide, all lobes irregularly dentate, bristly on both sides and along margin; petioles densely bristly, especially at base. Heads subglobose, 1–1.5 cm across; outer involucral bracts brownish green, ovate, bristly, the others longer, pubescent, with scattered bristles and ciliate margins; paleae subcartilaginous, pale brown, lanceolate, dorsally short-haired, with ciliate margin, 9 mm long, 3 mm wide; achene 5–7 mm long, tetrahedral, grooved, densely pubescent, angular teeth 1 mm, intermediate teeth much shorter; flowers pale yellow, pubescent outside, outer flowers 13 mm, the inner 9 mm long; calyx saucerlike, bristly, with unequal marginal prickles. Fl. July, Fr. August.

Pebbly slopes in the montane belt. – Caucasus: Dag. (Kara-Koisu valley, above Kula). Endemic. Described from the indicated locality. Type in Leningrad.

Section 2. LEUCOCEPHALAE Szabo in Math. Termtud. Ertes. XXXIX (1922) 274. – Paleae acuminate, dorsally pale, becoming brownish; teeth of outer calyx barely discernible.

The seven species of this section are circumpontine (in the broad sense); two of these are distributed outside the Soviet Union (Balkans, N. Anatolia).

Series 1. **Laevigatae** Bobr. – Outer calyx ribbed, the ribs terminating in tuberculate teeth; paleae acuminate; stems bristly below; leaves nearly coriaceous, pinnatifid, with lanceolate, entire lobes, sometimes cleft or subdentate.

Crimean-Pannonian plant; besides the Soviet species, C. l a e v i g a t a (W. et K.), distributed in Banat and Yugoslavia, is referred here.

39 **14. C. demetrii** Bobr. in Bot. zhurn. SSSR, XVII (1932) 484.

Perennial; stems 1 m, brownish in lower part, with spreading bristles, glabrous above, luminous, sulcate; leaves pinnatifid, the radicle sometimes lanceolate, entire; lower leaves and petioles bristly, short-haired on both sides, the upper less hairy or subglabrous, the lobes lanceolate, entire or dentate, sometimes slightly cleft. Heads subglobose, ca. 2 cm across; outer involucral bracts green, ovate, obtuse, with pubescent margins; paleae subcartilaginous, brownish, lanceolate, acuminate, nearly spinous, dorsally short-haired, with ciliate margins, 10–12 mm long, 3.5 mm wide; achene 6–8 mm long, pubescent, tetrahedral, grooved, tuberculate-ribbed above; flowers pale yellow, pubescent outside; outer flowers 10–13 mm, the inner 8–11 mm long; calyx saucerlike, bristly, with bristly-dentate margin. August. (Plate III, Figure 2.)

Pebbly mountain slopes. – European part: Crim. (E.). Endemic. Described from Karadag. Type in Moscow.

Note. This endemic Crimean species is vicarious with C. laevigata (W. et K.) Schrad. growing in Banat and Serbia. The Crimean plant clearly differs from the western species by the bristly stems and leaves, the character of the truncate lamina, the smaller heads, the pubescent involucral bracts and subcartilaginous, brownish, dorsally pubescent paleae. This species is named for D. P. Syreitshchikov, the only investigator of the Crimean flora to have collected the plant.

Series 2. **Coriaceae** Bobr. in Bot. zhurn. SSSR, XVII (1912) 485. – Outer calyx ribbed-tuberculate above; paleae acuminate; stems usually glabrous, sometimes bristles under heads and at base; leaves coriaceous, usually entire, rarely lyrate.

Pontic plant (Crim., W. Caucasus, Anatolia); C. paphlagonica Bobr. from E. Anatolia is also referred here.

15. C. coriacea (Willd.) Steud. Nomencl. (1821) 177; Litv. in Spiske rast. GRF, VI, 154; Bobrov in Bot. zhurn. SSSR, XVII, 485; Grossg., Fl. Kavk. IV, 50. – C. cretacea Roem. et Schult. Syst. III (1818) 51. – C. centauroides var. cretacea Coult. Mém. Dips. (1823) 26; Ldb. Fl. Ross. II, 449. – C. uralensis var. cretacea Boiss. Fl. or. III (1875) 123; Fedch. and Fler., Fl. Evrop. Ross. 929. – Scabiosa coriacea Willd. in Neue Schr. der Gesellsch. Naturf. Fr. zu Berlin, III (1801) 416; Enum. hort. reg. bot. Berol. I, 1809. – S. cretacea Pall. Tabl, phys. taur. (1795) 46, nom. nud.; M. B. Fl. taur.-cauc. I, 93. – Exs.: GRF, No. 1978; Dorfl. Herb. norm. No. 4105.

40 Perennial; stems 70–80 cm, glabrous, smooth, luminous, sometimes short-haired in lower part; radical leaves lanceolate, entire, long-petioled, rarely quinquefid, with appressed-bristly margin, other leaves pinnatifid, with lanceolate entire, sometimes barely cleft lobes, all leaves coriaceous, yellowish green, sometimes bristly above. Heads subglobose, 1.5–2 cm across; outer involucral bracts green, ovate, obtuse, pubescent; paleae subcartilaginous, brownish, oblanceolate, short-acuminate, dorsally short-haired, with ciliate margin, 9–11 mm long, 3–5 mm wide; flowers pale yellow, pubescent outside, outer flowers 11–14 mm, the inner 9–11 mm long; achene 6–8 mm long, tetrahedral, pubescent, furrowed, tuberculate-ribbed above. Fl. June, Fr. July. (Plate III, Figure 1.)

Dry pebbly foothill slopes. – European part: Crim. (mountainous part); Caucasus: Cisc. (in vicinity of Mineral'nye Vody), W. Transc. (only north of Novorossisk province). Endemic. Described from the Crimea. Type in Berlin.

Series 3. **Corniculatae** Bobr. in Bot. zhurn. SSSR, XVII (1932) 486. – Outer calyx crowned with small teeth; stems bristly below; leaves pinnatifid, more or less bristly.

Steppe plains of the European part of the USSR and mountain steppes of Ciscaucasia, Crimea and the Caucasus.

(41)

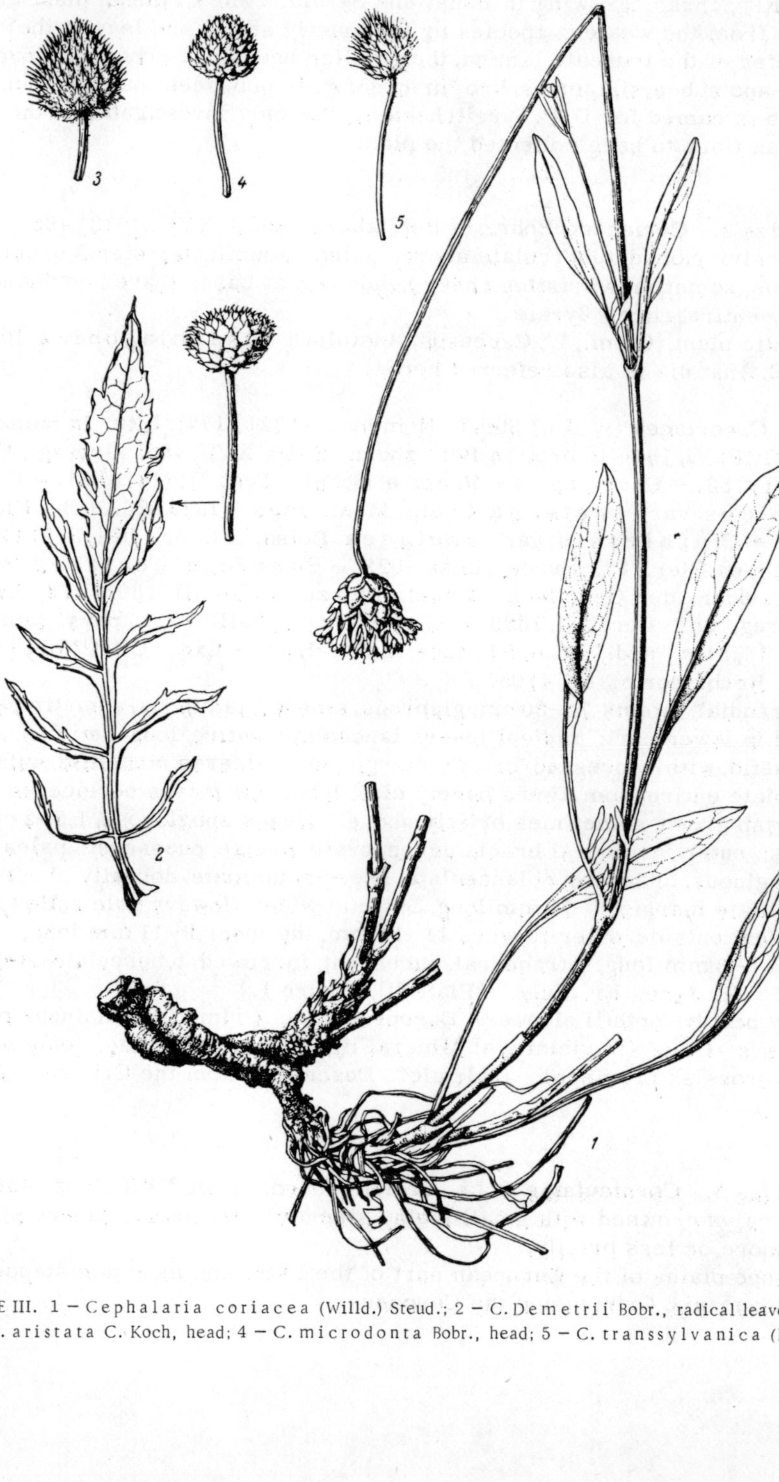

PLATE III. 1 – Cephalaria coriacea (Willd.) Steud.; 2 – C. Demetrii Bobr., radical leaves, head; 3 – C. aristata C. Koch, head; 4 – C. microdonta Bobr., head; 5 – C. transsylvanica (L.) Schrad., head.

16. **C. uralensis** (Murr.) Schrad. Cat. sem. hort. Götting. (1814); Shmal'g., Fl. II, 26; Litv. in Spiske rast. GRF, VI, 153; Fedch. in Fler., Fl. Evrop. Ross. 929; Bobrov in Bot. zhurn. SSSR, XVII, 486; yugo-vost. VI, 276; Grossg., Fl. Kavk. IV, 50. – C. centauroides var. corniculata Coult. Mém. Dips. (1823) 26. – C. centauroides var. uralensis DC. Prodr. IV (1830) 648; Ldb. Fl. Ross. II, 449; Korsh. Tent. Fl. ross. or. 197. – Scabiosa uralensis Murr. Comment. Goett. V (1782) 13 tab. 4; Rchb. Ic. Bot. tab. 314, No. 487. – S. corniculata W. et K. Pl. rar Hung. I (1802) 11, tab. 13; Rchb. l. c. tab. 315, No. 488. – Ic.: Murr. l. c.; Rchb. l. c.; W. et K. l. c.; Fl. yugo-vost. fig. 658. – Exs.: GRF, No. 1977.

Perennial; stems 0.5–0.8 m, in lower part cylindrical, pubescent and covered with spreading bristles, in upper part, ribbed, subglabrous, shiny, sometimes covered with long soft hairs under heads; radical leaves rarely entire, usually all pinnatifid, with decurrent lanceolate lobes, bristly on both
43 sides and on petioles; lobes of upper leaves sublinear, with few bristles. Heads subglobose, ca. 1.5 cm across; outer involucral bracts green, ovate, obtuse, the others longer, all pubescent, with ciliate margins; paleae subcartilaginous, brownish-yellow, pale, lanceolate, acuminate, dorsally pubescent with ciliate margin, 11 mm long, 3 mm wide; achene 6–8 mm long, pubescent, tetrahedral, furrowed, angular teeth small, 0.5–0.7 (1) mm long, intermediate teeth barely discernible; flowers pale yellow, pubescent outside, outer flowers 10–14 mm, the inner 8–10 mm long; calyx saucerlike, bristly, with irregular marginal teeth covered with bristles. Fl. June, Fr. July.

Steppes, pebbly slopes and sands, chalks. – European part: U. Dnp., Bes., M. D., V.-Don (extreme south), Transv. (S.), V.-Kama (S.), Bl., L. Don, Crim., L. V.; W. Siberia: U. Tob. (SW); Caucasus: Cisc. (Stavropol Plateau), W. Transc. (Novorossiisk District). **Gen. distr.**: Centr. Eur. (Transylvania, Dobrudja). Described from specimens grown from seeds obtained in the Urals. Type in Berlin?

17. **C. media** Litv. in Spiske rast. GRF, VI (1908) 155; Bobrov in Bot. zhurn. SSSR, XVII, 488; Grossg., Fl. Kavk. IV, 50. – C. leucantha var. albescens Ldb. Fl. Ross. II (1844) 449. – C. uralensis var. cretacea Somm. et Lev. in Tr. Peterb. bot. sada, XVI (1900) 215. – Succisa leucantha Hohen. Enum. Elisabethopol. (1832) 215. – Exs.: GRF, No. 1979; Pl. or exs. No. 43; Fl. Cauc. exs. No. 290.

Perennial; stems 70–80 cm, in lower part cylindrical, pubescent, with sparse retrorse hairs, in upper part subglabrous, ribbed, shiny; leaves pinnatifid, lobes decurrent, lanceolate or linear-lanceolate, acuminate, usually entire, with revolute margins; limbs with short scattered hairs and appressed bristles on both sides and along margins; lobes of upper leaves narrowly linear, the uppermost leaves linear, often entire. Heads subglobose, ca. 1.5 cm across; outer involucral bracts green, ovate, obtuse, the inner longer, all very short-haired; paleae subcartilaginous, pale, reddish-brownish above, lanceolate, acuminate, dorsally short-haired, with ciliate margins, 10 mm long, 3 mm wide; achene 6–8 mm long, pubescent, tetrahedral, when ripe fusiform, furrowed; outer calyx tipped with obsolete teeth; flowers pale yellow, pubescent outside, outer flowers 10–14 mm, the inner 8–10 mm long; calyx saucerlike, bristly, with irregular bristly marginal teeth. June.

44 Stony and pebbly slopes of middle and lower mountain belts. – Caucasus: Cisc. (N. Osetiya: Mikhailovskoe, Baryatinskaya), Dag., W. Transc. (S.), S. Transc. (N.), E. Transc. Endemic. Described from wild specimens in the Tbilisi botanical garden. Type published.

Note. D. I. Litvinov (GRF, No. 1979), who called it var. *pilosa*, points out that he collected this species in the Crimea, near the Georgievskii Monastery. His specimens have not been preserved and the only known collections from the Crimea should surely be referred to C. *uralensis*. Litvinov's determinations probably were erroneous and C. *media* does not seem to grow in the Crimea, nor has it ever been found in the NW Caucasus (Novorossiisk District).

18. C. velutina Bobr. in Bot. zhurn. SSSR, XVII (1932) 489; Grossg., Fl. Kavk., IV, 50.

Perennial; stems 40–70 cm, branching, densely covered with long retrorse bristles in lower part; leaves mostly radical, lyrate, sometimes subentire, long-petioled, lobes 1–5 on each side, decurrent, nearly opposite, 2–10 mm long, 1–5 mm wide, terminal lobe much larger, 4–6 cm long, 1–2 cm wide, acuminate, sometimes spatulate, rounded or slightly acuminate, all leaves entire or faintly dentate, with soft white bristles on both sides (dense at base) and along margin; cauline leaves narrowly lanceolate, the upper smaller, linear. Heads subglobose, ca. 1.5 cm across; outer involucral bracts brown, ovate, obtuse, the others longer, all densely covered with long silky hairs; paleae 6–8 mm long, nearly scarious, cuneate, brown, dorsally with soft long hairs, with ciliate margins, triangularly acuminate above, 4–5 mm wide in upper fourth; achene 6–8 mm long, pubescent, tetrahedral, furrowed at sides, tipped with obsolete teeth; flowers yellow, pubescent outside, outer flowers 11–13 mm, the inner 7–9 mm long, calyx saucerlike, bristly, margins with irregular bristly teeth. Fl. July, Fr. August.

Limestones in the montane belt of the eastern part of the Main Range of the Caucasus. – Caucasus: Dag., E, Transc. (Kuba). Endemic. Described from Kuba District, Karachai below Budug. Type in Leningrad.

Note. This species is distinguished from C. *media* Litv. by involucral bracts with dense long silky hairs; paleae nearly scarious, cuneate, acuminate; stems densely covered with long retrorse bristles below; leaves mostly radical, with ovate-lanceolate lobes, rather more densely
45 covered with white bristles beneath. The entire plant is pale, velutinous.

Section 3. ECHINOCEPHALAE Lange in Willk. et Lange, Prodr. Fl. Hisp. II (1870) 13. – Paleae aristately acuminate, rarely obtuse; awnless; outer calyx 8-toothed, with longer angular teeth, rarely teeth small, nearly equal or obsolete; annuals.

In addition to the Soviet species, this section contains 6–7 species, comprising 2–3 species series, mostly distributed in the eastern Mediterranean region.

Series 1. **Transsilvanicae** Bobr. h. l. – Teeth of outer calyx small or obsolete; paleae obtuse, lanceolate not tapering to awn. C. *sillingeri* Domin, from S. Slovakia, is also included here.

19. C. microdonta Bobr. in Izv. Glavn. bot. sada, XXVIII (1929) 391; Grossg., Fl. Kavk. IV, 49. — C. leucantha Ldb. Fl. Ross. II (1844) 448 p. p. non Schrad. — C. aristata Schmalh. Fl. II (1897) 26, non C. Koch. — C. caucasica Niem. nom. in herb. — C. centauroides var. Ledebouriana Trautv. nom. in herb. — Ic.: Bobrov, op. cit.

Annual; stems 0.5–1 m, ribbed, branching, glabrous above, bristly below; leaves bristly, the lower lyrate, with lanceolate acuminate dentate lobes, the terminal lobe larger, upper leaves much smaller, with subentire linear-lanceolate lobes. Heads 1–1.5 cm across, long-peduncled; outer involucral bracts ovate, obtuse, oblong, acuminate; paleae keeled in ripe fruits, ciliate reddish brown dorsally and along margins, appressed-hairy, sometimes silky; flowers heteromorphic, bluish, pubescent outside, the outer 2–3 times as long as and larger than the inner; achene fusiform; outer calyx pubescent, tetrahedral, furrowed, tuberculate-dentate, teeth small, obsolete, the 4 smaller ones angular; calyx tetrahedral, multidentate, pubescent. Fl. May–June, Fr. June–July. (Plate III, Figure 4.)

Steppe and shrubby slopes of the montane belt, rarely among crops. — Caucasus: Cisc., Dag., E. Transc. Described from Signakh District. Endemic. Type in Leningrad.

20. C. transsylvanica (L.) Schrad. Cat. sem. hort. Götting. (1814); Roem. et Schult. Syst. III, 45; Ldb. Fl. Ross. II, 448; Boiss. Fl. or III, 118; Shmal'g., Fl. II, 85; Bobrov in Izv. Glavn. bot sada, 388; Grossg., Fl.
46 Kavk. IV, 49. — Scabiosa transsylvanica L. Sp. pl. (1753) 98; M. B. Fl. taur.-cauc. I, 94. — Ic.: Hegi, III, Fl. VI, 1, 287. — Exs.: GRF, No. 1982; Herb. Fl. Cauc. No. 594; Schultz, Herb. norm. No. 1074; Fl. Hung. No. 254; Fl. exs. austro-hung. No. 3817; Fl. exs. reip. Boh.-Slov. No. 915.

Annual; stems erect, 0.3–1.5 mm [?], ribbed, branching, subglabrous above, bristly under heads and at base; lower leaves lyrate, the upper pinnatifid, with long lateral lobes and larger terminal lobe, limbs bristly, subdentate; heads globose or ovoid, 1–2 cm across; outer involucral bracts leaf-shaped, greenish, covered with long hairs; paleae ovate-lanceolate, aristately acuminate, the exterior wider than the interior, ciliate, reddish brown above and along awn, dorsally short-haired; flowers radial, the outer larger than the inner, pubescent outside, pale blue to nearly white; achene tetrahedral, outer calyx grooved, pubescent, crowned with 8 small, nearly equal teeth; calyx saucerlike, 1–1.5 mm across, with dentate bristly margin. Fl. May–June, Fr. June–July. (Plate III, Figure 5.)

Steppe slopes and shrub thickets, ditches, margins of arable land, vinyards and mountain slopes. — European part: Bes., Bl., L. Don (S.), Crim.; Caucasus: Cisc., Dag. W. Transc. (N.), E. Transc. **Gen. distr.:** W. Med., Centr. Eur., Bal.-As. Min. Described from Transylvania. Type in London.

Series 2. **Phalacrocarpus** Boiss. Fl. or. III (1875) 117, pro sect.; Szabo in Math. Termtud. Ertes. XXXIX (1922) 273, pro subgen. — Achene fusiform, outer calyx edentate.

C. sintenisii Freyn, from Transylvania, is also included here.

21. C. aristata C. Koch in Linnaea, XXIV (1851) 445; Boiss. Fl. or. III, 118; Bobrov in Izv. Glavn. bot. sada, XXVIII, 387; Grossg., Fl. Kavk. IV, 49. – C. calva Boiss. Diagn. sér. II, 6 (1854) 95. – C. gracilis Bordz. in Tr. Yur'evsk, bot. sada, XIII (1912) 22. – C. aristata f. gracilis Bordz. in Izv. Kievsk. bot. sada, VII–VIII (1928) 21.

Annual; stems 0.5–1.5 m, erect, branching, with scattered bristles; lower leaves oblong, lyrate-pinnate, lateral lobes dentate, the terminal much larger, upper leaves smaller, with small, lanceolate, entire lateral lobes; limbs pubescent. Heads 1.5–2.5 cm across, on long peduncles; outer involucral bracts rounded; paleae oblong, tapering into awn, midrib and awn reddish brown; paleae and involucral bracts densely covered with silky hairs; flowers pale yellow, hairy, distinctly heteromorphic, the outer twice as long as the inner and 3–4 times larger; achene fusiform, outer calyx with
47 bristly hairs, tetrahedral, furrowed, edentate; calyx tetragonal, multidentate, pubescent from base to tips of teeth. Fl. June, Fr. July. (Plate III, Figure 3.)

A weed of crops and fallow lands. – Caucasus: W. Transc. (extreme south, Chorokh valley). **Gen. distr.:** Arm.-Kurd. Described from Shateven in E. Anatolia. Type in Berlin.

22. C. armena Grossh. in Tr. Bot. sada, ser. 2, 1 (1920) 29; Bobrov in Bot. zhurn. SSSR, XVII, 5–6, 505. – C. hajastana Grossh. Fl. Kavk. IV (1934) 49, nom. nov.

Annual; stems 1–2 cm, ribbed, branching, short-bristly above; cauline leaves subsessile, pubescent, pinnatifid, with 3–5 lanceolate lateral lobes; terminal lobe much larger, 4 cm long; limbs with scattered bristles. Heads 1.5–2 cm across, long-peduncled; involucral bracts ovate, nearly white-tomentose, lilac-brown above; paleae ovate, acuminate, tapering to lilac awn, with ciliate margin, dorsally white-tomentose; flowers pale blue, the outer 13 mm, the inner 8 mm long; achene fusiform, bristly outer calyx edentate; calyx with numerous marginal bristles. August.

Weed among crops. – Caucasus: S. Transc. (near Akbash in Erevan District)? Endemic. Described from localities mentioned. Type in Tbilisi.

Series 3. **Syriacae** Bobr. h. l. – Paleae tapering to awn as long as paleae; angular teeth of outer calyx exceeding calyx.

C. stapfii Hausskn., from Kurdistan, is included here along with the Soviet species.

23. C. syriaca (L.) Schrad. Cat. sem. hort. Götting. (1814); Roem. et Schult. Syst. III, 22; Ldb. Fl. ross. II, 447; Boiss. Fl. or III, 120; Shmal'g., Fl. II, 26; Bobrov in Tr. po prikl. bot. XXI, 311; Grossg., Fl. Kavk. IV, 49. – C. Vaillantii Schott ex Roem. et Schult. Syst. III (1818) 46; M. B. Fl. taurc.-cauc. III, 99. – C. syriaca α pedunculata DC. Prodr. IV (1830) 648. – C. syriaca β sessilis DC. l. c. – C. boissieri Reut. in Boiss. Diagn. ser. II, 2 (1856) 122. – C. syriaca β boissieri Boiss. Fl. or. III (1876) 120. – C. syriaca var. persica Pau, Plantas de Persia (1918). – Scabiosa syriaca L. Sp. pl. (1791) 249; Encycl. Meth. VI,

704. — S. sibirica Lam. l. c. (1791) 249; Encycl. l. c. — Lepicephalus syriacus (L.) Lag. Gen. et sp. pl. (1816) 8; Litv. in Tr. Bot. muz. AN,
48 XV, 147. — L. syriacus var. turkestanicus Vved. in Sched. ad Herb. Fl. As. Med. XI (1927) 36. — Ic.: Sorn. rast. SSSR, IV, 165; Bobrov, op. cit. fig. 3, 4. — Exs.: HFAM, No. 296; Herb. Fl. Cauc. No. 593.

Annual; stems 20—70 cm, sulcate-ribbed, stiff-bristly, branches opposite, elongate, erect or horizontal, leaves opposite, amplexicaul, lanceolate, entire or with 1—2 lateral lobes at base, dentate, ciliate margin; limbs with sparse appressed hairs; in young plants radical leaves opposite, almost in rosette. Heads ovoid, long-peduncled at ends of branches or subsessile in bifurcations of branches; involucral bracts and paleae cartilaginous, short-haired, tapering to awn exceeding flowers; flowers pale lilac or pinkish; achene bristly; the 4 angular teeth of outer calyx rather longer, 4 intermediate ones shorter; calyx short, saucerlike, bristly, with 15—20 teeth. Fl. May, Fr. June.

Weed of cereal crops. — **Gen. distr.**: W. and E. Med., Bal.-As. Min., Arm.-Kurd., Iran. Described from Syria. Type in London.

ssp. transcaucasica Bobr. in Tr. po prikl. bot. XXI (1918) 315. — Stems and leaves densely bristly, leaves entire, rarely 1—2-lobed at base; angular teeth nearly half the length of the outer calyx, intermediate teeth one-sixth to one-fourth the length of the angular ones. — Caucasus: Dag., W. and E. Transc. **Gen. distr.**: Iran (NW).

ssp. turanica Bobr. l. c. — Stems with scattered bristles mainly in upper part and under heads; leaves with 1—2 lateral lobes at base; angular teeth nearly half the length of the outer calyx, intermediate teeth one-sixth to one-fourth the length of angular. — Caucasus: S. Transc.; Centr. Asia: T. Sh. (W.), Syr D., Amu D., Pam.-Al. (W.), Kara K. (in oases), Mtn. Turkm. **Gen. distr.**: Iran (N. Iran and Afghanistan).

Economic importance. In the eastern part of the Georgian SSR this species, known as makhobeli, is regarded as a rather useful plant and is not weeded out. The seeds accidentally ground with the wheat and baked give the bread a special aroma and taste and help to preserve its freshness. However, because of the oiliness of the seeds, flour, which contains makhobeli, goes bad rapidly. For this reason the local population grind flour only when necessary. Seeds of makhobeli contain 7.35% water, 16.25% nitrogenous
49 substances, 22.60% oil, 11.60% cellulose, 36.37% nitrogen-free extracts, and 5.85% ash; ripe fruits do not contain starch. The high oil content makes this an oil-plant suitable as fuel. (See Dzhaparidze and Kuparadze study (Tr. Tifl. bot. inst. I, 1933) on the biology and economic importance of makhobeli.)

Genus 1413. **PTEROCEPHALUS** * Vaill.
ex Adans. Fam. II (1763) 152; Coult. Mém. Dips. (1823) 31.

Flowers numerous, in heads subtended by 4—6 involucral bracts in 1—2 series; receptacle squamose; outer calyx 8-grooved for entire length

* From the Greek pteron — wing and kephale — head; at the end of flowering the capitate inflorescences are crowned by numerous plumous bristles.

crowned by small teeth or obsolete corona; calyx short-peduncled, with pappuslike appendage of 12–24 bristles; corolla quinquefid, irregular in marginal flowers; stamens 4; rarely 2–3. Perennials, with strongly branching caudex or annual herbs with opposite leaves; leaves crowded and subentire in perennials, in annuals the median and upper cauline leaves pinnatifid; heads on long leafless peduncles.

About 20 species distributed in Mediterranean countries.

1. Perennials, with strongly branching multicipital caudex . . . . 2.
\+ Annuals . . . . 4. **P. plumosus** (L.) Coult.
2. Flowers pinkish-lilac . . . . . 1. **P. afghanicus** (Aitch. et Hemsl.) Boiss.
\+ Flowers yellow . . . . 3.
3. Heads ca. 3 cm across, peduncles 3–5 cm long . . . . 2. **P. chorassanicus** Czerniak.
\+ Heads 8–9 mm across, pendunclesb 1–2 cm long . . . . 3. **P. fruticulosus** Korov.

1. **P. afghanicus** (Aitch. et Hemsl.) Boiss. Fl. or. Suppl. (1886) 286. – P. s a r a w s c h a n i c u s Lipsky in O. and B. Fedch., Perech. rast. Turk. 3 (1909) 154, nomen. – S c a b i o s a a f g h a n i c a Aitch. et Hemsl. in Journ. Linn. Soc. XVIII (1881) 67. – Ic.: Journ. Linn. Soc. XIX, tab. 15. – Exs.: HFAM, No. 448.

Perennial; loosely cespitose plant, sometimes spreading by virtue of its multicipital strongly branching caudex; leaves numerous, sessile, lanceolate,
50 rarely entire, usually narrowly lyrate-pinnatifid, 1.5–4(6) cm long, green, glabrous or subglabrous. Heads solitary, the terminal ca. 3 cm across, on 1.5–3(6) cm, appressed-hairy peduncles; involucral bracts 9–11, 2-seriate, lanceolate, colored, ca. 1 cm long; flowers lilac, 18–25, ca. 15 mm long, the inner nearly regular, 5-lobed, the outer nearly 2-lipped, corolla tube pubescent outside; receptacle pubescent; outer calyx pubescent, calyx with 17–18 pinnate, lilac bristles, slightly shorter than flowers, 10–12 mm long. Fl. June.

Stony plateaus and rock crevices of high mountains, 2.500–3,500 m. – Centr. Asia: Pam.-Al. (Zaravshan, Gissar, Kashka Darya basin to Kugitang Range, inclusively). **Gen. distr.**: Afghanistan. Described from SE Afghanistan (Kuram basin). Type in London, cotype in Leningrad.

2. **P. khorassanicus** Czerniak in Izv. Glav. bot sada, XXVI (1927) 260, nomen et in Fedde, Report. XXVII (1930) 281.

Perennial; cespitose-pulvinate perennials, with strongly branching woody multicipital caudex and thick knotty branches; leaves sessile, crowded, small, linear or sublinear-lanceolate, acuminate, entire, very rarely barely lyrate-pinnate, 1–2 cm long, 1–2 mm wide, densely covered with short hairs. glaucescent, thick. Heads solitary, ca. 3 cm across, peduncles 3–5 cm long, densely short-haired; involucral bracts nearly 2-seriate, lanceolate to half the length of flowers, acuminate, 10 mm long, entire, densely short-haired, the inner narrower; flowers pale yellow, 13–14 mm long; corolla 5-lobed, pubescent outside, tube long, narrowly cylindrical, 10–12 mm long, lobes ovate, irregular; outer calyx cylindrical, 8-grooved, with rigid hairs along ribs and very short corona; calyx sessile, reddish, with 15–19 long, brown, plumose bristles, 9–11 mm, shorter than flowers. Fl. July, Fr. August.

Shrubby sections of high mountains. – Centr. Asia: Mtn. Turkm. (may possibly be found in higher parts of Kopet Dagh). **Gen. distr.**: Iran. Described from Hazar-Mecheti in N. Iran. Type in Leningrad.

3. **P. fruticulosus** Korov. in Korov., Kult., Popov, Opisanie novykh vidov rast. II (1916), 74, fig. 21.

Loosely cespitose perennial, with strongly branching multicipital caudex; leaves numerous, sessile, linear, 10 mm long, 2 mm wide, entire or lyrate, with 1 pair of lobes at base, thick, short-haired. Heads few, 8–9 mm across,
51 on 1–2 cm peduncles; involucral bracts 2-seriate, lanceolate, densely short-haired, nearly as long as flowers; flowers ca. 1 cm long; corolla yellow, pubescent outside; outer lobe lanceolate, obtuse, larger than the lateral, inner lobe oval; outer calyx cylindrical, 1.5 mm long, grooved, rounded-dentate, pubescent; calyx bristles 13–16 shorter than corolla, appressed-plumose; stamens shorter than corolla. Fl. April–May.

Stony slopes. – Centr. Asia: Mtn. Turkm. (E. Kopet Dagh, Dar-Bulak Mountains at Iranian border near Meana). Described from locality mentioned. Type in Tashkent.

Note. Little is known about this species, which was collected only once (E. P. Korovin near Meana). The only reliable character distinguishing it from the preceding species is the small head with relatively long involucral bracts.

4. **P. plumosus** (L.) Coult. Mém. Dips. (1823) 31, tab. I, f. 15; Ldb. Fl. Ross. II, 451; Boiss. Fl. or. III, 147; Shmal'g., Fl. II, 27; Grossg., Fl. Kavk. III, 54. – Knautia plumosa L. Mant. (1771) 197; M. B. Fl. taur.-cauc. I, 99, III, 103. – Scabiosa plumosa Sibth. et Sm. Fl. graeca, II (1813) 111; Roem. et Schult. Syst. III, 70. – S. willichii Link in Roem. Collect. (1809) 2. – Cephalaria willichii Roem. et Schult, l. c. 53. – Ic.: Coult. l. c.; Sibth. et Sm. l. c. – Exs.: Fl. Cauc. exs. No. 47; GRF, No. 3798.

Biennial; root thin, branching; stems 15–40 cm, erect, branching, cylindrical, glandular-hairy, leafy, producing few heads; lower leaves oblong, entire, dentate, intermediate, lyrate, the upper nearly pinnate. Peduncles elongate, covered with soft hairs, heads solitary, 3 cm across, 15 mm long; involucral bracts 1-seriate, entire, lanceolate, pubescent, as long as flowers or longer; flowers meat-red to pinkish; corolla irregularly quinquefid, darker above, with enlarged outer lobe; stamens shorter than corolla, outer calyx ribbed, truncate and sparingly toothed above, with pubescent ribs; calyx with 11–13 reddish, plumose bristles; receptacle glabrous. Fl. April–May.

Submontane and montane belts in the semidesert and desert zones. – European part: Crim.; Caucasus: Dag., S. and E. Transc., Tal.; Centr. Asia: Mtn. Turkm. (from Kara-Kala to Gaudan). **Gen. distr.**: Arm.-Kurd., Iran. Described from the "East." Type in Geneva.

Tribe 3. **SCABIOSEAE** Van Tiegh. in Ann. Sc. Nat. sér. 9, X (1909). – Receptacle with involucral bracts; involucre sessile, 8-ribbed, with corona; calyx 5-merous; corolla 5-lobed; style duplex.

52 Genus 1414. **SUCCISA** * Neck.
Neck. Elem. I (1790) 109

Flowers in capitate inflorescence; involucral bracts 2-seriate, herbaceous, green; paleae scarious; outer calyx prismatic, tetrahedral, each facet with 2 grooves, contracted at base, tapering into short acuminate teeth at corners, pubescent; calyx saucerlike, with 5 setaceous rays or teeth; corolla 4-lobed.

Monotypic genus.

1. **S. pratensis** Moench, Meth. pl. (1794) 489; Boiss. Fl. or. III, 130; Shmal'g., Fl. II, 28; Grossg., Fl. Kavk. IV, 52. – S. praemorsa (Gilib.) Aschers. Fl. Brandenb. (1864) 285; Kryl., Fl. Zap. Sib. X, 2626. – S. vulgaris Presl, Fl. Čech. (1819) 31; Andrzh. in Kievsk. univ. izv. 7 (1862) 98. – S. palustris Sass, Phaner. Fl. Ösels (1860) 57. – Scabiosa succisa L. Sp. pl. (1753) 98; M.B. Fl. taur.-cauc. I, 94; Coult. Mém. Dips. 40; Ldb. Fl. Ross. II, 458. – S. praemorsa Gilib. Fl. lithuan. III (1781) 168. – Lepicephalus succisa Eichw. Skizze (1830) 151. – Ic.: Hegi, Ill. Fl. VI, Taf. 254, 1; Fl. yugo-vost. VI, Fig. 659. – Exs.: GRF, No. 769; Estonian plants, No. 178; Pl. Finl. exs. No. 371, 965; Fl. pol. exs. No. 735; Fl. exs. austro-hung. No. 3813; Fl. ital. exs. No. 1955; Fl. stir. No. 1258; Fl. exs. reip. Boh.-Slov. No. 1271.

Perennial; root fibrous, short, blackish, appearing as if cut off; stems 15–40(80) cm, bearing few peduncles above, rarely simple, glabrous below, appressed-hairy above; leaves ovate-elliptic, entire, acuminate, thick, shiny above, tapering into petiole, which in lower leaves equals half the length of the lamina; radical leaves with scattered appressed hairs; upper cauline leaves sessile. Heads on long peduncles, subglobose at first, becoming globose, ca. 2 cm across; involucral bracts 2-seriate, broadly lanceolate, acuminate, appressed-hairy, shorter, rarely longer than heads; bracts elliptic, scarious below, hairy above, 6–7 mm long; corolla 4-lobed, ca. 7 mm long, pubescent blue-violet outside, very rarely white, not elongate in marginal flowers; involucre ca. 5 mm long, tetrahedral, with 8 densely hairy longitudinal ribs; 4 basal ribs tapering above, acuminate; calyx teeth
55 brown-black, 2 mm long. Fl. July, Fr. September. (Plate IV, Figure 1.)

Swampy meadows, grass moors, shrubby formations, at forest edges beyond the northern forest and steppe zones, southeastern forest-steppe zone, sometimes solonetzic meadows. – European part: in the north to the extreme south of Kar.-Lap., in the south to Dv.-Pech. (Vologda, Syktyvkar), and in the south from C. Bes. through N. Bl., N. L. Don and N. Transv.; Caucasus: rarely in Cisc. (foothills), W. Transc.; W. Siberia: Ob (extreme south), U. Tob. (N.), Irt. (CW and CE); E. Siberia: Ang.-Say. (Minusinsk). **Gen. distr.**: Atl., Centr. and S. Eur., Bal.-As. Min. Described from Centr. Eur. Type in London.

Genus 1415. **SUCCISELLA** ** Beck
Beck, Fl. Nied. Oesterr. II, 2 (1893) 1145

Flowers in capitate inflorescence; involucral bracts not imbricate, ovate-lanceolate, acuminate, coriaceous, gradually passing into the lanceolate,

* From the Latin succidere – to cut from below, referring to the undercut fibrous root.

** Diminutive of Succisa.

(53)

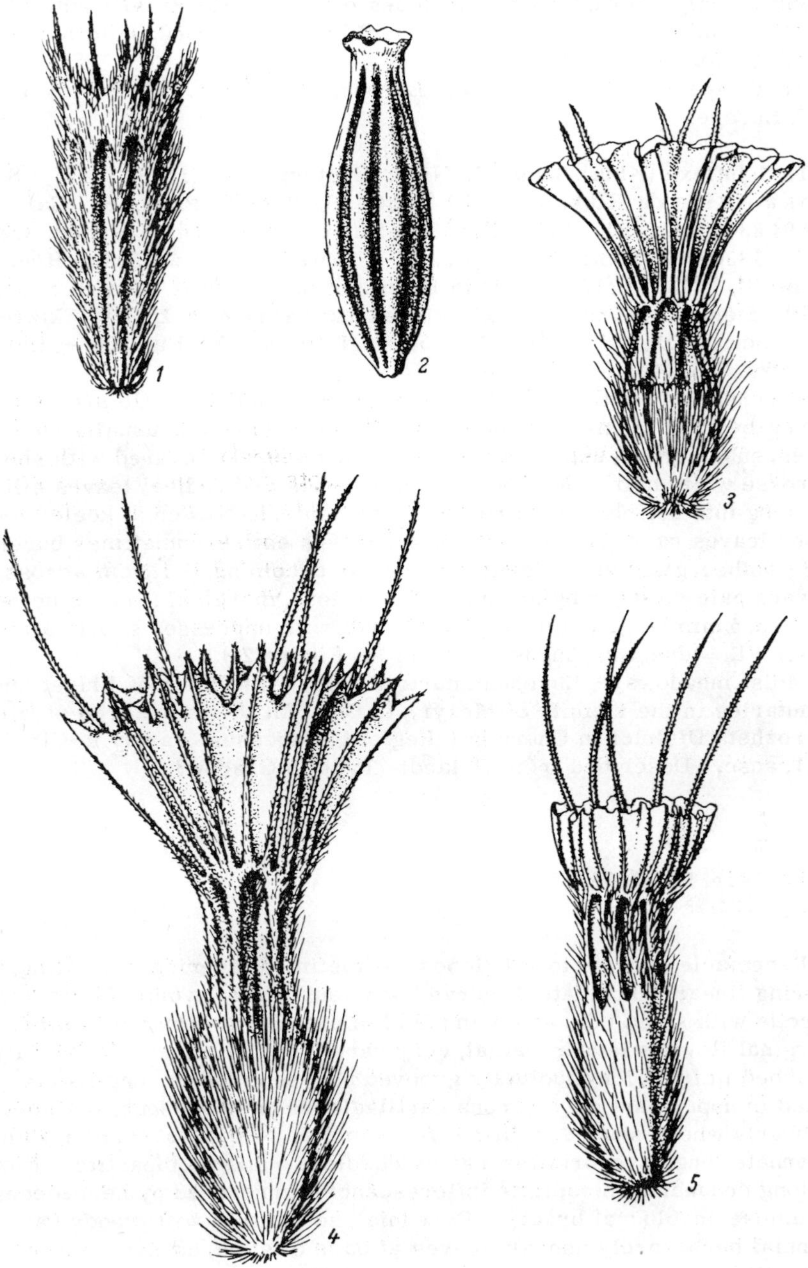

PLATE IV. Achenes: 1 – *Succisa pratensis* Moench; 2 – *Succisella inflexa* (Kluck) Beck; 3 – *Scabiosa alpestris* Kar. et Kir.; 4 – *S. soongorica* Schrenk; 5 – *S. caucasica* M.B.

acuminate paleae, which are also green but narrower; outer calyx pitcher-shaped, nearly terete, tapering above, with 8 deep grooves alternating with 8 ribs, shortly 4-lobed above, the lobes often connate, usually with alveolate venation, glabrous; calyx small, saucerlike, barely lobed, without teeth or bristles; corolla 4-lobed.

Three species distributed in S. Europe, from Spain to Yugoslavia, and in C. Europe.

1. **S. inflexa** (Kluk) Beck, Fl. Nied. Oesterr. II, 2 (1893) 1145. — Scabiosa inflexa Kluk, Dykc. III (1786) 56; Besser, Prim. Fl. Gal. II (1809) 336; Ldb. Fl. Ross. II, 458. — Succisa inflexa Jundz. Opis rosl. (1830) 56; Shmal'g., Fl. II, 28. — S. australis Wulfen in Roem. Arch. III, 3 (1805) 316; Schott in Roem. et Schult. Syst. III (1819) 61; Coult. Mém. Dips. 40. — Lepicephalus inflexus Eichw. Skizze (1830) 151. — Ic.: Beck, l. c. 1141, f. 154, 5; Hegi, Ill. Fl. VI, Fig. 160. — Exs.: Fl. pol. exs. No. 349; Fl. stir. exs. No. 390.

Perennial; root thin, creeping, branching, bearing sterile shoots and flower-bearing stems; stems ascending, 40—80(100) cm, usually rooting at nodes, subglabrous, upper part of peduncles densely covered with short retrorse white hairs; leaves of sterile shoots and radical leaves elliptic,
56 tapering into petioles; cauline leaves elongate, narrower at keeled base, upper leaves narrowly lanceolate; all leaves entire, sometimes barely 2—3-toothed, glabrous. Heads subglobose becoming 1—1.5 cm across; flowers pale violet, subglabrous, 4—5 mm long, marginal flowers not enlarged; fruit ca. 5 mm long; outer calyx glabrous, with appressed short hairs along ribs. Fl. June, Fr. August. (Plate IV, Figure 2.)

Moist meadows. — European part: U. Dnp. (floodplain of Pripet and its tributaries in the vicinity of Mozyr), U. Dns., M. D. (in the area of Kiev and Ostrozhets District in Chernigov Region); Caucasus: rarely in Cisc. (W.), W. Transc. Described from Poland. Type in Cracow?

Genus 1416. **SCABIOSA*** L.

L. Sp. pl. (1753) 98

Receptacle arcuate to subglobose, sometimes tapering at fruiting, terete, bearing linear-lanceolate, 1-nerved scarious white involucral bracts. Corolla with short tube and quinquefid bilaterally symmetrical limb; marginal flowers larger, radial, very rarely not enlarged. Outer calyx 8-ribbed in fruit, longitudinally grooved between ribs or longitudinally pitted in upper part only, corona cartilaginous-membranous, campanulate or nearly wheel-shaped, multinerved, dentate. Calyx saucerlike, with 5 connate long rays-bristles, rarely obsolete. Stigma bipartite. Flowers on long peduncles in capitate inflorescence, surrounded by herbaceous, not squamose, involucral bracts. Perennial, sometimes with woody base, or biennial herb, rarely annual; leaves at base of stem, connate opposite, pinnatifid or pinnatisect, very rarely entire.

* From the Latin Scabies — scabies; referring to early use in the treatment of this disease.

About 100 species, mostly of Mediterranean countries (in the broad sense); single species reach the Far East and the mountains of E. Africa.

Note. The annual Scabiosa atropurpurea L. (Spongostemma atropurpureum (L.) Van Tiegh.) is grown as an ornamental in the southern regions of the Soviet Union for its broad spectrum of flower colors, from dark purple to white. It is presumably descended from the annual S. maritima L. (Spongostemma maritimum (L.) Van Tieg.), from the Mediterranean countries of Europe. Of the Soviet species of Scabiosa,
57 several varieties of S. caucasica M. B. are cultivated in W. Europe. Perennial species of ornamental interest for their large, bright blue flowers are S. alpestris Kar. et Kir. (the roselike Tien Shan Scabiosa), S. comosa Fisch. (the E. Siberian species), and S. lachnophylla Kitag, a vicariant to the latter.

1. Outer calyx terete, grooved for entire length or pitted-grooved only in upper part . . . . 2.
+ Outer calyx subtetrahedral, with 4 distinct and 4 faint intermediate ribs, not grooved or pitted (Prismakena) . . . . 28.
2. Outer calyx terete in lower part, not ribbed, in upper part pitted-grooved, with 8 longitudinal pits (Asterocephalus) . . . . 3.
+ Outer calyx grooved, 8-grooved for entire length, 8-ribbed (Sclerostemma) . . . . 15.
3. Perennials, rarely biennials . . . . 4.
+ Annuals . . . . 12.
4. Radical and lower cauline leaves entire, rarely barely lyrate-pinnatifid, sometimes cauline leaves entire . . . . 8.
+ Radical and lower cauline leaves pinnatipartite, lobes linear or lanceolate, often cleft . . . . 5.
5. All leaves pinnatipartite or pinnatifid; heads 1.5–2 cm across in fruit; flowers yellowish or pinkish white or pale blue . . . . 6.
+ Upper cauline leaves linear, entire; heads 1–1.3 cm across in fruit; flowers pale yellow . . . . 7.
6. Corolla yellowish white or pinkish white; stems uniformly leafy; leaf lobes linear, 3–10 mm long, ca. 1 mm wide . . . . 1. **S. isetensis** L.
+ Corolla yellowish white; leaf lobes 5–20 mm long, 0.5 mm wide; stems and leaves white-haired . . . . 1a. **S. ulugbekii** Zak.
++ Corolla whitish, violet-blue; stems sparingly leafy; leaves mostly radical, leaf lobes lanceolate, 5–20 mm long, 2–3 mm wide . . . . 2. **S. austro-altaica** Bobr.
7. Perennials; stems 3–5(7), profusely branching, the entire plant appearing virgate; stems often densely white-bristly in lower part; upper branches horizontal; marginal flowers barely enlarged. Plants of stony outcrops in the mountain-steppe belt of Transcaucasia and the Crimea . . . . 8. **S. argentea** L.
+ Biennials and triennials; stems 1–3; upper branches diverging at 45°; marginal flowers much enlarged. Plants of sandy steppes and sands . . . . 9. **S. ucranica** L.
58 8. Stems leafless, very rarely with 1–2 leaves; leaves usually radical, entire; flowers yellow. Plants of limestones of Dagestan . . . . 7. **S. gumbetica** Boiss.

\+ Stems, apart from peduncles, more or less leafy; flowers from pinkish to blue . . . . 9.

9\. Stems profusely leafy, woody at base; all leaves entire, from lanceolate to oblong-ovate, acuminate; in rare cases the petiole bears a pair of small lobes, and thus leaves slightly lyrate; flowers grayish-blue. Plants of W. Transc. . . . . 6. **S. olgae** Albov.

\+ Stems slightly leafy; radical leaves entire, cauline leaves always pinnatifid . . . . 10.

10\. Flowers blue; corona 1.5–2 mm wide in fruit; calyx teeth and bristles, nearly twice as long as corona is wide. Plants of the Caucasus . . . . 3. **S. caucasica** M. B.

\+ Flowers pinkish, pinkish violet or yellowish violet; corona 5–6 mm wide in fruit; calyx teeth and bristles, exceeding width of corona by one-third or one-half. Plants of Centr. Asia . . . . 11.

11\. Flowers pinkish or pinkish violet; calyx teeth bristles, in fruit exceeding width of corona by one-third; stems with 1–3(4) nodes; petioles of radical leaves longer than lamina . . . . 4. **S. alpestris** Kar. et Kir.

\+ Flowers yellowish-violet; calyx teeth bristles, in fruit twice as long as corona is wide . . . . 5. **S. soongorica** Schrenk.

12\. Heads 1.5–3 cm across in fruit; calyx teeth and bristles, not more than twice as long as corona is wide . . . . 13.

\+ Heads 0.5–1 cm across, in fruit; calyx teeth and bristles 4–5 times as long as corona is wide . . . . 13. **S. olivieri** Coult.

13\. Heads globose in fruit; corona 5–8 mm wide; calyx teeth and bristles, barely longer or not more than twice as long as corona is wide . . . 14.

\+ Heads oblong in fruit; corona ca. 2 mm wide; calyx teeth and bristles, 2.5 to 3 times as long as corona is wide; median cauline leaves pinnatisect; heads in flower ca. 1 cm across . . . 10. **S. micrantha** Desf.

14\. Corona 6–8 mm wide; calyx teeth and bristles, exceeding the width of the corona; involucral bracts longer than the barely enlarged marginal flowers . . . . 12. **S. rotata** M. B.

\+ Corona ca. 5 mm wide; calyx teeth and bristles, nearly twice as long as corona is wide; involucral bracts shorter than marginal flowers,
59 the latter markedly enlarged . . . . 11. **S. persica** Boiss.

15\. Flowers yellow . . . . 16.

\+ Flowers pinkish, purple, violet or blue . . . . 21.

16\. Heads oblong in fruit; receptacle after abscission of fruits fusiform . . . . 17.

\+ Heads globose or subglobose in fruit . . . . 18.

17\. Stems simple or branching in middle, crisp-haired, especially under heads and at base; radical leaves entire or lyrate-cut, cauline leaves twice pinnatifid. Predominantly plants of plains in steppe zone . . . . 14. **S. ochroleuca** L.

\+ Stems simple or branching from base, with scattered spreading white bristles in lower part, especially at nodes; radical and cauline leaves 5–8 cm long, lanceolate, lyrate-pinnatifid, the terminal lobe reaching half the length of lamina. Plants of mountain steppes of E. Transc. and Dag. . . . . 15. **S. georgica** Sulak.

18. Calyx teeth and bristles dark brown, 5–7 mm long, 5–6 times as long as corona is wide . . . . 19.

\+ Calyx teeth and bristles brownish, half as long as corona is wide . 20.

19. Stems 50–80 cm, densely covered below with soft white hairs; heads 3–3.5 cm across at anthesis, in fruit ca. 1.5 cm; involucral bracts 3–4 times as long as corona is wide . . . . . . 16. **S. bipinnata** C. Koch.

\+ Stems 20–50 cm, glabrous in lower part; heads ca. 2.5 cm across at anthesis, in fruit ca. 2 cm; involucral bracts twice as long as corona is wide. (Verkhnyaya Svanetiya). . . 19. **S. correvoniana** Somm. et Lev.

20. Stems 40–80 cm, profusely branching; all leaves, except for the upper, with appressed white bristles to gray-tomentose hairs above and especially beneath; heads 1.5–2 cm across at anthesis. (Foothills and coastal belt of W. Transc.) . . . . . . 18. **S. sosnowskyi** Sulak.

\+ Stems 25–50 cm, simple or with 1–2 branches; leaves with short appressed white hairs on both sides, especially beneath; heads 2.5–3.5 cm across at anthesis. (Limestones in Abkhaziya) . . . . 17. **S. imeretica** (Somm. et Lev.) Sulak.

21. Heads 2–3(4) cm across at anthesis; marginal flowers enlarged 22.

\+ Heads ca. 1 cm across at anthesis; marginal flowers not enlarged . . . . 26. **S. meskhetica** Schchian.

22. Plants silvery-gray from short and dense pubescence; radical leaves lanceolate, 10–20 cm long, entire or barely lyrate-cut; the lamina comprising half the length of the leaf; all leaves velutinous-haired
60 on both sides, especially beneath, appearing nearly gray (W. Georgia, Lechkum) . . . . 22. **S. colchica** Stev.

\+ Plants subglabrous or sparingly pubescent, if canescent with short pubescence then plant deriving from Lenkoran lowland . . . . 23.

23. Calyx teeth and bristles somewhat expanding in lower part, keeled at cross section; cauline leaves glabrous, shiny above. (Transcarpathian Region) . . . . 27. **S. lucida** Vill.

\+ Calyx teeth and bristles orbicular; if glabrous above, then leaves not shiny , ... 24.

24. Calyx teeth and bristles 8–10 times as long as corona is wide. Plants of Talysh . . . . 25.

\+ Calyx teeth and bristles 5–6 times as long as corona, and if 6–8 times, then plant from W. Georgia . . . . 26.

25. Stems canescent from dense short pubescence, profusely leafy; terminal leaf lobes ovate, large, about one-third to one-half the length of leaves; lateral lobes obovate, 3–4-paired. Plants of Lenkoran lowland . . . . 21. **S. amoena** Jacq. f.

\+ Stems subglabrous, very sparingly leafy or even nearly leafless with very short hairs below; sometimes only lower cauline leaves with terminal lobes, not more than one-fourth the length of the leaves; lateral lobes 6–9-paired, lanceolate, cut into lanceolate or linear segments. Plants of Nagorny Talysh . . . . . . 20. **S. hyrcanica** Stev.

26. Root multicipital, producing 2–5 stems and sterile rosettes of leaves; stems 15–40 cm; leaves of sterile rosettes, sometimes also radical leaves entire, long-petioled; heads ca. 2 cm across at anthesis. (W. Georgia) . . . . 23. **S. velenovskyana** Bobr.

\+ Stems 40–80 cm, solitary or 2–3; leaves of sterile shoots, as well as the radical and lower cauline leaves, always lyrate-pinnatifid; heads ca. 3 cm across at anthesis . . . . 27.

27. Flowers pink; plant profusely branching; lower branches opposite, branches above branching once again, stems glabrous, only in lower part with few retrorse hairs, under heads hairs short-appressed. Plants of the eastern part of the Main Range, Caucasus . . . . 24. **S. overini** Boiss.

\+ Flowers pale blue or violet; stems with 1–2 branches at middle, glabrous below, sparse white hairs confined to the upper part of
61 peduncles . . . . 25. **S. columbaria** L.

28. Stems with short appressed hairs at base and under heads; leaves glabrous or subglabrous, cauline leaves pinnatipartite, long-lobed . . . . 29. **S. comosa** Fisch.

\+ Stems with spreading bristles and short crisp hairs below, these also in upper part sometimes with spreading bristles under heads; leaves covered on both sides with short hairs, sometimes rather densely so; cauline leaves lyrate-pinnatipartite, with linear-lanceolate lobes, sometimes cleft . . . . 28. **S. lachnophylla** Kitag.

Section 1. ASTEROCEPHALUS Coult. Mém. Dips. (1823) 33; Boiss. Fl. or. III, 131. — Outer calyx terete, in lower part not ribbed, in upper part pitted-grooved, with 8 longitudinal pits.

Series 1. **Isetenses** Bobr. — All leaves pinnatipartite with linear or lanceolate lobes, often cleft. Perennials; heads 1.5–2 cm across in fruit.

1. S. isetensis L. Mant. I (1767) 37; M. B. Fl. taur.-cauc. III, 100; DC. Prodr. IV, 656; Ldb. Fl. Ross. II, 455; Boiss. Fl. or. III, 137; Shmal'g., Fl. II, 29; Kryl., Fl. Zap. Sib. X, 2626. — S. rupestris M. B. Fl. taur.-cauc. I (1808) 95; Roem. et Schult. Syst. III, 64. — Ic.: Gmel. Fl. Sib. II, tab. 88, f. 1; Fl. yugo.-vost. IV, 281.

Perennial; root thick, woody, multicipital; stems 25–45 cm, 4–5, erect or ascending at base, with short crisp hairs, more densely pubescent above; leaves elliptic, radical leaves 5–10 cm long, petioles 1–2 cm, cauline leaves sessile, shorter, all leaves covered on both sides with appressed hairs, pinnatipartite, the lobes linear 3–10 mm long, 1 mm wide, often cleft, and then leaves bipinnate. Heads 2–3 cm across; involucral bracts oblong-ovate, tapering above, nearly one-third the length of the marginal flowers, densely tomentose; paleae keeled in upper third, narrowly lanceolate, filiform below, densely covered with thin hairs; outer calyx rimosely pitted above, terete, in lower half with appressed white bristles, 3 mm long in fruit; calyx 0.5 mm across, with setaceous, 4–5 mm long lobes with darkish tip; corolla yellowish white, rarely pink-white, pubescent outside; marginal flowers 13–15 mm long, nearly 2-lipped, with 2-lobed upper lip, median lobe
62 of lower lip larger than both the lateral lobes; corolla of median flowers 6–8 mm long, one of the 5 ovate lobes longer than the others; achene ovate, ca. 3.5 mm long, glabrous. Fl. July, Fr. August.

Steppes, rubbly steppe slopes, limestones and chalks, dry steppe and northern semidesert zone. – European part: V.-Kama (Urals and W. Ural area, south of Sverdlovsk), V.-Don (SE), Transv., L. Don (E.), L. V.; Caucasus: Cisc. (Stavropol Plateau); W. Siberia: U. Tob., Irt.; Centr. Asia: Ar.-Casp. (N.), Balkh. (N.). Endemic. Described from the Central Urals, Iset River. Type described by Gmelin.

1a. **S. ulugbekii** Zak. Sp. n. in Addenda XXIII, 457.

2. **S. austro-altaica** Bobr. sp. nov. in Addenda XXIII, p. 325. – S. isetensis auct. Fl. alt. non L.

Perennial; root thick, woody, multicipital; stems 1–4, with few leaves, 20–60 cm, erect, sparsely covered with short hairs, especially in upper part, with 1–3 heads; leaves mostly radical, elliptic, 5–10 cm long; petioles 1–2 cm; cauline leaves 2–3-paired, smaller; limbs short-haired above and beneath, more or less bristly along axis; all leaves pinnatipartite, the lobes lanceolate, 5–20 mm long, 2–3 mm wide, sometimes faintly pinnatifid. Heads 2–3 cm across, in fruit globose, 1.5–2 cm across; involucral bracts oblong-ovate, tapering above, one-third to one-half the length of the marginal flowers, densely covered with short, nearly tomentose hairs; paleae narrowly lanceolate, densely covered with thin hairs, filiform below, keeled in upper part; outer calyx rimosely pitted above, white-bristly mainly in lower part, 3–4 mm long in fruit; corona membranous, 2–3 mm wide; calyx 0.5 mm across, with bristly lobes darker at tip; 4–5 mm; corolla violet-blue, appressed-hairy outside; marginal flowers 15 mm long, nearly 2-lipped, median flowers 6–8 mm long. Fl. June, Fr. August. (Plate VII, Figure 1.)

Open rubbly slopes of foothills. – W. Siberia: Alt. (south of Zaisan basin). **Gen. distr.:** Dzu.-Kash. (Chernyi Irtysh River). Described from Kaldzhir. Type in Leningrad.

Note. This species is close to S. isetensis L., from which it is distinguished by its few-leaved stems, mostly radical leaves, with lanceolate lobes, 5–20 mm long, 2–3 mm wide, and violet-blue flowers.

Series 2. **Elegantes** Bobr. – Marginal flowers 2–3 cm long; membranous corona 16–20- or 24-nerved. Radical and lower cauline leaves long-petioled, lanceolate, entire, very rarely lyrate-pinnatifid. In addition to the Soviet species this section includes S. sulphurea Boiss. (As. Min.),
63 S. baliani Diraz. (pontic), and S. speciosa Royle (Himalayas).

3. **S. caucasica** M. B. Fl. taur.-cauc. I (1808) 98, III (1819) 102; Ldb. Fl. Ross. II, 1, 452; Boiss. Fl. or. III, 136; Shmal'g., Fl. II, 29; Grossg., Fl. Kavk. IV, 58. – S. connata Hornem. Hort. Hafn. (1813) 128. – S. elegans Spreng. Pl. min. cogn. pugill. sec. (1815) 24; Roem. et Schult. Syst. III, 84. – S. caucasica α heterophylla DC. Prodr. VI (1930) 654. – S. caucasica β elegans DC. l. c. – Asterocephalus elegans Lag. Gen. et sp. nov. (1816) 8. – A. caucasicus Spreng. Syst. I (1825) 381. – Ic.: Bot. Mag. No. 886. – Exs.: GRF, Nos. 111, 3240; Fl. Cauc. exs. No. 171.

Perennial; root woody, with 2–3 heads; stems 25–60(80) cm; with 3–4 nodes, sometimes subglabrous, usually short-haired in upper part, glabrous with few bristles below; lower radical leaves entire (very rarely lyrate-pinnatifid), lanceolate, with petioles longer than laminas; cauline leaves sessile, usually pinnatifid, with linear lobes; leaves with scattered bristles on both sides and along margin. Heads 3–4 cm across, in fruit short-ovate or subglobose, 3 cm across; involucral bracts lanceolate, acuminate, short-haired, in lower part long-bristly; flowers blue, with short appressed hairs outside, radiant; corolla of marginal flowers 2.5–3 cm long; outer calyx 8–9 mm long, in almost half of upper part pitted-grooved, with appressed white bristles along ribs; corona 1.5–2 mm wide, dentate-undulate, 16–20-nerved; calyx teeth and bristles half as long as corona is wide. Fl. August, Fr. September. (Plate IV, Figure 5.)

Subalpine and alpine meadows. – Caucasus: all regions except for Tal. **Gen. distr.**: E. Anatolia. Described from Kislovodsk District. Type in Leningrad.

4. S. alpestris Kar. et Kir. in Bull. Soc. Nat. Mosc. XV (1842) 536. – S. caucasica auct. fl. As. Med. p. p. non M. B., Fedch., Perech. rast. Turk. 3, 154. – S. caucasica β rosea Kar. et Kir. l. c. 375; Ldb. Fl. Ross. II, 452. – S. caucasica α typica Rgl. in Bull. Soc. Nat. Mosc. XL, 1 (1867) 16. – Exs.: HFAM, No. 297.

Perennial; root woody, with 2–3 heads; stems 30–50 cm, with 2–3(4) internodes, very short-haired, sometimes subglabrous, often faintly colored; petioles of radical and lower cauline leaves longer than laminas, the latter lanceolate, radical leaves always entire, very short-haired or subglabrous above and beneath, bristly along margin; cauline leaves opposite, in 1–3 pairs, pinnatisect below. Heads 3–4 cm across; involucral bracts
64 linear-lanceolate, white-bristly, often colored, 15–18 mm long; flowers pink or pink-violet, with crisp hairs outside, radial, the marginal flowers 25 cm long, heads in fruit globose, ca. 3 cm across; outer calyx 8–10 mm long, pitted-grooved for barely half the upper part, sparsely white-haired at surfaces and in lower part; corona 5–6 mm long, with 16–18 colored nerves and dentate-undulate margin; calyx teeth bristly, 1–2 mm longer than corona. Fl. July, Fr. August. (Plate IV, Figure 3.)

Subalpine and alpine belt. – Centr. Asia: Dzu.-Tarb., T. Sh. (mostly central slope but to the northern slope of Bardob in the Alai Range). **Gen. distr.**: Dzu.-Kash. (high mountain frontier of T. Sh.). Described from subalpine meadows of Sarkand in Dzungarian Ala-Tau. Type in Leinngrad.

5. S. soongorica Schrenk, Enum. pl. nov. I (1841) 55; Ldb. Fl. Ross. II, 453. – S. caucasica auct. fl. As. Med. p. p. non M. B.: Fedch., Perech. rast. Turk. 3 (1909) 154. – S. caucasica β songorica Rgl. in Bull. Soc. Nat. Mosc. XL, 1 (1867) 16. – Exs.: HFAM, No. 299.

Perennial; root woody, with 2–3 heads; stems 40–60(100) cm, with 4–7 internodes, short-haired throughout, densely covered with long hairs below; petioles of radical and lower cauline leaves shorter than laminas, the latter leaves lanceolate, entire, sometimes barely pinnate; cauline leaves opposite, pinnatifid, the terminal lobe much larger, lanceolate; leaves covered on both sides with short and long hairs. Heads 2.5–3 cm across; involucral bracts narrowly lanceolate, 1–1.5 cm long, densely long-bristled; flowers pale, yellowish violet, pubescent outside, radial, outer flowers 2 cm long, heads in fruit globose, 3–3.5 cm across; outer calyx 12–13 mm, more than half as long as tube pitted-grooved, appressed-bristly, the lower third densely covered with long white bristles; corona 5—6 mm wide, 20—24-nerved; calyx teeth and bristles twice as long as corona is wide. Fl. and Fr. June. (Plate IV, Figure 4.)

Herb and bunchgrass steppes and foothills. — Centr. Asia: Dzu.-Tarb. (Dzungarian Ala-Tau), T. Sh., Pam.-Al. (from Kugitang to Darvaza and Shugnan). **Gen. distr.**: Iran (Afghanistan). Described from meadows along the Tentek in Dzungarian Ala-Tau. Type in Leningrad.

Series 3. **Olgeanae** Bobr. — Stems profusely leafy, woody at base; corona of uniform width. Calyx bristles pale brown, longer than corona is wide.
65 The Albanian species S. epirota Hal. et Bald. also belongs to this series.

6. S. olgae Albov in Bull. Herb. Boiss. II (1894) 453; Grossg., Fl. Kav. IV, 57; Kolak., Fl, Abkh. IV, 163.

Perennial; root woody, multicipital, producing 3—10 stems; stems ascending, 25—45 cm, simple, woody in lower part, with 2—5 peduncles from axils of upper leaves, densely white-bristly throughout, outside peduncles profusely leafy; leaves entire, lanceolate to oblong-ovate, acuminate, tapering to short petiole, which rarely bears a pair of small linear lobes, leaves thus barely lyrate; leaves with appressed white hairs, densely hairy beneath, the entire plant silvery-gray. Peduncles 10—15 cm, densely covered with white hairs; heads at anthesis 3.5—5 cm across; involucral bracts lanceolate, attenuate-acuminate, with appressed white hairs, 1.5 cm long; marginal flowers radial, enlarged to 2 cm; corolla grayish blue, with densely white appressed hairs outside; heads in fruit (not counting bristles) ca. 2 cm across; outer calyx 5 mm long, 8-ribbed, longitudinally pitted-grooved between ribs in upper half, with white bristles in lower part, subglabrous above; corona ca. 5 mm wide, as long as tube, light brown, with 20—25 nerves and teeth; calyx teeth pale brown, 12 mm, 2.5 times as long as corona is wide. Fl. August. Fr. September. (Plate V, Figure 1.)

Limestones on bluff shores, crevices, taluses, in mountains to subalpine belt. — Caucasus: W. Transc. (from Tuapse to N. Abkhaziya). Endemic. Described from Asha (Radde collections). Type in Geneva, cotype in Tbilisi.

Series 4. **Gumbeticae** Bobr. — Stems usually simple, all leaves radical, narrowly lanceolate, entire, rarely stem with 1—2 undeveloped branches and pinnatifid leaves; heads large; flowers yellow.

At present this series is unknown outside Dagestan but presumably representatives will one day be found in Asia Minor, which is still rather poorly investigated.

7. S. gumbetica Boiss. Fl. or. III (1875) 137; Grossg., Fl. Kav. IV, 58. — Exs.: Herb. Fl. Cauc. No. 391.

Perennial; root woody, multicipital, covered near crown with remnants of petioles of past years; stems 2—5, 15—30 cm, simple, usually leafless, densely covered with crisp short hairs throughout, very rarely with

66 1 lateral branch developing in lower third from axil of one of the opposite cauline leaves; leaves usually all radical, 4–10 cm long, 3–10 mm wide, lanceolate, usually entire, tapering to petiole slightly larger than one-third of whole leaf, sometimes leaves nearly lyrate-pinnatifid, with very large terminal lobe and 2–4 pairs of small linear lateral lobes, covered with short appressed hairs, the entire plant appearing canescent; cauline leaves, if developed, always pinnatifid, with linear lobes. Heads 3–4 cm across at anthesis; involucral bracts 10–15 mm long, linear, acuminate, short-haired; marginal flowers to 1.5 cm long; corolla yellow, pubescent outside; outer calyx ca. 4 mm long, 8-ribbed, with very deep longitudinal pits between ribs in upper half, the upper quarter appearing glabrous; the lower part densely covered with long white bristles; corona membranous, light brown, 20–24-toothed, 2 mm long (ca. half length of the tube); calyx teeth and bristles dark brown, 4–5 times as long as corona is wide. Fl. June; Fr. July. (Plate V, Figure 2.)

Limestones in montane and subalpine belts. – Caucasus: Dag. Endemic. Described from Danuk in Gumbet District. Type in Leningrad.

Series 5. **Argenteae** Bobr. – Lower cauline leaves pinnatifid or bipinnatifid, with linear lobes and segments, very rarely lobes lanceolate; upper cauline leaves linear, entire. This series comprises S. eburnea Sibth. et Sm., S. pilosa Roem. et Schult., S. wulfenii Kern. and others, presumably ecological forms of S. argentea L. described from the E. Mediterranean.

8. S. argentea L. Sp. pl. (1753) 100. – S. ucranica auct. fl. transcauc. et taur. non L. – S. ucranica var. virgata Grossh. in Tr. Tifl. bot. sada, ser. II, 1 (1920) 31. – S. virgata Grossh. in Grossh. et Schischk. Sched. ad. Pl. or. exs. (1924) 32, No. 123; Grossgr., Fl. Kavk. IV, 58. – S. taurica Kotov in Ukr. Bot. zhurn. XIII, 3 (1956) 52.

Perennial; root woody, multicipital; stems 3–5(7), 30–70 cm, subglabrous in upper and median parts, with dense white bristles below, pale, brownish below, sometimes branching in lower third, the presence of several stems giving the plant a virgate appearance, upper branches divergent, sometimes nearly horizontal; lower cauline leaves sessile, pinnatifid, or bipinnatifid, with linear lobes and segments, bristly especially at base,
69 sometimes appearing silvery from pubescence; median leaves pinnatifid, with 1 or 2 pairs of lateral lobes, glabrous, the upper leaves linear, entire; peduncles 10–25 cm, pale, glabrous. Heads 1.5–2 cm across; at anthesis, in fruit 1–13 cm [?]; involucral bracts lanceolate, profusely pubescent, not longer than marginal flowers, marginal flowers small; corolla pale yellow; outer calyx ca. 2 mm long, 8-grooved, deeply pitted between grooves in upper half, white-bristly below; corona barely half as long as tube, pale brown, with 20 nerves and teeth; calyx teeth and bristles pale brown, one-fourth as long as corona is wide. Fl. June–July, Fr. July–August.

Rubbly slopes and gravelly soils in the mountain-steppe belt. – European part: Crim.; Caucasus: S. and E. Transc., Tal. **Gen. distr.**: Bal.-As. Min., Arm.-Kurd., Iran (NW). Described from the "East." Type in London.

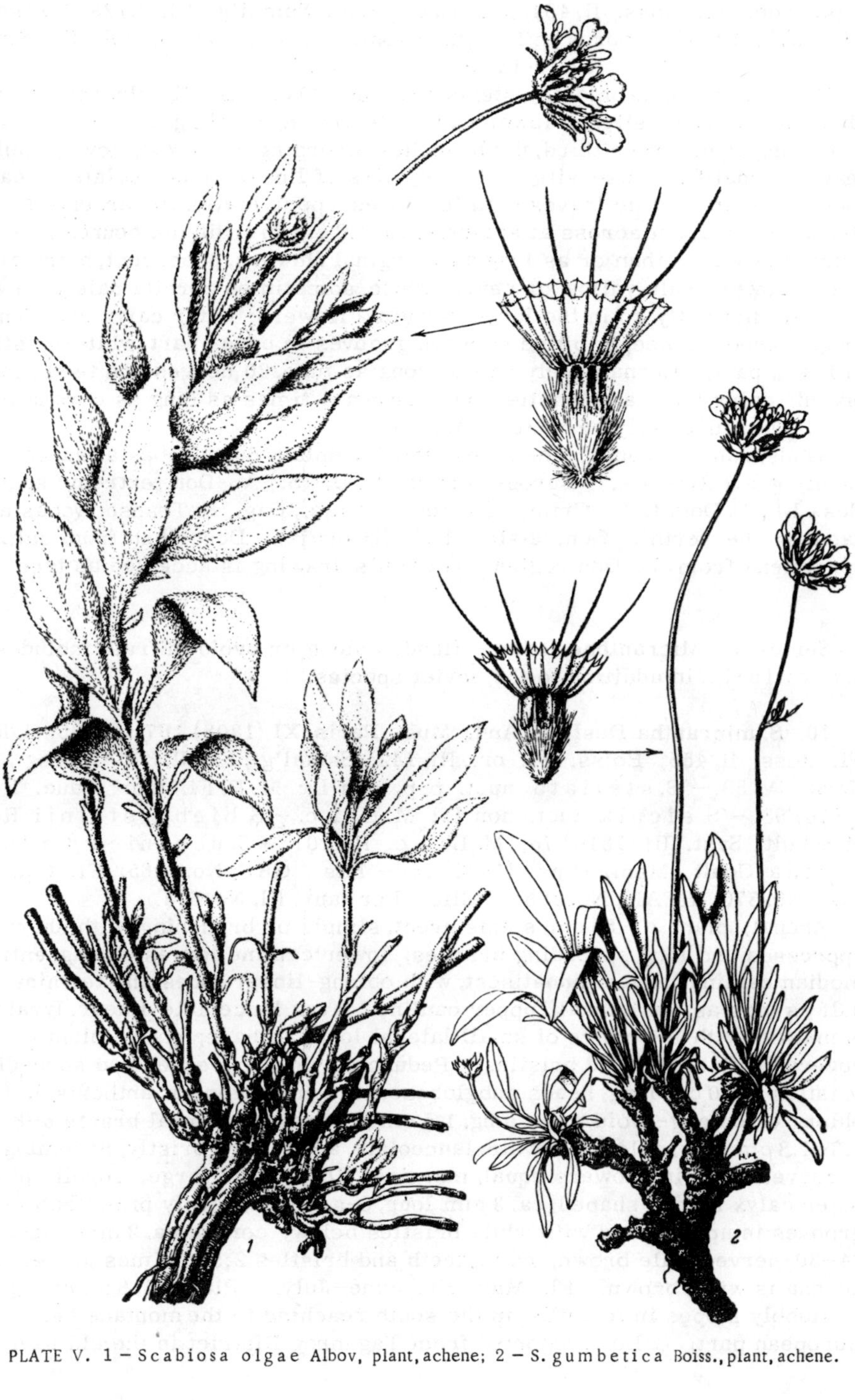

PLATE V. 1 – Scabiosa olgae Albov, plant, achene; 2 – S. gumbetica Boiss., plant, achene.

9. S. ucranica L. Sp. pl. (1762) 144; M. B. Fl. taur.-cauc. I, 97, III, 101, p. p.; Ldb. Fl. Ross. II, 454, pro max. parte; Shmal'g., Fl. II, 29; Grossg., Fl. Kavk. IV, 58; Bobrov in Fl. yugo-vost. VI, 281. – Ic.: Gmel. Fl. Sib. II, tab. 87. – Exs.: Herb. Fl. reip. Sov. Ucr. No. 97.

Biennial-triennial plants; stems 1–3, 30–70 cm, usually glabrous or with short hairs and bristles in lower part, pale brown, nearly glossy, profusely branching from lower third, the branches diverging at ca. 45°; lower cauline leaves pinnatifid, subsessile, with 3–4 pairs of linear or lanceolate-linear lobes, median cauline leaves with few lobes, upper leaves linear, entire. Heads 2–2.5(3) cm across at anthesis, ca. 1.3 cm in fruit, not counting bristles, shorter than or as long as marginal flowers, pubescent, hairy at base; flowers pale yellow or nearly white, sometimes corolla pale yellow distally, distinctly so in the large marginal flowers; outer calyx ca. 2.5 mm long, 8-grooved, deeply pitted between grooves in upper part, white-bristly in lower part; corona barely half as long as tube, 20 nerves and teeth, pale brown; calyx teeth and bristles pale brown, 4 times as long as corona is wide. Fl. June–July, Fr. July–August.

Sands, sandy steppes, pine forest and floodplain dunes, coastal sands, rarely gravelly soils. – European part: M. D. (S.), V.-Don (extreme south), Bes., Bl., L. Don, L. V., Crim.; Caucasus: Cisc., Dag., E. Transc. (coastal
70 sands to the north). **Gen. distr.**: Bal. (Dobruja). Described from cultivated specimens from the Don region. Gmelin's drawing is accepted as type.

Series 6. **Micranthae** Bobr. – Heads oblong or ovoid in fruit, includes S. sicula L., in addition to the Soviet species.

10. S. micrantha Desf. in Ann. Mus. Paris, XI (1808) 167, tab. 25; Ldb. Fl. Ross. II, 455; Boiss. Fl. or. III, 142; Shmal'g., Fl. II, 30; Grossg., Fl. Kavk. IV, 59. – S. stellata auct. p. p. non L.: M. B. Fl. taur.-cauc. I (1808) 98. – S. sicula auct. non L.: M. B. l. c. – S. biebersteinii Roem. et Schult. Syst. III (1818) 75; M. B. l. c. III, 101. – S. ucranica β micrantha Coult. Mém. Dips. (1823) 35. – Exs.: GRF, No. 1165; Fl. Cauc. exs. exs. No. 370; HFAM, No. 298; Callier, Iter taur. III, No. 628.

Annual; plant 20–60 cm, stems erect, simple or branching, with short appressed and long spreading bristles; lower cauline leaves oblong, entire; median cauline leaves pinnatisect, with oblong-linear lobes, the terminal half as long as the lateral; upper cauline leaves lanceolate-linear, lyrate-pinnatifid, with 1–2 pairs of small lateral lobes at base; entire plant covered with appressed bristles. Peduncles with appressed and spreading bristles 5–30 cm long, heads subglobose, ca. 1 cm across at anthesis, in fruit oblong or oblong-ovoid, 2 cm long, 1.5 cm across; involucral bracts 6–8, 1.3–1.8 cm long, oblong to ovate-lanceolate, acuminate, bristly, becoming recurved in fruit; flowers equal, marginal flowers not large; corolla pink; outer calyx funnel-shaped, ca. 3 mm long, 8-grooved, deeply pitted between grooves in upper third, with white bristles below; corona ca. 2 mm wide, 24–30-nerved, pale brown; calyx teeth and bristles 2.5–3 times as long as corona is wide, brown. Fl.. May, Fr. June–July. (Plate VI, Figure 2.)

Rubbly slopes in foothills, in the south reaching to the montane belt. – European part: L. Don (reported from Taganrog District in the old literature),

Crim.; Caucasus: Dag., W., E. and S. Transc.; Centr. Asia: T. Sh. (W.), Syr D., Pam.-Al. (vicinity of Stalinabad), Mtn. Turkm. **Gen. distr.**: Bal.-As. Min., Arm.-Kurd., Iran (N.). Described from the "East." Type in Paris.

Series 7. **Rotatae** Bobr. — Heads globose in fruit. Corona rotate, 5–8 mm wide.

11. S. persica Boiss. Diagn. I ser. 10 (1849) 81; Grossg., Fl. Kavk. IV, 59. — S. setulosa Fisch. et Mey. in Ann. Sc. Nat. IV ser. 1 (1854) 31. —
71 S. palaestina β microcephala Boiss. Fl. or. III (1875) 145. — S. palaestina ϵ persica Boiss. l. c. (1875). — S. talyschensis Fomin in Vestn. Tifl. bot. sada, X (1908) 34. — S. microcephala Grossh. Fl. Kavk. IV (1934) 59, non Nym. 1855. — Ic.: Tchihat. Bot. As. Min. Atlas, tab. 28. — Exs.: Pl. or. exs. No. 70.

Annual; plant 15–40 cm, stems erect or branching, with appressed hairs and white bristles; leaves lyrate-pinnatifid, with large linear-lanceolate lobes and 1–2 pairs of lateral linear lobes, sometimes pinnatifid, with 2–3 pairs of larger linear lateral lobes, very rarely 4–5 pairs of lateral lobes, and then irregularly dentate-cleft; all leaves with appressed white bristles. Peduncles long, 10–25 cm; heads ca. 3 cm across at anthesis, globose, ca. 2.5 cm across in fruit; involucral bracts lanceolate-linear, 10–12 mm long, bristly, shorter than marginal flowers; marginal flowers very long, 12–14 mm; corolla pale blue, pink in herbarium, lobes cut into linear segments, the corolla of outer flowers larger, cut into 4–5 segments, of the others into 3; outer calyx funnel-shaped, 5–6 mm long, 8-grooved, deeply pitted between groves in upper half, with white bristles in lower part; corona ca. 5 mm wide, 30-nerved, in fruit nearly spreading, 12–13 mm across; calyx teeth and bristles nearly twice as long as corona is wide. Fl. May, Fr. May–June.

Gravelly and sandy soils of foothill plains and submontane belt in desert zone. — Caucasus: S. and E. Transc., Tal.; Centr. Asia: Mtn. Turkm. **Gen. distr.**: As. Min., Arm.-Kurd., Iran. Described from Iran. Cotype in Leningrad.

Note. 1) Allied forms distributed in the E. Mediterranean countries and the Near East and included in this and the preceding species have been the source for 20 species, the overwhelming majority of which are but ecotypes of three to four real species.

2) S. schizantha Bordz. and S. palaestina ssp. armena Bordz. (Vestn. Kievsk. bot. sada. VII-VIII (1928) 22, 23), described by E. I. Bordzilovskii, presumably refer to S. persica Boiss., the type specimen of which has apparently been lost.

12. S. rotata M. B. Fl. taur.-cauc. III (1819) 102; Ldb. Fl. Ross. II, 454; Boiss. Fl. or. III, 145; Shmal'g., Fl. II, 30; Grossg., Fl. Kavk. IV, 59. — S. monspeliaca Georgi, Beschr. Russ. R. III, 4 (1801) 727, non Jacq. — S. stellata auct. p. p. non L.: M. B. Fl. taur.-cauc. I, 98. — S. linifolia C. Koch in Linnaea, XVII (1844) 36; Ldb. l. c. — S. rotata β linifolia Boiss. l. c. — Asterocephalus rotatus Spreng. Syst. I (1825) 381. — Ic.: Fl. Cauc. exs. No. 371; Woron. Herb. Fl. Cauc. No. 540.

72 Annual; plant 15–30 cm; stems erect or branching, with appressed stiff hairs and white bristles; lower cauline leaves oblong-lanceolate, tapering into petiole, entire, very rarely barely dentate-cleft; median cauline leaves lyrate-pinnatifid, with 1–2 pairs of lateral linear lobes at base; upper leaves usually entire, linear-lanceolate, rarely barely lyrate; all leaves with appressed bristles. Heads at anthesis subglobose, 1.5–2 cm across, on rough bristly 3–15-cm long peduncles, in fruit globose, 2 cm across; involucral bracts 1.5–2 cm long, longer than marginal flowers, lanceolate, bristly; marginal flowers rather large; corolla pink; outer calyx funnel-shaped, 4–5 mm long, with white bristles at base, 8-grooved, deeply pitted between grooves in upper half; corona 6–8 mm wide, 28–30-nerved, often spreading in fruit, nearly flat, rotate, ca. 2 cm across; calyx teeth and bristles brown, barely as long as corona is wide. Fl. April–May, Fr. May–June. (Plate VI, Figure 5.)

Dry rubbly slopes of foothills and rubbly soils of foothill plains. – European part: Crim. (Feodosiya District); Caucasus: Dag., E. and S. Transc.; Centr. Asia: Mtn. Turkm. **Gen. distr.:** As. Min., Arm.-Kurd., Iran. Described from vicinity of Tbilisi. Type in Leningrad.

Series 8. **Flavidae** Bobr. – Heads 0.5–1 cm across in fruit; calyx teeth and bristles 4–5 times as long as corona is wide. In addition to the Soviet species, S. flavida Boiss., from S. Iran, is also included here.

13. S. olivieri Coult. Mém. Dips. (1823) 36, tab. II, f. 10; DC. Prodr. IV, 656; Endl. et Fenzl, Sertum Cabul. 7, tab. IV; Ldb. Fl. Ross. II, 455; Boiss. Fl. or. III, 141; Grossg., Fl. Kavk. IV, 59. – S. rhodantha Kar. et Kir. in Bull. Soc. Nat. Mosc. (1841) 435; Nevsk. in Tr. Bot. inst. AN SSSR, ser. 1, 4, 277. – Ic.: Coult. l. c.; Endl. et Fenzl, l. c.

Annual; plant 10–40 cm, canescent from appressed pubescence; stems pale, furcately branching nearly from base, with short appressed hairs, upper part nearly white, with slender subfiliform peduncles; leaves borne in branching stems, oblong, 2–5 cm long, 0.3–1 cm wide, entire or appearing lyrate, with 1–2 pairs of basal auricles, pubescent on both sides. Peduncles slender, 3–10 cm; heads 0.5–1 cm, in fruit, ignoring calyx teeth, ca. 1.cm
75 across; involucral bracts 3–5, oblong or ovate, acuminate, pubescent, 2–4 mm long, shorter than flowers; flowers 5–12, the marginal flowers barely larger than the inner; corolla pale, nearly white, sometimes pink or bluish, pubescent outside; outer calyx ca. 2 mm long, funnel-shaped, 8-grooved, deeply pitted between grooves in upper half, with white bristles at base; corona ca. 3 mm wide, nearly spreading, pubescent with 20–24 nerves outside, with irregularly toothed margin; calyx teeth and bristles pale brown, 4 to 5 times as long as corona is wide, calyx grayish green, a 5-pointed star, with very long tapering rays. Fl. April, Fr. May–June.

Plains and foothills in deserts and semideserts, in southern mountains to the montane belt. – Caucasus: E. (Karabakh), S. Transc.; Centr. Asia: all regions except for W. Ar.-Kasp., where it occurs from Turgai District through Pam. Al. in the southeast to Khorog District. **Gen. distr.:** E. Med. (Arabia, Palestine, Syria), As. Min., Arm.-Kurd., Iran. Described from Iran. Type in Geneva.

(73)

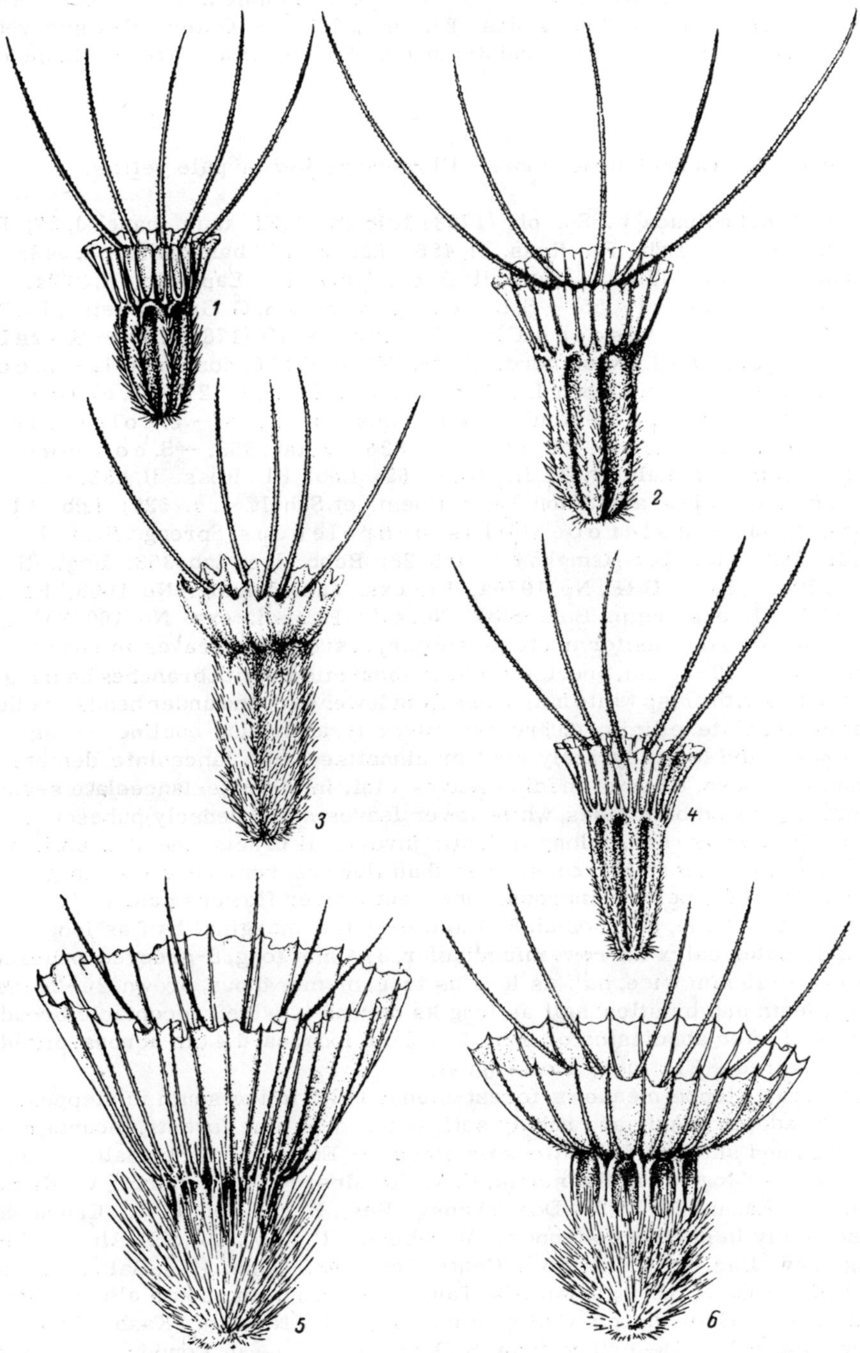

PLATE VI. Achenes: 1 – Scabiosa bipinnata C. Koch; 2 – S. velenovskyana Bobr.; 3 – S. comosa Fisch.; 4 – S. colchica Stev.; 5 – S. rotata M.B.; 6 – S. micrantha Desf.

Section 2. SCLEROSTEMMA Koch in Röhl. Deutsch. Fl. I (1823) 749; Koch, Synops. ed. 3, 1, 295; Boiss. Fl. or. III, 131. — Outer calyx grooved for entire length, 8-grooved and 8-ribbed. 2n = 16 (Kachidze in Planta 7, 4, 1929).

Series 1. **Ochroleucae** Bobr. — Flowers yellow or pale yellow.

14. S. ochroleuca L. Sp. pl. (1753) 101; M. B. Fl. taur.-cauc. I, 97; DC. Prodr. IV, 658; Ldb. Fl. Ross. II, 456; Turcz. Fl. baic-dahur. I, 543; Culak. in Vestn. Tifl. bot. sada. III, 3, 63; Kryl., Fl. Zap. Sib. X, 2624; Grossg., Fl. Kavk. IV, 55. — S. heterophyllos S. G. Gmel. Reise, I (1770) 150, tab. 28. — S. lutescens Gilib. Fl. lithuan. III (1781) 169. — S. stellata Schangin in Pall. Neue nord. Beitr. VI (1793) 76, nom. non L. — S. columbaria auct. p. p. non L.: Ldb. Fl. alt. I (1829) 127. — S. columbaria γ ochroleuca Coult. Mém. Dips. (1823) 38. — S. columbaria β ochroleuca Rchb. Ic. pl. rar. IV (1826) 43, tab. 353. — S. commutata Roem. et Schult. Syst. III (1818) 65; Ldb. Fl. Ross. II, 453. — S. ceratophylla auct. non Ten.: Roem. et Schult. l. c. 521; Ldb. Fl. Ross. II, 456. — Asterocephalus ochroleucus Spreng. Syst. I (1825) 383. — Ic.: S. G. Gmel. l. c. tab. 28; Rchb. l. c. tab. 353; Hegi, Ill. Fl. VI, 1, 306. — Exs.: GRF, No. 1976a; Fl. exs. austro-hung. No. 1005; Fl. stir. No. 679; Fl. exs. reip. Boh.-Slov. No. 477; Fl. pol. exs. No. 160, 384 a, b.

Biennial; root fusiform, stems solitary, rarely few, leaves in rosette at
76 first; stems 30—80 cm, erect, simple or about middle with branches bearing solitary heads with crisp white hairs mostly in lower part and under heads; radical leaves petiolate, elliptic, entire, dentate or lyrately cut; cauline leaves 2—5 pairs, the lower lyrately cleft or pinnatisect into lanceolate, dentate or pinnatifid lobes, lobes of median leaves cleft into lyrate-lanceolate segments, short-haired on both sides, while lower leaves more densely pubescent. Heads 2—3 cm across, oblong in fruit; involucral bracts linear, acuminate, green, usually short-haired, shorter than flowers, very rarely as long; bracts tapering below, glabrous, pubescent above; flowers pale yellow, pubescent outside, the median 5—7 mm long, the marginal half as long, radial; outer calyx narrow, infundibular, 3—4 mm long, 8-grooved, pubescent; corona 1—15 mm wide, half as long as tube, membranous, brown, 20—24-nerved; calyx teeth and bristles half as long as corona is wide; receptacle broadly fusiform after abscission of fruit; 1—1.5 cm long, ca. 0.4 cm across, pitted, short-haired. Fl. June, Fr. August.

Steppes, steppe meadows, forest-steppe belt; in the south in steppes, dry meadows, sandy and rubbly soil, sometimes pine forests, mountain steppes and shrubby formations on slopes. — European part: Balt. (Lithuania), U. Dnp. (to Mogilev in the north), U. V. (to Moscow in the north), V.-Kama (south of Kazan), M. D., V.-Don, Transv., Bes., Bl., L. Don, L. V.; Caucasus: Cisc. (only in northern steppes); W. Siberia: U. Tob., Ob (SE), Alt.; E. Siberia: Ang.-Say., Dau. (extreme SW); Centr. Asia: Ar.-Casp. (N.), Balkh. (N. and SE); Dzu.-Tarb. (Dzungarian Ala-Tau), T. S. (E., in the south also in Issyk-Kul basin). **Gen. distr.**: Centr. Eur. (S.), Bal. (N.), Dzu.-Kash. (W.), Mongolia (NW). Described from S. Germany. Type in London.

15. S. georgica Sulak. in Vestn. Tifl. bot. sada, III, 3 (1927) 64; Grossg., Fl. Kavk. IV, 54.

Perennial; root fusiform, stems usually solitary, 30–80 cm, erect, in lower part, especially at nodes, with scattered white spreading bristles, under heads with white retrorse appressed hairs; radical leaves petiolate; cauline leaves lyrate-pinnatifid, 3–4-paired, sessile, lanceolate, 5–8 cm long, 2 cm wide, with large terminal lobe half the length of the lamina; in lower part of leaf 3–5-pairs of oblanceolate lateral segments, with appressed white bristles
77 on both sides, more beneath, irregularly cleft; leaves in upper part lyrate-pinnatifid to entire, lanceolate-linear. Heads long-peduncled, ca. 2 cm across, becoming ovoid in fruit; involucral bracts lanceolate-linear, with appressed white hairs, shorter than marginal flowers; flowers yellow, with short appressed hairs outside, marginal flowers larger; outer calyx narrowly grooved, white-haired; corona ca. 1 mm wide, one-third to one-half the length of the tube, membranous, brown, 20–24-nerved; calyx teeth and bristles half as long as corona is wide; receptacle fusiform after abscission of fruits, ca. 1 cm long, ca. 2 mm across, tuberculate, pubescent, with numerous membranous bracts. June–August.

Slopes of mountain steppes. – Caucasus: Dag., E. Transc. Endemic. Described from Temir-Khan-Shura (now Buinaksk). Cotype in Leningrad.

16. S. bipinnata C. Koch in Linnaea, XVII (1843) 37, XXIV, 446; Ldb. Fl. Ross. II, 457; Sulak. in Vestn. Tifl. bot. sada, III, 3, 73; Grossg., Fl. Kavk. IV, 55; S. ochroleuca auct. fl. cauc. p. p. non L.: M. B. Fl. taur.-cauc. I, 97; Ldb. l. c. 456. – S. maritima auct. fl. cauc. non L.: Boiss. Fl. or III, 134, quoad pl. Szov.; Grossg., op. cit. IV, 57. – S. svanica Schchian in Zam. po sist. i georg. rast. 16 (1951) 92. – ?S. pinnata Güld. Reise, II (1791) 10, nom. – Exs.: GRF, No. 1976 b, sub. S. ochroleuca L.

Perennial; plant 50–80 cm high, especially in lower part, densely covered with soft white hairs; stems erect, branching, very leafy; rosetted leaves of sterile shoots and lower cauline leaves long-petiolate, 10 cm long, lanceolate, entire or lyrate-cut; cauline leaves sessile, smaller, bipinnatifid, with narrowly lanceolate or linear, entire or dentate lobes. Heads on long hairy peduncles, in fruit globose, ca. 1.5 cm across, at flowering 3–3.5 cm across; involucral bracts recurved nearly as long as flowers, densely covered with soft hairs; flowers pale yellow, pubescent outside, marginal flowers larger, radiant; outer calyx deeply furrowed for entire length; corona membranous, toothed, one-fourth to one-third the length of the tube; calyx teeth and bristles dark brown, 7 mm long, five times as long as corona. July–September. (Plate VI, Figure 1).

Meadows in high-mountain belt, rubbly slopes and shrubby formations. – Caucasus: Cisc., Dag., W. Transc., E. Transc. (W.), S. Transc. **Gen. distr.**: E. Anatolia. Described from Kartalinia and Imeretia. Type in Berlin.
78 Note. Described by A. S. Shkhiyan from Upper Svanetia, this species presumably is the shade form of the Koch species.

17. S. imeretica (Somm. et Lev.) Sulak in Vestn. Tifl. bot. sada. III, 3 (1927) 75. – S. ochroleuca var. imeretica Somm. et Lev. in Tr. Peterb. bot sada, XVI (1900) 217. – S. calcarea (Alb.) Sulak. op. cit. non S. calcarea Tecl. 1896; Grossg., Fl. Kav. IV, 56; Kolak., Fl. Abkh.

IV, 160. — S. ochroleuca var. calcarea Alb. Prodr. Fl. colch. (1895) 127; Lipskii, Fl. Kavk. 340. — S. schaorica Kem.-Nath. in Vestn. Tifl. bot. sada., V (1931) 15. — Ic.: Kolak., op. cit., Plate XV.

Perennial; stems 25—50 cm, solitary or 3—4, erect, usually simple or branched once or twice, covered with white bristles at base, with appressed white bristles and retrorse hairs above; radical leaves petiolate, obovate or oblanceolate, entire, rarely lyrate-pinnatifid, with 2—3 pairs of small lateral lobes and large terminal one, both sides, especially the lower, with short appressed white hairs; cauline leaves 2—3-paired, the lower pinnatisect into 2—3 cm long, ca. 1 cm wide, irregularly cleft-dentate lobes, upper cauline leaves with lanceolate lobes. Heads solitary or 2—3, long-peduncled, at flowering 2.5—3.5 cm across, globose, ca. 1.5 cm across in fruit; involucral bracts lanceolate, white-haired, shorter than marginal flowers, the latter 2 cm long; outer calyx 3.5—4 mm long, longitudinally grooved; corona nearly half the length of the tube; calyx teeth and bristles twice as long as corona is wide. Fl. July, Fr. August.

Limestone in ranges of the alpine belt. — Caucasus: W. Transc. Endemic. Described from alpine pastures of Mingrelia. Type in Leningrad.

18. S. sosnowskyi Sulak. in Vestn. Tifl. bot. sada, III, 3 (1927) 74; Grossg., Fl. Kavk. IV, 55; Kolak., Fl. Abkh. IV, 160.

Perennial; stems 40—80 cm, erect, very leafy, with short white appressed retrorse hairs above, in lower part with short white bristles, more densely bristly at nodes; radical leaves petiolate, entire or lyrate-pinnatifid, terminal lobe large, rhombic or ovate, lateral lobes cuneate, all lobes cleft-dentate; lower cauline leaves sessile, with narrower lobes; leaves on both sides, especially beneath, with appressed white bristles to densely gray-tomentose hairs; upper cauline leaves pinnatifid, with less pubescent linear
79 lobes. Heads long-peduncled, at anthesis 1.5—2 cm, in fruit ovoid, ca. 1 cm across; involucral bracts shorter than marginal flowers, lanceolate-linear, appressed-hairy; flowers pale yellow, sometimes nearly white, the marginal slightly enlarged; outer calyx ca. 3 mm long, funnel-shaped, longitudinally grooved, 8-ribbed, appressed-hairy; corona ca. 1.5 mm wide, membranous, 20—24-nerved; calyx teeth and bristles brownish, twice as long as corona is wide. Fl. July, Fr. August—September.

Rubbly slopes and outcrops of the coastal belt and foothills. — Caucasus: W. Transc. **Gen. distr.**: E. Anatolia (coast). Described from Novorossiisk District. Type in Tbilisi.

19. S. correvoniana Somm. et Lev. in Tr. Peterb. bot. sada., XII (1892—1893) 152, XVI (1900) 218; Grossg., Fl. Kavk. IV, 55.

Perennial; stems 20—50 cm, solitary or 2—3, usually simple or bifurcated above, glabrous below, with appressed short retrorse hairs above, especially under heads; radical leaves oblanceolate, lyrate-pinnatifid, with large irregularly crenate terminal lobe, and oblong entire lateral lobes; leaves very short below, pubescent mostly along nerves, upper leaves sessile, ovate, pinnatifid, the terminal lobe not larger than the lateral, or all lobes sometimes shallowly cleft. Heads solitary, long-peduncled, ca. 2.5 cm, fruit globose, ca. 2 cm across; involucral bracts lanceolate, acuminate, short-haired, not longer than the median flowers; flowers pale yellow, the marginal radiant; outer calyx 4–5 mm long, longitudinally grooved; corona half the length of the tube; calyx teeth and bristles dark brown, 5–6 times as long as corona is wide. Fl. July, Fr. September.

Alpine meadows. – Caucasus: W. Transc. (Upper Svanetia). Described from between the Nenskvyr and Seken rivers. The type was grown in Geneva from seeds collected there. Type in Florence.

Note. The independence of this species is very doubtful. It closely resembles S. bipinnata C. Koch as well as S. imeretica (Somm. et Lev.) Sulak, from which it is distinguished by the often entire radical leaves and shorter calyx teeth.

Series 2. **Columbariae** Bobr. – Flowers pale blue, pale violet or pink.

20. S. hyrcanica Stev. in Bull. Soc. Nat. Mosc. XXIX, IV (1856) 371 in textu; Trautf. in Tr. Peterb. bot. sada, VIII, I, 430; Sulak. in Vestn. Tifl.
80 bot. sada, III, 3, 69; Grossg., Fl. Kavk. IV, 57. – S. pyrenaica auct.: C. A. M. Verzeichn. (1831) 48; Ldb. Fl. Ross. II, 458; non All. (1785). – ?S. columbaria γ M. B. Fl. taur.-Cauc. I (1808) 96, cfr. Beschr. d. Länder zwischen Terek u. Kur (1800) 132.

Perennial; plant 30–70 cm high; stems 3–5, subglabrous, in lower part with very short hairs, slightly canescent, cylindrical, bearing from middle few-leaved or even nearly leafless opposite branches; leaves deeply pinnatisect; lower cauline leaves barely lyrate-pinnatisect, terminal lobe ca. one-fourth the entire leaf, ovate-lanceolate, deeply cleft-dentate, with 7–8 lateral pinnatifid lobes with lanceolate to linear segments; median and upper cauline leaves all pinnatisect, with linear terminal lobe, and lanceolate lateral lobes, cleft into linear segments; leaves sparingly pubescent above, hairy beneath, lower cauline leaves rather more hairy. Peduncles 10–30 cm, subglabrous or with scattered hairs, more densely pubescent under heads; heads ca. 3 cm across at anthesis, ca. 1.2 cm in fruit (not counting bristles); involucral bracts linear, short-haired; flowers pink-violet; outer calyx 2 mm long, covered for two-thirds with appressed hairs; corona not more than one-third the length of the tube; calyx teeth and bristles very short pubescent, 9–11 mm long. Fl. June, Fr. July.

Mountain steppe slopes. – Caucasus: Tal. (highlands). **Gen. distr.**: Iran (Gilyan). Described from mountainous Talysh. Cotype (Tellavakh, Meier collections) in Leningrad.

21. S. amoena Jacq. f. Eclogae, I (1811–1816) 86, tab. 59 et access. I, f. 4; Boiss. Fl. or. III, 132, quoad pl. talyschensem. – S. longipedunculata Fisch. Cat. jard. Gorenki, 1808 (1812) 43, nomen. – S. grisea Sulak. in Vestn. Tifl. bot. sada., III, 3 (1927) 68, excl. syn. quoad pl. lenkoranicam; Grossg., Fl. Kavk. IV, 57, pl. lenk. – Ic.: Jacq. f. l. c.

Perennial; plant 30–70 cm high; stems 3–5, canescent with dense short hairs, (these longer and more dense in lower part), from middle with opposite profusely leafy branches; internodes 3–7 cm; lower cauline leaves subsessile, with ovate crenate-dentate terminal lobe not more than one-third the length of the lamina, and 3–4 lateral obovate lobes; median cauline leaves larger, the terminal lobe ca. half the length of the lamina, more deeply

cleft-dentate; in upper cauline leaves terminal lobe more than half the
81 length of the lamina; all leaves lyrate-pinnatifid, pubescent, more densely so beneath, lower leaves with more prominent nerves. Peduncles 10–20 cm, short-haired; heads 2–2.5 cm across, in fruit globose, ca. 1 cm across (not counting bristles); involucral bracts lanceolate, short-haired; flowers pale violet; outer calyx 2–2.5 mm long, 8-ribbed, deeply grooved, with scattered appressed hairs in lower part and in depth of groove; corona one-third to one-half the length of tube, brownish; calyx teeth and bristles dark brown, short-haired; 8–10 mm long. Fl. June, Fr. July–August.

Plains, submontane belt, seashores, shrubby formations. – Caucasus: Tal. (Lenkoran lowland, seashores and adjacent islands). **Gen. distr.:** Iran (Astrabad Province). Described from specimens cultivated in Vienna. The first published description serves as type.

Note. The authors of the Caucasian flora have ignored Ledebour's doubt about the distribution of this species in the USSR (Fl. Ross. II, 459).

S. amoena was described and beautifully drawn by Jacquin's son from a plant grown in Vienna, where it was grown from collected seeds in the gardens of Count Razumovskii, near Moscow, who had received the seeds from England. These facts were not known to Ledebour.

Close study of Jacquin's drawing shows that the plants grown in the gardens of Moscow, Vienna and England in the first decade of the last century most nearly resemble Scabiosa of the Columbaria group, distributed in the Lencoran lowland, and it is there that the above-mentioned seeds were presumably first collected. Marschall von Bieberstein is known to have traveled through the Caspian area of the Caucasus (between Terek and Kura) from 1796 to 1797. His account of his travels (Beschreibung der Länder zwischen den Flüssen Terek und Kur am Kaspischen Meere, Frankfurt. 1800, p. 132) reveals his special interest in the Scabiosa of this group. It is only natural to assume that he collected seeds of one of them in the Kura lowland (where he sojourned) and dispatched them to botanical gardens for cultivation. It is for these reasons that we submit that the type material of this species derives from material collected by Bieberstein in the Lenkoran lowland.

22. S. colchica Stev. in Bull. Soc. Nat. Mosc. XXIX, 4 (1856) 371 in textu; Trautf. in Tr. Peterb. bot. sada., VIII, 3 (1883) 430. – S. letschchumensis Kem.-Nath. in Vestn. Tifl. bot. sada, 5 (1931) 13; Grossg., Fl. Kavk. IV, 56. – Ic.: Fl. Gruzii, VIII, Figure 367.

82 Perennial; plant silvery-gray from short dense pubescence; stems 50–80 cm, solitary or 2–3, simple or with 1–2 branches, erect or ascending, with appressed white retrorse hairs throughout, hairs longer in lower part; radical leaves lanceolate, 10–20 cm long, the laminas half the length of the leaf, entire or barely lyrate and then with very large terminal lobe, shallowly crenate-dentate, with teeth turned upward; lower cauline leaves short-petioled, the upper sessile, upper and lower leaves opposite, lyrate-pinnatifid, with cleft-dentate lobes of which the terminal often more than half the length of the leaf; all leaves canescent with velutinous pubescence on both sides, especially beneath. Heads solitary, on long peduncles, ca. 3 cm across at anthesis, globose, 1.5–2 cm in fruit; involucral bracts nearly 2-seriate, with densely appressed short hairs, lanceolate-acuminate,

the outer bracts ca. 1 cm long, half the length of the marginal flowers; corolla pink or light meat-red, white-haired outside, enlarged in outer flowers; outer calyx ca. 4 mm long, 8-ribbed, longitudinally deeply grooved, white-haired along ribs, especially in their lower half; corona one-third to one-half the length of the tube, pale brown, 16–20-nerved; calyx teeth and bristles dark brown, 4–5 times as long as corona is wide. Fl. and Fr. July–August. (Plate VI, Figure 4.).

Limestone slopes. – Caucasus: W. Transc. (W. Georgia, Lechkhum). Endemic. Described from Lechkhum. Type in Helsinki.

Note. S. holophylla Bordz., described by E. I. Bordzilovskii (Ucr. Bot. Rev. II (1926) 51, 52). The type, presumably lost, should probably be referred to Koch's Scabiosa, because of its entire leaves. The only doubtful data in Bordzilovskii's report concerns the flowers, described as pale yellow.

23. S. velenovskiana Bobr. nom. nov. – S. dubia Vel. Fl. Bulg. (1891) 243 243, non Moench (1777); Sulak. in Vestn. Tifl. bot. sada., III, 3, 71; Grossg., Fl. Kavk. IV, 56.

Perennial; root woody, multicipital, producing stems 2–5 and rosettes of next year; stems 15–40 cm, usually simple or few-branched, white-haired, more densely pubescent in lower part and under heads; leaves of sterile rosettes, sometimes the radical leaves as well, entire, oblong-spatulate, long-petioled, irregularly dentate-crenate, densely hairy on both sides, especially beneath; lower cauline leaves larger, lyrate, with large obovate terminal lobe, crenate-dentate, with 2–3 pairs of small lateral lobes, median and upper cauline leaves nearly bipinnatifid, with lanceolate or linear lobes,
83 glabrous, with sparse bristles along margins and nerves. Heads elongate, ca. 2 cm across at anthesis, in fruit ca. 1.5 cm; involucral bracts lanceolate, slightly broadened at base, tapering, shorter than flowers, appressed-hairy; flowers pink-violet, the marginal larger, pubescent outside; outer calyx 3 mm long, deeply grooved, pubescent; corona half as long as tube, 20–24-nerved; calyx teeth and bristles dark brown, 7–9 mm long. Fl. August, Fr. September. (Plate VI, Figure 2.)

Herbaceous slopes, shrubby formations, forest edges of upper mountain belt. – Caucasus: W. Transc. (Adzhariya), E. Transc. (W.), S. Transc. (SW – Georgia). **Gen. distr.**: Bal.-As. Min. Described from Bulgaria. Type in Prague.

24. S. owerini Boiss. Fl. or. III (1875) 133; Sulak. in Vestn. Tifl. bot. sada., III, 3, 66; Grossg., Fl. Kavk, IV, 55. – S. crinita auct. p. p. non Ky. et Boiss. ex Boiss. (1875); Sulak., op. cit. 70; Grossg., op. cit. 56. – S. lagodechiana Sosn. in Vestn. Tifl. bot. sada., 32 (1914) 17. – S. alexeenkoana Sulak. op. cit.; Grossg., op. cit.

Perennial; plant 40–80 cm high; stems 2–3, glabrous, with few retrorse bristles confined to lower part, and short appressed hairs above, cylindrical; branches bifurcate in lower third and even lower, twice bifurcate above, yielding profusely branching plants; leaves of cauline rosettes, sometimes also the lowermost cauline leaves, petiolate, lyrate-pinnatifid, and obovate, with shallowly crenate-dentate terminal lobe, one-third to one-half the length of the leaf, and 3–4 pairs of small lateral cleft-dentate lobes; median cauline leaves sessile, nearly bipinnatifid, with narrowly lanceolate or linear lobes and linear segments; lobes and segments of upper leaves linear and glabrous. Peduncles 15–30 cm, with retrorse hairs beneath heads; heads at anthesis ca. 3 cm across, in fruit (not counting bristles) globose, slightly elongate, 1.5 cm; involucral bracts ovate, 10–13 mm long, attenuate-acuminate, pubescent; flowers pink, the marginal enlarged. Fl. July, Fr. September.

Light forests in montane and high mountain belts, shrubby formations and meadows. – Caucasus: Dag. (S. Dagestan, Kuba, Kusary), E. Transc. (southern slope of Main Range, Lagodekhi, Zakataly, Shemakha). Described from the foot of Saridag, at the sources of the Samur River. Endemic. Type in Leningrad.

84 Note. S. alexeenkoana, described by T. Sulakadze, appears to represent a somewhat impoverished form, more at home in the more arid montane belt.

25. S. columbaria L. Sp. pl. (1753) 99; M. B. Fl. taur.-cauc. I, 96, III, 101, p. p.; DC. Prodr. IV, 658; p. p.; Ldb. Fl. Ross. II, 457; Boiss. Fl. or. III, 131, p. p.; Shmal'g., Fl. II, 28; Grossg., Fl. Kavk. IV, 57. – S. bellidifolio Gilib. Fl. lithuan. III (1781) 169; ej. Exerc. phytol. 143. – S. polonica Piotr. in Verh. bot. Ver. Brand. XXXIX (1898). – ?S. coerulea Güld. Reise, II (1791) 10, nom. – S. purpurea Sulak. in Vestn. Tifl. bot. sada., III, 3 (1927) 72; Grossg., op. cit.; – S. crinita auct. p. p. non Ky. et Boiss. ex Boiss. 1875; Sulak., op. cit. 67; Grossg., op. cit. 56. – Asterocephalus columbaria Spreng. Syst. I (1825) 382. – Exs.: Fl. exs. austro-hung. No. 1004; Wol. Fl. pol. exs. No. 953a; Pl. pol. exs. No. 158 (sub S. columbaria var. polonica Piotr.). – Ic.: Hegi, III. Fl. VI, 1, Taf. 254, 4.

Perennial; stems 40–80 cm, usually solitary, 1–2 times bifurcately branching from middle, stems glabrous below, sparingly white-haired only in upper part; lower radical leaves and leaves of sterile shoots lyrate-pinnatifid, with slightly larger rounded crenate-cleft terminal lobe and 5–7 pairs lateral lobes, from obovate to oblong, short-cleft; lower cauline leaves nearly bipinnatifid, with lanceolate or linear-lanceolate lobes; upper cauline leaves pinnatifid, with linear lobes; radical and lower cauline leaves sparsely white-haired mostly along nerves and margins, sometimes radical and lower cauline leaves, gray, tomentose-hairy (var. grisea Somm. et Lev. in Tr. Peterb. bot. sada, XVI (1900) 217). Heads ca. 3 cm across, long-peduncled, in fruit globose, ca. 1.5 cm; involucral bracts 1–1.5 cm long, lanceolate, acuminate, short-haired and long-white-bristly; bracts membranous, keeled, pubescent; flowers pale blue or violet, white-haired outside; marginal flowers radiant, larger; outer calyx 3 mm long, 8-ribbed, grooved, white-bristly; corona 1 mm wide, half the length of the tube, pale brown, 16-toothed; calyx teeth and bristles dark brown, 5–6 times longer than corona is wide. Fl. June–July, Fr. August.

Dry meadows and herbaceous slopes, shrubby formations and forest edges. – European part: Balt. (Ezel, Dago, Moon, SW Lithuania), U. Dnp. (Brest-Litovsk, Mogilev), U. Dns. (Carpathian region), Bes., Bl., L. Don (Veliko-Anadol'skoe forest and, introduced, in Kupyansk District to the north), Crim.; Caucasus: Dag., E. Transc., rarely W. Transc. (coast of Adzhar ASSR); Centr. Asia: Mtn. Turkm. (W. Kopet Dagh, Khasardag Mountain).
85 **Gen. distr.:** Centr. and S. Eur., Bal.-As. Min. Described from the mountains of Centr. Eur. Type in London.

26. S. meskhetica Schchian in Zam. po sist. i georg. rast. Tbil. bot. inst. 16(1951)92. – Ic.: Fl. Gruzii, VIII, Plate 368.

Perennial; plant densely covered with appressed short hairs; stems 35–50 cm, short-haired, woody at base, leafy in lower half; radical leaves ovate, entire, crenate, long-petioled; median cauline leaves lyrate, with large ovate terminal lobe and 3 pairs of lanceolate lateral lobes; upper cauline leaves pinnatifid or bipinnatifid, with narrowly lanceolate dentate lobes. Peduncles elongate, densely pubescent under heads; heads at anthesis 1 cm across, in fruit globose, 1.3 cm; involucral bracts lanceolate, acuminate, appressed-hairy, as long as flowers or shorter; flowers pink; marginal flowers not prominent, making heads discoidal; outer calyx ca. 1.3 cm across, ca. 2.mm long, deeply grooved, bristly; corona half the length of the tube; calyx teeth and bristles pale brown, 2–3 times as long as corona is wide. Fl. August.

Gypsiferous outcrops in montane belt. – Caucasus: S. Transc. (Meskhetiya, SW Georgia). Endemic. Described from between Adigeni and Mlashe. Type in Tbilisi.

Note. Shkhiyan's plant apparently represents the discoidal form of S. columbaria. With small marginal flowers, such plants are common in the asteroid Compositae. In the Dipsacaceae the discoidal form occurs in Knautia arvensis. As very little is known about Shkhiyan's plant (we have never seen it), it deserves special study. Though here treated as a separate species, it should probably be called S. columbaria f. discoidea Bobr.

27. S. lucida Vill. Prosp. (1779) 18; ej. Hist. Dauph. II (1789) 293; Hegi, Ill. Fl. VI, 308; Vizn. rosl. URSR, 323; Szaf., Kulcz., Pavl. Rosl. Polskie, 633. – S. columbaria α lucida Coult. Mém. Dips. (1823) 38. – S. columbaria ssp. lucida (Vill.) Vollm. Pl. pol. exs. No. 159, a, b. – Ic.: Hegi, l. c. Fig. 166 d, e. – Exs.: Fl. exs. austro-hung. No. 1003; Fl. exs. reip. Boh.-Slov. No. 372; Fl. stir. No. 676, 1260; Pl. pol. exs. No. 159, a, b.

Perennial; root erect, producing stem and rosette of next year; stems 10–60 cm, erect or ascending, simple or divided above into simple elongate branches, glabrous, sometimes with scattered hairs below, with 3–7 pairs of
86 leaves in lower part; radical leaves ovate or rhombic, large-toothed, barely lyrate with small lateral lobes; median cauline leaves lyrate, with long rhombic terminal lobe and lanceolate or linear, often acutely toothed lateral lobes; upper leaves from entire to bipinnatifid, with linear lobes, glabrous, rarely ciliate along margin and nerves; lower leaves subglabrous, with scattered hairs beneath, nearly shiny above. Peduncles elongate, short-haired above; heads 2–4 cm across; involucral bracts shorter or longer than heads; flowers reddish violet, pubescent outside; marginal flowers larger and radiant; outer calyx 2–3 mm long, pubescent outside; marginal flowers larger and radiant; outer calyx 2–3 mm long, pubescent, deeply grooved for entire length; corona ca. 1.5 mm wide; calyx teeth and bristles 8 mm, 4–5 times as long as corona is wide, shiny, purple-black, keeled, somewhat broadening in lower part. Fl. July, Fr. September.

Stony and shrubby formations on slopes in upper mountain belt. – European part: U. Dnp. (E. Carpathian Mountains, Carpathian area).

PLATE VII. 1 – Scabiosa austro-altaica Bobr., part of plant with radical and lower cauline leaves; 2 – S. comosa Fisch., part of plant with sterile rosette and lower cauline leaves and flowering head.

Gen. distr.: mountains of S. Europe (Pyrenees, Vosges, Jura, Alps, mountains of Illyria, Tatras, Carpathians), Bal. Described from the French Alps. Type in Paris?

Section 3. PRISMAKENA Bobr. sect. nov. in Addenda XXIII, p. 325. – Outer calyx 4-faceted, without grooves or pits, with 4 distinct and 4 faint intermediate ribs.

In addition to the 2–3 Soviet species, this section includes 3–5 species recently described from N. China, Korea and Japan (S. hopeiensis Nakai, S. hairalensis Nakai, S. austromongolica Hurusawa, S. mansanensis Nakai, S. togashiana Hurusawa, S. zukoensis Nakai), as well as the old Japanese species S. japonica Miq. S. superba Grünn (Shansi) and T. tschiliensis Grünn (Chihli) apparently refer to one and the same high mountain species, whose description may have been copied by the Japanese authors.

28. S. lachnophylla Kitag. ex Nakai, Honda et Kitag. Contr. cogn. Fl. Mansh. (1935) 33. – S. fischeri auct, fl. or. extr. non DC. – Ic.: Kom. and Alis., Opred. rast. Dal'nevost. kr. II, Plate 297; Kitagawa, l. c. tab. X; Gard. Chron. ser. 3, XCVIII (1935) 179.

Perennial; root multicipital; stems ca. 60 cm, erect or ascending, few-branched, rarely simple, in lower part with spreading bristles and crisp short hairs, in upper part usually with crisp short hairs, sometimes with
89 spreading bristles under heads; rosetted leaves of sterile shoots 7–15 cm long, 2–4 cm wide, elliptic, nearly entire or lyrate-pinnate, long-petioled; cauline leaves short-petioled or sessile, lyrate-pinnate, with uniform color on both sides of leaves, rather fleshy, with short, often rather dense crisp hairs, sometimes mixed with long bristles. Heads long-peduncled, solitary, at anthesis ca. 3.5 cm across, in fruit 1–1.5 cm, subglobose; involucral bracts linear, 1 cm, membranous in lower part, greenish, cowl-shaped, ciliate above, ca. 5 mm long in fruit; outer calyx 5 mm long in fruit, tube obconical, densely pubescent, with pale membranous 16-nerved corona, hairy outside along nerves, crenate-dentate, with 16 unequal teeth; calyx cylindrical at anthesis, in fruit nearly flat, with 5 setaceous teeth, much longer than corona is wide; flowers blue; corolla of median flowers 8 mm long, 5-lobed, tube of outer flowers ca. 9 mm long, 2-lipped, the upper lip with 2, ca. 5-mm long lobes, the lower 3-lobed, lobes 10 mm long; fruit cylindrical, ca. 3 mm long, with persistent calyx, bearing 3 mm long bristles above. Fl. July, Fr. August.

Forest edges, shrubby thickets, forest meadows, sandy banks of rivers and lakes, meadows. – Far. East: Ze.-Bu. (S.), Uss. (excluding north). **Gen. distr.**: Manchuria, N. Korea. Described from Manchuria. Type in Japan.

29. S. comosa Fisch. ex Roem. et Schult. Syst. III (1818) 84. – S. fischeri D. C. Prodr. IV (1830) 658; Ldb. Fl. Ross. II, 456; Turcz. Fl. baic-dahur. I, 542. – S. dahurica Pall. in herb. Fisch.

Perennial; root multicipital; stems 25–45 cm, erect, sometimes ascending at base, simple or few-branched, lower part colored, with very

short appressed retrorse hairs, at middle subglabrous, with very short crisp hairs under heads; rosetted leaves of sterile shoots 5–12 cm long, 0.5–2 cm wide, limb lanceolate, entire or incised, acutely toothed, 2–4 cm long; petiole 3–8 cm; cauline leaves short-petioled or subsessile, pinnatifid, with linear lobes; leaves nearly the same color on both sides, glabrous, very rarely margin with sparse cilia or bristles. Heads long-peduncled, solitary, 3–4 cm across at anthesis, in fruit subglobose, 1.5 cm; involucral bracts linear, with broadened base, with appressed short hairs, 0.8–1.5 cm long; paleae strongly keeled, membranous below, greenish in upper third, acuminate,
90 with appressed hairs, 6 mm long in fruit; flowers blue-violet, rarely white (albinism!); corolla of median flowers regularly 5-lobed, in radiant marginal flowers irregularly 2-lipped, upper lip 2-lobed, lower 3-lobed, with middle lobe broadly obovate, 1 cm long; outer calyx ca. 5 mm long in fruit, tube obconical, densely pubescent, with pale brown membranous corona, 16(18) nerves pubescent, terminating in irregular teeth; calyx with colored setaceous teeth, 4 mm long. Fl. August, Fr. September. (Plate VI, Figure 3. Plate VII, Figure 2.)

Dry meadows, grass plots in light coniferous forests, steppes, rubbly mountain slopes, riparian sands. – E. Siberia: Ang.-Say. (SE), Dau. **Gen. distr.:** N. Mongolia. Described from specimens cultivated by Besser in Transbaikalia (Kremenets Garden) from seeds of plants collected by Pallas and preserved in Leningrad.

Note. 1) Formally, the type of this species are the specimens grown in the Kremenets Garden. In fact, however, Fischer erroneously distributed them as S. comosa Fisch.; the same name is entered in his own handwriting on the labels of S. dahurica, collected and named by Pallas. The Fischer herbarium in Leningrad includes Pallas' specimens, which are the proper type specimens.

Priority obviously belongs to S. comosa Fisch. Ledebour knew this name from the description and specimens sent to him in Tartu by Fischer. On Fischer's labels on Scabiosa comosa m. alba, he noted that the epithet comosa should be discarded as absurd (nomen ineptum). Out of respect for Ledebour, Russian botanists refer to the species as S. fischeri DC.

2) A form with leaves dissected into subfiliform lobes grows in the steppe region of Transbaikalia, near its frontier with SE Mongolia and SW Manchuria (lower reaches of Kerulen River, basin of Upper Hailar (Argun')) and in the steppes of SE Transbaikalia, along the Onon and Argun' rivers, apparently in the most arid habitats. Lomonosov's collections of this form between Kulusutaev and Hulun Nor (1870) were labeled by Trautvetter as var. tenuiloba Trautv. It is very likely that S. hairalensis Nakai (Journ. of Jap. Bot. XIX (1943) 273) refers to this form. Because of ths lack of type specimens and the ambiguities of the description, this question cannot be decided at present.

Excluded from the "Flora of the USSR" are

1. Scabiosa gramuntia L., erroneously reported for the Crimea.
2. Scabiosa reversa Ldb. in Ind. hort. Acad. Dorpat. (1824)
91 Suppl. II, 6, erroneously reported for Siberia (Index Kewensis). This has been based on specimens cultivated from achenes of the order Columbariae Bobr., from S. Europe.

Family CLVII. **CUCURBITACEAE*** HALL.

Flowers usually diclinous, monoecious or dioecious (rarely hermaphroditic), regular, solitary, axillary or in raceme, panicles, umbels or axillary fascicles; corolla and calyx 5- (rarely 4–6-) merous, connate below into a slightly thickened basal ring, free above; corolla gamopetalous, usually 5-lobed or 5-fid, to dissected, yellow or white, rarely greenish or red; stamens 1–5, (mostly 5), usually 4 connate in pairs, the fifth free, thus giving the superficial impression of only 3 stamens, filaments or anthers of all stamens sometimes adnate. Inferior ovary (sometimes half-inferior) predominantly trilocular, with accrescent fleshy placentas. The pistillate flowers sometimes have 2–5 staminodes. Style with thickened fleshy stigmas. Fruit a berry (sometimes dry) or a berrylike many-seeded pepo with a juicy pulp and coarse (sometimes woody) cortex; seeds without endosperm and embryo erect with broad fleshy cotyledons. Annual or perennial herbs (sometimes shrubs or even arbuscula), often with climbing or prostrate stems, the leaves alternate, simple, entire or lobed to divided and dissected.

Economic importance. The Cucurbitaceae ranks among the more important families of edible and officinal plants. It also comprises all kinds of gourds used as water-jars and the sponges – substitutes used for bathing and kitchen purposes.

The literature on this family is enormous, but the most comprehensive study remains Cogniaux's monograph (in DC. Monogr. Phanerogam. III). Engler's "Pflanzenfamilien" comprises a study by Müller and Pax (1894). Unfortunately the treatment by Cogniaux (1916) confirmed by Cogniaux and Harms (1924) in Engler-Prantl's "Pflanzenreich" has not been completed, although the basic principles were published. The tribe Cucurbiteae and subtribe Cucumerinae were published in 1924. For Cucurbitaceae grown in the USSR and adjacent countries, attention is drawn to the many works by
92 K. I. Pangalo et al., published for the most part between 1925 and 1956, and widely known in the USSR.

Key to the Genera

1. Fruit a globose berry not more than 7–8 mm across . . . . 1422. **Bryonia** L.
+ Fruit different . . . . 2.
2. Staminate flowers with 2 membranous distally connate bracts forming cuplike involucre in flower . . . . 1420. **Momordica** L.
+ Involucre obsolete . . . . 3.
3. Stems rather slender; flowers small, white or greenish-white; fruit 1–1.5(2) cm, when ripe, thin, coriaceous, with or without bristles, if larger (4–5 cm long, 3–4 cm wide) then dry, with coriaceous walls beset with long prickles-bristles . . . . 4.
+ Stems usually thick, fleshy; flowers yellow or orange, if white then large (petals 3–4 cm long, 2–3 cm wide); fruit much larger, if small (ca. 5 cm) then smooth, with woody walls . . . . 7.

* Treatment by I.T. Vasil'chenko.

4. Fruit large (4–5 cm long, 3–4 cm wide); corolla usually 6-lobed . . . . 1430. **Echinocystis** Torr. et Gr.
\+ Fruit not as large; corolla 5-lobed . . . . 5.
5. Fruit distally dehiscent by operculum . . . 1417. **Actinostemma** Griff.
\+ Fruit dehiscing by 3 longitudinal valves turning outward, or indehiscent . . . . 6.
6. Fruit dehiscent by 3 longitudinal valves, 3-seeded, glabrous or subglabrous . . . . 1419. **Schizopepon** Maxim.
\+ Fruit indehiscent, 1-seeded, densely covered with bristly hairs . . . . 1429. **Sicyos** L.
7. Tendrils present, simple or branching . . . . 8.
\+ Tendrils absent . . . . 1423. **Ecballium** A. Rich.
8. Tendrils simple . . . . 9.
\+ Tendrils branching, very rarely simple (see Citrullus colocynthis (L.) Schrad.) . . . . 11.
9. Root tuberiform-thickened; stamens free; leaves usually entire, cordate-ovate, acute . . . . 1418. **Thladiantha** Bge.
93 + Root not tuberiform-thickened, thin; 4 stamens connate in pairs, the fifth free; leaves usually lobed (to divided) or more or less angular . . . . 10.
10. Ovary densely covered with soft hairs; corolla sulphur-yellow; leaf lobes (segments) rounded . . . . 1426. **Melo** Adans.
\+ Ovary covered with sparse bristles; corolla vitelline-yellow; leaf lobes (segments) triangular-mucronate . . . . 1425. **Cucumis** L.
11. Corolla white, dissected to base; petioles with 2 dentate glands below limb . . . . 1427. **Lagenaria** Ser.
\+ Corolla yellow to orange, lobed or divided, rarely dissected; petioles without distal glands . . . . 12.
12. Fruit dry, tapering; staminate flowers borne in racemes, on long peduncles . . . . 1421. **Luffa** Adans.
\+ Fruit juicy, variously shaped; staminate flowers solitary or clustered in axils of leaves . . . . 13.
13. Leaves pinnatipartite or pinnatisect; corolla rotate or broadly campanulate, rather small, deeply 5-fid . . . . . 1424. **Citrullus** Forsk.
\+ Leaves 5-lobed; corolla more than 5–6 cm long, campanulate, lobed for not more than half its length . . . . 1428. **Cucurbita** L.

Genus 1417. **ACTINOSTEMMA** * Griff.

Griff. Account. Bot. Call. Cantor (1837) 24, tab.3.

Flowers monoecious, dioecious or hermaphroditic; staminate flowers in racemes or panicles, very rarely solitary, receptacle cupulate, squamous; sepals linear-lanceolate; petals whitish, lanceolate or elongate-ovate, long-acuminate; stamens 5, free. Pistillate flowers solitary, in groups or proximal in mixed racemes with staminate flowers, with flask-shaped receptacle and 3–5 glandlike staminodes; calyx and corolla similar to those

* From the Greek actinos – radius and stemma – corona (the flowers resembling a corona).

of staminate flowers. Ovary ovoid, usually verrucose, unilocular; style short, fleshy, with 2 stigmas. Fruit ovoid, conically tapering above, with 2–4 seeds, dehiscing by operculum above middle. Perennial herbs, their stems climbing by means of tendrils, glabrous or with scattered hairs, leaves alternate, hastately cordate; tendrils simple or bipartite.

About 8–10 species in E. Asia and the Himalayas.

94 **1. A. lobatum** Maxim. in Franch. et Sav. Enum. pl. Jap. I (1875) 175; Cogniaux in DC. Monogr. Phanerogam. III, 921 et in Pflanzenr. H. 66, 33; Kom. in Tr. Peterb. bot. sada, XXV, 2, 348; Kom. and Alis., Opred. rast. Dal'nevost. kr. II, 988. – A. lobatum a. longiloba et b. subintegra Kom. in Fl. Man'chzh. III, 2 (1907) 548. – Mitrosicyos lobatus Maxim. Prim. Fl. Amur. (1859) 112. – Ic.: Maxim. l. c. tab. 7.

Perennial; stems 1 m, glabrous, sparingly branching, prostrate or climbing by means of numerous bifurcate tendrils, sparsely pubescent, becoming glabrous; leaves thin, cordate- or ovate-triangular-hastate, 3–5-lobed, to divided (usually terminal lobe long, lateral ones shorter (f. longiloba Kom.)) or entire, more or less undulantly emarginate (f. subintegra Kom.), with deep trapeziform or rounded notch, thin-petioled, (0.5)1–3(4) cm, with very short hairs along nerves and margins, (1.5–2)3–8(12) cm long, (1.5–2)3–6(8) cm wide. Staminate flowers solitary or in racemes, on slender short-haired peduncles; calyx linear-lanceolate, subulate-acuminate, sparingly pubescent, with thin denticles, ca. 3–5 mm long; petals subulate or filiformly acuminate, ovate-lanceolate, denticulate at base, ca. 3–4 mm long; pistillate flowers axillary, solitary, the peduncles long, jointed about middle; fruit ovoid, greenish, brown when dry, (1)1.5–2 cm long, (0.8)1–1.5 cm wide, 2(3–4)-seeded, covered in lower part with soft bristles (later falling), dehiscing distally; seeds (8)10–13 mm long, (6)7–9 mm wide, flattened (often planoconvex), pitted-rugose. Fl. July–September, Fr. from August.

Banks and valleys of rivers and lakes. – Far East: Uss. **Gen. distr.**: Jap.-Ch. Described from Japan. Type in Leningrad.

Genus 1418. **THLADIANTHA*** Bge.

Bge. Enum. pl. chin. (1834) 29

Flowers dioecious; the staminate in racemes or umbels, sometimes solitary; receptacle short-campanulate or nearly rotate; sepals 5, linear or lanceolate; corolla campanulate, 5-fid to base; stamens 5, free, unequal, there is a rudimentary pistil in the shape of a gland. Pistillate flowers solitary or in fascicles, with 5 staminodes (4 in approximate pairs, as often
95 are the stamens). Ovary usually oblong, with 3 placentas and numerous ovules. Fruit oblong-ovoid, indehiscent, many-seeded, juicy; seeds obovate, smooth, flattened, immarginate. Perennial herbs, their stems climbing by simple or bifurcate tendrils; roots flattened into tubers; leaves of the Soviet species alternate, entire, cordate.

About 25 species in SE Asia.

* From the Greek tlaein – press, squeeze, anthos – flower. Bunge saw only one specimen with deformed flowers.

1. T. dubia Bge. Enum. pl. chin. (1834) 29 et Mém. sav. etr. II (1835) 103; Naud. in Ann. Sc. Nat. 4 sér. XII, 150, XVI, 186, 5 sér. VI, 11; Cogniaux, in DC. Monogr. Phanerogam. III (1881) 422 et in Pflanzenr. H. 66 (1916) 42; Pax in Pflanzenfam. IV, 5, 13; Kom. and Alis., Opred. rast. Dal'nevost. kr. 988; Kom., Fl. Man'chzh. III, 2, 551; Lapin in Opred. rast. Tashk. oazisa, I, 277; Visyulina in Viznach. rosl. URSR, 503; Bobrov in Fl. Maevskogo (1954) 324. – Ic.: Naud. (1854) tab. 10; Cogniaux (1916) f. 13; Pax, l. c. f. 11; Michurin, Soch. IV, Plate III.

Perennial; stems with spreading hairs, 0.5–1.5 m, weakly branching, climbing by simple pubescent tendrils; leaves broadly ovate-cordate, short-acuminate, with deep lyrate notch at base, sometimes slightly dentate, 5–10 cm long, 4–9 cm wide, more or less densely finely pubescent on both sides (to tomentose beneath), with stiff bristles and short hairs along nerves and petioles, nerves branching furcately from base of limb; tendrils thin, simple, sitff-haired. Flowers solitary; corolla yellow; peduncles of staminate flowers slender, with long soft hairs, 1–3 cm; receptacle shortened, nearly rotate, pubescent, 7–8 mm across; sepals recurved, linear-lanceolate, 12–13 mm long; petals elongate-ovate, acute, short-haired, 2.5 cm long; pistillate flowers on thickish peduncles, 0.5–1.5 cm long; staminodes short; ovary oblong, pubescent; style with 3 diverging stigmas; fruit oblong-ovoid, green or red, with 10 longitudinal furrows, covered with spreading short hairs, 4–5 cm long, 2.5 cm across; seeds obovate, tapering at base, smooth, 4–4.5 mm long, 2.5–3 mm wide, flattened. Fl. from June to September, Fr. from August.

Riparian meadows, sandy coasts, as a weed in gardens and orchards, sometimes grown as an ornamental. – European part: sometimes weed of gardens and orchards, locally cultivated as a climber, to Kalinin, Gorki in the north; Far East: Uss. (known only along the Suchan River). **Gen. distr.**: northeast regions of China. Described from Peking. Type in Paris.

96 Note. The distribution of this interesting plant in the USSR, beyond its natural area, is worthy of attention. Apparently introduced into the European part of the USSR as an ornamental, it escaped locally to become a weed of gardens and orchards. Lapin (op. cit.) claims that it has successfully acclimatized to Tashkent, but after long years of growing on the embankments of the Ankor Canal, it was eradicated through the introduction of earth-moving equipment. I. V. Michurin (1948), who considered Thladiantha eminently suitable for experiments in remote hybridization, successfully crossed it with Cucurbita pepo, Cucumis sativus and Melo sativus. On 18 July 1913 A. A. Bulavkin collected a specimen of T. dubia near the Suchan River (S. Maritime Territory). This differed from the type in its smaller flowers, with sepals longer than petals and markedly pubescent beneath. Komarov referred this plant to T. dubia var. parviflora Kom. (in herb.).

Genus 1419. **SCHIZOPEPON** * Maxim.

Maxim. Prim. Fl. Amur. (1859) 110

Flowers small, hermaphroditic or diclinous, axillary, solitary, rarely in few-flowered racemes; receptacle broadly campanulate; sepals 5; corolla

* From the Greek schizein – split, opening, pepon – melon; referring to the dehiscing fruits, and their melonlike shape.

whitish, rotate, deeply 5-fid; stamens 5, 4 connate in pairs, producing bilocular anthers, free, with unilocular anther. Ovary broadly ovoid, trilocular, with short fleshy 3-partite style. Fruit ovoid, fleshy, tapering, 3-seeded, dehiscing by 3 recurved valves; seeds ovoid, flat, crenately undulant. Perennial herbs with alternate cordate-ovate, long-petioled leaves, covered with finely scattered hairs; tendrils 2-partite, long.

Monotypic genus.

1. **S. bryoniifolius** Maxim. Prim. Fl. Amur. (1859) 111 et in Ann. Sc. Nat. 4 sér. XIII (1860) 95; Cogniaux in DC. Monogr. Phanerogram. III, 917 et in Pflanzenr. 66 (1916) 255; Pax in Plfanzenfam. IV, 5 (1889) 21; Kom., Fl. Man'chzh III, 2 (1907) 548; Kom. and Alis., Opred. rast. Dal'nevost. kr. II, 988. – Ic.: Maxim. l. c. (1859) tab. 6; Cogniaux, l. c. (1916) f. 60; Kom. and Alis., op. cit. Plate 298.

Perennial; stems thin, branching, pubescent, becoming glabrous, 2 mm, climbing by bifid tendrils; leaves thin, cordate-ovate, with large long-acuminate
97 terminal lobe and short broadly triangular lateral lobes, 3–7-angled, the margin undulant, small-toothed, 5–10(14) cm long, (3)4–8(11) cm wide, both sides (especially beneath) with short bristly hairs, petioles 5–12 cm long, thin, sparsely haired or glabrous. Flowers on short slender suberect peduncles which subsequently elongate to 2–3 cm and turn downward, in fruit elongating to 6–10 cm; receptacle subglabrous, ca. 1 mm across; sepals erect, 1.5–2 mm long, corolla lobes 2 mm long, 1–1.2 mm wide, acute, glandular-verrucose on both sides; ovary glandular-verrucose, globular-ovoid, 2–3 mm across; style 1 mm long, with diverging stigmas; fruit green, glabrous or with sparse hairs, more or less tuberculate-rugose, 10–15 mm long, 8–12 mm wide, brown when dry, 3-seeded; seeds dirty gray, tapering upward, ca. 9–10 mm long, 6–7 mm wide at base, tuberculate in upper part. Fl. July–August, Fr. August–September.

Shrubby thickets in mixed forests, along streams and rivers. Far East: Uss., Sakh. **Gen. distr.**: Jap.-Ch. Described from Ussuri. Type in Leningrad.

Note. Cogniaux's diagnosis of Schizopepon (1881 and 1916) does not quite agree with the characters of the species. For example, the diagnosis mentions solitary flowers, hermaphroditic, but Figure 60 shows them in racemes (which are also shown by Komarov and Alisova, Figure 298). Moreover, Cogniaux himself, writing of S. bryoniifolius var. paniculatus Kom., mentions declinous flowers arranged in racemes – many-flowered in the case of staminate flowers and few-flowered in the case of pistillate ones. Study of herbarium specimens of S. bryoniifolius revealed that in addition to solitary axillary flowers there also occurred racemes usually of not more than 2–3 flowers. Hence, we have made the necessary corrections in the generic diagnosis.

Genus 1420. **MOMORDICA** * L.

L. Sp. pl. (1753) 1009

Flowers monoecious, rarely dioecious; staminate flowers solitary or in racemes or corymbs; receptacle embraced by membranous connate bracts;

* From the Latin mordere (perfectum momordi) – to bite, gnaw, (referring to the seeds of this genus, which often appear gnawed at the margin).

sepals 5; corolla yellow or white, 5-partite or 5-lobed, rotate or broadly campanulate; stamens (2)3–5. Pistillate flowers solitary; ovary oblong or
98 flask-shaped, with 3 placentas; style thin, with 3 stigmas. Fruit oblong, flask-shaped or cylindrical, sometimes subglobose, often prickly or verrucose, fleshy; seeds inflated or flattened. Annual or perennial herbs with prostrate and climbing stems, simple or branching tendrils, lobed or palmately compound and leaves with 3–9 leaflets.

More than 60 species in the tropics and subtropics of both hemispheres, mainly in Africa.

1. **M. balsamina** L. Sp. pl. (1753) 1009; Willd. Sp. pl. IV, 601; Ser. in DC. Prodr. III, 311; Schrad. in Linnaea, XII, 420; Naud. in Ann. Sc. Nat. 4 sér. XII, 132, 5 sér. V, 20; Boiss. Fl. or. II, 757; Cogniaux in DC. Monogr. Phanerogam. III, 439; Pax in Pflanzenfam. IV, 5, 23; Cogniaux and Harms in Pflanzenr. H. 88. 28. – Ic.: Gaertn. De fruct. et seminibus pl. tab. 88; Rev. Hort. f. 75.

Annual; stems strongly branching, subglabrous, 0.5–1.5 m, climbing by simple filiform glabrous tendrils; leaves rounded, glabrous or sparsely coarse haired, palmately 3–5 lobed or divided, with acute dentate, rhombic lobes, more or less constricted at base, 2–5(7) cm wide, outer lobes (segments) smaller, the median larger than the lateral ones; petioles 1–3 cm puberulent. Staminate flowers solitary, on 1.5–5(7) cm peduncles; bracts sessile, rounded-cordate, dentate, subglabrous, 6–12 mm wide; sepals ovate, pubescent, mucronate, ca. 5–6 mm long; corolla orange-yellow, dark brown in lower part, slightly irregular (asymmetrical), lobes obovate, 15 mm long, 7–10(12) mm wide; pistillate flowers ebracteate or with bracts sessile at base of peduncle, this shorter than in staminate flowers; fruit red-orange, broadly ovoid or subglobose, angular and tuberculate, elongating into a fleshy beak, 3–6 cm across, irregularly dehiscing; seeds ovoid, flattened, finely tuberculate, with nearly entire margin, 10–11 mm long, 7 mm wide.

Rarely cultivated. – European part: L. V.; Centr. Asia: Syr D. **Gen. distr.**: tropics and subtropics of the Old and New Worlds. Described from India. Type in London.

Note. Cultivated in the tropics for food (edible fruits – "Poma Hierosolymitana") or medicine, and as an ornamental. In the herbarium of the Botanical Institute of the Academy of Sciences of the USSR there is a part of a staminate specimen, collected by S. I. Korzhinskii in Astrakhan, as well as a branch without flowers or fruit collected in Centr. Asia. On
99 the label there is a comment by O. A. Fedchenko: "Momordica balsamina – "Karillya" – Indian squash culta." The cultivation of Momordica was undoubtedly introduced into the USSR from the south, probably from India, Afghanistan or Iran. Today this rare plant is being grown by amateur gardeners in the republics of Central Asia and, possibly, in other southern regions of the USSR.

Genus 1421. **LUFFA** * Adans.

Fam. II (1763) 138

Flowers diclinous, monoecious. Staminate flowers in racemiform inflorescences; receptacle campanulate or turbinate; sepals 5, triangular or lanceolate; petals 5, nearly free, spreading, yellow or white; stamens 3–5, free, attached to receptacle tube. Pistillate flowers solitary, with 35– staminodes; ovary elongate, furrowed, angular or cylindrical, with 3 placentas; style columelliform, with 3 bilobate stigmas. Fruit dry, elongate or cylindrical, smooth or with acute ribs, glabrous or prickly, fibrous inside, trilocular, many-seeded. Annual herbs climbing by means of tendrils, with 5–7-lobed leaves (rarely leaves nearly entire).

About 10 species, all in warm zones.

Economic importance. Both species described below are cultivated in the southern part of the USSR as ornamentals and for the sponges obtained from the dry fruits used in the manufacture of women's slippers, men's summer hats and service caps.

1. Fruit ribless; seeds smooth, with 0.5–1 mm wide margin . . . . 1. **L. cylindrica** (L.) Roem.

\+ Fruit 10-ribbed; seeds rugose, emarginate . . . . 2. **L. acutangula** (L.) Roxb.

1. L. cylindrica (L.) Roem. Fam. II (1846) 63; Naud. in Ann. Sc. Nat. 4 sér. XII, 119; Cogniaux in DC. Monogr. Phanerogam. III, 456; Pax in Pflanzenfam. IV, 5, 25; Visyulina v Viznachn. rosl. URSR, 54; Cogniaux u. Harms in Pflanzenr. H. 88, 62. – Momordica cylindrica L. Sp. pl. (1753) 1009; Willd. Sp. pl. IV, 63; Ser. in DC. Prodr. III, 311. – M. luffa L. l. c. – Ic.: Pflanzenfam. IV, 5, f. 18; Pflanzenr. H. 88, f. 8.

Annual; stems pentahedral, scabrous along ribs, 3–6 m; leaves palmately 5-lobed, 15–25 cm across, with triangular or lanceolate, more or less acute
100 lobes, scabrous above and beneath; peduncles 10–12 cm, thickened scabrous; tendrils short-pubescent, usually trifid. Raceme of staminate flowers 13–15 cm long, with 15–20 flowers on 1–2 cm peduncles; pistillate flowers on 2–10 cm peduncles, borne in the axil of the same leaf as the staminate inflorescence; receptacle broadly campanulate, more or less pubescent; sepals lanceolate, slightly longer than receptacle; corolla 4 cm across; petals oblong-cuneate, rounded or slightly notched at apex, bright yellow, 2–3 cm long, 1–1.5 cm wide; ovary cylindrical, glabrous; fruit fusiform, cylindrical or elongate-clavate, ribless, 50 cm long or longer, 6–10 cm across; seeds ovoid, smooth, ca. 12 mm long, 8–9 mm wide, with 0.5–1 mm wide margin. Fl. from June, Fr. from August.

Cultivated in the southern part of the USSR. **Gen. distr.**: cultivated in all warm zones, sometimes wild. Described from China and Ceylon. Type in London.

2. L. acutangula (L.) Roxb. Hort. Beng. (1814) 70; Ser. in DC. Prodr. III, 302; Ldb. Fl. Ross. II, 1, 141; Roem. Fam. II, 66; Naud. in Ann. Sc. Nat. 4 sér. XII, 122; Cogniaux in DC. Monogr. Phanerogam. III, 459; Pax in

* From loofah, Arabic name for L. cylindrica.

Pflanzenfam. IV, 5, 25; Cogniaux u. Harms in Pflanzenr. 88, 68; Visyulina v Viznachn. roslin URSR, 505; Nikitin in Fl. Turkm. VI, 362. – Cucumis acutangula L. Sp. pl. (1753) 1011. – Ic.: Gartenflora XLVIII, Abb. 30; Pflanzenfam. IV, 5, f. 18.

Annual; stems pentahedral, scabrous along ribs, 3–6 m; leaves rounded, pale green, scabrous above and beneath, 5–7-angled or nearly lobed, 15–20 cm across, with 8–12 cm petioles; tendrils slightly scabrous, usually trifid. Staminate inflorescence 10–15 cm with 17–20 flowers on 1–4 cm peduncles; inflorescence borne in axil of same leaf, 5–10 cm long; receptacle short, campanulate, pentahedral, with short hairs; sepals lanceolate, keeled, slightly longer than receptacle; corolla pale yellow; petals obcordate, notched at apex, 2 cm long, 2–2.5 cm wide; ovary clavate, 10-ribbed, obtuse or short-acuminate, 15–30 cm long, 6–10 cm across; seeds ovoid, flattened, without border, rugose, yellowish-black or black, bidentate at base, 11–12 mm long, 7–8 mm wide. Fl. from June, Fr. from August.

Cultivated in southern part of the USSR. **Gen. distr.**: probably a native of India, cultivated in warm zones of both hemispheres, sometimes wild. Described from specimens of unknown origin ("Hab. in Tataria, China"). Type in London.

101 Genus 1422. **BRYONIA** * L.

Flowers diclinous in corymbs or umbels or 1–2, sometimes sessile, dirty- (greenish-) white or yellowish. Corolla in staminate flowers more or less campanulate, 5-fid, with oblong-ovate or divided lobes, or rotate; sepals 5; stamens 5, 2 connate pairs and 1 free, giving impression of 3 stamens. Pistillate flowers with globose inferior ovary and 3–5 obsolete staminodes; corolla and calyx as in staminate flowers but usually smaller, stigmas 3. Fruit a globular few-seeded berry, with thin, black, red or yellow (to orange), sometimes yellowish skin; seeds obovoid, flattened, usually with narrow circular border. Perennial herbs, climbing by means of simple tendrils, with thickened root (as thick as a man's hand), white inside; leaves pentagonal or 5-lobed to deeply divided, sometimes entire, usually cordate or pubescent only along margin.

Ten to fifteen species distributed in the Mediterranean area, S. Europe and Near Asia.

Economic importance. Poisonous, medicinal (Radix Bryoniae), and ornamental plants.

Note. Bryonia is one of the most interesting plants in Russian flora because of variations in sex. Thus, monoecious or polygamous flowers occur in the dioecious species. Botanists have extensively studied variations in the shape of leaves, in particular the differences in the shape of the leaves of the staminate and pistillate specimens of the dioecious species, but so far to no avail. Many interesting peculiarities, such as parthenocarpy, have been observed in the fertilization and in the development of the fruit. A great deal has been written about the medicinal properties of Bryonia, but we shall limit ourselves here to Cogniaux and Harms' survey of the more important investigations of the genus in "Pflanzenreich" (H. 88, 1924).

* From the Greek bryein – to shoot, from its rapid growth.

In the USSR Bryonia has not been thoroughly investigated and the above-mentioned authors therefore touch upon Russian species only superficially and the present treatment is indeed not complete. More material
102 must be collected, principally from the Caucasus, Crimea and Central Asia. Such collections should comprise staminate inflorescences of the monoecious species, which are often confined to the upper part of the plant, and pistillate flowers prevalent in the lower part. It is desirable that entire plants be collected. Since this is difficult because of long stems, lower and upper parts of the plant with the flowers and fruits should be collected separately. Collections of dioecious species must comprise staminate and pistillate specimens. Finally, the color of the corolla and berries must be recorded on the spot before drying. The habitat should also be described.

1. Monoecious plants; stigma usually smooth . . . . 2.
+ Dioecious plants; stigma usually scabrous . . . . 4.
2. Fruits black . . . . 1. **B. alba** L.
+ Fruits red or orange . . . . 3.
3. Staminate and pistillate flowers in mixed racemes: the former distal to the latter; linear-filiform, as long as petals or longer. Leaves lobed . . . . 2. **B. monoica** Aitch. et Hemsl.
+ Staminate flowers in separate racemes in upper part of plant, pistillate flowers in racemes found in lower part of plant; sepals shorter than petals; leaves divided, with long attenuate terminal lobe . . . . 4. **B. transoxana** Vass.
4. Leaves entire, large, 9–13 cm long, 6–9 cm wide, resembling leaves of burdock; seeds semirounded, ca. 5 mm long, 4 mm wide; flowers sessile or subsessile (S. Tadzhikistan) . . . . 5. **B. lappifolia** Vass.
+ Leaves lobed to divided, or shape different; seeds more elongate, obovoid; flowers in stalked racemes, peduncled (sometimes stalks and peduncles rather short) . . . . 3. **B. dioica** Jacq.

1. B. alba L. Sp. pl. (1753) 1012; Willd. Sp. pl. IV, 621; M. B. Fl. taur.-cauc. II, 410; Ser. in DC. Prodr. III, 307; Ldb. Fl. Ross. II, 1, 140; Boiss. Fl. or. II, 762; Cogniaux in DC. Monogr. Phanerogam. III, 476; Pax in Pflanzenfam. IV, 5, 25; Shmal'g., Fl. I, 371; Hegi, Ill. Fl. VI, 1, 317; Cogniaux u. Harms in Pflanzenr. H. 88, 84; Grossg., Fl. Kavk. IV, 60; Fedch. in Fl. yugo-vost. VI, 282; Visyulina v Viznachn. roslin URSR, 503; Nikitin v Fl. Turkm. VI, 364. – B. nigra Gilib. Fl. lithuan. I (1781) 60. – Ic.: Rchb. Ic. Fl. Germ. XIX, tab. 1620; Hegi, l. c.: Pflanzenr. 88, f. 11, H-R.; Nikitin, op. cit. Plate L.
103 Perennial; monoecious plant; stems rather thin, 2–4 m, faceted, glabrous or covered with small prickles and bristles sessile on white tubercles; leaves glabrous or pubescent, 3–6(10) cm across (sometimes larger), the limb broadly cordate, ovate or triangular-ovate, more or less deeply notched at base, 5–7-lobed or divided, lobes triangular or ovate, irregularly dentate or undulant. Staminate flowers usually (5)7–15 in racemes with long peduncles (5–20 cm) in upper part of stem; pedicels of staminate flowers filiform, protuberant, 0.5–2 cm; sepals 1.5–2 mm long, much shorter than corolla, the latter 4–5 mm, bluish-yellow or yellowish-white, with green nerves, like receptacle more or less pubescent outside, the lobes ovate or

oblong-ovate; pistillate flowers in corymbiform or umbelliform inflorescences, usually in lower part of plant, on shorter (2–20 cm) stalks, sepals as long as corolla, the latter the same color as in the staminate flowers; stigma glabrous; fruit black, globose, 7–8 mm across; seeds pale brown, rufous or cinnamon brown, sometimes grayish-brown, thinly rugose, with narrow border, 5 mm long, 3–4 mm wide, laterally flattened, obovoid. Fl. May–July, Fr. June–August.

Shrubby formations, forest edges, ravines, as a weed in orchards, sometimes as ornamental climber in gardens and parks. — European part: Balt., Lad.-Ilm. (Pskov), U.V., V.-Kama, U. Dnp., M.D., V.-Don, U. Dns., Bes., Bl., Crim., L. Don; Caucasus: Cisc., W., E. and S. Transc.; Centr. Asia: Mtn. Turkm., Ar.-Casp., Balkh., Dzu.-Tarb., T. Sh. **Gen. distr.**: Scand., Centr. Eur., Bal.-As. Min., Arm.-Kurd., Iran (N.). Described from W. Europe. Type in London.

Note. Bieberstein (l. c.) cites the variety β for B. alba with the following diagnosis (p. 411); "caule petiolisque hispido-aculeatis, pedunculo communi racemi petiolo breviore." He adds that though he never saw the fruits it might be dioecious. Later, he referred the variety β to B. dioica as having dioecious flowers, but again notes that he never saw the fruits. He tentatively considered them to be red ("videbantur attamen rubrae euasurae").

At about the same time, Steven (in Catal. Horti Gorenk. (1812) 32) described B. aspera from the Caucasus, with the following diagnosis: "follis cordatis palmato-quinquelobis calloso-asperis; lobis acutis inaequaliter remoto dentatis integerrimisve racemis axillaribus subcorymbosis, floribus dioicis, baccis nigris." B. aspera also appears in the work
104 of K. A. Meier (Ind. pl. cauc. 83) and Ledebour (l. c.). Later it was discarded in favor of B. alba L. However, the issue is still obscure and we would not claim B. aspera as identical with B. alba (or B. dioica) nor that it is a separate species, distinguished from B. alba by dioecism and shortened stalks, and from B. dioica by the black color of its fruits.

2. B. monoica Aitch. et Hemsl. in Transact. Linn. Soc. sér. 2, III (1888) 65; Cogniaux u. Harms in Pflanzenr. H. 88, 87; Nikitin in Fl. Turkm. VI, 365. — Ic.: Aitch. et Hemsl. l. c. tab. 10; Nikitin, op. cit. tabl. L.

Perennial; monoecious plant; stems 2–4 m, climbing high, thin, glabrous, or with short scattered prickles and bristles; leaves with (1)1.5–3 cm petioles, their limb triangular-ovate, deeply notched at base, usually faintly 5-lobed, the lobes triangular, more or less acuminate, the terminal much longer than the lateral, the latter obtuse, sometimes weakly expressed; margin slightly undulant, subglabrous above, very short bristly and scabrous along margin and beneath; sometimes leaves deeply notched to divided, the lamina varying from (2.5–3)5–9 cm long and from 3–7 cm wide. Flowers greenish-yellowish, with short papillae on the outside, in 6–10-flowered racemes usually pistillate, the upper staminate; pedicels filiform, 4–8(10) cm; sepals narrowly linear, divergent, 8–10 mm long, as long as or longer than lanceolate petals; receptacle of staminate flowers broadly campanulate, 4–5 mm long and as wide, globose in pistillate flowers; stigma filiform, glabrous; fruit smooth, globose, red. Fl. April–May, Fr. May–June.

Dried-out riverbeds (often with species of Tamarix), stony slopes, banks of lakes. – Centr. Asia: there are reports of the occurrence of this species in Mtn. Turkm. (Badkhyz). **Gen. distr.:** Afghanistan (Badkhyz), from where it was described. Type in London.

3. **B. dioica** Jacq. Fl. Austr. II (1773) 59; Willd. Sp. pl. IV, 621; M. B. Fl. taur.-cauc. Supplem. 625; Ser. in DC. Prodr. III, 307; Ldb. Fl. Ross. II, 1, 140; Boiss. Fl. or II, 760, Suppl. 242; Cogniaux in DC. Monogr. Phanerogam. III, 470; Shmal'g., Fl. I, 371; Pax in Pflanzenfam. IV, 5, 25; Hegi, Ill. Fl. VI, 1, 318; Cogniaux u. Harms in Pflanzenr. H. 88, 77; Grossg., Fl. Kavk. IV, 60; Visyulina in Viznachn. rosl. URSR, 503; Nikitin in Fl.
105 Turkm. VI, 364. – B. dioica β subsessilis Boiss. Fl. or. II (1872) 760. – B. ruderalis Salisb. Prodr. (1796) 159. – B. nitida Link. Enum. II II (1822) 404; Ser. l. c. – B. lutea Bast. ex. Ser. l. c. – B. subsessilis (Boiss.). Bornm. in Beih. Bot. Centralbl. VIII (1938) 271. – Ic.: Jacq. l. c. tab. 199; Hegi, l. c. tab. 255; Pflanzenr. H. 88, f. 10, 11 A–G; Nikitin, op. cit. Plate L. – Exs.: Bornm. Iter. pers.-turc. No. 3561 (sub B. subsessilis).

Perennial; dioecious plant; stems 2–4 m, thin, scabrous from acute prickles and bristles, with basal tubercles, or subglabrous; leaves like stems pubescent, limb broadly cordate or angular-ovate, 5–7-angled or lobed to divided and nearly dissected, more or less deeply notched at base; lobes usually acute, triangular, their margin undulantly dentate to lobed. Flowers in specimens with staminate flowers long-peduncled (10–20 cm), with yellowish-white or bluish-yellow corolla, the sepals shorter than corolla; in specimens with pistillate flowers, these form corymbiform or umbelliform inflorescences on short stalks (1–4 cm, sometimes shorter), pubescent outside, sometimes sparingly so, as in staminate flowers; sepals half as long as corolla; stigma scabrous; receptacle 5–6 mm across, pubescent; fruit 7–8 mm across, globose, scarlet-red or yellow; seeds yellowish or yellow, with brown speckles, thinly rugose, flattened, obovate. Fl. May–June, Fr. June–August.

Habitat as in preceding species. – European part: U. Dns., U. Dnp., Bes., M. D., Bl., Crim.; Caucasus: Cisc., E. and S. Transc.; Centr. Asia: Mtn. Turkm., Pam.-Al., Syr D., Amu D., Balkh., T. Sh. **Gen. distr.:** Centr. and Atl. Eur., Bal.-As. Min., Med., Arm.-Kurd., Iran. Described from Austria. Type in Vienna.

Note. Boissier (Fl. or. II (1872) 760) refers the S. Iranian variety which appears in the work of Cogniaux (l. c. 472) and β subsessilis Cogniaux and Harms (l. c. 78) to B. dioica, distinguished from the type species by the markedly shorter, sometimes obsolete stalks of the pistillate racemes, the almost sessile flowers on short pedicels borne in the leaf axil, and the more strongly developed pubescence. Later, Bornmüller (l. c.) described it as B. subsessilis (Boiss.) Bornm. This southern race of B. dioica is surely worthy of notice, although it does not differ sharply from the type of B. dioica, neither geographically nor morphologically. Until more material becomes available, Boissier's view, according to which B. subsessilis is a variety of B. dioica, is accepted. In Central Asia the extreme northern habitats of this variety are the sands
106 of Muyun-Kum and the Sary-Su River valley (Lake Balkhash area).

4. B. transoxana Vass. sp. nov. in Addenda XXIII, p. 326.

Perennial; dioecious plant; stems glabrous or subglabrous; leaves hastate-triangular, 5-lobed or 5-fid, with triangular-lanceolate or oblong-ovate lobes, the median markedly longer than the obtuse, acuminate lateral lobes, leaves glabrous above, with stiff short hairs sessile on white tubercles beneath, deeply notched at base, margin undulant, sometimes with teeth and lobes. Pistillate flowers in very loose, elongate racemes, at base of stalks often sessile, the stalks (2)3–7 cm, including rhachis; fruit red, seeds dingy yellow, with narrow border, 4.5–5 mm long, 3.5–4 mm wide; staminate flowers unknown. Fl. April–May, Fr. May–June. (Plate VIII, Figure 2.)

Shrubby formations on lake and river banks. – Centr. Asia: Pam.-Al. (W.-Kashka Darya River basin). Endemic. Described from Shakhrisyabz. Type in Leningrad.

Note. This plant, with lower part missing, was collected by V. I. Lipskii near Chopuk, Shakhrisyabz District; its dioecism must therefore be verified.

The herbarium of the Botanical Institute of the Academy of Sciences of the USSR comprises specimens collected by V. Pel'ts in S. Kyzyl-Kum (7 April 1912), Shavir-Kuduk (5 April 1912), and near Kuldja (11 April 1912). These are monoecious, with red berries in loose racemes, and differ from B. monoica by their deeply divided, nearly dissected, leaves, the dense glandular pubescence of all parts of the flowers, and the sepals which are shorter than the petals; they differ from B. transoxana by the profuse pubescence of the flowers and the shape of the leaves. Whereas the leaves of B. transoxana possess a markedly elongate terminal lobe, the Kyzyl-Kum Bryonia has leaves which are uniformly, deeply 5-fid in the middle part of the plant, and terminal leaves with a markedly elongate, acute, linear-lanceolate median lobe and much smaller, acute, lateral lobes, considerably more pubescent than those observed in B. transoxana. As Pel'ts' collections consist of incomplete specimens, we refrain for the time being from describing the Kyzyl-Kum Bryonia as a new species (B. kyzilkum Vass. h. l.). The collections of V. I. Smirnov (23 June 1919, No. 123),
109 from south of Karsakpai, at the edge of the desert steppe Bet-Pak-Dala, are highly interesting. In shape and pubescence of leaves they resemble B. transoxana, which they differ from by the crowded fruit and dark-spotted, dingy background of the seeds. Dried herbarium material does not give a reliable idea of the color of the fruits but part of the ripe fruits is grayish-greenish. The fruits dehisce when ripe (the red fruits are not dehiscent). The Karsakpai specimens are here referred to B. betpakdalensis Vass. h. l., as the description of a new species required more material, especially plants in flowers.

5. B. lappifolia Vass. sp. nov. in Addenda XXIII, 458.

Perennial; dioecious plant; stems faceted, glabrous or with sparse pubescence; leaves entire, sometimes with obsolete lobes, and undulant margin, similar in shape to the reduced leaves of Lappa, 9–13 cm long, 6–9 cm wide, deeply notched at base, glabrous above, beset with white prickles and bristles, sessile on white tubercles beneath and along margin. Petioles 3.5–4.5 cm with spreading bristles. Fruit red, globose, crowded in groups 5–7 in axils of leaves, sessile or on 1.5–2 mm long, pubescent pedicels; seeds greenish-grayish, bordered, finely rugose, ca. 5 mm long,

(107)

PLATE VIII. 1 – Bryonia lappifolia Vass., part of the plant, seeds; 2 – B. transoxana Vass., part of the plant, seeds.

4 mm wide. Plants with staminate flowers are not described, as they have never been collected. Fl. May (?), Fr. June. (Plate VIII, Figure 1.)

Savin juniper slopes.— Centr. Asia: Pam.-Al. (known only from Gardani-Ushti Range in S. Tadzhikistan). Endemic. Described from the range indicated. Type in Leningrad.

Genus 1423. **ECHBALLIUM** * A. Rich.

A. Rich. in Dict. class. hist. nat. VI (1825) 19

Flowers diclinous, monoecious, very rarely dioecious; staminate flowers in racemes; receptacle short-campanulate; sepals 5, linear-lanceolate; corolla broadly campanulate or nearly rotate, deeply 5-fid; stamens 5, 2 connate pairs and 1 free. Pistillate flowers solitary, usually in axil of
110 the same leaf as racemes of staminate flowers, with 3–5 short liguliform staminodes, their perianth the same as in staminate flowers. Ovary oblong, bristly, with 3 placentas; style short, with three 2-fid stigmas. Fruit juicy, many-seeded, separating from pedicel when ripe and violently ejecting mucus and seeds. Perennial plants with ascending or prostrate stems, without tendrils, with alternate cordate leaves.

Monotypic genus.

1. E. elaterium (L.) A. Rich. in Dict. class. hist. nat. VI (1825) 19; Ldb. Fl. Ross. II, 1, 141; Boiss. Fl. or II, 760; Cogniaux in DC. Monogr. Phanerogam. III, 467; Pax in Pflanzenfam. IV, 5, 26; Cogniaux u. Harms in Pflanzenr. 88, 94; Hegi, Ill. Fl. VI, 1, 313; Grossg. Fl. Kavk. IV, 50; Fedch. in Fl. yugo-vost. VI, 283; Visyulina v Viznachn. rosl. URSR, 504. — Momordica elaterium L. Sp. pl. (1713) 1010.

Perennial; stems 0.5–1.5 mm thick, with short branches, more or less scabrous; root thickened, fleshy, white; leaves cordate-ovate or slightly lobed, crenate-dentate, gray-tomentose beneath, rugose, hard-scabrous, with thickened prominent nerves beneath, 5–10(20) cm long, 4–8(15) cm wide; petioles fleshy, terete, more or less bristly-scabrous, 5–15 cm long. Pedicels of staminate inflorescences villous or bristly, shorter than, as long as or longer than petioles; pedicels 1–3 cm, with linear bracts at base; sepals bristly-hairy; corolla pale yellow, lobes oblong-ovate, acute; fruit glaucous-green, oblong, or oblong-ovoid, bristly, obtuse at both ends, (3)4–6 cm long, 1.5–2.5(3) cm wide, when ripe detaching from pedicel and violently ejecting seeds and mucus via aperture thus formed; seeds small, elongate, flattened, smooth, narrowly bordered, ca. 4 mm long. Fl. from June, Fr. from August.

Weedy places in hedges, walls, roadsides, etc. — European part: Bes., Bl., Crim., L. V., L. Don, Transv. (Ural'sk); Caucasus: all regions; Centr. Asia: Syr D. (known from Tashkent). **Gen. distr.:** Med., Bal,-As. Min. Described from S. Europe. Type in London.

Note. One of the more interesting plants of the Russian flora; it provides an example of seed dispersal by ejection, due to the structure of the fruit walls (e.g., Körner's Pflanzenleben, II (1891) 771). This phenomenon and the chemism of the fruits have been studied by Harms (l. c.).

* From the Greek ecballeio — I am ejecting, referring to the ejection of the seeds.

111 Genus 1424. **CITRULLUS** * Forsk.
Forsk. Fl. aegypt.-arab. (1775) 167

Flowers diclinous, monoecious or dioecious, solitary, rarely in fascicles. Staminate flowers with broadly campanulate receptacle, 5 narrow sepals and rotate or broadly campanulate, deeply 5-fid yellow corolla with obtuse oblong-ovate lobes; stamens 5, 4 connate in pairs; with glandlike rudiment of ovary. Pistillate flowers with perianth similar to staminate and 3 short bristles or ligules as staminodes. Ovary ovoid, with short style and 3 fleshy, nearly 2-lobed stigmas. Fruit globose or elongate, juicy or dry, many-seeded, indehiscent; seeds obovoid, flattened, often bordered. Annual or perennial herbs, with prostrate or climbing stems; leaves alternate, rounded or triangular-ovate, deeply 3–5-lobed or divided, lobes (segments) lobed, divided or dissected; tendrils 3–5-fid, rarely simple.

There are 5–7 species, in Africa and Asia.

Economic importance. Citrullus is grown for food, forage, and medicinal purposes (see: Pangalo. Arbuzy severnogo polushariya, Tr. prikl. bot. gen. i sel. XXIII, 3 (1929–1930)).

1. Shoots in upper part villous-lanate; leaves slightly scabrous; ovary with long lanate-tomentose hairs; cultivated plant . . . . 1. **C. vulgaris** Schrad.
+ Shoots stiff-haired; leaves coarsely scabrous; ovary stiff-haired; weed . . . . 2. **C. colocynthis** (L.) Schrad.

1. C. vulgaris Schrad. in Eckl. et Zeyher, Enum. pl. (1836) 279; Linnaea, XII, 412; Naud. in Ann. Sc. Nat. 4 sér. XII (1859) 100; Cogniaux in DC. Monogr. Phanérogam. III, 508; Pax in Pflanzenfam. IV, 5, 27; Cogniaux u. Harms in Pflanzenr. 88, 103; Fedch. in Fl. yugo-vost. VI, 283; Visyulina in Viznachn. rosl. URSR, 505. – Cucurbita citrullus L. Sp. pl. (1753) 1010; Willd. Sp. pl. IV, 610. – Cucumis citrullus Ser. in DC. Prodr. 3 (1828) 190. – Citrullus edulis Spach, Hist. nat. veg. Phaner. VI (1838) 214. – C. edulis Pang. in Byull. Mosk. obshch. isp. pripr. 5–6 (1945) 81. – C. colocynthoides Pang., op. cit. 82. – Ic.: Pangalo in Tr. prikl. bot. gen. i sel. XXIII, 3, 45, 48–51, 61, 65, 67.

112 Annual; stems thin, flexible, creeping or climbing, usually cylindrical-5-faceted, to 4 mm and longer, branching; when young stem densely covered with soft spreading hairs; tendrils 2-fid, pubescent; leaves long-petioled, pubescent, coarse, triangular-ovate, cordate, stiff-scabrous above and beneath, deeply 3-fid, the lobes once or twice pinnatifid, with acute median lobe elongate at apex, lateral lobes usually rounded, sometimes leaves entire, more or less lobed, laminas 8–10 to 20–22 cm long and 5–10 to 15–18 cm wide. Flowers with boat-shaped bractioles; staminate flowers solitary, 2–2.5 cm across, on villous stalks; receptacle broadly campanulate, pubescent; sepals narrowly lanceolate to subulate-filiform; ovary greenish, villous, broadly infundibular; stamens 5, 2 connate pairs and 1 free; pistillate flowers solitary, slightly larger than the staminate; ovary more or less pubescent; style thin, ca. 5 mm; stigma 5-lobed, greenish; fruit a many-seeded, juicy pepo, variable in shape, size and color, flesh edible; seeds flat, often bordered, with hilum variously colored, Fl. June, Fr. August.

* Diminutive of the word citrus (referring to the shape of the fruit and the consistency of its flesh).

A valuable cultivated food plant in the southern regions of the USSR. **Gen. distr.:** originating from Africa, grown in all warm zones.

Note. The so-called forage Citrullus (C. vulgaris var. colocynthoides Schweinf. = C. colocynthoides (Schweinf.) Pang.), grown in the south for forage, deviate from type by their slightly larger flowers (3 cm across), acuminate petals, yellow rounded stigma, compact thick rind, tasteless, white-greenish flesh of the fruit and the seeds lacking a hilum. Also of interest are C. vulgaris var. fistulosus (Stocks) Duthie et Fuller (= C. fistulosus Stoks) with small, globosely flattened sweet fruits, with fibrous, dryish placentas of loose, interwoven, thick fibers. In the hot summers of the southern regions, C. vulgaris is a vital, easily transportable, dessert and thirst-quenching fruit. It is also recommended in the treatment of kidney ailments, though this potential has received very little recognition. In the USSR the best watermelons are grown in the sandy regions of the southeastern European part of the USSR and N. Caucasus, and on unirrigated land in the foothills of Central Asia. Some plants produce
113 gigantic fruits, of 20 kg and more; these are grown in the Syr Darya river valley in Kazakhstan and elsewhere. Besides being eaten raw, watermelons are also used for pickling (fermentation) and in the making of a syrup (called "nardek" in the southeast), outstanding in taste and odor.

2. C. colocynthis (L.) Schrad. in Linnaea, XII (1838) 414; Naud. in Ann. Sc. Nat. 4 sér. XII, 99; Boiss. Fl. or. II, 759; Clarke in Hook. Fl. Brit. Ind. II, 620; Cogniaux in DC. Monogr. Phanerogam. III, 510; Pax in Pflanzenfam. IV, 5, 27; Cogniaux u. Harms in Pflanzenr. 88, 109; Nikitin in Fl. Turkm. VI, 366. – Colocynthis vulgaris Schrad. Ind. sem. hort. Götting. (1833) 2; Thellung, Fl. advent. Montpellier, 494. – C. officinalis Schrad. l. c. 421. – Cucumis colocynthis L. Sp. pl. (1753) 1011; Ser. in DC. Prodr. III, 302. – Ic.: Wight, Ic. pl. Ind. or II, tab. 498; Pflanzenfam. IV, 5 f. 17; Pflanzenr. 88, f. 13; Pangalo in Tr. prikl. bot. gen. i sel. XXIII, 3, 71, 75.

Annual; stems 1–2 m, thin (ca. 3 mm across at base), prostrate or climbing, with well-defined facets, stiff-haired, with dense white recurved hairs; leaves elongate-triangular, (3)5–8 cm long, (2.5)4–5 cm wide, usually deeply notched or short-cuneate at base, deeply pinnatifid, with 5–7 pinnatilobate lobes, gray-green above, with scattered glandular hairs and bristles, ash-gray, pubescent with glandular and simple hairs beneath; tendrils simple or bifid, more or less coarse-haired, sometimes subglabrous. Stalks of staminate flowers short, villous, stiff-haired; receptacle broadly campanulate with spreading hairs; sepals subulate, apically reflexed, as long as receptacle; corolla lobes ovate, acute; ovary obovoid, stiff-haired; fruit pale green, yellowish when ripe, with longitudinal yellowish-green striae, pubescent at first, later glabrous, globose, 7–10(12) cm across, flesh dry, whitish-yellowish, hard, very bitter; seeds brownish, not bordered, ca. 6 mm long. Fl. August– September, Fr. September–October.

A weed in fields and along edges of sandy deserts. – Centr. Asia: Kara K. (known from several localities in the Tedzhen Oasis and near Geok-Tepe, in the Turkmen SSR). **Gen. distr.:** Africa, Near Asia (including India), Med. Introduced into W. Europe. Described from specimens of unknown derivation. Type in London.

Note. A perennial, Colocynthis has become an annual in the Turkmen SSR, where it reproduces solely by seeds. Especially in Near Asia, the bitter fruits have been used for medical purposes since ancient times. (For a survey of the literature, see Harms "Pflanzenreich," (1924).) In
114 some parts of North Africa the fruits are eaten cooked, they also are a favorite food of ostriches (see: Pangalo in Tr. prikl. bot. gen. i sel. XXIII, 3 (1929–1930) 71).

Genus 1425. **CUCUMIS** * L.
L. Sp. pl. (1753) 1012

Flowers diclinous, monoecious, rarely dioecious, 2.5–3(4) cm across, short-peduncled. Staminate flowers axillary, usually in sessile clusters; receptacle cup-shaped or subcylindrical; sepals lanceolate-subulate; corolla broadly funnel-shaped, deeply dissected, usually with 5 (rarely 6–8) lobes; stamens 5, 2 connate pairs and 1 free. Pistillate flowers solitary (rarely few in clusters); ovary fusiform, cylindrical, sometimes curved, sparsely covered with prickles and tubercles. Fruit elongate, of varying size and shape, with edible juicy flesh; seeds whitish, acute, elongate, not bordered. Annual plants, with alternate leaves and simple tendrils.

About 30 species in Africa and Asia

Economic importance. In the USSR grown almost exclusively as a valuable food plant (see: Gabaev. Ogurtsy Azii, Tr. prikl, bot. gen. i sel. XXIII, 3 (1929–1930); Filov. Ogurtsy mira, Byull. po plodovodstvu, vinogr. i sadov, 10, Stalinabad. 1948).

1. **C. sativus** L. Sp. pl. (1753) 1012; Willd. Sp. pl. IV (1805) 615; Ser. in DC. Prodr. III, 300; Naud. in Ann. Sc. Nat. 4 sér. XI, 27; Cogniaux in DC. Monogr. Phanerogam. III, 498; Hegi, Ill. Fl. VI, 1 321; Cogniaux u. Harms in Pflanzenr. 88, 143; Fedch. in Fl. yugo-vost. VI, 284; Visyulina in Viznachn. rosl. URSR, 504; Nikitin vo Fl. Turkm. VI, 367. — C. esculentus Salisb. Prodr. (1796) 157. — C. muricatus Willd. l. c. 613. — Ic.: Hegi, l. c. f. 171.

Annual; stems prostrate or climbing by simple tendrils, bristly-haired, faceted, 0.5–4 m; leaves on thickened stiff-haired petioles, cordate-ovate, 3–5-lobed or pentagonal, lobes nearly equal, acute, crenate or dentate, bristly-haired above and beneath, (8)12–18(20) cm across; corolla vitelline yellow, 2.5–4 cm across, sometimes larger; staminate flowers with densely pubescent (villous) receptacle; sepals nearly as long as receptacle; corolla lobes oblong-lanceolate, acute; ovary often cylindrical or fusiform, 2 cm long when flower dehiscing, tuberculate, with black, white or rufous prickles;
115 fruit usually cylindrical or short-cylindrical, sometimes angular or rounded, dark or yellow-green, smooth or tuberculate, sometimes with prickles, ripe fruit brown or cinnamon-brown, with netted surface; seeds numerous, oblong, 8–14 mm long. Fl. and Fr. May–June.

Cultivated nearly through the USSR, with the exception of the extreme north, where it can be grown in hothouses. **Gen. distr.:** originally from India, cultivated in all warm and moderate zones. Described from a specimen of unknown origin. Type in London.

* Ancient name for cucumber, used by Roman authors.

Genus 1426. **MELO** Adans.

Adans. Fam. II (1763) 138

Flowers diclinous, staminate, pistillate, sometimes also hermaphroditic, on one and the same plant, rarely all flowers hermaphroditic. Staminate flowers in umbelliform inflorescences, the pistillate solitary. Calyx campanulate, densely pubescent, usually with 5 linear or subulate lobes. Corolla broadly infundibular, deeply dissected into 5 (sometimes more) lobes, sulfur-yellow. Stamens 5, 4 connate in pairs, the fifth free, with flexuous short-pediceled anthers. Inferior ovary ovoid or globose, densely lanate-tomentose, with 3–5 placentas and numerous ovules. Style short; stigma 3–5-lobed. Fruit a sweet juicy pepo in the cultivated forms, sour or bitter in the weedy forms. Seeds ovoid or obovoid, flat. Annual herbs with creeping cylindrical to pentahedral stiff-haired branching stem, and simple tendrils; leaves rounded, cordate or reniform, entire or lobed to dissected, stiff-haired.

About 10 species in the warm zones of the globe.

Economic importance. A food plant, sometimes a weed.

Pangalo (see: Dyni kak samostoyatel'nyi rod Melo Adans. in Bot. zhurn. 35, 6 (1950) 371) has restored the old genus Melo, described by Adanson (l. c.) but subsequently usually united with Cucumis.

According to Pangalo the differences between these genera are as follows:

	Melo	Cucumis
Number of chromosomes	2n = 24	2n = 14
Form of chromosomes	Unsegmented	Segmented
Form of ovary	Ovoid or globose	Usually cylindrical
Ovary pubescence	Long-tomentose or short, appressed, thick	Prickly, sparse
116 Color of corolla	Sulfur-yellow	Vitelline yellow
Lobes of leaf laminas	Rounded	Triangular-mucronate
Length of main branch	1.5–2.5 m, rarely shorter	0.5–1.5 m, rarely slightly shorter
Sugar content of fruit	1.5–12%	1–2%
Cellulose content of fruit	1.3–9%	0.4–0.5%

To this should be added the considerable individualization of Melo vis-à-vis Cucumis. Though we agree with Pangalo, we do not accept his division of cultivated Melo into a series of species, which, to us, are but groups of varieties (variety types) produced in the course of selection for cultivated plants. (See: Harms in Pflanzenr. 88 (1924) 123, "Zusatz über Melonen-Sorten," and the more important literature. Pangalo's best works on Melo of the USSR are "Sornopolevye dyni" (Field Weed Melo) and "Kriticheskii obzor osnovnoi literatury po sistematike, geografii i proiskhozhdeniyu kul'turnykh i chast'yu dikikh dyn" (Critical Survey of the Literature on the Classification, Geography and Derivation of Cultivated and Wild Melo) in Tr. prikl. bot. gen. i sel. XXVII, 3, 1929–1930).

1. Fruit large (10 cm and longer); cultivated food plants . . . . 2.
\+ Fruit small (2–5 cm); weeds or ornamentals . . . . 3. **M. dudaim** (L.) Sageret.

2. Fruit cylindrical-clavate, 1–2 cm, often wavy-curved or flexuous . . . .
. . . . 2. **M. flexuosus** (L.) Sageret.
\+ Fruit shorter, straight . . . . 1. **M. sativus** Sageret.

1. M. sativus Sageret ex Roem. Fam. II [1846) 72. – M. vulgaris Moench ex Cogniaux in DC. Monogr. Phanerogam. III (1881) 484; Nikitin in Fl. Turkm. VI, 370. – M. adzhur Pang. in Bot. zhurn. XXXV, 6 (1950) 375. – M. cassuba Pang. l. c. – M. adana Pang. l. c. 576. – M. cantalupa Pang. l. c. – M. ambiguus Pang. l. c. 577. – M. chandalak Pang. l. c. – M. ameri Pang. l. c. 578. – M. zard Pang. l. c. 579. – M. chinensis Pang. l. c. 580. – M. conomon (Thbg.) Pang. l. c. – M. monoclinus Pang. l. c. – Cucumis melo L. Sp. pl. (1753) 1011; Willd. Sp. pl. IV, 613; Ser. in DC. Prodr. III, 300; Cogniaux, l. c. 482; Shmal'g, Fl. I, 377; Ldb. Fl. Ross. II, 1, 142; Hegi, Ill. Fl. VI, 1, 323; Fedch. in Fl. yugo-vost. VI, 284; Visyulina in Viznachn. rosl. URSR, 504. – C. melo β cultus Kurz in Journ. As. Soc. Beng. XLVI, 2 (1877) 102; Cogniaux, l. c. 484;
117 Cogniaux u. Harms in Pflanzenr. 88, 121. – Ic.: Hegi, l. c. f. 172.

Annual; stems usually prostrate, covered with spreading hairs, cylindrical-pentahedral or ribbed; leaves rounded-ovate or subreniform, entire, angular or more or less lobed, with rounded dentate lobes, cordate at base, stiff-haired, on rather long petioles. Flowers ca. 2–2.5 cm across, the staminate flowers often few in axil of leaves, the pistillate solitary, on short pedicels; corolla campanulate, the lobes obtuse; sepals subulate; ovary long-tomentose; fruit with typical "melon" odor very variable in form, size, taste and color; seeds flat, yellowish, sessile on fibrous placentas. Fl. May–August, Fr. June–October.

Many varieties cultivated in the more southern regions of the USSR for their valuable fruits. **Gen. distr.:** Apparently from Near Asia, grown in all parts of the world. Described from a specimen of unknown origin. Type in London.

Note. Melons rank among the more valuable cultivated plants with some, particularly the Central Asian varieties considered unequalled in the total quality of their fruits. These are sometimes rightly called "northern pineapple." Some melons surpass even the pineapple in taste and fragrance. V. P. Drobov, himself a lover of melons and one of the top experts on nature and economy of Central Asia believes that the best melons of Central Asia – possibly of the whole world (see: V. V. Nikitin, 1954) – grow on transient islands of muddy river deposits in the Khorezm Oasis at the estuary of the Amu Darya. In addition to being eaten raw, melons yield an excellent dried product, for use in confectionary.

2. M. flexuosus (L.) Sageret ex Roem. Fam. II (Peponiferarum) (1846) 75; Pangalo in Bot. zhurn. 6 (1950) 577. – Cucumis flexuosus L. Sp. pl. ed. 2 (1763) 1437; Willd. Sp. pl. IV, 615; Ser. in DC. Prodr. III, 101; Cogniaux in DC. Monogr. Phanerogam. III (1881) 484. – C. melo flexuosus (L.) Naud. in Ann. Sc. Nat. 4 sér. XI (1859) 34; Cogniaux u. Harms in Pflanzenr. H. 88, 125.

Annual; stems rather thin, prostrate, scabrous, flexuous, with elongate internodes; leaves cordate-ovate, petioled, dentate, entire or lobed, median lobe much larger and longer than the lateral. Flowers monoecious,

diclinous, sessile in groups in axils of leaves; sepals densely pubescent; corolla yellow; pistillate flowers long-stalked, with long fusiform or cylindrical ovary, 10–12 cm long at flowering, and rudimentary staminodes;
118 fruit 0.5–1(2) m, cylindrical-clavate, often serpentinely curved or flexuous, smooth or furrowed, gray or blackish-green, when ripe usually yellow or white, with reddish or pale pink loosely fibrous flesh of unpleasant odor. July–August.

Rarely cultivated in the southern regions of Central Asia. **Gen. distr.**: cultivated in the countries of Near Asia. Described from India. Type in London.

Economic importance. The young, 5–7-day old ovaries, are used as food.

3. M. dudaim (L.) Sageret in Ann. Sc. Nat. VIII (1826) 313. – M. agrestis (Naud.) Pang. in Bot. zhurn. 6 (1950) 580; Nikitin in Fl. Turkm. 370; Cucumis dudaim L. Sp. pl. (1753) 1011; Willd. Sp. pl. IV, 615; Ser. in DC. Prodr. III, 301; Boiss. Fl. or. II, 759; Cogniaux, in DC. Monogr. Phanerogam. III, 484. – C. melo dudaim Naud. in Ann. Sc. Nat. 4 sér. XI (1859) 69; Cogniaux u. Harms in Pflanzenr. 88, 126. – C. melo var. agrestis Naud. in Ann. Sc. Nat. 4 sér. XI (1859) 73, XII (1859) 110; Cogniaux, l. c. 483; Cogniaux u. Harms, l. c. 120; Grossg., Opred. rast. Kavk. 414. – C. odoratissimus Moench, Meth. pl. (1794) 654. – Melo microcarpus (Alef.) Pang. in Bot. zhurn. 6 (1950) 575; Nikitin, in Fl. Turkm. VI, 368. – C. melo var. microcarpus Alef. Landw. Flora (1866) 198. – Ic.: Nikitin, op. cit. Plate LI; Pangalo in Tr. prikl. bot. gen. i sel. XXIII, 3, 238–239, 242–243. – Exs.: Fl. Cauc. exs. No. 324.

Annual; stems 30–100 cm, branching, stiff-haired, prostrate, ascending or erect, slightly faceted; leaves rounded or ovate, 3–5-angled or lobed, lobes rounded, cordate, bristly-haired, 3–6 cm across, sometimes larger. Flowers monoecious, diclinous or hermaphroditic, slightly smaller than in the cultivated plant; receptacle pubescent; sepals and petals pubescent outside; fruit oblong-ovoid or ovoid to globose, 1–3(5) cm, variously colored (green, yellow or yellow-orange, purple-brown, etc.), spotted or striped, flesh white-yellow or pale pink, sweet or sour to bitter; seeds usually white-yellow, small, ovoid. Fl. June–September, Fr. July–October.

In the southern regions of Central Asia, M. dudaim is grown as an ornamental; in Transcaucasia it occurs as a weed of irrigated crops. **Gen. distr.**: Iran, As. Min., Arabia, Med., introduced in America and other countries. Described from Egypt. Type in London.

Note. Among the typical M. dudaim are plants mostly cultivated as ornamental, with aromatic fruits which easily detach from the peduncles. We tentatively regard M. dudaim with odorless nonabscising fruits as
119 M. dudaim var. agrestis (Naud.) Vass. h. l. The difficulty of the problem derives from Naudin (l. c.) who interpreted Cucumis melo var. agrestis too broadly, including in it M. dudaim. Pangalo (op. cit.) restricted Naudin's conception of var. agrestis to include only M. dudaim of Central Asia, Iran and Afghanistan, and at the same time promoted this variety to specific rank. Unable to carry out a critical study of this question and pending critical revision, we refer M. agrestis (Naud.) Pang. to M. dudaim, as interpreted by Pangalo. Neither Naudin nor Pangalo have left any data on the authenticity of M. agrestis or on its whereabouts.

M. dudaim is a weed, mainly of cotton. It naturally hybridizes with cultivated specimens of Melo (Pangalo, op. cit.).

Genus 1427. **LAGENARIA** * Ser.

Ser. in Mém. Soc. Genève, III, 1 (1825) 25

Flowers diclinous, monoecious, solitary; staminate flowers on long peduncles with infundibular receptacle; sepals 5; corolla dissected to base; petals 5, white; stamens 5, 4 connate in pairs and 1 free; a gland represents the rudimentary pistil. Pistillate flowers in cup-shaped receptacle, on shorter peduncles; ovary ovoid or cylindrical, with 3 placentas; style short, thick, with 3 bilobate stigmas and 3 poorly developed staminodes. Fruit of many shapes, fleshy inside, with lignified outer shell, indehiscent, with numerous obovoid flattened seeds. Annual plants, with climbing or prostrate stems, finely pubescent, with bifid tendrils and alternate dentate leaves with a musky odor.

Monotypic genus.

1. L. vulgaris Ser. in Mém. Soc. Genève, III, 1 (1825) 25 et in DC. Prodr. III (1828) 299; Ldb. Fl. Ross. II, 1, 141; Naud. in Ann. Sc. Nat. 4 sér. XII, 91 et 5 sér. V, 8; Cogniaux in DC. Monogr. Phanerogam. III, 416; Cogniaux u. Harms in Pflanzenr. 88, 201; Visyulina in Viznachn. rosl. URSR, 504;
120 Nikitin in Fl. Turkm. IV (1954) 371. – Cucurbita lagenaria L. Sp. pl. (1753) 1010. – Ic.: Ser. l. c. tab. 2; Pflanzenr. 88, f. 24; Kobyakova in Tr. prikl. bot. gen. i sel. XXIII, 3, pp. 479–481.

Annual; plant covered with soft sticky hairs; stem thick, angular; leaves finely pubescent, rounded-cordate, angular or faintly 3-lobed, 10–40 cm across, on semicylindrical rigid, often hollow, petioles provided with 2 dentate apical glands, and 5–7 palmately diverging glands at tip of petiole. Staminate flowers on peduncles longer than petioles; receptacle 2–3 cm long; sepals narrowly triangular, one-third to one-half as long as receptacle; petals crisp, pubescent or tomentose, more or less emarginate, mucronate, 3–4 cm long, 2–3 cm wide; pistillate flower peduncles as long as or shorter than petioles; ovary densely long-villous; fruit glabrous, of many shapes and sizes; seeds white or brown, elongate-obovoid or triangular, apically dissected or bidentate, rarely rounded, 7–20 mm long. Fl. from June, Fr. from August.

Cultivated in the southern regions of the USSR as an ornamental and for its fruits. Gen. distr.: grown in all warm zones. Described from America. Type in London.

Note. Lagenaria ('bottle' or 'calabash' gourd) is among the older plants, many varieties of which are cultivated throughout the world. Cogniaux and Harms (l. c.) surveyed the extensive literature on Lagenaria. In the USSR the fruits are made into flasks and water bottles. The large-fruited plants yield flask-shaped fruits, the small-fruited bear cylindrical, conical or pear-shaped fruits which are made into handsome tobacco containers by careful processing and polishing (Kobyakova, op. cit.).

* From the Latin lagena – a bottle, referring to the shape of the fruit.

Genus 1428. **CUCURBITA** * L.

L. Sp. pl. (1753) 1010

Flowers unisexual (plants with bisexual flowers are excluded), large, axillary, solitary or (the staminate) in fascicles. Staminate flowers with campanulate receptacle; sepals 5(4–7); corolla campanulate, large, yellow, 5(4–7)-lobed to divided, lobes spreading or erect at apex; stamens 5, 4 connate in pairs, 1 free; anthers S-shaped centrally connate. Pistillate
121 flowers with 3–5-celled inferior ovary and 3 small triangular staminodes; style 1, with 3–5 bilobate stigmas. Fruit usually a large fleshy gourd with numerous flattened seeds. Annual or perennial herbs with prostrate stems, or climbing by means of 2 multifid tendrils, and alternate, entire or palmatilobate leaves.

About 15 species throughout the world.

Economic importance. A food plant with oleaginous seeds also grown as fodder and ornamental (see: Zhiteneva. Mirovoi sortiment kul'turnykh tykv., Tr. prikl. bot. gen. i sel. XXIII (1929–1938)).

1. Stems sharply faceted, with coarse conical prickles; leaves usually deeply lobed to divided; corolla lobes erect . . . . 1. **C. pepo** L.
+ Stems cylindrical or obtusely faceted, pubescent; leaves usually faintly lobed or entire; corolla lobes recurved . . . . 2.
2. Stems obtusely faceted; sepals flat, often with broadened leaf-shaped appendages; corolla lobes acuminate; fruit elongate, constricted at middle; fruiting pedicels faceted . . . . 2. **C. moschata** Duch.
+ Stems cylindrical; sepals linear or filiform; corolla lobes rounded; fruit of different shape; fruiting pedicels cylindrical . . . . 3. **C. maxima** Duch.

1. **C. pepo** L. Sp. pl. (1753) 1010; Willd. Sp. pl. IV, 609; Ser. in DC. Prodr. III, 316; Naud. in Ann. Sc. Nat. 4 sér. VI, 29; Cogniaux in DC. Monogr. Phanerogam. III, 545; Pax in Pflanzenfam. IV, 5, 33; Hegi, Ill. Fl. VI, 1, 389; Fedch. in Fl. yugo-vost. VI, 285; Visyulina in Viznachn. rosl. URSR, 505; Whitaker and Bohn. in Econ. bot. IV, 57; Nikitin in Fl. Turkm. VI, 372. – C. verrucosa L. l. c. – C. melopepo L. l. c. – C. ovifera L. Mant. pl. I (1767) 126. – C. subverrucosa Willd. l. c. – C. aurantia Willd. l. c. 607. – Ic.: Pflanzenfam. IV, 5, f. 18; Hegi, l. c. f. 172–173; Zhiteneva in Tr. prikl. bot. gen. i sel. XXIII, 3, 161–169, 172–176.

Annual; stems prostrate or climbing, 2–10 m, ribbed, stiff-haired, like leaves densely covered with stiff, somewhat spiny hairs, acutely scabrous to the touch; leaves 5-lobed, erect, cordate, their lobes acute, more or less lobed, with rounded incisions between lobes. Pedicels abruptly pentahedral; staminate flowers in axillary fascicles with linear or subfiliform sepals; pistillate flowers solitary; corolla 7–10 cm across, with erect acuminate lobes, golden yellow or yellow-orange; fruit large, very variable in shape, size, color and surface texture; flesh fibrous, the easily separating placentas
122 bearing numerous yellowish-white, distinctly bordered seeds; fruiting pedicels abruptly faceted. Fl. June–July, Fr. September–November.

* From cucumis – cucumber and orbis – circle.

Grown for food and forage, rarely as an ornamental, in the more southern regions of the USSR. **Gen. distr.**: Originally from America, it is grown in all parts of the world. Described from a cultivated specimen of unknown origin. Type in London.

Note. Cucurbita rates among the most popular cultivated plants in the USSR where it is represented by many varieties and groups (Citrulin-Cucurbita, vegetable marrow, bush squash, crookneck squash, small-fruited ornamental gourds, etc.). Its fruits are massive, some weighing 100–150 lbs. The Ukrainian "khan'ki" with small exquisite, varicolored fruits are favorite window decorations in the Ukraine. Cucurbita produces the largest fruits of any plant in the USSR.

2. **C. moschata** Duch. ex-Poir. in Dict. Sc. Nat. XI (1818) 234; Ser. in DC. Prodr. III, 317; Roem. Syn. fam. 2, 83; Naud. in Ann. Sc. Nat. 4 sér. VI, 41; Cogniaux in DC. Monogr. Phanerogam. III, 546; Visyulina in Viznachn. rosl. URSR, 505; Whitaker and Bohn in Econ. bot. IV, 59; Nikitin in Fl. Turkm. VI, 373. – Ic.: Naud. l. c. tab. 2; Zhiteneva in Tr. prikl. bot. gen. i sel. XXIII, 3, 160, 163, 165, 166–168.

Annual; stems 4–5 m, usually prostrate, obtusely faceted, finely pubescent, often black-spotted at nodes; leaves rounded-reniform, crenate-dentate, 5–7-lobed, intensively green, often with white spots and pendent margins, lobes acute, sometimes entire, velutinous, with incisions between them. Staminate flowers on more or less terete peduncles with flat receptacle and linear sepals, often broadened at apex; corolla 7–10 cm across, pale yellow, campanulate, its lobes acuminate, recurved; peduncles of pistillate flowers pentahedral, coarse, broadened at apex; fruiting pedicels faceted; fruit variously colored, usually cylindrical, constricted at middle, with easily separating placentas; seeds dingy white, distinctly bordered. Fl. June–July, Fr. August–September.

Rarely grown for food in the southern regions of the USSR. In Central Asia (local name "pilav-kadu"), it is one of the ingredients of pilav. **Gen. distr.**: originally from S. America it is grown in moderate and tropical zones. Described from America. Type in Paris.

3. **C. maxima** Duch. in Lam. Encycl. méth. 2 (1786) 151; Ser. in DC. Prodr. III, 316; Naud. in Ann. Sc. Nat. 4 sér. VI, 17; Cogniaux in DC.
123 Monogr. Phanerogam. III, 544; Visyulina in Viznachn. rosl. URSR, 506; Whitaker and Bohn in Econ. bot. IV, 62; Nikitin in Fl. Turkm. VI, 373. – C. pepo var. α L. Sp. pl. 2 (1763) 1435. – C. turbaniformis M. Roem. Synops. 3 (1847) 87. – Pepo macrocarpus Rich. ex Spach, Hist. nat. veg. Phaner. 6 (1838) 202. – Ic.: Naud. l. c. tab. 1; Zhiteneva in Tr. prikl. bot. gen. i sel. XXIII, 3, 159–164, 166–167.

Annual; stems cylindrical, prostrate, 4–5 m thick, like leaves rather finely pubescent; leaves more or less reniform, serrate-dentate, 5-lobed, with short obtuse rounded lobes, sometimes not lobed, only undulantly notched at margin, deeply notched at base, petioled, covered with stiff hairs. Staminate flowers on cylindrical peduncles, with cup-shaped or obconoid receptacle and linear or filiform sepals; corolla campanulate, intensively yellow, its lobes wide, recurved; column of anthers conoid, longer than filaments; peduncles of pistillate flowers thickened, fleshy, cylindrical or clavate; fruit usually globosely flattened, placentas spongy, not easily separating; seeds ovoid, large, smooth, white or yellowish; pedicels when ripe cylindrical. Fl. from June, Fr. from August.

Grown mostly in the southern regions of the USSR. **Gen. distr.**: originally from S. America, grown in all warm zones. Described from America. Type in Paris.

Note. There is a subspecies, C. maxima ssp. turbaniformis (Roem.) Vass. h. l. (= C. turbaniformis Roem.), the fruits of which have a rounded, turbanlike apical appendage, with trifid style and stigma, the lobes of the latter being recurved to one side and appressed to ovary.

Genus 1429. **SICYOS** * L.

L. Sp. pl. (1753) 1012

Flowers diclinous, monoecious; staminate flowers in racemes or corymbs; sepals 5, subulate; corolla rotate or subcampanulate, deeply 5-fid; stamens (2)3–7, the filaments distally adnate; pistillate flowers axillary, in fascicles; corolla greenish- or dingy-white. Ovary 1-locular, with short style and 2–3 stigmas. Fruit 1-seeded, elongate-ovoid, tapering, usually bristly-prickly, with coriaceous pericarp. Annual herbs, their stems prostrate or climbing by means of (2)3-multifid tendrils, with alternate, more or less lobed or angular thin leaves.

124 An Australian-S. American genus of 30–40 species.

1. S. angulata L. Sp. pl, (1753) 1012; Willd. Sp. pl. IV, 625; Ser. in DC. Prodr. III, 309; Ldb. Fl. Ross. II, 142; Cogniaux in DC. Monogr. Phanerogam. III, 872; Pax in Pflanzenfam. IV, 5, 38; Shmal'g., Fl. I, 376; Hegi, III. Fl. VI, I, 327; Visyulina in Viznachn. rosl. URSR, 506; Bobrov in Maevskii, Fl. (1954) 324. Ic.: Hegi, l. c. f. 155.

Annual; stems angularly furrowed, climbing, 3–6 m, with spreading stiff hairs and 3–5-fid tendrils; leaves rounded or ovate-cordate, 3–5-angled or lobed, with triangular lobes, denticulate, deeply notched at base, finely scabrous above and beneath, (3)5–12(20) cm long and almost as wide, on long (sometimes short) pubescent petioles. Staminate flowers in racemes or capitate corymbs, on 10–20 cm pubescent peduncles; pedicels thin, pubescent; receptacle nearly rotate, pubescent, 4–5 mm; sepals ca. 1 mm, lanceolate-subulate; corolla 9–14 mm across, dingy yellow, with green veins, more or less pubescent, the lobes triangular-lanceolate, ca. 3–4 mm long; pistillate flowers smaller, crowded in heads, 10–15 sessile on common 1–2 cm peduncles; fruit oblong-ovoid, acuminate, 10–15 mm, short-glandular, with long bristly hairs, yellowish or dingy gray (dark), indehiscent, with 1 seed. Fl. July–August, Fr. August–September.

A weed of gardens, hedges and walls. – European part: U. Dns., Bes., U. Dnp., M. D., V.-Don. **Gen. distr.**: N. Am., introduced into W. Eur., Australia and other countries. Described from America. Type in London.

Note. Ledebour's (l. c.) record for the Caucasus was never confirmed, nor is the species represented in the herbarium of the Botanical Institute of the Academy of Sciences of the USSR.

* Sicyos – ancient Greek name of Cucumis sativus, referring to the similarity of the leaves of both plants.

Genus 1430. **ECHINOCYSTIS*** Torr. et Gr.

Torr. et. Gr. Flora of North-America, I (1840) 542

Flowers diclinous, monoecious, small, white; staminate flowers in racemes or panicles; sepals 5–6, subulate or filiform; corolla deeply 5–6-fid, usually rotate, with lanceolate or linear verrucose lobes; stamens 3, with filaments adnate to column. Pistillate flowers solitary or in fascicles in
125 axils of the same leaves as staminate flowers, sometimes bearing staminodes. Ovary ovoid or globose, prickly, 1- or nearly 2-locular; style very short; stigma semiglobose, 2–3-lobed or divided. Fruit long-prickled, fibrous inside; seeds ovoid or oblong. Perennial or annual herbs, climbing by means of branch tendrils and alternate, 5–7-angled or lobed, sometimes entire, leaves.

An American genus of about 30 species.

1. **E. echinata** (Muhl.) Vass. comb. nov. h. l. – E. lobata (Mich.) Torr. et Gr. l. c.; Naud. in Ann. Sc. Nat. 4 sér. XVI, 187; Cogniaux in DC. Monogr. Phanerogam. III, 815; Pax in Pflanzenfam. IV, 5, 39; Igoshina in Bot. mat. Gerb. AN SSSR, XVII, 515. – Sicyos lobata Mich. Fl. Bor.-Amer. 2 (1803) 217. – Momordica echinata Mühl. in Trans. Am. Phil. Soc. III (1793) 180; Willd, Sp. pl. IV, 605; Ser. in DC. Prodr. III, 310. – Micrampelis lobata (Mich.) Creene in Pittonia, II (1890) 127; Kom. and Alis., Opred. rast. Dal'nevost. kr. 988. – Ic.: Torr. Fl. St. of New-York, 1 (1843) tab. 30.

Annual; roots fibrous, numerous; stems thin, 5–6 m, strongly branching, angular-furrowed, short-haired at nodes, climbing by means of 3–4-fid tendrils; leaves thin, pale green above and beneath, finely tuberculate-scabrous, suborbicular or ovate, more or less deeply notched at base, 5-lobed to divided, with acute triangular lobes, the terminal larger and longer than the lateral, 5–10(15) cm long and nearly as wide, on slender, rather long, short-haired, sometimes subglabrous, petioles. Peduncles of staminate racemes slender, with sparse short hairs, densely bearing sessile flowers on slender short pedicels nearly to base; receptacle whitish, 2–3 mm wide; sepals spreading, capilliform, 1–1.5(2) mm long; corolla white, usually 6-lobed, the lobes linear-lanceolate, acuminate, pubescent-glandular, 5–7 mm; pistillate flowers slightly larger than the staminate, axillary, solitary (or in two's), peduncles shorter than those of racemes bearing staminate flowers; fruit glaucous green, broadly ovoid, ca. 4–5 cm long, 3–4 cm wide, covered with numerous (3)5–10 mm prickles and bristles, slightly broadened toward base, 2-locular, dry, irregularly indehiscing at apex, 2-seeded; seeds oblong-ovoid, flattened, large, black-brown, truncate, slightly swollen, 15–17 mm long, 7–8 mm wide. Fl. July–September, Fr. from August.

River and lake banks, seldom in shrubs and as weed in gardens, sometimes grown as an ornamental. – European part: U. Dns.; Far East: Uss. (gardens of Vladivostok). **Gen. distr.**: N. Am., introduced into Eur. Described from N. America. Type in London.

* From the Greek echinos – prickle, kuste – cyst.

126 Family CLVIII. **CAMPANULACEAE** JUSS.*

Flowers staminate-pistillate (amphisporangiate, hermaphroditic), regular (actinomorphic). Calyx tube adnate to ovary, with 3–4–5–6-lobed limb. Calyx teeth or lobes alternating with corolla lobes with imbricate or valvate aestivation. Corolla sympetalous, usually 5-fid, rarely 6–10- or 3–4-fid, tubular, campanulate, infundibular, sometimes nearly totate, with valvately arranged lobes, lobes usually short or half the length of the corolla, sometimes longer, rarely lobes free below at anthesis, but compactly united apically. Stamens alternating with corolla lobes and as many attached to base of corolla, sometimes to a disk (nectary). Very rarely (Sphenoclea) stamens nearly epipetalous. Filaments usually broadened at base. Anthers free, at times cohering to form a tube or only proximally, 2-chambered, dehiscing by longitudinal clefts. Ovary inferior, half-inferior, rarely superior (alien genus Cyananthus), 2–10-locular, often 3-locular, when ovary 1- or 3-locular, the perianth usually 5-, rarely 4-merous. Style usually 3-fid, rarely 2-multifid or stigma stellate, very rarely capitate (for example, Sphenoclea), usually long and thin, sometimes strongly exserted from flower, rarely very short and thick (Sphenoclea). Sometimes disk (nectary) forming an inflated ring, hollow cylinder or tube, around the base of the style. Placentas axial, rarely apical and basal. Ovules usually numerous, anatropous. Fruit a variously dehiscing capsule, sometimes a berry (alien genus Canarina), often with persistent calyx teeth. Seeds with erect embryo and abundant protein. Annual, monocarpic or perennial herbs, semishrubs, rarely shrubs (alien genus), often with milky juice, with alternate sometimes opposite leaves, rarely partly in whorls. Stipules absent. Stems erect, ascending or creeping, sometimes pendent or climbing, leafy. Inflorescences spicate, racemiform or paniculate; often flowers solitary, terminal or axillary, bracteate and often bracteolate, chasmogamic, rarely cleistogamic, protandrous or proterogynous. Roots often thickened and fleshy, the rhizome sometimes producing shoots.

127 Type genus Campanula L.

The Campanulaceae comprise about 40 genera with some 600 species, mainly in the northern hemisphere; they are almost completely absent in the tropics. The largest concentration of species is in the Mediterranean area. In the southern hemisphere, Campanula is mainly replaced by Wahlenbergia. Only Cephalostigma and Sphenoclea are found in the tropics. In the north, Campanula alone reaches to the Arctic. In the USSR there are 20 genera and 224 species of Campanulaceae.

Besides the monograph on Lobeliaceae, there are no up-to-date works on the Campanulaceae s. str. Some of the genera have been investigated (see Notes to characteristics of the genera). The first monograph on the taxonomic rank of the tribe — excellent for its time, but now obsolete — is due to Alphonse de Candolle (1830). Brief surveys were made by Schönland (Engler — Prantl, Pflanzenfam.) and Baillon (Hist. des plantes. VIII (1881)). Lobeliaceae,as a tribe, were described only recently (1943). In the "Flora of the USSR," the Lobeliaceae are treated as a distinct family. The Caucasian Campanulaceae are described by Fomin in "Flora caucasica critica" (1903–1907).

* Treatment by A.A. Fedorov.

De Candolle was the first to draw up the taxonomy of the family Campanulaceae (Monograph. Campan.), initially dealing only with the tribe Campanuleae s. l., but later establishing the subfamily Campanuloideae, without Lobeliaceae, Sphenocleaceae and Cyphiaceae. In this system only two subtribes were recognized, separated by manner of dehiscence of the capsule (from the apex vs. laterally from the base). Later, in his "Prodromus," de Candolle gave the former tribe the status of a family, regarding Sphenocleacea, Lobeliaceae and Cyphiaceae as of the same taxonomic rank. For the subtribes now promoted to tribes de Candolle used the names Wahlenbergieae and Campanuleae of the former authors.

De Candolle's later changes revealed his tendency to reunite the true Campanulaceae with Lobeliaceae, Sphenocleaceae and Cyphiaceae in one family. Phylogenetically this is wrong, since historically and geographically the three groups developed in isolation, as reflected in marked morphological differences. In this respect we fully support de Candolle's position;
128 we consider the Lobeliaceae and the Sphenocleaceae as distinct families. In our treatment the Sphenocleaceae were ranked as a subfamily, because the original Key to Families of Sympetalae did not list them as a family (see Note to subfamily Sphenocleoideae). In spite of de Candolle's correct ideas, the taxonomy of Campanulaceae s. l., of the Engler school, was widely accepted in the literature, with variations due to Schönland (l. c.), Dalla Torre and Harms (Genera siphonogam.) and by Engler in "Syllabus." Wimmer's monograph includes the Lobeliaceae in the Campanulaceae, as in Engler's taxonomy.

We are dividing the subfamily Campanuloideae (or Campanulaceae s. str.) into 8 tribes, of which 6 are new, based primarily on the different modes of dehiscence of the capsule, the shape of the corolla and aggregate characters, determining general similarity. Three new genera are added to the Campanuleae s. str., in addition to Campanula, Symphyandra and Adenophora: Brachycodon, Popovicodonia and Astrocodon. The only species of Brachycodon has been excluded from Campanula (C. fastigiata Duf.); Popoviocodonia is established for two bell-flowers intermediate between Campanula and Adenophora, for Adenophora uyemurae Kudo and Campanula stenocarpa Trautv. et Mey. The new genus Astrocodon includes Campanula kruhseana Fisch. ex Rgl. et Tiling and C. expansa Rud. The new tribes Peracarpeae, Ostrovskieae, Michauxieae, Edrajantheae and Jasioneae were created for the Euro-Asiatic genera Peracarpa, Ostrowskia, Michauxia, Edrajanthus and Jasione, which have nothing in common taxonomically and cannot be included in the Campanuleae or Wahlenbergieae. The large new tribe Phyteumateae includes Phyteuma, Asyneuma, Cylindrocarpa, Legouzia, Sergia and Cryptocodon. The latter two genera were newly established for some taxonomically distinct Campanulaceae of Central Asia — Campanula sewerzowii Rgl., Phyteuma regelii Trautv. (genus Sergia) and Campanula monocephala Trautv. (genus Cryptocodon), Platycodon and Codonopsis are included in the Wahlenbergieae.

Many members of the family Campanulaceae are important ornamentals long cultivated in gardens, in particular botanical gardens. Campanula includes species which yield vitamin C and minor food plants. Species of

129 Asyneuma and especially Michauxia contain rubber; indeed all campanulas contain small quantities of rubber; (see: Grossgeim. Rastitel'nye bogatstva Kavkaza. Izd. 2-e. 1952). In the People's Republic of China, some species of Codonopsis are used in folk medicine. The species of Adenophora are minor food plants. (See relevant Notes.)

Key to Genera

1. (Nectary) disk annular or tubular, surrounding base of style; corolla usually tubular-campanulate, shallowly lobed; style often markedly exserted; leaves dentate or entire; flowers in racemiform or paniculate inflorescences, with clavate aestivation; root usually thickened . . . . 1434. **Adenophora** Fisch.
+ (Nectary) disk absent . . . . 2.
2. Capsule short, obconical, 2-locular, with very small seeds dehiscing by planoconical cap; style very short, with capitate stigma; flowers very small, greenish, with short obtusely lobed calyx and short tubular corolla and spreading limb; thick-stemmed aquatic plant with lanceolate leaves and compact spicate inflorescences . . . . 1450. **Sphenoclea** Gaertn.
+ Capsule dehiscing by small proximal holes, valves or slits in lower part, at sides or at apex, or nearly indehiscent; style long, with distinct 2–3–5-fid stigma, rarely with faintly 2-fid clavate stigma . . . . 3.
3. Calyx-tube short-cylindrical, at sides with numerous elongate oval parallel notches, teeth long, linear-lanceolate, usually 7, rarely 5–9; corolla very large, to 10 cm long, infundibular to campanulate, with shallow lobes, bluish or pink-violet; leaves whorled, dentate, obovate; plant more than 1 m high, with massive thick stem . . . . 1438. **Ostrowskia** Rgl.
+ Calyx-tube without elongate notches; corolla much smaller and if 7–8 cm long, then a very rare garden plant . . . . 4.
4. Corolla and calyx usually divided into 7–10 lobes; corolla lobes narrow, linear; style with stellate stigma; plant high, strong, with thick white stem; leaves stiff glaucescent, long, spicate, racemiform,
130 interrupted; inflorescence; abundant milky juice . . . . 1439. **Michauxia** L'Her.
+ Corolla calyx 5-, very rarely 4-merous . . . . 5.
5. Corolla obconical, small, much shorter than calyx, pale blue; small, 5. annual, furcately branching, with small irregularly dentate slightly cuneate leaves . . . . 1433. **Brachycodon** Fed.
+ Corolla longer than calyx, if shorter, then limb distinctly rotate . . . 6.
6. Small, delicate plant, 5–10 cm high, with thin stems, flowers small, not more than 3 cm long, pyriform drooping capsule, thin leaves; in habit resembling small specimens of Circaea alpina. In the USSR confined to Sakhalin and Kamchatka . . . . 1437. **Peracarpa** Hook f. et Thoms.
+ Plant with different characters; flowers usually very large . . . . . . 7.

7. Ovary half-inferior, capsule dehiscing at apex by 3 valves; stems climbing or flexuous, weak, sometimes very long; corolla flask-shaped to campanulate, rather large, pink, slightly violet or greenish; calyx teeth broad, herbaceous; stigma lobes short; root thick, grumous, club-shaped . . . . 1446. **Codonopsis** Wall.

\+ Ovary inferior, if half-inferior then capsule dehiscing by 5 valves; stems erect, ascending, not climbing, sometimes slightly prostrate or drooping; shape of corolla different . . . . 8.

8. Capsule splitting from tip to base into uneven fragments, forming a funnel; corolla tubular, tapering above, shallowly lobed; stems much shortened, slightly columelliform; flowers terminal, sessile; plant forming dense cushions due to crowded stems, densely covered with narrow short imbricate procumbent leaves . . . . 1448. **Edrajanthus** A. DC.

\+ Capsule never dehiscing from tip to base; corolla not tapering at apex if slightly constricted below apex, then not narrowly tubular; stems only rarely columelliform, and if so then only in lower part; median and upper cauline leaves always more or less remote, not imbricately procumbent . . . . 9.

9. Ovary half-inferior; capsule dehiscing distally at top by 5 recurving
131 valves, opposite the persistent sepals; corolla large, nearly rotate; leaves broad, serrate; root thickened . . . . . 1447. **Platycodon** A. DC.

\+ Ovary inferior; capsule dehiscing by several apertures, clefts or valves, if by apical aperture then anthers connate at base . . . . . 10.

10. Corolla limb flat, rotate; calyx teeth longer than limb lobe; calyx long, narrow; capsule elongate, prismatic, dehiscing by distal clefts or pores . . . . 1445. **Legouzia** Durand.

\+ Corolla limb usually not rotate, if rotate then much longer than calyx with its teeth, corolla often campanulate or tubular-campanulate, sometimes dissected nearly to base; calyx teeth shorter than or as long as corolla; shape of capsule different . . . . 11.

11. Petals of deeply dissected corolla tightly joined distally at beginning of anthesis to form a tube around the protruding style, remaining distinct proximally; inflorescence compactly spicate or capitate; leaves usually with broad laminas and long petioles; root thickened . . . . 1440. **Phyteuma** L.

\+ Corolla lobes at anthesis divergent, never distally concrescent . . 12.

12. Anthers connate at base; stigma clavate, faintly 2-fid; the very small blue flowers arranged in compact regular spherical inflorescences; corolla lobes and calyx narrow, linear; style exserted; plant few-stemmed, leafy in lower half . . . . 1449. **Jasione** L.

\+ Anthers free, if connate then over entire length; stigma distinctly 2–3-fid; flowers not in compact spherical inflorescences . . . . . 13.

13. Capsule narrow, cylindrical, with neck-shaped constriction, cut allowing seeds to scatter; many-stemmed perennials, with narrow leaves and thick hard tuberous root . . . . 1444. **Cylindrocarpa** Rgl.

\+ Shape of capsule different . . . . 14.

14. Corolla opening widely at anthesis, with stellately spreading broadly lanceolate lobes, divided nearly to base, or at least to middle . . . 15.

\+ Corolla campanulate, with simple or curved, not stellately spreading lobes, the lobes usually short; if corolla dissected to base then lobes liguliform . . . . 16.

132 15. Capsule dehiscing by valves at base; leaves linear, broadly lanceolate or shortly rhombic-lanceolate, entire or acutely toothed; corolla 3 cm across . . . . 1436. **Astrocodon** Fed.

\+ Capsule dehiscing by distal pores; leaves oblong or broadly lanceolate, obtusely toothed; corolla 1–1.5 (2) cm across . . . . 1435. **Popoviocodonia** Fed.

16\. Corolla divided to base . . . . 17.

\+ Corolla divided for one-third of its length, rarely more deeply . . 19.

17\. Corolla lobes narrow, linear-liguliform not longer than 1 cm, usually shorter; flowers in spicate-racemiform inflorescences; root usually not thick . . . . 1441. **Asyneuma** Griseb. et Schenk.

\+ Corolla lobes broadly liguliform; flowers in paniculate or capitate inflorescences; root very thick, tuberous, hard . . . . 18.

18\. Stigma 2-fid; capsule pentagonal at cross section, not ribbed, dehiscing by 3 slits in middle part; flowers crowded in heads enveloped by terminal leaves; low plant, with very thick tuberous weedy root; leaves ovate-elliptic, serrate . . . 1443. **Gryptocodon** Fed.

\+ Stigma 3-fid; capsule ribbed, constricted above, dehiscing by pores at apex or middle; flowers in many-flowered paniculate inflorescences not covered by terminal leaves; root thick, tuberous; leaves linear or lanceolate, entire or dentate . . . . 1442. **Sergia** Fed.

19\. Anthers cohering as if into a tube or connate; plant many-stemmed, often with drooping flowers; stems leafy, often faintly pendent; leaves with short cordate, markedly dentate, sometimes lobed lamina . . . . 1432. **Symphyandra** A. DC.

\+ Anthers all or partly free; corolla campanulate or infundibular, sometimes slightly tubular; all other characters different . . . . 1431. **Campanula** L.

Subfamily 1. **CAMPANULOIDEAE** Schoenland in Pflanzenfam. IV, 5(1889)48. – Stamens attached to base of petals or to epigynous disk. Fruit a capsule dehiscing by valves or opening by pores at base on apex or by apical aperture, rarely dehiscing by valves extending all along the fruit. Calyx lobes, petals and stamens usually 5, rarely 6–10 or 3–4. Corolla most often campanulate or infundibular, sometimes tubular, rotate, not divided
133 deeper than one third of its length, sometimes cut nearly to base, lobes sometimes divided at base but not united by apices. Anthers free, rarely connate as if to form a tube, or proximally connate. Filaments usually broadened. Nectary disk developed, or often absent. Ovary 2–3–5-, rarely multi-locular. Mostly monocarpic perennials, or annual herbs with some alien semishrubs and shrubs. Leaves, stems, root and rhizome very varied.

Type species: Campanula L.

Tribe 1. **CAMPANULEAE** Rchb. Consp. (1828) 194, s. str. – Corolla 5-merous like calyx, campanulate, infudibular or tubular-campanulate, rarely nearly rotate, divided for one-quarter to one-half its length, very rarely more deeply, into more or less curved lobes. Calyx with or without appendages in notches between teeth. Ovary 3- rarely 5-locular. Stamens free or united by anthers. Filaments broadened proximally. Anthers oblong. Style elongate, with 3, rarely 5 stigmas. Nectary disk sometimes developed and then usually cylindrical, hollow. Capsule dehiscing by pores or small valves at sides of apex or at base, rarely in middle. Annual or perennial herbs, with alternate leaves, often with rosettes of radical leaves, rarely with some whorled cauline leaves, with rhizomes or more or less thickened roots.

Genus 1431. **CAMPANULA*** L.

L. Sp. pl. (1753) 163; A. DC. Monogr. Campan. (1830) 213; idem in DC. Prodr. VII, 2, 457.– Roucela Dum. Comm. bot. (1829) 14.– Medium Fisch. ex A. DC. l.c. (1830) 213, pro syn.– Quinqueloculaґia C. Koch in Linnaea, XXIII (1850) 360.– Marianthemum Schrank in Denkschr. bot. Ges. Regensb. II (1822) 34.– Erinia Noulet, Fl. bassin sous pyrén. (1837) 407.– Drymocodon Fourr. in Ann. Soc. Linn. Lyon, N. S. XVII (1869) 111.– Rapunculus Fourr. l.c.– Syncodon Fourr. l.c.– Čenekia, Nenningia, Sykoraea, Weitenwebera Opiz, Seznam (1852) 36–68.

Calyx 5-toothed, with tube closely appressed to ovary. Corolla 5-lobed, campanulate, infundibular, tubular-campanulate, rarely flat or nearly rotate, usually not deeply divided. Stamens with free, sometimes contiguous anthers, the filaments usually broadened at base. Style with 3-fid or 5-fid stigma. Lobes of stigma usually arcuately curved or spirally twisted. Nectary disk absent. Ovary 3–5-locular. Capsule membranous,
134 erect or pendulous, 3–5-locular, usually crowned by persistent dry calyx teeth, dehiscing by pores or small valves at apex and at base. Seeds numerous. Mostly perennial herbs, often with more or less fleshy thickened rhizomes or roots, rarely annuals or monocarpic herbs with fibrous root. Leaves alternate, sometimes in basal rosettes, petioled or sessile. Stems simple or branching. Inflorescences paniculate, many-flowered, racemiform, or plant 1-flowered.

Lectotype: C. latifolia L.

More than 300 species, mostly in the temperate parts of the Caucasus, Near Asia, the mountains of W. Europe, and in N. America. There are 150 species of Campanula in the USSR.

The species of Campanula are marked by a variety of habitats (forests, meadows, slopes, steppes and deserts); they are rather more abundant and diverse in the subalpine and alpine zones of high mountains. Campanula has not been adequately studied, nor is there a monograph devoted to this genus alone. Therefore, Alphonse de Condolle's work (l. c. and "Prodromus") is still important. The arrogantly-named "Monographia generis Campanula" (Linnaea, XXVI, 1853), an article by Farkas-Vukotinovic, was a poor show, by comparison with de Candolle's splendid work. Proceeding from exaggerated and confused views on the taxonomic value of the shape of the leaves, that author arbitrarily renames

* Diminutive of campana, a pre-Linnean name referring to the shape of the corolla.

some species of Campanula without attempting a monograph. All
subsequent works (particularly the invaluable treatment by Boissier in
"Flora orientalis") are fragments or short articles. Fomin's survey for
the "Flora caucasica critica" is very important. A. L. Kharadze has also
just published a valuable study of Caucasian Campanula.

There are undoubtedly some natural groups of species in Campanula
which should be recognized as sections, subsections, and orders. However,
existing classifications, besides that given by Kharadze, reflected the original
division of the genus into two large sections, established on different grounds
135 by de Candolle and Boissier. De Candolle's division of the genus into the
sections Medium and Eucodon based on the presence or absence of calyx
appendages fails to do justice to accomodate many new species. This
character proved to have little relation to all other characters. Boissier
proposed for part of section Eucodon the new, more natural section, Rapun-
culus, merging the rest of Eucodon with section Medium. His new classi-
fication was founded on a very stable character — the dehiscence of the
capsule in the upper part (Rapunculus) and in the lower (Medium). Many
years later Kharadze promoted section Medium to subgenus, dividing it into
two subgenera, Quinqueloculares and Triloculares, which were
in turn subdivided into sections.

We are adopting the Boissier division of Campanula into two large
sections, divided into subsections and series. This system of subsections
has no strict "hierarchy" of characters, which becomes lost in the division
of many small, similar subsections (by the totality of their characters).
At the same time, our classification certainly approaches the real phylo-
genetic relationship.

Species of Campanula have for a long time been grown in gardens for
their beatiful flowers. The following have been adopted for cultivation:
C. abietina, C. allionii, C. alpina, C. alsinoides, C. argyro-
tricha, C. betulifolia, C. carpatica, C. cashmiriana, C. cenisia,
C. collina, C. excisa, C. formanekiana, C. garganica, C. glo-
merata, C. hawkinsiana, C. hercegovina, C. heterophylla,
C. incurva, C. isophylla, C. laciniata, C. lactiflora, C. lasio-
carpa, C. latifolia, C. longistyla, C. lyrata, C. medium, C. mi-
rabilis, C. morettiana, C. peregrina, C. persicifolia, C. cha-
missonis, C. piperi, C. portenschlagiana, C. poscharskyana,
C. punctata, C. pyramidalis, C. raddeana, C. rotundifolia,
C. rupestris, C. sartorii, C. saxatilis, C. stevenii, C. triden-
tata, C. tubulosa, C. versicolor, C. vidalii, C. waldsteiniana,
C. zoysii. Many of these species grow wild in the USSR, most of them
are also cultivated in Europe and N. America. In this context Clifford
Crock's efforts to grow in England rare, endemic Caucasian species are
136 worthy of note and in particular C. autraniana, C. choziatowskyi,
C. hypopolia, C. kachetica, C. kemulariae, C. kolenatiana,
C. petrophila (Campanulas, London-New York. 1951). Until half
a century ago, many Caucasian Campanulas were cultivated in the
Tiflis Botanical Garden. There is a huge collection of live wild
species and numerous garden forms at the Botanical Garden of the
V. L. Komarov Botanical Institute of the Academy of Sciences in Leningrad.

Sometimes eaten as salad, with vinegar and oil from the roots of Campanula rapunculus, C. latifolia and C. persicifolia. The leaves of the last two species are also edible. According to Grossgeim (Rastitel'nye bogatstva Kavkaza, 2nd ed. (1952)), the leaves of C. glomerata contain 39.0 mg% vitamin C and those of C. latifolia even 144.5 to 200 and up to 400 mg%.

1. Capsule dehiscing by small holes or valves at apex or along upper half of lateral walls, erect, usually elongate. Appendages between calyx teeth absent . . . . 2.
\+ Capsule dehiscing by small holes or valves at base, nearly always pendulous, usually shortened; appendages between calyx teeth usually well developed, rarely small, hardly discernible, appearing as small folds between teeth, slightly tapering . . . . 20.
2. Leaves acutely dentate, nearly aristate, oblanceolate or lanceolate-spatulate, radical leaves in rosette; plant not higher than 3–10 cm; stems 1-flowered; corolla glabrous, 3 cm long; calyx teeth with acute dentoid lobes; capsule cylindrical, broadened distally; appendages between calyx teeth none; rhizome with shoots . . . . 150. **C. lasiocarpa** Cham.
\+ Leaves entire or dentate, but not aristate, sometimes with very small hamate prickles along dentate margin . . . . 3.
3. Corolla ca. 1 cm long, divided nearly to base into suberect lanceolate obtuse lobes, slightly shorter or nearly as long as calyx teeth, if longer they are not more than one and a half times as long; at anthesis and especially later calyx-tube inflated at apex, clavate, bluish or dark violet or nearly black, sparingly pubescent; shape of capsule similar; stems 10–15 cm, leafy, 1-flowered; leaves glabrous, obovate or oblanceolate, entire or faintly crenate . . . . 149. **C. uniflora** L.

137 + Corolla 2–3–5 cm long, divided for not more than one-third to one-half its length, broadly campanulate or narrower, often with curved lobes, sometimes nearly rotate; calyx-tube not clavately inflated distally; capsule globose-obconical or oblong . . . . 4.
4. Lower cauline leaves long-petioled, with short ovate-cordate, irregularly capitate-dentate laminas; stems branching; flowers large, with long pedicels and nearly rotate corolla; capsule longitudinally striated by 10 prominent nerves, ovoid-cylindrical, elongate-oblong or obconical; calyx teeth with lateral dentiform lobes; root thickened . . . . 148. **C. carpatica** Jacq.
\+ Lower cauline leaves short-petioled, others sessile, if cauline leaves tapering to petiole, then leaf base elongate-cuneate . . . . 5.
5. Stems thick, branching, covered at base with fibrous remnants of radical leaves; corolla narrowly infundibular to campanulate, 4 cm long, glabrous, divided for one-third into narrowly ovate, acute lobes with sparsely ciliate margin . . . . 135. **C. turczaninovii** Fed.
\+ Stems without fibrous remnants of leaves at base; corolla broadly infundibular or campanulate, margin not ciliate . . . . 6.
6. Cauline leaves ovate-oblong or very broadly lanceolate, crenate or dentate, commonly stiff-haired or margin with hamately curved bristles . . . . 7.

\+ Cauline leaves lanceolate, lanceolate-linear or linear, glabrous or subglabrous, obscurely crenate or dentate . . . . 9.

7\. Plant glabrous throughout, but for small bristles along leaf teeth; calyx-tube covered with white papillae and verrucae; root collar with squamous remnants of leaves; stems 50–70 cm, rather thick, longitudinally furrowed; leaves ovate-oblong, crisp-crenate, with hamate fine bristles along teeth; inflorescence paniculate; corolla broadly campanulate; calyx teeth lanceolate acute with broad base . . . . 134. **C. pontica** Alb.

\+ Plant more or less covered with stiff hairs; calyx-tube bristly pubescent . . . . 8.

8\. Plant 20–30 cm high; flowers intensively blue-violet, in small few-
138 flowered raceme or corymb, densely covered with short bristles; leaves irregularly and obtusely serrate-dentate . . . . 133. **C. hieracioides** Kolak.

\+ Plant 100–150 cm high; flowers nearly white or pale blue, in broad, few-flowered panicle, covered with short stiff, usually scattered hairs; leaves biserrate . . . . 132. **C. lactiflora** M. B.

9\. Calyx teeth markedly declinate, narrowly lanceolate or nearly subulate, sometimes broadly lanceolate but then plant low-growing, 8–10 cm high; inflorescence usually branching, paniculate . . . . . 10.

\+ Calyx teeth appressed to corolla, lanceolate or narrowly lanceolate, rarely nearly subulate; flowers usually solitary, sometimes in few-flowered, mainly unbranched inflorescence . . . . 16.

10\. Flowers few, 4–5 cm long, broadly campanulate, bright blue; corolla with slightly inflated base; leaves glabrous, shiny, crenate, the cauline linear-lanceolate; plant 50–70 cm high . . . . . 138. **C. persicifolia** L.

\+ Flowers 2–3 cm, infundibular; base of corolla never inflated, bluish-violet or almost white or bluish; inflorescence paniculate, many-flowered, if 1-flowered then calyx lobes usually broad . . . . 11.

11\. Small plant, 8–10 cm high; stems 1-flowered; calyx lobes broadly lanceolate; rhizome short-creeping . . . . . . 142. **C. vaidae** Penzes.

\+ High plant, with branching inflorescences, rarely 1-flowered; calyx lobes narrowly lanceolate or subulate; root thin or thickened . . . 12.

12\. Corolla white or nearly so; calyx teeth narrow, almost bristly, nearly half the length of the corolla; inflorescence paniculate; root fusiform or thickened, fleshy, radishlike . . . . 13.

\+ Corolla blue-violet; calyx teeth half the length of the corolla; root thin, fibrous . . . . 14.

13\. Root thickened, fleshy, radishlike . . . . 136. **C. rapunculus** L.

\+ Root fusiform, hard . . . . 137. **C. lambertiana** A. DC.

14\. Stem acutely faceted, alate from decurrent extensions of petioles; flowers in compressed panicle . . . . . . 141. **C. hemschinica** C. Koch.

\+ Stem cylindrical or obtusely angled; inflorescence spreading . . . 15.

15\. Stem pubescent in lower part, without sterile shoots at base; calyx
139 lobes small-toothed . . . . 139. **C. patula** L.

\+ Stem glabrous, thin, weak, with thin sterile, often creeping shoots at base; calyx lobes edentate; entire plant thin, delicate; leaves becoming semitransparent upon drying . . . . 140. **C. abietina** Griseb. et Schenk.

16 Rhizome firm, branching in upper part, where densely covered with long coarsely bristled grayish-brown remnants of stems and petioles; entire plant glabrous; stems developing from branching rhizome, usually in dense fascicles, simple, thin, stiff, after flowering, strict, sparingly leafy, 1-flowered; leaves linear, obscurely and remotely dentate or nearly entire; flowers violet, with deeply incised corolla; calyx with subulate linear lobes; capsule 10-nerved, slightly ribbed . . . . 143. **C. alberti** Trautv.

\+ Rhizome thin, creeping, with creeping shoots . . . . 17.

17\. Calyx teeth rather broad, to 1.5 mm, oblong-lanceolate, obtuse, distally boat-shaped, curved inward, slightly longer than or as long as tube; rhizome usually oblique; stems solitary, often branching in upper part, glabrous; rosetted leaves broadly lanceolate; corolla broadly infundibular, blue, 4(5) cm long . . . . 144. **C. altaica** Ldb.

\+ Calyx teeth narrow, 1 mm wide, acute or subulate, flat, not boat-shaped . . . . 18.

18\. Calyx teeth covered with spongy whitish cells, narrowly lanceolate, acute; corolla blue, glabrous, divided for one-third; style blue, divided up to middle; plant glabrous or sparingly pubescent, with thin creeping rhizome and thin stems, bearing 1–3 flowers; radical leaves ovate or oblong . . . . 145. **C. beauverdiana** Fom.

\+ Calyx teeth green, without superficial spongy layer; corolla blue-violet . . . . 19.

19\. Calyx teeth short, obconical at anthesis, much longer than calyx . . . . 147. **C. wolgensis** P. Smirn.

\+ Calyx teeth elongate, nearly cylindrical at anthesis, usually as long as calyx . . . . 146. **C. stevenii** M. B.

20 Ovary 5-locular; stigmas 5 . . . . 21.

\+ Ovary 3-locular; stigmas 3 . . . . 23.

140 21. Rosetted leaves none; cauline leaves sessile, ovate or ovate-lanceolate; flowers to 7 cm long; stems high, often to 1 m thick . . . . ★**C. medium** L.

\+ Rosetted leaves developed, irregularly lyrate-incised or subcordate, petiolate; cauline leaves sessile; flowers smaller, 2–3 cm long . . . . 22.

22\. Lower leaves short-petioled, cordate, irregularly incised-lobed-dentate; calyx teeth triangular, with lateral teeth; corolla covered with soft hairs . . . . 1. **C. crispa** Lam.

\+ Lower leaves long-petioled; petioles lobed-dentate; limb irregularly crenate; calyx teeth ovate-lanceolate, lateral teeth none; corolla sparingly pubescent along nerves . . . . 2. **C. lyrata** Lam.

23\. Annuals, with thin root and furcate stems; style distinctly thickened at base . . . . 24.

\+ Perennials, with rhizomes or more or less thickened roots; style not thickened at base . . . . 25.

24\. Calyx teeth slightly shorter than corolla, lanceolate, acute, erect, becoming accrescent and declinate; flowers subsessile; leaves sessile, ovate or obovate . . . . 23. **C. erinus** L.

\+ Calyx teeth one-half to three-quarters the length of the corolla, lanceolate, declinate; flowers short-pediceled; lower leaves short-petioled, spatulate; cauline leaves sessile, oblong-lanceolate . . . . 24. **C. propinqua** Fisch. et Mey.

25. Corolla dingy white, outside and even more inside purple punctate and maculate, pubescent in lower part, bearded inside, short-cylindrical to campanulate, slightly inflated at middle, globose at base; flowers pendulous; calyx with ovate acute appendages, ciliate . . . . 21. **C. punctata** Lam.

\+ Corolla never maculate-punctate, regularly blue, violet, reddish, white, sometimes paler inside in lower part . . . . 26.

26. Calyx teeth and calyx appendages and leaves with large transparent marginal prickles; flowers blue, large, in dense pyramidal paniculate inflorescences; rosetted leaves oblong-lanceolate, with winged petioles; cauline leaves sessile . . . . 3. **C. mirabilis** Alb.

\+ Calyx teeth and appendages and leaves never with the same prickles,
141 cilia or other excrescences along margin . . . . 27.

27. Plant scabrous with fine bristly hairs, sessile on verrucae; calyx appendages markedly accrescent after flowering, appearing inflated-broadened, prominently netted from protruding nerves, concealing tube; flowers small, sessile, in interrupted spicate inflorescence; leaves stiff, ovate-oblong . . . . 45. **C. stricta** L.

\+ Plant glabrous or variously pubescent, hairs never sessile on verrucae . . . . 28.

28. Stems furcate, flexuous, leafy; rosetted leaves none; cauline leaves subrhombic, subsessile; flowers not more than 1 cm long; corolla pale blue, twice as long as calyx; rhizome hidden in rock crevices . . . . 20. **C. imeretina** Rupr.

\+ Stems simple or branching, not furcate . . . . 29.

29. Style pubescent at apex; corolla campanulate, with globose base and attenuate-acuminate lobes; flowers on long, sometimes nearly radical pedicels on stem which branches nearly from base, hence inflorescence often prominently branched, rarely few-branched; leaves linear-lanceolate; root thick, fusiform, spongy . . . . 22. **C. alpina** Jacq.

\+ Style glabrous or covered with tiny papillae, not pubescent . . . . . 30.

30. Capsule suberect, flowers pendulous; all leaves cauline, linear, green above, densely gray-tomentose beneath; stems branching, flexuous, weak, thin; rosette none; calyx teeth lanceolate or sublinear, arcuate, much longer than small tube and triangular appendage half the length of the corolla . . . . 125. **C. hypopolia** Trautv.

\+ Capsule pendulous, if erect then flowers erect; leaves ovate, cordate, lanceolate or of another shape, if linear then glabrous or regularly pubescent above and beneath . . . . 31.

31. Flowers sessile in compact heads, whorls or clusters; appendages in notches between calyx teeth absent or inconspicuous; stems leafy, often high, simple or with few branches; rosettes none or weakly developed; cauline leaves sessile . . . . 32.

\+ Flowers solitary, pediceled, rarely subsessile, in more or less branching or racemiform inflorescence, or stems 1-flowered . . . 43.

142 32. Inflorescence long, spicate, appearing interrupted, composed of nearly equally sized, dense, sometimes secund clusters of rather small flowers or small whorls, with bracts at base; pubescence of stiff hairs; stems coarse, high, sometimes as thick as a small finger, usually simple; leaves sessile or very short-petioled . . . . . . 33.

\+ Inflorescence not long, compactly capitate for a good part or interrupted in lower part, whorled, sometimes branching, with long, broad or shorter bracts; lower leaves petioled, the upper sessile; pubescence of stiff, spreading or soft appressed hairs, rarely plant glabrous . . . . 34.

33. Inflorescence subcapitate at apex, interruptedly spicate, composed of subsessile flowers in dense whorls; corolla pubescent along nerves outside; seeds obliquely ovoid, yellowish . . . . . . 43. **C. cervicaria** L.

\+ Inflorescence composed for entire length of equal secund clusters or whorls of 6–7 flowers; corolla glabrous; seeds nearly cylindrical, obtuse, dark brown . . . . 42. **C. macrostachya** Waldst. et Kit.

34. Inflorescence terminal, capitate, entirely covered by oblong obtuse bracts, longer than flowers, forming broad infundibular envelope around inflorescence; calyx teeth white-ciliate; seeds flattened, spongy-verrucose, covered with small bristles . . . . 44. **C. machrochlamys** Boiss. et Huet.

\+ Inflorescence capitate or interrupted below, whorled, sometimes branching; bracts shorter, not concealing flowers; plant with soft and stiff hairs, sometimes glabrous . . . . 35.

35. Inflorescence elongate, interrupted, usually branching, nearly paniculate, with clusters of sessile flowers united at tip of stem in compact, more or less leafy head; radical leaves very long-petioled, large, oblong-ovate, with distinct, sometimes deeply cordate base, lamina 10–15 cm long, 7 cm wide; stems to 1 m more; corolla pubescent outside . . . . 33. **C. cephalotes** Nakai.

\+ Inflorescence not branching; radical leaves petiolate but laminas never reaching 15 cm in length and 7 cm in width; stems usually ca. 50 cm, rarely 1 m; corolla usually subglabrous outside . . . . . 36.

36. Inflorescence terminal, compact, capitate . . . . 37.

\+ Inflorescence capitate in upper part, slightly interrupted and whorled below . . . . 40.

37. Inflorescence head small, the size of a hazel nut or slightly larger,
143 5–10-flowered; rhizome thin, horizontal; stems straight, thin, 50 cm, nearly square in cross section, with spreading hairs; lower leaves long-petioled, oblong, acute, with rounded base; flowers not longer than 1.5–1.7 cm; corolla violet, glabrous, deeply divided into lanceolate lobes . . . . 36. **C. subcapitata** M. Pop.

\+ Inflorescence head large, 4–5 cm and more across; plant usually with shortened stem, 20–30 cm, sometimes to 50 cm . . . . 38.

38. Plants with spreading stiff hairs, slightly bristly; stems thick, strong, longitudinally striate; radical leaves petiolate, with truncate or cuneate base, irregularly and finely crenate-dentate; cauline leaves sessile, semiamplexicaul; inflorescence covered with bracts which conceal the blue flowers for about two-thirds; calyx teeth white-bristly . . . . 39. **C. maleevii** Fed.

\+ Plants with soft, dense, long hairs or subglabrous 39.

39. Rather sparingly pubescent or subglabrous plants; stems usually short, 10–20, rarely 30 cm, straight, simple, with few leaves; radical leaves petiolate, ovate-oblong, acute, with obtuse or cuneate base; flowers rather large, 2–4 cm long; corolla tubular-campanulate, barely pubescent outside . . . . 41. **C. trautvetteri** Grossh.

\+ Densely pubescent plant with soft spreading hairs; stems 25–30 cm; radical leaves petiolate, oblong-lanceolate, gradually and cuneately tapering to base; corolla glabrous . . . . 40. **C. panjutinii** Kolak.

40\. Radical leaves with orbicular or slightly cuneate base, oblong or elliptic, acute, leaves scabrous, with fine short hamately curved hairs on leaves; inflorescence interruptedly spicate, with compact, few-flowered whorls; calyx whitish, densely pubescent . . . . 37. **C. oblongifolia** (C. Koch) Charadze.

\+ Radical leaves with obtuse or shallowly cordate base; calyx greenish, not strongly pubescent . . . . 41.

41\. Bracts longer than flowers; scabrous pubescence of short, rather dense hairs; leaves cauline, cordate, amplexicaul base; calyx teeth with ciliate margin . . . . 38. **C. symphytifolia** (Alb.) Kolak.

\+ Bracts shorter than flowers; plants usually with sparse, soft or stiff
144 hairs or glabrous . . . . 42.

42\. Leaves gray-tomentose beneath, sparingly pubescent above; stem pubescent; inflorescence elongate, narrow, interrupted . . . . 35. **C. farinosa** (Roch.) Andrz.

\+ Leaves green on both sides, more densely pubescent beneath; stems usually sparingly pubescent; inflorescence a compact terminal head, and not far below it whorls or axillary clusters of flowers . . . . 34. **C. glomerata** L.

43\. Calyx without appendages or folds in notches between teeth, its teeth linear-lanceolate, acute, more or less declinate; stems densely leafy; cauline leaves lanceolate or linear; radical leaves, if present, with broad lamina, abruptly set off from petioles; plant usually subglabrous, with thin rhizome or more or less thickened fleshy root . . . . . . 44.

\+ Calyx with appendages or small folds between teeth, set at an angle, if appendages absent then cauline and radical leaves nearly equal in shape . . . . 48.

44\. Branching root thickened, fleshy; cauline leaves elliptical-oblong or lanceolate, 7–8 mm wide; stems thick, densely leafy; flowers in secund racemes . . . . 131. **C. napuligera** Schur.

\+ Roots or rhizome not fleshy or thickened; cauline leaves usually narrowly lanceolate or linear, not more than 2–4 mm wide; stems thin, sometimes nearly filiform, often sparingly leafy; flowers solitary or in spreading paniculate inflorescences . . . . 45.

45\. Stems 5–10 cm, usually 1-flowered; flowers large, 2–3 cm long, pendulous or drooping; cauline leaves linear, broadening slightly at apex, becoming slightly lobed . . . . 128. **C. giesekiana** Vest.

\+ Stems 10–40(50) cm, sometimes 1–2-flowered, often many-flowered; flowers smaller, not more than 1.5–2.5 cm long; cauline leaves always tapering distally, never spatulate . . . . 46.

46\. Inflorescence many-flowered, paniculate; corolla not more than 1.5 cm long; calyx teeth short, not recurved, subulate . . . . 127. **C. rotundifolia** L.

\+ Flowers 1–2, rarely more; calyx teeth appressed or markedly recurved . . . . 47.

47\. Calyx teeth strongly declinate, usually long, only half the length of the corolla or shorter, green ("herbaceous"), rarely shorter and narrower, darkish; calyx and capsule with prominent riblike nerves; stems thin but not weak, usually to 20 cm and more. Cauline leaves long,

145 sometimes to 5 cm, narrowly lanceolate or linear . . . . 126. **C. langsdorffiana** Fisch.

\+ Calyx teeth appressed to corolla or slightly declinate, always short, one-third the length of the corolla, like calyx-tube dark-colored; calyx and capsule not ribbed; stems thin, often filiform, weak, often pendulous; cauline leaves narrowly linear, always entire . . . . 129. **C. kladniana** (Schur) Witasek.

++ Calyx teeth appressed to corolla, long, only one and a half times as long or rarely half as long as corolla, like tube, dark-colored; calyx not ribbed; stems thin, not weak, arcuately ascending; upper cauline leaves broadly lanceolate, linear; lower leaves dentate . . . . 130. **C. polymorpha** Witasek.

48\. Calyx teeth and appendages covered with whitish slightly curved bristly cilia, appendages resembling teeth in size and shape, radical leaves with cuneate base; stems more or less branching, many-flowered . . . . 49.

\+ Calyx without whitish bristles but if bristly, then leaves cordate, appendages usually distinctly different from calyx teeth, or absent . . . . 64.

49\. Calyx appendages very short and narrow, not concealing even part of tube; capsule broadened above, abruptly set off at base; stems branching from base; corolla tubular-campanulate, glabrous; inflorescence branches few-flowered . . . . . 5. **C. fedorovii** Charadze.

\+ Calyx appendages as long as or slightly shorter than tube which they cover partly or entirely; capsule oblong or ovate . . . . 50.

50\. In addition to ascending low stems, the roots develop weak, prostrate and usually also radicant, rather long shoots; plant glaucous, with appressed hairs; radical leaves and leaves of sterile shoots obovate or rounded, long-petioled, small, slightly crenate; flowers not more than 1.2–1.5 cm long; calyx appendages half the length of the tube . . . . 4. **C. caucasica** M. B.

\+ Stems straight or ascending, sometimes partly decumbent, but not radicant, prostrate; plant green or grayish, rarely glaucous; radical leaves not rounded; calyx appendages as long as tube, sometimes concealing it . . . . 51.

51\. Stems not more than 20 cm, ascending and partly decumbent, branching, with spreading white hairs; lower leaves broadly lanceolate, obtuse,
146 crenate-serrate, petiolate; flowers few, 2.5 cm long, violet, long-pediceled; corolla glabrous, infundibular . . 17. **C. darialica** Charadze.

\+ Stems ascending or erect, not decumbent . . . . 52.

52\. Calyx appendages nearly as long as sepals, broad, obtuse, with overlapping margins, completely concealing tube; many-stemmed plant, the stems densely covered with crisp retrorse hairs . . . . 53.

\+ Calyx appendages shorter than sepals, margins not overlapping; stems few or stem solitary, variously pubescent, not crisp . . . . . 54.

53\. Stems appressed, crisp pubescent, numerous, arcuately ascending; flowers medium-sized, ca. 2.5 cm long; corolla tubular-campanulate, violet . . . . 7. **C. daghestanica** Fom.

\+ Stems numerous, ascending, covered with white stiff spreading hairs recurved below; flowers 3–4 cm long; corolla narrowly campanulate, bright violet-blue . . . . 6. **C, komarovii** Maleev.

54\. Plant usually low-growing, 2–5 cm, very rarely to 20 cm high; stems solitary or few, arcuate, stiff-haired; flowers 3.5 cm long, tubular-campanulate, rather dark violet; calyx teeth and appendages with coarse bristly cilia . . . . 13. **C. talievii** Juz.

\+ Plant not low-growing, usually exceeding 20 cm, sometimes to 80 cm and more; stems solitary or numerous; other characters different . . . . 55.

55. Stems solitary or else usually not more than 2–3 . . . . 56.

\+ Stems always numerous and equal, or central stems coarser and thicker . . . . 60.

56. Stems sturdy, 50–120 cm; flowers usually 3–4–5 cm long, blue-violet, with broad infundibular to campanulate corolla; plant densely covered with short stiff hairs; style as long as corolla, elongating only after flowering . . . . 8. **C. schischkinii** Kolak. et Sachok.

\+ Stems usually 50 cm, rarely longer; flowers 1.5–2, rarely 3 or 4 cm long, but then style markedly esxerted at anthesis . . . . 57.

57. Style markedly exserted at anthesis; flowers 2–3–4 cm long, bright blue-violet . . . . 58.

\+ Style as long as corolla or inconspicuously exserted; flowers large, 1.5–2 cm long, bright blue violet . . . . 59.

58. Flowers 3–4 cm long; corolla urceolate-campanulate, constricted above middle, glabrous outside, bright violet; stems 50 cm, 3–5 mm
147 thick, branching slightly in upper part, with sparse stiff pubescence . . . . 9. **C. longistyla** Fom.

\+ Flowers 2 cm long; corolla infundibular to campanulate; stems 30–40 cm, 1–1.5(3) cm across, branching above, sparsely stiff-haired; species similar to preceding but smaller and thinner in all parts . . . . 10. **C. charadzae** Grossh.

59. Calyx appendages with prominent reticular nerves, covered with stiff spreading hairs; flowers 1.5–2 cm long; inflorescence broad, branching often paniculate . . . . 14. **C. sibirica** L.

\+ Calyx appendages with prominent reticular nerves, pubescence sparse; flowers 2.5–3 cm long; inflorescence narrow, usually racemiform . . . . 12. **C. elatior** (Fom.) Grossh.

60. Central stems distinctly thicker and coarser than the lateral or all stems thick, coarse, profusely pubescent or bristly-pubescent, scabrous . . . . 61.

\+ Stems rather thin, not coarse, similar, pubescence usually short or longish, not bristly . . . . 62.

61. All stems, especially the central ones 20–40 cm long, 0.5–0.7 cm across, coarse, arcuate, with stiff bristly pubescence; flowers 3 cm long, in densely branching inflorescence, short-pediceled, bright blue-violet; root oblique or vertical, thick, the collar densely covered with brown remnants of petioles . . . . 11. **C. charkeviczii** Fed.

\+ Median stems coarse, thick, 3 mm across, the lateral thin, shorter (rarely stem single), pubescence dense, declinate; flowers on rather long pedicels, (1.5)2–2.5(3) cm long, violet-blue, arranged in loose inflorescences . . . . 15. **C. taurica** Juz.

62. Corolla broadly campanulate, with globose base, broadly ringent, bright blue, 2.5–3 cm long; calyx small, with narrow lobes and appendages; stems many, arcuate, thin, few-branched, like whole plant short-haired . . . . 19. **C. schelkownikowii** Grossh.

\+ Corolla tubular-campanulate or infundibular, with nearly cylindrical base, 1.5, rarely 2 cm long; stems prominently branched . . . . . 63.

63. Radical leaves 5–10 cm long, ovate-oblong, markedly crenate-lobate proximally . . . . 18. **C. brassicifolia** Somm. et Lev.

\+ Radical leaves oblanceolate or nearly spatulate, acute, obscurely crenate . . . . 16. **C. hohenackeri** Fisch. et Mey.

148 64. Flowers and capsule always pendulous; radical and lower cauline leaves mostly large, sometimes rather small, with cordate or obtuse, very rarely cuneate base, long-petioled, rarely sessile, always more or less distinctly large-toothed; inflorescence usually many-flowered, racemiform or slightly paniculate, sometimes plants 1-flowered . . . . 65.

\+ Flowers erect but capsule pendulous, rarely nearly erect; radical leaves mostly small, cuneately tapering to petioles, if base obtuse or cordate, then leaf not more than 1 cm long, usually denticulate or crenate, often nearly entire, plants with 1 or few flowers . . . . 118.

65. Stems many, forming dense fascicles on thick, branching, grumose rhizome, rather weak, sometimes decumbent and pendulous, regularly and rather densely leafy . . . . 66.

\+ Stems usually solitary, straight, not decumbent or pendulous . . . 83.

66. Radical leaves cordate, elongate, with very long and thin petioles; cauline leaves with the exception of the upper, also long-petioled; calyx teeth prostrate . . . . 67.

\+ Radical leaves, if present basally obtuse, with suborbicular lamina, with broadened, alate lobate-dentate petioles; cauline leaves short-petioled or sessile; calyx teeth not spreading . . . . 77.

67. Flowers 2.5–3 cm long; radical leaves with conspicuously attenuate, cordate or ovate, acute or obtuse, acutely or obtusely toothed laminas (3)4–7 cm long . . . . 68.

\+ Flowers 1–1.5(2) cm long; radical leaves with short, rounded or shortly ovate, dentate laminas, 2–3(4) cm . . . . 74.

68. Flowers white, reddish, 4 cm long; plant with sparse grayish curly hairs; stems 20 cm, flexuous; leaves stiff, thin, ovate or cordate or basally obtuse . . . . 73. **C. finitima** Fom.

\+ Flowers violet or bluish; pubescence different . . . . 69.

69. Calyx teeth with scattered marginal teeth and cilia; corolla glabrous outside and along margins; flowers violet, ca. 3 cm long; leaves twice large-toothed or partially incised-dentate; radical leaves 10 cm long, 3–5 cm wide, ovate or ovate-oblong, acute; stems thick, declinate or pendulous; inflorescence loose; rhizome thick, dark brown, grumose . . . . 76. **C. ossetica** M. B.

149 + Calyx teeth with transparent basally broadened, marginal bristles, corolla pubescent outside and along margins or only along margins, rarely entirely glabrous . . . . 70.

70. Corolla pubescent inside, narrowly campanulate; glaucescent green plant; leaves indurate, slightly fleshy, large-toothed; pedicels shorter than calyx . . . . 78. **C. betulifolia** C. Koch.

\+ Corolla glabrous or subglabrous, sometimes with bearded margins . . . . 71.

71. Radical leaves obtuse, incised-dentate at apex . . . . 72.

\+ Radical leaves acute or attentuate-acuminate, dentate . . . . 73.

72. Radical leaves obstuse, incised, crenate, 7 cm long; corolla glabrous, broadly campanulate, violet; calyx green, acutely toothed; subglabrous plant with few, usually branching stems . . . 79. **C. kolenatiana** C. A. M.

\+ Radical leaves of the same shape, 4–5 cm long, cuneate; corolla narrowly infundibular; calyx glaucous, with thinly acuminate teeth . . . . 80. **C. akhverdovii** Charadze.

73. Anthers orange; flowers violet; leaves acutely toothed, triangular-hastate, cordate . . . . 77. **C. raddeana** Trautv.

\+ Anthers yellow; flowers bluish; corolla bearded at margin; leaves twice large-toothed, cordate-ovate, attenuate . . 81. **C. kemulariae** Fom.

74. Corolla like rest of plant densely covered with short bristles; leaves rounded-ovate, deeply cordate, acutely large-toothed; small plant, with branching rhizome and weak stems, not more than 20 cm long . . . . 84. **C. takhtadzhianii** Fed.

\+ Corolla glabrous or sparingly pubescent; glabrous or subglabrous plant . . . . 75.

75. Corolla covered outside with small sparse papillae, ciliate along nerves, pubescent inside, blue, 2.5 cm long; stems 10–12 cm, branching, prominently flowered; leaves rounded-ovate, large-toothed . . . . 85. **C. elegantissima** Grossh.

\+ Corolla glabrous outside, bearded inside . . . . 76.

76. Corolla infundibular, constricted at base, 2 cm long; leaves ovate-cordate with erect acuminate teeth . . . . . 83. **C. choziatowskyi** Fom.

\+ Corolla campanulate, not constricted at base, 1.5 cm long; leaves with teeth inclined forward, nearly hamately curved, rounded-ovate, often
150 nearly reniform . . . . 82. **C. bayerniana** Rupr.

77. Calyx with developed appendages in notches between teeth, margins of teeth and appendages with stiff setaceous cilia; radical and lower cauline leaves with lobed-dentate petioles; all leaves broadly ovate, broadly rhombic or suborbicular, partly irregularly incised, obtusely toothed; pubescence villous, grayish, stems many, corymbiformly branching, developing from branching grumose thick rhizome . . . . 86. **C. kantschavelii** Zagareli.

\+ Calyx without or with very short appendages, pubescent, without marginal setaceous cilia . . . . 78.

78. Calyx appendages one-quarter the length of the teeth; leaves indurate, slightly fleshy but dry and very brittle, ovate, cuneately tapering to short petioles, with large acute teeth pubescent along margin; inflorescence corymbiform-paniculate; corolla 2 cm long, pale blue or whitish; plant with markedly scabrous dense stiff hairs hence leaves and stems glaucescent . . . . 87. **C. radula** Fisch.

\+ Calyx appendages absent or inconspicuous; leaves with truncate rounded or slightly cordate base . . . . 79.

79. Radical leaves with laminas ca. 2 cm long, 2 cm across; the cauline leaves smaller but usually longer than 1 cm, with stiff and scabrous pubescence, without tomentum beneath; stems whitish, arcuate; flowers 2 cm long; calyx lobes dentate; rhizome covered with white remnants of petioles . . . . 88. **C. kachetica** Kantsch.

\+ Radical and cauline leaves rarely 1 cm, usually not more than 5–7 mm long and wide; leaves densely tomentose, white and gray, sometimes sericeous-silvery beneath; flowers ca. 1 cm long, often much smaller . . . . 80.

80\. Leaves few, usually subelliptic-rhombic, elongate, obscurely toothed, often nearly entire; stems reddish, thin, with subcapilliform branches, pubescence thinly tomentose, partly sericeous . . . . 90. **C. evolvulacea** Royle.

\+ Leaves short and broadly ovate, distinctly large-toothed; densely tomentose, white or grayish beneath . . . . 81.

81\. Leaves flat, not crisp-undulant, gray-pubescent above, more densely so beneath, large-toothed or slightly crenate; corolla one and a half times as long as calyx teeth, pink or blue; sometimes in addition to
151 chasmogamic flowers also much smaller cleistogamic flowers on filiform branches . . . . 89. **C. incanescens** Boiss.

\+ Leaves crisp-dentate or strongly undulant, bright green and pubescent above; corolla 2–4 times as long as calyx teeth . . . . 82.

82\. Calyx teeth short-triangular, half the length or as long as tube . . . . 92. **C. ketzkhovelii** Sosn.

\+ Calyx teeth elongate, triangular, twice as long as tube . . . . 91. **C. massalskyi** Fom.

83\. Calyx teeth markedly growing after flowering, much longer than tube, prostrate, appendages nearly absent; radical leaves and leaves of sterile shoots long or very long, thin-petioled, triangular, with cordate, obtuse or broadly cuneate base; stems 1-flowered, simple or slightly branching, branches terminating with flowers; rhizome branching, usually with sterile shoots . . . . 84.

\+ Calyx teeth slightly accrescent after flowering, not spreading . . . 89.

84\. Calyx appendages filiform, reflexed, small, shorter than tube; plant with scattered rigid hairs; rhizome rather thin; stems subglabrous, 1–2-flowered; radical leaves cordate-ovate or elongate-ovate, acute, irregularly incised, obtusely toothed; calyx teeth glabrous, triangular-lanceolate, accrescent in fruit, spreading; corolla 3.5 cm long, dark violet . . . . 71. **C. kolakovskyi** Charadze.

\+ Calyx appendages barely discernible, dentiform, or obsolete . . . . 85.

85\. Stems one- or few-flowered . . . . 86.

\+ Stems with many-flowered branching inflorescence . . . . 87.

86\. Radical leaves as long as or longer than cauline leaves, with long, subfiliform petioles; limb much shorter than petioles, broadly ovate, with truncate or broadly cuneate base and large obtuse teeth; calyx teeth lanceolate, much longer than tube, spreading; corolla broadly campanulate, blue-violet; rhizome branching, with sterile shoots . . . . 70. **C. jadvigae** Kolak.

\+ Radical leaves distinctly shorter than cauline leaves; limb as long as or two to three times as long as petioles, ovate-triangular, with cordate base and large, obtuse teeth; calyx teeth triangular-lanceolate, spreading, much longer than tube; corolla campanulate, pale blue-violet with ciliate margin; plant covered with small setaceous hairs
152 throughout . . . . 69. **C. dzyschrica** Kolak.

87\. Plant covered with sparse small transparent hairs; leaves glaucescent beneath, green above; radical leaves asymmetrical, long-petioled, ovate-triangular, rarely some lanceolate, with faintly notched, mostly truncate base, irregularly dentate; calyx teeth triangular-lanceolate, spreading; corolla pale blue . . . . 74. **C. engurensis** Charadze.

\+ Plant quite glabrous . . . . 8.

88\. Radical leaves with faintly cordate, truncate or cuneate base, long-petioled, acutely incised-dentate, elongate ovate-acuminate, semi-coriaceous; calyx teeth spreading, longer than tube, narrowly lanceolate . . . . 73. **C. autraniana** Alb.

\+ Radical leaves deeply cordate with long, thin petioles, obtuse and irregularly bidentate, triangular-ovate, with acute and attenuate apex, thin; calyx teeth lanceolate, spreading, longer than tube . . . . 72. **C. suanetica** Rupr.

89\. Calyx appendages long, as long as tube, usually longer, recurved below, lanceolate; leaves mostly triangular-cordate, with short, sometimes elongate lamina . . . . 90.

\+ Calyx appendages obsolete or very short and dentiform or longer, filiform . . . . 98.

90\. Leaves deeply incised to lobed-dentate, broadly ovate-triangular, acute, with cordate or obtuse base, lamina to 10 cm long, 8 cm wide, thin, green beneath and above, scabrous; plant 70 cm; stems sulcate, reddish, covered with thin fine bristly hairs; calyx teeth and appendages covered with whitish, slightly curved, acute bristles; corolla ca. 2 cm long, slightly longer than the lanceolate calyx teeth; flowers in short few-flowered apical raceme . . . . 46. **C. sclerotricha** Boiss.

\+ Leaves usually dentate, crenate or obscurely crenate, not incised-dentate or lobed-dentate . . . . 91.

91\. Corolla blue; plant nearly completely glabrous, pale green; leaves the same color above and beneath, glabrous, acute or rather large-toothed with hardish cilia confined to margin . . . . 47. **C. brotheri** Somm. et Lev.

\+ Corolla white, yellow, pink or wine-red; plant more or less pubescent; leaves always pubescent beneath, rarely differing in color – gray or
153 whitish – obtusely toothed, crenate or barely crenate . . . . 92.

92\. Radical leaves elongate, ovate-oblong, with small basal notch or slightly cordate, acute, obtusely crenate-serrate; flowers in raceme, white-yellow, not longer than 1.5 cm, plant with appressed pubescence . . . . 52. **C. kirpicznikovii** Fed.

\+ Radical leaves with short broadly cordate-ovate, sometimes nearly triangular lamina; flowers pink, reddish-violet, white or yellow, sometimes large . . . . 93.

93\. Flowers 4 cm and longer, white, becoming yellow when dry; radical leaves very long-petioled, with petioles 25 cm long, reniform, rounded-cordate or acuminate-cordate, irregularly crenate-dentate; plant pubescent, with stem branching nearly from base; rhizome oblique, rather thin . . . . 48. **C. dolomitica** E. Busch

\+ Flowers not more than 2 cm long; radical leaves not as long-petioled . . . . 94.

94. Flowers pink or reddish-violet . . . . 95.

\+ Flowers white or yellow . . . . 96.

95. Flowers wine-pink or pink-violet; leaves white-tomentose beneath, subglabrous above . . . . 54. **C. letschchumensis** Kem.-Nath.

\+ Flowers pale pink, sometimes nearly white; leaves gray-green beneath, pubescent, not tomentose . . . . 53. **C. makaschvilii** E. Busch.

96. Stems regularly and densely leafy distally; apical leaves 4 cm wide, slightly tapering, grayish-green, obscurely crenate, the lower reniform; flowers white, bluish . . . . 51. **C. leskovii** Fed.

\+ Stems few-leafed; lower leaves much larger than others, tapering abruptly at stem apex, distinctly crenate or dentate; flowers white, yellowing or ocher-yellow . . . . 97.

97. Live flowers white or straw-colored, becoming slightly yellow when dry; leaves tomentose beneath . . . . 49. **C. alliariifolia** Willd.

\+ Flowers, even when dry, ocher-yellow; leaves above and beneath slightly pubescent, not tomentose . . . . 50. **C. ochroleuca** Kem.-Nath.

98. Calyx devoid of appendages, with recurved plicate notches between teeth; high or low plants with poorly developed or obsolete rosette and
154 rather densely leafy stem; leaves cordate, dentate; cauline and radical leaves similar . . . . 99.

\+ Calyx with very small dentiform or longer filiform appendages, or without appendages, and then with pliciform recurved notches between teeth; rosette of radical leaves usually well developed; in shape and especially in size radical leaves differ distinctly from the cauline and are always much larger; stem with rather few leaves . . . . 105.

99. Calyx teeth one-third to one-half as long as corolla, erect; flowers usually 3–5 cm long . . . . 100.

\+ Calyx teeth at least one-fifth to one-fourth the length of the corolla, recurved or spreading; flowers 2–3 cm long . . . . 102.

100. Lower leaves triangular-cordate, twice large-toothed at margin; calyx subglabrous or hairy, blackish; teeth green; stem slightly faceted, reddish, with white bristly hairs . . . . 28. **C. trachelium** L.

\+ Lower leaves oblong-ovate, with rounded or cordate base, irregularly serrate or biserrate; calyx glabrous or teeth with short curly cilia . . . . 101.

101. Cauline leaves all petioled, the petiole half the length of the lamina; calyx teeth dentate, with curly, ciliate margin . . . . 27. **C. odontosepala** Boiss.

\+ Cauline leaves partly sessile, partly very short-petioled, 5–10 cm long; robust plant, 100 cm and more; calyx glabrous, its teeth not dentate; stem many-flowered . . . . 25. **C. latifolia** L.

\+\+ Cauline leaves of similar shape, 2–3(4) cm long; plant 10–12 cm, with hairy calyx; stem 1-flowered . . . . 26. **C. megrelica** Mand. et Kuthath.

102. Leaves more or less densely gray-haired beneath; flowers in short raceme or narrow panicle; sepals spreading after flowering . . . . 32. **C. bononiensis** L.

\+ Leaves often subglabrous, with scattered hairs, or more densely pubescent but not gray-haired beneath; flowers in secund raceme; sepals turning downward after flowering . . . . 103.

103. Lower leaves with obtuse or faintly cordate base; flowers 3 cm and longer; corolla broadly campanulate . . . 30. **C. grossheimii** Charadze

\+ Lower leaves cordate; flowers 2–2.5 cm long; corolla infundibular, campanulate . . . . 104.

104. Leaves from lower to medium cauline deeply cordate; calyx and
155 stem subglabrous . . . . 31. **C. cordifolia** C. Koch.

\+ Lower leaves cordate; median and upper cauline leaves with obtuse base; calyx and stem scabrous-hairy . . . . . . 29. **C. rapunculoides** L.

105. Notches between calyx teeth nearly without appendages or with a small fold; limb of radical leaves usually oblong, ovate or elliptic . . . . 106.

\+ Notches between calyx teeth with distinct, short, sometimes dentiform or subulate, elongate and hamately curved appendages; limb of radical leaves usually elongate-ovate, with acutely attenuate apex . . . . . 113.

106. Stems ca. 10 cm, rarely 15 cm long and more; rhizome usually producing many sterile shoots; leaves small, the lamina usually not more than 2–3–4 cm long . . . . 107.

\+ Stems usually 20–30(50) cm; rhizome not producing many sterile shoots; leaves with lamina 4–8 cm long . . . . 109.

107. Calyx teeth 1 cm long or more, 2–3 mm wide; corolla with acute strongly attenuate lobes; capsule ovoid; plants resembling Viola in habit . . . . 60. **C. sphaerocarpa** Kolak.

\+ Calyx teeth ca. 5 mm long, narrowly lanceolate or sublinear, sometimes falcate, corolla with acute, not attenuate lobes . . . . 108.

108. Canescent plant; radical leaves broadly elliptic, obtuse or rounded, with truncate or barely cuneate base, crenate; rhizome with its branches covered with membranous remnants of petioles; calyx lobes sublinear, falcate, half the length of the corolla . . . . 62. **C. irinae** Kuthath.

\+ Plant short-haired, green; radical leaves ovate, rounded or acute, crenate; calyx teeth lanceolate, short, only one-quarter the length of the ciliate-margined corolla . . . . 61. **C. fonderwisii** Alb.

109. Corolla 4 cm long, four times as long as calyx teeth, with narrow attenuate, acute lobes . . . . 57. **C. schistosa** Kolak.

\+ Corolla 2.5–3 cm long or much shorter, not more than twice as long as calyx teeth, lobes acute, not long-attenuate . . . . 110.

110. Plant, especially the stems, sparingly pubescent, subglabrous; radical leaves with distinct winged petiole; calyx teeth appressed to corolla,
156 hamately curved distally; stem thin, often geniculately bent, few-leafed . . . . 58. **C. kluchorica** Kolak.

\+ Plant more or less densely pubescent; calyx teeth erect . . . . . . 111.

111. Stem slender, 60 cm, distinctly flexuous, sparingly leafy, arcuate below, tapering above, glabrous above, with spreading hairs beneath; flowers 1.5–2 cm long; calyx with bristly hairs . . . . 59. **C. albovii** Kolak.

\+ Stems thick, 20–30–40(50) cm, erect or with arcuate base; cauline leaves well developed, pubescent for entire length; calyx not bristly-haired . . . . 112.

112. Stems coarse, very often with long branches; corolla broadly campanulate, 3 cm long; strong, partly bristly-haired plant . . . . 56. **C. annae** Kolak.

\+ Stems rather slender, simple; corolla 2–2.5, rarely 3 cm long, campanulate; plants with sparse spreading hairs, not bristly . . . . 55. **C. collina** M. B.

113. Calyx appendages longer than tube, subulate, acuminate, hamately recurved; rhizome creeping, covered with squamous remnants of petioles; sparsely pubescent or subglabrous plant, with slender, leafy, 1- or 2-flowered stems, sometimes branching from middle; radical leaves ovate-cordate, small; capsule pendulous . . . . 68. **C. calcarata** Somm. et Lev.

\+ Calyx appendages much shorter than tube, acuminate, not curved at apex, often reduced, dentiform; glabrous or pubescent plants . . . . 114.

114. Plants nearly quite glabrous; rhizome with thick shoots, densely covered with remnants of petioles; stems rather slender, simple; radical leaves oblong-ovate, acutely biserrate; capsule short, globose-conical, with prominent riblike nerves . . . . . 66. **C. siegizmundii** Fed.

\+ Plants more or less densely pubescent . . . . 115.

115. Stems sturdy, to 80 cm thick, coarse, like leaves long white-haired; calyx white-haired, bristly; corolla blue, 3.5 cm; capsule short, semi-globose, pendulous . . . . 63. **C. calcarea** (Alb.) Charadze.

\+ Stems not more than 50 cm, pubescent, with short or appressed hairs; calyx not usually bristly . . . . 116.

116. Stems branching from base with thin branches; corolla narrow, infundibular, violet; lower leaves long-petioled, ovate, irregularly dentate; plants with appressed pubescence . . . . 67. **C. sommieri** Charadze.

157 + Stems simple or slightly branching in upper part; corolla campanulate, whitish or pale blue . . . . 117.

117. Capsule short, globose-obconical, subglabrous . . . . 64. **C. sarmatica** Ker.-Gawl.

\+ Capsule narrowly cylindrical, white-lanate . . . . 65. **C. woronovii** Charadze.

118. Calyx appendages obsolete or very small, dentiform; stems often few, rarely 1-flowered, but then appendages absent or calyx teeth falcately curved; radical leaves spatulate, ovate or lanceolate, abruptly or sometimes gradually attentuate into petiole; cauline leaves often lanceolate, sometimes reniform, short-petioled; rhizome cespitose or slightly so . . . . 119.

\+ Calyx appendages developed, usually as long as tube or longer, rarely shorter; stems nearly always 1-flowered, very rarely few-flowered, but then calyx appendages longer than tube; like cauline leaves radical leaves always with narrow cuneate base, if broadly cuneate then appendages short; rhizome usually densely cespitose; often forming dense cushions . . . . 125.

119. Cauline leaves reniform, dentate, short-petioled, small; all leaves transparent, prominently nerved beneath; plant not more than 5–10 cm high, canescent; stems filiform, rather densely leafy, rhizome thin; flowers 1.5–2 cm long; corolla violet, pubescent outside . . . . 120. **C. andina** Rupr.

\+ Cauline leaves spatulate, lanceolate, linear, never reniform . . . . 120.

120\. Radical leaves ovate, the short lamina passing abruptly into thin petiole; rhizome loosely cespitose, partly with prostrate shoots; stems 7–10 cm, rarely slightly longer, 1–2–3-flowered, leafy; corolla 2 cm, blue, sparingly pubescent outside along nerves . . . . 119. **C. petrophila** Rupr.

\+ Radical leaves tapering cuneately into petiole; rhizome thick, with columelliform branching, or relatively thin without prostrate shoots . . . . 121.

121\. Plant acutely scabrous, with small, short rigid bristly hairs, glaucescent; rhizome with columelliform branching, densely cespitose, producing cushions, with rosettes of leaves; these spatulate, acutely toothed or crenate; stems ca. 10 cm, arcuate, corymbiformly branching, leafy; calyx stiff-haired; corolla tubular to infundibular, blue . . . . 121. **C. karakuschensis** Grossh.

158 + Plant subglabrous or with faint scattered pubescence; stems 1-flowered or slightly branching, not corymbiform . . . . 122.

122\. Rosetted leaves remotely acutely toothed, rather broad, 1 cm wide, sometimes wider, with rather short petioles, cuneate-obovate, acuminate . . . . 122. **C. lehmanniana** Bge.

\+ Rosetted leaves obscurely and remotely toothed, nearly entire, narrowly lanceolate, ca. 3, rarely 5 mm wide . . . . 123.

123\. Calyx stiff-haired, with distinct triangular pubescent appendages, the teeth narrowly lanceolate, slightly falcately curved; stems and petioles filiform; radical leaves lanceolate, obscurely and remotely denticulate or nearly entire; flowers 1.5 cm long, pale blue . . . . 116. **C. aldanensis** Fed. et Karav.

\+ Calyx subglabrous or very sparingly pubescent, appendages obsolete or inconspicuous . . . . 124.

124\. Rosetted leaves scabrous, grayish, the petioles shorter than or as long as laminas . . . . 123. **C. capusii** (Franch.) Fed.

\+ Rosetted leaves smooth, green, their long filiform petioles twice as long as laminas . . . . 124. **C. eugeniae** Fed.

125\. Flowers 0.5 cm long; small plants, not higher than 2–5 cm; rhizome with columelliform branching, forming small cushions; calyx teeth short-ciliate; leaves short-cuneate, with few rather large apical teeth . . . . 118. **C. minsteriana** Grossh.

\+ Flowers 2–3–4-cm long; plant 5–10–15-cm high. . . . . 126.

126\. Leaves covered with reflexed bristles above, glabrous beneath; calyx with ciliate margin; stems usually 1–4-flowered . . . . 117. **C. ledebouriana** Trautv.

\+ Leaves pubescent or glabrous, if pubescent only above then calyx white-tomentose inside; stems always 1-flowered . . . . 127.

127\. Leaf margins with spreading, thin rather long cilia or thick reflexed prickles . . . . 128.

\+ Leaves glabrous or pubescent, their margin without long cilia or prickles . . . . 129.

128\. Rosetted leaves glandular-crenate, with long spreading cilia along margin; calyx teeth and appendages ciliate . . . . . 93. **C. ciliata** Stev.

\+ Rosetted leaves with thick reflexed prickly cilia along margin,
159 coriaceous, shiny; calyx teeth and appendages glabrous . . . . 94. **C. dzaaku** Alb.

129\. Leaves, especially those of rosetted, cuneately spatulate with 3 distinct apical teeth, the sides usually edentate, corolla glabrous outside; calyx rather densely pubescent . . . . 130.

\+ Leaves dentate, obscurely dentate, crenate or entire, not apically 3-toothed . . . . 131.

130\. Corolla 2 cm, campanulate, with wide hemispherical base; calyx teeth linear-lanceolate, acute, calyx fine-haired . . 95. **C. tridentata** Schreb.

\+ Corolla 3–4 cm, campanulate, with narrow base; calyx teeth broadly linear-lanceolate, obtuse, calyx usually densely lanate with tangled hairs . . . . 96. **C. biebersteiniana** Roem. et Schult.

131\. Receptacle stiff-haired; corolla pubescent or glabrous . . . . . . 132.

\+ Receptacle and corolla glabrous, the latter sometimes pubescent along margin and below . . . . 143.

132\. Rosetted leaves tapering into more or less long petiole . . . . . . 133.

\+ Rosetted leaves shortened, spatulate, lanceolate or linear-lanceolate, short-petioled . . . . 138.

133\. Corolla pubescent or tomentose . . . . 134.

\+ Corolla glabrous . . . . 136.

134\. Leaves elongate-oblanceolate, gradually tapering into petiole, dentate or nearly entire; corolla 4 cm long, bright blue-violet; densely cespitose plant . . . . 97. **C. aucheri** A. DC.

\+ Leaves rounded ovate or short-elliptic, tapering rather abruptly into petiole . . . . 135.

135\. Radical leaves rounded or ovate-rounded, more or less distinctly serrate, with long thin petioles; rhizome thick, ligneous . . . . 103. **C. argunensis** Rupr.

\+ Radical leaves elliptic or obovate-elliptic, rounded, irregularly crenate, with long petioles; rhizome branching, with elongate columelliform bundles of branches . . . . 105. **C. doluchanovii** Charadze.

136\. Leaves elongate-spatulate, especially at apex acutely incised-dentate, subglabrous; plant pulvinate, loosely cespitose . . . . 106. **C. fominii** Grossh.

\+ Leaves with short, obovate or broadly spatulate crenate or dentate lamina, sometimes nearly entire, not incised-dentate . . . . 137.

137\. Leaf spatulate, slightly dentate or nearly entire . . . . 107. **C. bellidifolia** Ad.

160 + Leaf broadly spatulate, distinctly dentate . . . . 108. **C. sosnowskyi** Charadze.

138\. Corolla pubescent . . . . 139.

\+ Corolla glabrous . . . . 142.

139\. Rhizome thick, densely cespitose, with clustered thick columelliform 1 cm thick branches; stems arcuately curved, partly prostrate; leaves oval-spatulate, crenate or serrate, subglabrous or gray-pubescent, 0.7–1 cm wide; corolla 3 cm long . . . . 104. **C. meyeriana** Rupr.

\+ Rhizome branches ca. 0.3 cm across; stems erect; leaves not more than 0.3–0.5 cm wide . . . . 140.

140. Branching rhizome with crowded silvery-white remnants of leaf petioles; stems rather densely leafy; cauline and radical leaves spatulate, rounded at apex, entire plant densely covered with soft tangled hairs . . . . 98. **C. radčhensis** Charadze.

\+ Branching rhizome without silvery-white remnants of petioles; remnants of petioles gray or brown; stems with sparse leaves . . 141.

141. Rosetted leaves broadly oblanceolate or spatulate, remotely dentate, glabrous above, shiny; calyx whitish, densely pubescent, appendages as long as or longer than tube . . . . 99. **C. armazica** Charadze.

\+ Rosetted leaves spatulate, irregularly serrate, appressed-hairy on both sides, dull; calyx with fine appressed hairs, appendages twice as long as tube, which they conceal . . . . 100. **C. alpigena** C. Koch.

142. Calyx appendages triangular, acute, without filiform tips; leaves sublinear or linear-lanceolate, obscurely crenate or entire beyond middle; rhizome branching, sometimes with profuse rather long shoots; corolla 3 cm long . . . . .101. **C. saxifraga** M. B.

\+ Calyx appendages gradually acutely attenuate from broad base, with nearly filiform tips, and ciliate margin; leaves narrowly obovate or narrowly spatulate, few-toothed at apex, pubescent; rhizome thick, branching, multicipital; corolla 1–1.3 cm long . . . . 102. **C. ruprechtii** Boiss.

143. Calyx appendages shorter than tube . . . . 144.

\+ Calyx appendages longer than tube . . . . 146.

144. Leaves lanceolate, shiny, with appressed retrorse hairs above, quite glabrous beneath; calyx teeth glabrous outside, white-tomentose inside . . . . 113. **C. besenginica** Fom.

161 + Leaves more or less pubescent above and beneath, more strongly so beneath or subglabrous; corolla with ciliate margin . . . . 145.

145. Radical leaves ovate, the lamina tapering rather abruptly into petiole, subglabrous . . . . 115. **C. chamissonis** Fed.

\+ Radical leaves lanceolate or linear-lanceolate, pubescent; corolla hairy below and along margins . . . . 114. **C. dasyantha** M. B.

146. Radical leaves narrowly linear or narrowly linear-lanceolate; calyx teeth narrowly linear, nearly subulate . . . . 111. **C. ardonensis** Rupr.

\+ Radical leaves with broad oval or spatulate lamina and short or long petiole . . . . 147.

147. Limb of radical leaves oval-rounded, rather abruptly tapering into long nearly filiform petiole . . . . 112. **C. kryophila** Rupr.

\+ Limb of radical leaves spatulate, gradually tapering into rather short petiole . . . . 148.

148. Rhizome thick, with thick branches, densely covered with remnants of petioles; leaves with reflexed teeth in upper part; calyx teeth obtuse; flowers 4 cm long . . . . 109. **C. anomala** Fom.

\+ Rhizome with thin glabrous branches and shoots; leaves finely serrate at apex; calyx teeth acute; flowers 2.5 cm long . . . . 110. **C. circassica** Fom.

Section 1. MEDIUM DC. in Lam. et DC. Fl. Franc. ed. 3, III (1805) 706 s. latiore; A. DC. Monogr. 216. — Sect. Eucodon A. DC. l. c. 214, p. p. —

Capsule dehiscing by basal pores or valves; calyx with more or less strongly developed appendages or with small reflexed folds between teeth, representing reduced appendages, rarely without appendages; all other characters different.

Type of section: lectotype of genus.

Note. Of the two sections of Campanula here accepted one must be based on the type species of the genus (in this case, the lectotype), C. latifolia L. However, A. de Candolle included this species in section Eucodon A. DC and this name should be retained in spite of the new delimitation. Nevertheless, as accepted by us, i. e., also including section Medium DC., Medium should be selected over Eucodon, for reasons of priority even though in de Candolle's view the type species of section Medium was C. medium L., not C. latifolia L.

162 Subsection 1. QUINQUELOCULARES Boiss. Fl. or. III (1875) 893. — Gen. Quinquelocularia C. Koch in Linnaea, XXIII (1850) 63. — Subgen. Quinqueloculares (Boiss.) Charadze in Zam. po sist. i geogr. rast. Tbil. bot. inst. 15 (1949) 16. — Ovary 5-locular; stigmas 5; calyx with reflexed appendages between teeth; capsule dehiscing by valves at base. Monocarpic plants, with high stems, usually with many-flowered inflorescences. Radical leaves entire or irregularly lyrate-incised.

Type species: C. medium L.

Series 1. **Lyratae** Fed. — Rosetted leaves lyrate; petioles lobate-dentate; cauline leaves sessile.

1. C. crispa Lam. Encycl. Meth. I (1783) 581; A. DC. Monogr. 216 idem in DC. Prodr. VII, 458; Boiss. Fl. or III, 895; Trautf. in Tr. Peterb. bot. sada, IV, 164; Lipskii, Fl. Kavk. 337; Fomin in Mater. Fl. Kavk. IV, 6, 19; Grossg., Fl. Kavk. IV, 61; Grossg., Opred. rast. Kavk. 415; Kharadze in Fl. Gruzii, VIII, 132. — Quinquelocularia crispa C. Koch in Linnaea, XXII (1850) 631. — Ic.: Tourn. Relat. d'un Voy. du Lev. II (1717) 378.

Biennial; glabrous or subglabrous plant; stems rather thick, elongate, usually straight, branching; lower leaves long-petioled, smooth, subcordate; cauline leaves acute, cordate, the petioles shorter than lamina; upper leaves subsessile, serrate-incised, crisp; petioles with lobes or teeth. Flowers axillary, solitary, very short-pediceled. Calyx more or less scabrous, with triangular, acute, remotely small-toothed teeth and turned up margins, appendages small, dentiform, one-third the length of the tube; corolla pubescent, broadly campanulate, pale blue, the ovate mucronate lobes half the length of the corolla; capsule semiglobose, flattish above, scabrous; seeds smooth, brown. June–July.

Basalt and andesite rock crevices, riparian ravines. — Caucasus: S. Transc. (Aragats, Mantash, Kara-Kala), E. Transc. (Borzhomi, Akhalkalaki). **Gen. distr.**: E. Anatolia (Kars, Erzerum, Sarykamysh, Ol'ty, Ardagan). Described after a specimen of Tounefort collected between Kars and Erzerum. Type in Paris, cotype in Leningrad.

2. C. lyrata Lam. Encycl. Meth. I (1783) 581; Roem. et Schult. Syst. V, 148; A. DC. Monogr. 219; idem in DC. Prodr. VII, 2, 457; Boiss. Fl. or. III, 899; Trautf. in Tr. Peterb. bot. sada, VI, 1, 57; Lipskii, Fl. Kavk. 377; Fomin in Mater. Fl. Kavk. IV, 6, 19; Grossg., Fl. Kavk. IV, 62; C. stricta auct. non L.: Ldb. Fl. Ross. II, 2 (1845–1846) 877. – Ic.: H. Clifford Crook,
163 Campanulas, 122. – Exs.: Sintenis, Iter or. 4192; Balansa, Pl. d'Orient, 335; Orphanides, Fl. Graeca, No. 656.

Biennial; stems 30–50(70) cm, straight, stiff, slightly virgate, branching, with spreading stiff hairs, sometimes stems few, lateral stems low, partly prostrate; leaves covered with short scabrous hairs; radical and lower cauline leaves with ovate, obtuse or cordate lamina, irregularly crenate, with slightly winged, finely lobate-dentate petiole; upper cauline leaves sessile, ovate-lanceolate or oblong, irregularly and largely serrate-dentate. Flowers erect, axillary, terminal, short-pediceled, blue, 2–3 cm long, in many-flowered spicate-racemiform, sometimes branching inflorescence; teeth of pubescent calyx straight, ovate-lanceolate, slightly acuminate, broad, flat, herbaceous, half the length of the tube, with sparingly pubescent nerves; calyx appendages recurved, broad, ovate, obtuse, nearly as long as teeth, adnate after flowering, reticular, herbaceous; stamens as long as calyx lobes, filaments membranous, broadened below, ciliate; anthers half as long again as filaments; corolla divided into ovate acute nearly straight lobes; style slightly shorter than corolla tube, with 5 stigmas; capsule ovoid, 5-locular, crowned with persistent calyx teeth and surrounded by its post-florally developing appendages, dehiscing by 5 small basal holes; seeds ovoid, very small. June–July.

Mountain slopes, shrubby formations. – Caucasus: possibly near the Turkish border. **Gen. distr.**: Bal.-As. Min. Described from Asia Minor. Type in Paris.

Note. The record for the Caucasus concern a specimen in Ledebour's herbarium, collected by Nordmann in the vicinity of Rize, near the Turkish border but erroneously determined by Ledebour as C. stricta; an error subsequently rectified by Boissier. Ruprecht recommended (on the label) that C. lyrata be excluded from the flora of the USSR, inasmuch as Rize is in Turkey. If we are not following this recommendation it is because we feel that one day it will probably be found in the southern and southwestern parts of Transcaucasia.

Cultivated from 1797 in gardens in stony hills.

Series 2. **Mediae** Fed. – Rosetted leaves none, cauline sessile, ovate or lanceolate.

★C. medium L. Sp. pl. (1753) 236; A. DC. Monogr. 225; idem in DC. Prodr. VII, 2, 460; Ldb. Fl. Ross. II, 875. – C. grandiflora auct. non L.: Lam. et DC. Fl. France (1778) 334. – Marianthemum medium Schur
164 in Verh. Siebenburg. Ver. Naturw. IV (1853) 48. – Sykoraea hortensis Opiz, Seznam (1852) 36. – Ic.: St.-Hil. Pl. de Fr. tab. 72 et icon. auct. alior. mult.; H. Clifford Crook, Campanulas, 133. – Exs.: Baenitz, Herb. Europ. sine numero.

Biennials; root cylindrical, thick, white, branching slightly; stems straight, 1 m, cylindrical, rather thick, longitudinally striate, with spreading stiff hairs, branching in upper part; rosetted leaves none; cauline leaves sessile, ovate-lanceolate, crenate, ca. 7 cm wide; stiff, with stiff, white hairs along nerves, especially profuse beneath. Flowers 7 cm long, erect in bud, then sometimes slightly declined, short-pediceled, in broad many-flowered raceme or paniculate inflorescence; calyx-tube obconical, stiff-haired, with 5 prominent nerves; teeth straight, broad, ovate, acuminate, short-haired along midrib and margins, appendages ovate, reflexed, obtuse, broad, with profusely white erect bristly hairs; corolla three times as long as calyx teeth, poculiform-campanulate, inflated below middle, tapering toward base of lobes, blue, white or pink, with glabrous or slightly pubescent nerves, and short ovate acute recurved lobes; filaments membranous, ciliate; anthers slightly longer than filaments; style cylindrical, nearly as long as corolla, pubescent nearly from top to bottom, the hairs forming 10 longitudinal rows; stigmas 5, filiform, subglabrous beneath, at first erect, later spirally or arcuately curved; capsule pendulous, ovoid, ca. 1.5 cm long, surrounded by adnate calyx appendages, 5-locular or 5-valvular; seeds ovoid, flat, ca. 2 mm long, brown in center, lighter toward margins. June–August.

A garden plant in the USSR. Described from Germany and Italy. Type in London.

Note. Pallas, Güldenstaedt, Beber and Georgi consider this species as native to the Ukraine, Crimea and Ciscaucasia (Terek River); therefore, Ledebour (l. c.) included it in "Flora Rossica," although with the comment that it has not subsequently been observed in the regions indicated. Steven (Bull. Soc. Nat. Mosc. IV (1856) 416) noted that Pallas had mistaken C. medium from the Crimea and Transcaucasia for C. sibirica s. l., i. e. probably C. taurica Juz. and C. elatior Grossh. which often have rather large flowers. The same mistakes were perhaps repeated by others, or else they may have found wild specimens of the garden species C. medium. In floras of Western Europe C. medium is very often cited, but
165 de Candolle (l. c.) had observed that this Campanula, apparently a native of Greece, Italy and France, had escaped from gardens.

C. medium is widely cultivated in the USSR and probably grows wild in places, for which reason it is included here. Many varieties are known, differing in size and color of the corolla from white and blue to pink, double flowers, with two overlapping corollas, as well as flowers with rotate corollas (var. monstrosa hort.), and other ornamental "monstrosities." In the past C. medium was used in folk medicine and seasoned with oil and vinegar, eaten as a salad.

Linnaeus took the epithet "medium" from earlier authors who had used it in describing the bellflower. This epithet translated into Russian is "medium," although the ending of the word does not grammatically fit the generic name Campanula, indicating that the species epithet derives from the former generic name.

Subsection 2. SPINULOSAE (Fom.) Fed. comb. nov. – Sect. Spinulosae (Fom.) Charadze in Zam. po sist. i geogr. rast. Tbil. bot. inst. 15 (1949) 17. – Ser. Spinulosae Fom. in Mater. Fl. Kavk. IV, 6 (1904) 21. – Calyx with appendages in notches between teeth as long as tube; teeth, appendages and leaves with transparent prickles along margins; flowers large, in dense pyramidal paniculate many-flowered inflorescences. Stems sometimes few, branching from base. Rosetted leaves oblong-spatulate, crenate-dentate, with winged petioles; cauline leaves sessile. Capsule dehiscing by 3 small basal pores.

The only species of the subsection is C. mirabilis Alb.

3. C. mirabilis Alb. in Bull. de l'Herb. Boiss. III (1895) 228; ej. Prodr. Fl. Colch. 158; Lipskii, Fl. Kav. 377; Fomin in Mater. Fl. Kavk. IV, 6, 22; Medv. in Tr. Tifl. bot. sada, XVIII, 2, 334; Grossg., Fl. Kavk. IV, 64; Grossg., Opred. 417; Kharadze in Fl. Gruzii, VIII, 135; Kolak., Fl. Abkh. IV, 166. – C. mirabilis var. aureo-marginata Kolak. l. c. 166. – Ic.: Alb. Prodr. Fl. Colch. tab. 1; Kharadze, op. cit. tabl. 370; H. Clifford Crook, Campanulas, 132. – Exs.: Herb. Fl. Cauc. No. 293.

Biennial or perennial; rather sturdy plant, 50–70 cm high, subglabrous, with short, thick, fusiform root; stems pyramidally branching from base, glabrous, 1 cm thick; radical leaves oblong-spatulate, tapering into winged
166 petiole; cauline leaves sessile, ovate-oblong, with cordate base, the upper cordate; all leaves thick, coriaceous, crenate-dentate, bearing translucent prickly cilia, smooth above, scabrous with small papillae beneath. Flowers large, with rather short pedicels, bracts small; teeth of the glabrous calyx broadly lanceolate, obtuse, half the length of the calyx; calyx pale lilac or blue, broadly campanulate, divided for one-third into acuminate oval lobes, furrowed in rictus; calyx appendages triangular-lanceolate, acute, as long as tube, like teeth with prickly translucent marginal cilia; style enclosed in corolla; broadened lower part of filaments finely papillate; receptacle glabrous; capsule broadly ovoid; seeds oblong-elliptic, narrowly winged. July–August (September). (Plate IX, Figure 3.)

Limestone slopes, mountain river ravines, forest belt. – Caucasus: W. Transc. (Abkhaziya, mouth of the Gegi River before falling into the Vzyb', Zhoekvar near Gagry, Arbik Range, Tsikherv ravine, Khipst River). Endemic. Described from Abkhaziya, Type in Geneva, cotype in Tbilisi.

Note. A rare, very beautiful plant, cultivated since 1898. Al'bov has called it "empress of the Abkhazian flora." This bellflower is taxonomically unique and probably represents a Tertiary relict.

In the Leningrad Herbarium there are specimens labeled as collected in the mountains of Karabakh, presumably because of a conscious attempt by the collector to mystify, rather than as a result of a mix-up.

Subsection 3. TRILOCULARES Boiss. Fl. or. III (1875) s. valde restr. – Campanulastra Nym. Consp. Fl. Europ. 476, p. p. – Sect. Sibiricae Charadze in Zam. po sist. i geogr. rast. Tbil. bot. inst. 15 (1949) 24 excl. spec. un. – Ser. Sibiriace Fom. in Mater. Fl. Kavk. IV, 6 (1905) 22, p. p. – Subsect. Intermediae Fom. op. cit. 20. – Gen. Nenningia Opiz, Seznam (1852) 36. – Notches between calyx teeth with long, broad appendages curved downward, appendages and teeth with setaceous ciliate margins. Rosette of radical leaves weakly developed. Stems straight, ascending, rarely prostrate. Perennial of biennial herbs, with hard, partially woody roots.

Type species of subsection: C. sibirica L.

Series 1. **Caucasicae** Charadze in Zam. po sist. i geogr. rast. Tbil. bot. inst. 15 (1949) 24. — Calyx with appendages in notches between teeth half the length of the tube. Stems prostrate or ascending, weak. Plant
167 developing long, rooting or reduced shoots. Corolla small, pale blue. Radical leaves short, obovate or suborbicular or obsolete.

4. **C. caucasica** M. B. Tabl. Prov. Casp. (1798) 112 (diagn. brev.); ej. Beschr. Länder zw. Terek u. Kura Append. (1800) 140 (descript.); ej. Fl. taurc.-cauc. I, 156; A. DC. Monogr., 245; idem in DC. Prodr. VII, 469; Rupr. in Bull. Acad. Sc. Pétersb. XI, 193; Boiss. Fl. or. III, 907; Lipskii, Fl. Kav. 378; Fomin in Mater. Fl. Kav. IV, 6, 39; Medv. in Tr. Tifl. bot. sada, XVIII, 2, 334; Grossg., Fl. Kavk. IV, 65; Grossg., Opred. rast. Kavk. 418. — C. sibirica auct. p. p. non L.; Ldb. Fl. Ross. II, 2, 879. — C. sibirica var. caucasica Trautv. in Tr. Peterb. bot. sada, VI, 1 (1879) 61.

Perennial; glaucous plant, entirely covered with appressed hairs, very rarely subglabrous; fusiform root producing thin, brittle, prostrate branching often rooting, sterile shoots and ascending, 3–4-flowered stem ca. 10 cm; radical leaves obovate or rounded, tapering into petiole, petiole slightly longer than lamina; cauline leaves shortened, short-petioled, all leaves slightly crenate, with densely and crisply white-tomentose margin. Pedicels longer than flowers, bracteolate; flowers 1.2–1.5 cm long; teeth of spreading ciliate calyx triangular, one-quarter the length of the corolla, the latter tubular-obconical, glabrous or with sparingly pubescent nerves; appendages nearly half the length of the calyx teeth; capsule pendulous. July–August.

Stony mountain bluffs. — Caucasus: E. Transc., Dag. Endemic. Described from Azerbaidzhan (Kurt-Bulak). Type in Leningrad.

5. **C. fedorovii** Charadze in Zam. po sist. i geogr. rast. Tbil. bot. inst. 19 (1956) 36. — C. sibirica var. divergens f. corymbosa Fom. in Mat. Fl. Kavk. IV, 6 (1904) 26.

Perennial; stem base slightly woody, with reduced lateral shoots, stems branching from base, flexuous, ascending, with reduced internodes in upper part, 15–40 cm, covered with scattered bristly hairs; lower cauline leaves lanceolate or broadly so, 2.5–3.5 cm long, 5–8 mm wide, tapering at base; median leaves subsessile, slightly amplexicaul, crisp-crenate at margin, attenuate at apex, obtuse, gradually diminishing toward apex of stem, covered above and beneath with scattered bristly hairs, green; branches 3–7-flowered, elongate. Pedicels longer than calyx, declined, deflected in fruit, with bracteoles ciliate at margin; flowers 2.5 cm; teeth of narrow and
168 cuneately tapering calyx elongate-linear-lanceolate, sparse-ciliate at margin, much longer than the narrow and short appendages, one-third as long as tube; corolla glabrous, nearly three times as long as calyx, shallowly divided into acute lobes, tubular-campanulate; capsule elongate, broadening above, abruptly tapering to base, with persistent wilting and markedly declined calyx teeth and appendages. August.

Mountain steppes in subalpine zone. — Caucasus: Dag. (Danukh and Salatau Range). Endemic. Described from Danukh in Dagestan. Type in Tbilisi, isotype and paratype in Leningrad.

N o t e . Well-distinguished from all species affined to C. s i b i r i c a s. l. by the very short and narrow calyx appendages and capsule broadening in upper part and much constricted below.

Series 2. **Komarovianae** Fed. – Ser. Elatae Charadze in Zam. po sist. i geogr. rast. Tbil. bot. inst. 15 (1949) 24, p. min. p. – Calyx appendages broad in notches between teeth, overlapping at margins, completely concealing tube, slightly smaller and shorter than teeth; corolla large, tubular-campanulate, bright violet-blue. Stems many, faintly branching in upper part.

6. C. komarovii Maleev in Izv. Glav. bot. sada. XXIX, 3–4 (1930) 424; Maleev in Zap. Nikitisk. bot. sada, XIII, 2, 163; Grossg., Fl. Kavk. IV, 65 (err. "Comarovi"); Grossg., Opred. rast. Kavk. 418; Kolak in Spiske gerb. fl. SSSR, XII 113. – C. s i b i r i c a var. h o h e n a c k e r i f. l o n g e a p p e n d i c u l a t a h i s p i d a Fomin in schedis, non C. h o h e n a c k e r i Fisch. es Mey. – Ic.: Maleev, op. cit. (1930) 425 (photo). – Exs.: GRF, No. 3698.

Perennial; root ligneous; stems usually many, ascending, 15–20(45) cm, woody at base, covered with stiff white unequal hairs declined or recurved below; leaves 2–4 cm long, 0.5–1 cm wide, lower leaves oblong-spatulate, crenate at margin, with lamina decurrent in narrowly winged petiole; upper cauline leaves and bracts lanceolate, crenate, sessile, all scabrous with white bristles. Inflorescence more or less compressed; pedicels 1–2-flowered, the lower 2–3 times as long as flowers, the upper equal to or shorter; flowers 3–4 cm; calyx covered with very stiff white and more or less spreading bristles, the teeth broadly lanceolate or more often triangular-ovate, approximate, the appendages slightly shorter than teeth, bristly-stiff-hairy, ovate or triangular-ovate, recurved below, contiguous with each other; corolla bright violet-blue, pubescent outside along nerves, with acute more or less deflexed lobes, 4–5 times as long as calyx; style hidden in corolla or barely protruding. June. (Plate X, Figure 3.)

169 Dry limestone cliffs, in Savin Juniper open woodland. – Caucasus: W. Transc. (extreme west). Endemic. Described from the Black Sea coast, between Novorossiisk and Gelendzhik. Type in the herbarium of the Nikitskii Botanical Garden in the Crimea, cotype in Leningrad.

N o t e . This handsome and unique plant is narrowly endemic, growing only between Novorossiisk and Gelendzhik, where it stretches from the sea along the mountain slopes up to the summit of Markhot Range. Before Maleev's discovery there were only very poor specimens in the older herbaria which had been collected near Novorossiisk by Litvinov and Lipskii. One of them (cited above) bears an inscription by Fomin who, because of the poor state of preservation of the older herbaria, could not have been ingenious enough to surmise that a completely specific species were to be found in the Novorossiisk area. In vain did Maleev belittle the value of his discovery when he partly (p. p.) identified his species with var. d i v e r g e n s Trautfetter (and Fomin). In the list of habitats for var. d i v e r g e n s Fomin there is mention en passant of a specimen collected by Medved in the area of Novorossiisk. Hence we cannot identify C. k o m a r o v i i with any of the earlier described and published varieties of C. s i b i r i c a. Taxonomically, C. k o m a r o v i i can only be related to the

Dagestan endemic C. daghestanica Fomin, as Maleev himself believed. Stankov's error in reporting C. komarovii for the Crimea has been correctly noted by Yuzepchuk (Mater. gerb. Bot. inst. AN SSSR, XIV, 1951, p. 39).

7. C. daghestanica Fom. in Mater. Fl. Kavk. IV, 6 (1904) 29; Grossg., Fl. Kavk. IV, 65; Grossg., Opred. rast. Kavk. 418; Fler., Spisok rast. Sev. Kavk. i Dag. 534.

Perennial; glaucous plant, particularly the calyx; rhizome woody; stems ascending, simple or branching, flexuous, with crisp and appressed hairs; radical leaves spatulate; cauline leaves oblong-lanceolate, attenuate into petiole, upper leaves sessile; all leaves rather densely covered with crisp and appressed hairs, obtusely crenate and sparingly pubescent at margin. Flowers medium-sized, with slender pedicels furnished with narrowly lanceolate bracteoles, one-fourth to one-third as long as corolla; teeth of the glaucous and glabrous calyx triangular-lanceolate, obtuse, one-fourth to one-third as long as the corolla, the latter blue, tubular, glabrous or slightly pubescent along nerves; calyx appendages obovate, concave, nearly as long as teeth, like teeth spreading-pectinate-ciliate, cilia with hamately curved apex; style hidden inside corolla; receptacle glabrous. May—July.

Limestone and pebbly slopes in montane and submontane belts. — Caucasus: Dag., E. Transc. Endemic. Described from Chirkei near the north-
170 western border of Dagestan. Type in Leningrad.

Series 3. **Longistylosae** Fed. — Ser. Elatae Charadze in Zam. po sist. i geogr. rast. Tbil. bot. inst. 15 (1949) 24, p. p. — Calyx appendages distinctly shorter than teeth; corolla large, campanulate or infundibular-campanulate, much longer than calyx, sometimes with constriction in lower part; style usually exserted; inflorescence paniculate. Plants with high stiff-haired stems.

8. C. schischkinii Kolak. et Sachok. in Spiske rast. Gerb. fl. SSSR, XII (1953) 113. — Exs.: GRF, No. 3699.

Perennial; stems sturdy, solitary, densely covered with short stiff hairs, 50—120 cm; lower leaves oblong-lanceolate, gradually tapering into petiole, irregularly serrate-dentate at margins, (3)5—10 cm long, 5–12 mm wide; upper leaves small, semiamplexicaul at base, 2—3 cm. Inflorescence broadly paniculate, with declined or arcuate branches, 5—20 cm; pedicels 5—12 mm; calyx teeth triangular, 7—12 mm long, 2—4 mm wide, bristly at margin, with ovate or obovate recurved appendages expanding after flowering and bristly at margin; corolla blue-violet, infundibular or campanulate-infundibular, (2.5)3—4(5) cm long, cut for one-third into ovate-triangular, short-acuminate lobes sparingly pubescent at margin; style 3-fid, exceeding corolla after flowering. June.

Pebbly slopes of forest belt. — Caucasus: W. Transc. (extreme west). Endemic. Described from near Tuapse, between Ol'ginsk and Dzhubga. The type is an exsiccate in Leningrad, the rest of the set are isotypes.

Note. This species is closest to C. longistyla Fomin, with which it has contiguous distribution areas. It is slightly reminiscent of C. komarovii

(171)

PLATE IX. 1 – Campanula punctata Lam.; 2 – C. alpina Jacq.; 3 – C. mirabilis Alb.

Maleev and other species, related to C. sibirica s.l. In spite of Kolakovskii's views on the matter (op. cit.) it is not related to C. imeretina Rupr. Judging by Fl. Abkhazii, he seems to have taken for C. imeretina a quite different plant, related to C. sibirica s. l., from Abkhaziya where C. imeretina has not thus far been recorded (see Note to C. imeretina).

9. C. longistyla Fom. in Tr. Tifl. bot. sada, VI, 3 (1904) 37; "longestyla"; Grossg. Opred. rast. Kavk. 417; Kolak. Fl. Abkh. IV, 166; Kharadze in
173 Fl.. Gruzii, VIII, 163; Fomin in Mater. Fl. Kavk. IV, 6, 20 quoad pl. Colch.; Medv. in Tr. Tifl. bot. sada, XVIII, 2, 334; Grossg., Fl. Kavk. IV, 64, p. p. — C. sibirica var. majo Albov Prodr. Fl. Colch. (1895) 154, non Boiss. — Ic.: Kharadze, op. cit., tabl. 374; H. Clifford Crook, Campanulas, 121 (photo). — Exs.: GRF, No. 3241; Fl. Cauc. exs. No. 275.

Perennial; pubescence of sparse stiff hairs; stems rather thick, 50 cm, 3—5 mm across, straight and ascending, pyramidally branching; lower leaves oblong, sessile, tapering into broad, to 10 cm long petiole; upper leaves oblong-lanceolate, irregularly orbicular-denticulate. Flowers violet, 3—4 cm long, drooping, with pedicels half the length of the corolla, sometimes even much shorter; calyx teeth reflexed, lanceolate, acute, one-fifth to one-fourth the length of the corolla; corolla urceolate-campanulate, glabrous outside, more or less barbate in rictus, with inflated base, constricted above middle; calyx appendages strongly recurved toward pedicels, ovate-triangular, obtuse, with curved margins, these and teeth ciliate; style exserted, with 3, rarely 4—5 stigmas; broadened part of filaments chartaceous; receptacle glabrous. May—July.

Limestone cliffs in middle and subalpine mountain zones. — Caucasus: W. Transc. (Abkhaziya and the Tuapse-Sochi District). Endemic. Described from the mountain pass between Novomikhailovsk and Olginsk, near Tuapse. Type in Leningrad.

Note. This species was reported by its author for W. and E. Transcaucasia, and by Grossgeim (Fl. Kavk.) for Dagestan. Subsequently Grossgeim described a bellflower from the eastern part of the Main Range, accepted as C. longistyla, as C. charadzae Grossh., thus reducing C. longistyla Fom. to its original true limits. According to Fomin (op. cit.) this "very handsome plant" was successfully grown for a long time in the Tiflis (now Tbilisi) Botanical Garden, where it is probably preserved to this day; in England it has been cultivated since 1907 (see: Clifford Crook, l. c.).

10. C. charadzae Grossh., in Dokl. AN Armyansk SSR, VII (1947) 168; Grossg., Opred. rast. Kavk. 418; Kharadze in Fl. Gruzii, VIII, 154. — C. longistyla Fom. f. parviflora Fom. in Mater. Fl. Kavk. IV, 6, (1904) 21.

Perennial; plant covered with rather sparse stiff hairs; stems 30—40 cm, comparatively thin, 1—1.5(3) mm thick, straight or ascending, usually with few branches; lower leaves oblong-spatulate, the upper oblong-lanceolate, finely rounded, crenate-serrate. Flowers on long, slender pedicels, violet, 2 cm long, rather small, drooping; calyx teeth declined
174 narrowly lanceolate, acute, not more than one-third the length of the

urceolate-infundibular corolla, the latter glabrous outside, bearded in rictus; calyx appendages markedly recurved, like teeth ciliate, obtuse; style prominently exserted. May–August.

Subalpine meadows. – Caucasus: Dag., E. Transc. Endemic. Described from Tushetia (Dzhvari-Vozeli). Type in Leningrad.

Note. This species is a vicariant of the preceding. In addition to its morphology C. charadzae is also distinguished from C. longistyla by its pollen grains, which are half the size of those of C. longistyla. It is also distinguished by its habitat, meadows of the subalpine belt, not rocks.

Series 4. **Elatae** Charadze in Zam. po sist. i geogr. rast. Tbil. bot. inst., 15 (1949) 24, s. restr. – Calyx appendages broadly lanceolate, nearly as long as or longer than tube; corolla large, tubular-campanulate, violet-blue. Inflorescence few-branched, sometimes densely so or flowers solitary. Stems tall, thick, usually solitary, rarely reduced, low. Radical leaves usually large, broadly spatulate.

11. C. charkeviczii Fed. sp. nov. in Addenda XXIII, p. 327.

Perennial; multicaulescent plant, with thick indurate vertical or oblique root, 1.5–2 cm across, with rather long grayish-brown remnants of dead petioles at root neck; stems 20–40 cm long, 0.5–0.7 cm thick, ascending from arcuate base, becoming erect, densely leafy, scabrous from short and dense simple hairs, usually slightly purple and longitudinally striated; lamina of rosetted leaves obovate-spatulate or elliptic, 10–12 cm long, 3–4 cm wide, gradually and cuneately decurrent on winged petioles, nearly as long as lamina; leaves conspicuously, unequally or indistinctly crenate, obtuse, covered above and beneath with scabrous hairs or setulose, often purple; lower cauline leaves similar in shape to the radical but smaller; upper leaves narrower and subsessile. Flowers in dense, slightly branching raceme, ca. 3 cm long, bluish-violet, erect or drooping, short-pediceled; lobes of obconical calyx triangular-lanceolate, acute, bristly-ciliate, one-third to one-half the length of the tubular-campanulate corolla, appendages reflexed, lanceolate, setulosely ciliate, concealing tube, corolla glabrous or
175 finely papillate, divided for one-third into ovate acute lobes; style not exserted. June.

In rock-crevices, subalpine mountain belt. – European part: Crim. (Baidar Yaila). Described from locality indicated. Type in Leningrad.

Note. C. charkeviczii differs from other species of the subsection Triloculares, by its large, coarse, thick and rather short stems, dense inflorescences, very large rosetted leaves and the dense, scabrous and slightly bristly pubescence of all parts, with the exception of the corolla. It is also notable for the very thick roots and the abundant brown, rather long, remnants of petioles at the root neck. The most closely related species is the (not the mountain) lowland species of Hungary and neighboring countries.

12. C. elatior (Fom.) Grossg., Opred. rast. Kavk. (1949) 418, absque diagn., descr. nostra in Addenda XXII, 460; Kharadze in Fl. Gruzii, VIII, 160 descr. georgica compl. – C. sibirica f. elatior Fomin in Mater. Fl. Kavk. IV, 6 (1904) 23.

Biennial; root fusiform, indurate, light brown; stems solitary, rather thick, straight, simple or slightly branching in upper part; sulcate-angular, 60 cm high, 4, rarely 5 mm thick, sometimes rather densely covered with short, slightly setulose scabrous hairs, especially in lower part; radical and lower cauline leaves elliptic-spatulate, obtuse, gradually tapering into broad winged petioles, sometimes with small obsolete lobes or rounded teeth, irregularly sometimes bicrenate, with petiole about the length of the leaf, 8–10 cm long, 2 cm wide; median and upper cauline leaves narrower, small, narrowly oblong-lanceolate or sublinear, subsessile, semiamplexicaul, usually obtuse. Inflorescence narrow, racemiform, often somewhat paniculate; flowers 2.5–3 cm long, short-pediceled, bluish-violet, declined; teeth of the very short calyx narrowly lanceolate, acute, like entire calyx bristly-ciliate, usually only one-third to one-half the length of the narrow, tubular, infundibular, subglabrous or very sparingly pubescent corolla, the latter divided for one-fourth into acute, often recurved lobes; calyx appendages obtusely lanceolate, one-third to one-half the length of the calyx lobes; style not exserted. May–July.

Meadows, forest edges, herbaceous edges of fields, submontane plains,
176 sometimes in mountains. – European part: Bl., L. Don (S.); Caucasus: Cisc., Dag., E. Transc. (Kazbek). Endemic. Described from Kropotkin (Kavkazskaya Stanitsa). Type and paratype in Leningrad.

Note. This species is closest to C. sibirica L. s. str., from the North and East. In the Caucasus it grows together with C. taurica, from which it differs by a solitary stem, an elongated inflorescence, much longer calyx teeth, nearly half the length of the large corolla. It is not found with C. taurica in the mountains of the Crimea. From the north-eastern race, C. sibirica s. str., it differs by sparser and shorter hairs and by the absence of prominent netted veins in the calyx appendages.

13. C. talievii Juz. in Bot. mat. Gerb. Bot. inst. AN SSSR, XIV (1951) 39. – C. sibirica auct. non L.; Taliev in Tr. Mosk. Obshch. ispyt. prir. XLII (1908) 80. – C. sibirica var. divergens Popl., Spisok rast. Krymsk. zapov. (1931) 92, non Trautv. – C. sibirica var. hohenackeri Fomin in schedis, non Trautv. – C. komarovii Stankov in Stankov and Taliev, Opred. rast. (1949) 747, p. p. non Maleev.

Perennial; root woody, vertical; usually low-growing plant, 2.5–20 cm high, rarely solitary, rather thick stem higher, usually ascending, arcuately curved; rosetted leaves spatulate, 2–7 cm long, 0,7–2 cm wide, both sides densely covered with stiff hairs, markedly scabrous; cauline leaves oblong-lanceolate, obtuse, sessile. Inflorescence usually few-flowered, sometimes flowers solitary; flowers 2–3.5 cm long, tubular-campanulate, rather narrow, dark violet; calyx teeth rather broadly lanceolate, with broadly ovate rounded appendages not quite contiguous, calyx one-fourth to one-third the length of the corolla; corolla glabrous; style not exserted. June–July.

Dry, mostly stony habitats in the subalpine zone. – European part: Crim. (mountains of Crimean Yaila: Roman-Kosh, Ai-Petri, Ekliza-Burun, Demerdzhi). Endemic. Described from Roman-Kosh Mountain. Type in Leningrad.

Note. According to Yuzepchuk (op. cit.) this distinct Yaila (subalpine) species is related to C. sibirica s. l., only qualitatively different from

C. taurica, also from the Crimea, but in the forest belt. Stankov (op. cit.) has incorrectly identified it with the W. Caucasian endemic C. komarovii Maleev, to which it bears no relation. Neither can it be identified with
177 C. sibirica var. divergens Trautv., as suggested by Popavskaya, op. cit. and Taliev, op. cit.

Series 5. **Sibiricae** Fom. in Mater. Fl. Kavk. IV, 6 (1905) 22, excl. sp. nonnull. – Ser. hohenackeri Charadze in Zam. po sist. i geogr. rast. Tbil. bot. inst. 17 (1953) 104, p. p. – Ser. Hemschinicae Charadze, op. cit. 15 (1949) 24, nom. impropr. – Calyx appendages lanceolate, shorter or longer than tube but not concealing it; corolla small, infundibular, or tubular-campanulate, pale, rarely large but if so then bright blue and broadly campanulate; inflorescence prominently branching, loose. Stems rather thin, usually numerous, rarely solitary.

14. C. sibirica L. Sp. pl. (1753) 236; Roem. et Schult. Syst. V, 145; A. DC. A. DC. Monogr. 244, p. p.; idem in DC. Prodr. VII, 465, p. p.; Ldb. Fl. Ross. II, 2, 879, p. p.; Trautf. in Tr. Peterb. bot. sada, VI, 1, 61, promin. p.; Shmal'g., Fl. II, 178, p. p.; Kryl., Fl. Zap. Sib. XI, 2632; Korsh. Tent. Fl. ross. or. 270; Fedch. and Fler., Fl. Evrop. Ross. 937; 0. and B. Fedch., Perech. rast. Turk. III, 356; Szaf., Kulcz., Pawl. Rosl. Polskie, 641; Pavl. Fl. Tsentr. Kazakhst. III, 201. – C. racemosa Gmel. Reise, I (1774) 153. – C. sibirica var. typica Trautv. in Tr. Peterb. bot. sada, V, 2 (1878) 453, p. p., VI, 1 (1879) 61. – C. rapunculoides var. neglecta Rgl. in Bull. Soc. Nat. Mosc. XXVI (1853) 674, p. p. non C. rapunculoides L. nec var. neglecta Ldb. – C. rapunculoides auct. non L.; Trautf. in Tr. Peterb. bot. sada, VI, 1 (1879) 75, quoad pl. ex Bukoni (pr. 1. Saissan). – Marianthemum sibiricum Schur in Verh. Naturw. IV (1853) 48. – Nenningia paniculata Opiz, Seznam (1852) 68. – Ic.: Gmel. Fl. Sibir. III (1768) tab. 29 (manca); ej. Reise, tab. 34 (mala); Cyreishch., Ill. Fl. Mosk. gub. III, 229. – Exs.: GRF, No. 422; Tarachkov and Poganko, Gerb. Orlovsk. gub. No. 214; Fl. Gall. et Germ. exs. No. 696.

Biennial; root thin, fusiform, branching, sometimes rather thick, indurate; entire plant covered with spreading stiff hairs; stems (15)20–60 cm, straight, ribbed, more or less branching, with slightly declined branches, rarely simple; leaves faintly orbicular-dentate with sparingly pubescent margins, lower leaves petioled, elliptic or oblong, sometimes narrowly spatulate, obtuse, 3–10 cm long, 0.5–2 cm wide; upper leaves nearly entire or faintly crenate, sessile, sometimes slightly amplexicaul, lanceolate or linear-lanceolate, acuminate, becoming narrower and shorter toward stem tip. Flowers 15–20 cm [?] long in narrow paniculate inflorescence, pendulous or drooping, bracteate, short-pediceled; teeth of stiff-haired bristly-ciliate
178 calyx linear-lanceolate or lanceolate, one-fourth to one-third the length of the bluish-lilac, narrowly campanulate, sometimes nearly white, corolla, the latter sparingly pubescent outside along nerves, divided for one-third into ovate acute lobes, with glabrous margins, calyx appendages reflexed, rather broad, oblong, like rest of calyx with stiff bristles and cilia; capsule grayish brown, dehiscing at base; seeds oblong, shiny, grayish-brown or yellowish, often slightly angular, 1 mm long. May–July.

Dry forests, grass steppes, birch outliers, fallows, herbaceous edges of fields, precipitous riverbanks, sometimes as a weed in crops. — European part: Lad.-Ilm., U. V., V.-Kama (S.), U. Dns., U. Dnp.; W. Siberia: throughout the region, but in the east only slightly beyond Yenisei; Centr. Asia: Dzu-Tarb., Ar.-Casp. (N.), Balkh. (N.). **Gen. distr.:** Centr. Eur. Described from Siberia. Type in London.

Note. It appears that the single species, i. e. C. sibirica s. str. is distributed over the entire Russian Plain as well as Western Siberia and Central Europe. A multitude of local races, which grow in separate areas and are peculiar to the upper and middle mountain zones as well as to the alpine zone, are encountered in the Caucasus, Crimea, N. Iran, Asia Minor and in the Mediterranean area and Eastern Europe. In the Caucasus there grow several distinct species. The closest relatives of true C. sibirica are two rather similar species — C. taurica and C. elatior. The latter grows in the extreme south of the plain of the European part of the USSR and in Ciscaucasia.

15. C. taurica Juz. in Bot. mat. Gerb. Bot. inst. AN SSSR, XIV (1951) 36. — C. sibirica var. divergens Trautv. in Tr. Peterb. bot. sada, IV, 1 (1876) 164; Fomin in Mater. Fl. Kavk. IV, 6, 24; Grossg., Fl. Kavk. IV, 65, pro min. parte, non C. divergens Willd. — C. sibirica var. hohenackeri f. umbrosa Fomin in schedis, non C. hohenackeri Fisch. et Mey. — C. sibirica var. taurica Trautv. l. c. — C. komarovii auct. p. p. non Maleev: Stankov in Stankov and Taliev, Opred. rast. (1949) 747, p. p. — C. divergens auct. pro max. parte, non Willd.: Grossg., Opred. rast. Kavk. (1949) 418. — C. ciscaucasia Charadze in Zam. po sist. i geogr. rast. Tbil. bot. inst. 17 (1953) 100; Charadze in Fl. Gruzii, VIII, 161, descr. georgica. — Ic.: Kharadze, op. cit. (1953) tabl. 1 (mediocr.).

Biennial or perennial; plant (15)25–45(50) cm high; stems solitary, often few; main stem straight, usually twice as thick as the others (3 mm); stems cylindrical, indurate in lower part, not sulcate, slightly sulcate above, covered with stiff retrorse whitish hairs, in upper part with spreading elongate branches forming loose paniculate inflorescence; rosetted leaves spatulate, rounded at apex, crenate-dentate, like cauline leaves covered with scattered or rather densely spreading curved or declined hairs, soon drying
179 up; cauline leaves remote, the lower elongate-obovate or sublanceolate, obtuse, tapering to petiole, the upper lanceolate or narrowly lanceolate, acute, sessile, slightly amplexicaul, acutely serrate. Flowers 1–3 on each branch of inflorescence, horizontally spreading, (1.5)2–2.5(3) cm long, elongate-tubular-campanulate, violet-blue; calyx several times shorter than corolla, with lanceolate teeth and narrow ovate appendages covered, like calyx lobes, with spreading bristly hairs. June–July.

Dry plots in leafy forests. — European part: Crim. (S.); Caucasus: Cisc. (south to Kislovodsk). Endemic. Described from the Crimea (Uspenskii Monastery near Bakhchisarai). Type in Leningrad.

Note. This species does not differ from C. ciscaucasica described below by Charadze from N. Caucasus (Klukhori) and, as correctly stated by Yuzepchuk (op.cit.), is found in Ciscaucasia, from Taman' to Klukhori and Mineral'nye Vody, where it admixes with C. elatior.

16. C. hohenackeri Fisch. et Mey. in Suppl. ad Ind. sem. hort. Petrop. IX (1844) 9; C. A. M. in Beitr. z. Pflanzenr. d. Russ. Reichs, VI, 24; Rupr. in Bull. Acad. Sc. Pétersb. XI, 217; Grossg., Fl. Kavk. IV, 65; Kharadze in Fl. Gruzii, VIII, 162. – C. sibirica var. major Boiss. Fl. or. III (1875) 901, p. p.; Somm. et Lev. Enum. pl. Cauc. 314. – C. sibirica var. hohenackeri Trautv. in Tr. Peterb. bot. sada, II, 2 (1873) 564, VI, 1 (1879) 62, p. p. – C. sibirica var. hohenackeri Fomin in Mater. Fl. Kavk. IV, 6 (1904) 26 et f. parviflora Fomin, l. c. 27, non var. hohenackeri Somm. et Lev. l. c. – C. sibirica var. multicaulis Rupr. in sched. – Exs.: Herb. Fl. Cauc. No. 292.

Biennial; root rather thick, light brown, indurate, fusiform; stems many, 15–45 cm, rather thin, usually reddish, longitudinally striated, slightly ribbed, like entire plant covered with short scabrous hairs, usually not long, straight at center, arcuately ascending at sides, branching in upper part and sometimes at base, terminating in many-flowered paniculate inflorescence with spreading branches; lower leaves oblanceolate or broad and nearly oblong-ovate, sometimes slightly spatulate, usually acute, somewhat crenate, tapering gradually into rather distinct, not long petiole; cauline leaves becoming gradually fewer toward tip of stem, sessile, lanceolate, linear-lanceolate,
180 crenate-serrate, sparingly pubescent along margins. Flowers 1.5, rarely 2 cm long, usually much shorter, pale violet-blue; teeth of short obconical calyx suberect, lanceolate, one-fourth to one-third the length of the tubular-campanulate glabrous corolla, appendages broadly lanceolate, obtuse, like lobes covered with coarse sparse bristly curved cilia especially at margins, markedly netted with prominent nerves, corolla divided for one-fourth into acute lobes; style as long as or slightly exserted from corolla. June–July. (Plate X, Figure 1).

Dry rocky places in supalpine mountain zone. – Caucasus: Cisc. (Main Range), Dag., W. Transc. (E.), S. Transc., E. Transc. **Gen. distr.**: As. Min., Iran (N.). Described from Kobi, Georgian SSR. Type in Leningrad.

Note. This species has long been known for its hardiness. Ruprecht (l. c.) has observed that under cultivation in St. Petersburg it did not transform into typical C. sibirica. According to him, the many-stemmed C. hohenackeri resembled C. sibirica, a subalpine plant growing among rocks. The numerous stems, short pubescence, small and numerous flowers in markedly branching inflorescences, the low-growing habit, shape of the leaves and part of the calyx serve to distinguish this species from C. sibirica, C. elatior and C. taurica, which it replaces in the Caucasus. C. hohenackeri somewhat resembles C. darialica, but the flowers of the latter are much larger, the calyx appendages are not reticular, the pubescence is spreading and the leaves have a different shape.

After being grazed by cattle, this species, like many other bellflowers, grows again, producing thin branches late in the fall, and blossoming profusely with smaller flowers.

17. C. darialica Charadze in Zam. po sist. i geogr. rast. Tbil. bot. inst. 17 (1953) 104; Charadze in Fl. Gruzii, VIII (1952) 160, descr. georg. – C. sibirica var. hohenackeri Somm. et Lev. Enum. pl. Cauc. (1900) 314, non Trautv. nec Fomin, neque C. hohenackeri Fisch. et Mey. – Ic.: Kharadze, op. cit. (1953) tabl. 2 (mediocr.).

Biennial; stems 20 cm, not thick, spreading-branching from base, branches converging toward stem and reaching the same level but diverging laterally, often even prostrate, but then again ascending and branching at apex, covered with spreading stiff white hairs; lower leaves broadly lanceolate or elongate-ovate, obtuse or acute, short-petioled, crenate-serrate; upper leaves smaller and narrower, sessile, semiamplexicaul. Flowers few, 2–2.8 cm long, violet, drooping or declined with long, slender pedicels;
181 calyx teeth rather long, acutely lanceolate, erect, one-fourth to one-third the length of the infundibular, glabrous, hairy-nerved corolla; appendages spreading, not reticular, like lobes with bristly white cilia along margins, corolla divided for one-fourth into acute teeth; style not exserted; capsule obconical, brown; seeds brown, small. July.

Dry pebbly slopes in montane belt. – Caucasus: E. Transc. (C. Caucasus). Endemic. Described from the Kazbek district near Chmi, Georgian SSR. Type in Tbilisi, topotype in Leningrad.

Note. Sommier and Levier presumably described this species (from Dar'yal Range between Lars and Kazbek), erroneously referring it to the previously established var. hohenackeri (from C. sibirica). C. darialica differs from C. hohenackeri by the quite glabrous, nonreticular calyx appendages, large corolla, spreading-haired stem and very long pedicels as well as by the solitary stem, branching strongly from base, and partly prostrate branches. Sommier and Levier grew their plant in Florence, but judging by their description of the cultivated specimens it does not seem to have changed its characters. Kharadze (op. cit.) brings this species into relation with C. daghestanica Fom., but in our view this is a mistake. Rather, this species is one of those closely related to the common C. sibirica.

18. C. brassicifolia Somm. et Lev. in Nuov. Giorn. Bot. Ital. (1895) 94; Somm. et Lev. in Tr. Peterb. bot. sada, XVI (1900) 320; Lipskii, Fl. Kavk. 378, Dopoln. I, 69. – C. imeretina auct. p. p. non Rupr.: Fomin in Mater. Fl. Kavk. IV, 6 (1904) 29, p. p.; Kharadze in Fl. Gruzii, VIII, 162. – C. sibirica var. imeretina Trautv. in Tr. Peterb. bot. sada, IV, 2 (1876), also VI, 1 (1879) 62, non C. imeretina Rupr. – Ic.: Somm. et Lev. l. c. (1900) tab. XXX (optime delin.).

Perennial; green, glaucescent, subglabrous plant; root thick, stems numerous, many-flowered, ascending or inclined or leafy, sterile; basal leaves in rosette numerous, large, chartaceous when dry, indurate, long-petioled; lamina 5–10 cm long, ovate-oblong, obtuse, obtusely crenate or bicrenate, with short-ciliate margin, at base crenate-lobate, tapering and decurrent along petiole, lamina completely glabrous or slightly scabrous, short haired above, with white nerves; petiole as long as or half the length of the lamina, more or less narrowly winged and ciliate for entire length;
182 stems with strict elongate branches nearly from base, leafy, sometimes flexuous and spreading, furcate, covered with sparse short downward-turned hairs; cauline leaves much smaller than in rosette, the lower tapering to petiole as long as lamina; leaves on branches small, elliptic, oblong, subsessile, in shape similar to bracts; branches thin, numerous, forming several corymbiform panicles. Flowers elongate, thin, spreading pubescent or glabrous; teeth of calyx coarse, long-ciliate accrescent, otherwise

subglabrous; calyx triangular-lanceolate, nearly three-fourths the length of the blue corolla, the latter tubular-campanulate; glabrous outside, usually sparsely hairy inside, incised for one-fourth or one-third its length, appendage triangular, acute or ovate, one-third to one-half the length of the calyx teeth but at flowering as long as or slightly longer than tube; stigmas 3, becoming hamately curved, slightly exserted from corolla or barely as long; capsule nearly semiglobose, 3-locular, dehiscing by small holes at base; seeds flattened, slightly rimmed. July.

Forest belt, slopes of riverbanks. — Caucasus: W. Transc. Endemic. Described from right bank of Rion River in Kutaisi. Type in Florence.

Note. This unjustly neglected species, formerly incorrectly considered synonymous with C. imeretina Rupr. (see Note to C. imeretina), has here been restored. In fact, it has nothing in common with C. imeretina. It seems that most authors found themselves hypnotized by the new species of Sommier and Levier. Thus, Lipskii commented (op. cit.): "Certe nil aliud est nisi C. imeretina Rupr." Fomin (op. cit.) also lumped these contrasting species, C. imeretina and C. brassicifolia, together. To be sure, Sommier and Levier themselves are partly to blame for this, as they specifically compared their new species with C. imeretina, obviously only because C. brassicifolia was described from the same locality as C. imeretina (Kutaisi). A comparison of C. brassifi-folia with C. imeretina, C. betulifolia and even C. autraniana, such as was made by the above-named authors, is quite useless, as these species belong to the most varied groups. In fact, C. brassicifolia is only related to C. sibirica s.l. and especially to C. hohenackeri Fisch. et Mey.

19. C. schelkownikowii Grossg., Opred. rast. Kavk. (1949) 418; absque diagn.; descr. nostra in Addenda XXIII, p. 328.

Biennial; root fusiform, brown, sturdy; stems numerous, partly straight,
183 partly prostrate or arcuately ascending, 25—30 cm long, 1—2 mm thick, slightly branching in upper part, leafy, covered with short white scabrous hairs; radical and lower cauline leaves oblanceolate, obtuse, acute, obsoletely crenate-serrate, scabrous, 5 cm long, 0.8 cm wide, gradually tapering into a narrowly winged petiole; upper and median leaves diminishing in size, sessile, similar to lower leaves or lanceolate, with rounded base. Flowers 2.5—3 cm long, pale, violet-blue, with narrow filiform pedicels, drooping; teeth of very short calyx linear-lanceolate, acute, sometimes nearly subulate-linear, ciliate, shorter than broadly campanulate, glabrous corolla, the latter divided for one-fourth, its lobes acute, appendages narrow, obtuse, as long as tube but one-third to one-half as long as lobes, ciliate; style not exserted. June.

Dry mountain slopes, to upper zone. — Caucasus: S. Transc. (Karabakh). Endemic. Described from Ziarat Mountain in Azerbaidzhan. Type in Leningrad.

Note. This species differs from the relatives of C. sibirica, in its broadly campanulate corolla, very narrow, nearly subulate calyx teeth, very slender, nearly filiform pedicels and inflorescent branches. A very beautiful plant because of its abundant large flowers; it seems to be suitable for parks and gardens in stony hills.

Subsection 4. PHASIDIANTHE Fed. subsect. nov in Addenda XXIII, p. 329. – Subsect. Triloculares ser. rupestres Boiss. Fl. or. III (1875) 901, p. p. – Sect. Sibiricae Charadze in Zam. po sist. i geogr. rast. Tbil. bot. inst. 15 (1949) 24, p. p. – Ser. Sibiricae Fom. in Mater. Fl. Kavk. IV, 6 (1904) 22, p. p. – Calyx with appendages in notches between teeth; flowers axillary, opaque, corolla small, only twice as long as calyx, blue, glabrous. Rosetted leaves apparently absent, cauline leaves short-petioled or sessile, subrhombic. Stems furcate, borne on tip of rhizome, which grows in rock crevices. Endemic in the Caucasus.

Only species of subsection: C. imeretina Rupr.

This subsection is intermediate between subsections Triloculares and Annuae. Ruprecht, who recorded C. imeretina for the Caucasus, pointed out its affinity with the Mediterranean C. dichotoma L. and the Abyssinian C. rigidipila Hochst. et Steud. var. heterophylla. However, in spite of this, C. imeretina should be included in a separate subsection endemic to the Caucasus, as here described (see also Addenda).

184 **20. C. imeretina** Rupr. in Bull. Acad. Sc. Pétersb. XI (1867) 217; Boiss. Fl. or. III, 908; Lipskii, Fl. Kavk. 378; Fomin in Tr. Tifl. bot. sada, VI, 2, 9; Fomin in Mater. Fl. Kavk. IV, 6, 29, quoad specim. Rupr.: Grossg., Fl. Kavk. IV, 65, p. p. (calami lapsu: "imeretica"), Kharadze in Fl. Gruzii, VIII, 162, pro minima parte.

Perennial; rhizome ligneous, growing in rock crevices, with numerous thin flexuous furcate, 1 mm thick stems, 15–20 cm long, covered with short, spreading stiff hairs; rosettes apparently absent; lower cauline leaves broadly ovate or obovate-spatulate, slightly rhombic, tapering to winged petiole, as long as or much shorter than lamina; upper leaves sessile, broadly elliptic or ovate, often subrhombic, faintly crenate, all leaves 1–3 cm long or less, thin, covered with short stiff hairs; flowers solitary, the lower, sometimes also the upper, opposite to leaves, on glabrous pedicels 3–5(10) mm; calyx 5 mm long, 3–4 mm wide, incised for two-thirds, appendages one-third to one-half the length of the teeth and like the latter stiff-haired along margin; corolla blue, campanulate, only half the length of the calyx, glabrous inside and outside; style as long as corolla, stigma hamately curved at apex. November. (Plate X, Figure 2.).

In rock crevices of lower forest mountain belt. – Caucasus: W. Transc. Endemic. Described from the shady limestone banks of the Tsiteli-Tskhale River, near Kutaisi. Type in Leningrad.

Note. Only Boissier, following Ruprecht, correctly interpreted this species. All other authors confused it with species related to C. sibirica s. l., in spite of the fact that it is only remotely related to it. Neither Fomin (op. cit.) nor Trautfetter understood this species. Fomin (op. cit.) merged it with campanules of the C. sibirica type from near the "locus classicus" of C. imeretinae, even though they were quite different. Trautfetter described var. imeretina from a foreign specimen without knowing the type. Grossgeim's description, in turn, is so short that it is difficult to decide whether he had seen the type specimen. Kharadze (op. cit.) followed Fomin, but in a note he adds that the species is ambiguous and requires further investigation. Kolakovskii (Fl. Abkh. IV, 167) describes as C. imeretina, a quite unknown plant; his record for the

alpine zone is incredible. (Ruprecht reported that the species was found in the "Imeretia demissior".) All these errors are due to the neglect of Ruprecht's statement (l. c.) on the close affinity of C. imeretina with C. dichotoma L. and C. rigidipila Hochst. et Steud. Ruprecht refers
185 only casually to the similarity to the Caucasian species, which he considers as related to C. sibirica s. l. merely because all other Caucasian campanules do not resemble C. imeretina at all. Unfortunately, there are only two specimens of the latter in the Leningrad Herbarium, Ruprecht's type and a plant later collected by Sredinskii, probably in the same locality (the label, written by Sredinskii, merely states: "Kutaisi"). The discovery by Sredinskii of C. imeretina identical with Ruprecht's specimens refutes any attempt to present Ruprecht's specimens as disfigured, or exceptional. Both specimens prove that C. imeretina does not have the characters attributed to it by modern botanists. It is distinguished by the complete absence of rosettes, thin furcate flexuous stems, small flowers with corolla about twice the length of the calyx, leaves with very short, very broad, subrhombic lamina, in general habit rather more reminiscent of Stellaria media or Asperugo procumbens than in campanulas of the C. sibirica type.

Subsection 5. TULIPELLA Fed. subsect. nov. in Addenda XXIII, 461. — Appendages in notches between calyx teeth ovate, acute, calyx tube, appendages and teeth covered with whitish declined cilia; corolla large, cylindrical-campanulate, slightly inflated in middle, divided into straight lobes, with purple dots inside. Flowers drooping, pedicels rather long. Capsule pendulous, opening at base by three small holes.

Type species: C. punctata Lam.

21. **C. punctata** Lam. Encycl. Meth. I (1783) 586; Roem. et Schult. Syst. V, 138; A. DC. Monogr. 242; idem in DC. Prodr. VII, 465; Ldb. Fl. Ross. II, 2, 876; Turcz. Fl. baic.-dahur. II, 476; Kom., Fl. Man'chzh. III, 2, 552; Kom. and Alis., Opred. rast. Dal'nevost. kr. II, 993. — C. trachelium auct. non L.: Thunb. Fl. Japon. (1784) 88. — Ic.: Bot. Mag. tab. 1723; Gmel. Fl. Sibir. III, tab. 30 (mediocr.); Kom., op. cit., tabl. 301, Fig. 2—3 (bona); H. Clifford Crook, Campanulas, 167. — Exs.: Karo, Pl. Amur. et Zeaënsae, No. 363, 138.

Perennial; root rather thin, fibrous; stem straight, ca. 50 cm, pubescent, scabrous, cylindrical, paniculately branching in upper part; leaves rather numerous, pubescent, pale beneath; lamina of radical leaves cordate, ovate, acute, 7 cm long, 4 cm wide, crenate, with reddish pubescent petioles; cauline
186 leaves remote, ovate, acute or lanceolate, twice crenate-serrate, short-petioled or sessile, sometimes petioles winged. Flowers 1—5, drooping; pedicels rather long, pubescent; petioles covered with spreading white cilia; calyx teeth lanceolate, acute, with reddish tips, straight, with reflexed ovate acuminate appendages; corolla 6 cm long, poculiform-campanulate, slightly inflated at middle, dingy white, covered outside and especially inside with purple dots, lanately pubescent outside, bearded inside, with ovate, acute erect lobes, much longer than calyx with teeth; stamens half the length of the corolla, filaments membranous, scabrous, hairy in lower part; style not exserted, with 3 filiform stigmas; capsule 3-locular, pendulous, opening by three small holes at base. July. (Plate IX, Figure 1.).

(187)

PLATE X. 1 – Campanula hohenackeri Fisch. et Mey.; 2 – C. imeretina Rupr.; 3 – C. komarovii Maleev.; 4 – C. propinqua Fisch. et Mey.

Forests and shoals of forest riverbanks. — E. Siberia: Dau.; Far East: Ze.-Bu., Uda, Uss., Sakh. **Gen. distr.:** Korea, Jap.-Ch., NE China. Described from Japan. Type possibly in Paris.

Note. Not reported for Sakhalin, but in the herbarium of the Botanical Institute of the Academy of Sciences of the USSR in Leningrad there is a specimen collected by Krishtofovich on the western shore of the island.

Subsection 6. DASYSTIGMA Fed. subsect. nov. in Addenda XXIII, p. 329. — Appendages in notches between calyx teeth small, lanate, calyx-tube short, obconical, the teeth long-acuminate, much longer than tube; corolla campanulate, with hemispherical base, divided into attenuate-acuminate lobes; style pubescent at tip and along stigmas. Flowers long-pediceled, in branching, few- or many-flowered, nearly paniculate inflorescences. Root robust, fusiform, spongy.

Type species: C. alpina Jacq.

22. C. alpina Jacq. Enum. stirp. Vindobon. (1762) 36 et 210; A. DC. Monogr. 248; idem in DC. Prodr. VII, 466; Sogorski et Schneid. Fl. Carpat. Centr. II. 372; Szaf., Kulzc., Pawl. Rosl. Polskie, 640. — Marianthemum alpinum Schur in Verh. Siebenburg. Ver. Naturw. IV (1853) 48. — Ic.:
189 Sims in Bot. Mag. tab. 957; H. Clifford Crook, Campanulas, 28. — Exs.: Fl. pol. exs. No. 362; Fl. exs. austro-hung. No. 2979.

Perennial; root fusiform, thick, spongy; stem 5—10 cm, simple or slightly branching, sulcate, lanate-pubescent, surrounded by rosetted leaves; leaves yellowish, linear-lanceolate, notched-crenate, or entire, densely lanate along margin and nerves; radical leaves crowded, 2—4(5) cm long, 0.5—0.7 cm wide, tapering at base; cauline leaves rather numerous, reduced, straight, linear, acute. Flowers 3—4, or many drooping in pyramidal inflorescence; pedicels 1-flowered, 8 cm, sometimes from just above root; calyx-tube small, obconical, subglabrous, the teeth straight, long-acuminate, much longer than tube, lanate, appendages ovate, acute, ca. 1 mm long, lanate; corolla bright blue, campanulate, 2.5—3 cm long, with ovate acute lobes; filaments broadend at base, ciliate; anthers narrow, cylindrical; style trifid, pubescent at tip. June—July. (Plate IX, Figure 2.),

Open ground and rocks in alpine mountain belt. — European part: U. Dns. (Carpathians). **Gen. distr.:** Centr. Eur. (Austria, Alps, Carpathians, Tatra Mountains, Transylvanian Alps). Described from the Austrian Alps. Type in Vienna.

Note. Cultivated since 1779 for its beatiful flowers; in the USSR only in botanical gardens.

Subsection 7. ANNUAE (Boiss.) Fed. comb. nov. — Sect. Annuae Charadze in Zam. po sist. i geogr. rast. Tbil. bot. inst. 15 (1949) 32. — Annuae Boiss. Fl. or. III (1875) 895, nom. — Gen. Erinia Noulet. Fl. Pyr. (1837) 467. — Gen. Roucela Dumort. Comm. bot. (1822) 14. — Appendages between calyx teeth present or absent; corolla barely or 2—2.5 times as long as declined calyx teeth; capsule pendulous, dehiscing at base by small valves; style with conspicuously thickened base. Stems low, furcate, leafy. Leaves not rosetted. Flowers subsessile in angles of branching stem or terminal.

Type species: C. erinus L.

23. C. erinus L. Sp. pl. (1753) 240; Roem. et Schult. Syst. V, 150; A. DC. Monogr. 295; idem in DC. Prodr. VII, 473; Boiss. Fl. or. III, 932; Grossg. in Izv. Azerbaidzh. fil. AN SSSR, 1–2, 118. – Erinia campanulata Noulet, Fl. Pyr. (1837) 407. – Roucela erinus Dumort. Comm. bot. (1822), 14. – Ic.: Columna, Pl. nov. hist. 29, f. 31; J. Bauhin, Hist. 799.

190 Annual; root whitish, thin, fibrous; stem furcate, 30 cm, often not more than 10 cm, with stiff spreading hairs, slightly faceted, sulcate; leaves alternate, rarely opposite, stiff-haired, ovate or obovate, sessile, obtuse, 1–2 cm, crenate-dentate, sometimes teeth rather large becoming small lobes. Flowers terminal or sessile, in angles of furcate stem, later drooping slightly, the upper crowded in terminal groups of 2 or 3, opening after lower flowers; lobes of the stiff-haired obconical calyx straight, acute, becoming accrescent and curved, slightly shorter than the tubular, pale blue or whitish basally pubescent corolla, divided into straight elongate lobes; stamens half the length of the corolla, the filaments white glabrous, slightly broadened at base; anthers as long as filaments; style villous, not branching, with 3-lobed stigma, conspicuously thickened at base; capsule turbinate, pendulous, dehiscing at base by valves; seeds very small, elongate:ovoid, not flattened, acuminate, shiny. May.

Dry limestone cliffs. – Caucasus: S. Transc. (S. Karabakh). **Gen. distr.:** Med. (also Canary Islands, Madeira Island), Bal.-As. Min., Arm.-Kurd., Iran. Described from Italy and France. Type in London.

Note. Rare in the USSR; first recorded in 1911 by Voronov for Dashkesan, Karyaginsk District of Azerbaidzhan. In 1937 Grossgeim (op. cit.) found it not far from there near Pirchevan. We had no new collections at our disposal. Mostly distributed in the Mediterranean area this campanula is uniform throughout its area of distribution.

24. C. propinqua Fisch. et Mey. in Ind. sem. hort. Petrop. II (1835) 32; DC. Prodr. VII, 462; Boiss. Fl. or. III, 930; Fomin in Mater. Fl. Kavk. IV, 6, 31; Grossg., Fl. Kavk. IV, ed. 1-e, 62. – C. pestalozzae Boiss. Diagn. sér. I, 11 (1849) 62. – C. hispidissima Hochst. in Lorent, Wander. im Morgenlande (1845) 331. – Exs.: Herb. Fl. Cauc. No. 144.

Annual; plant with spreading scabrous hairs; stem 10–15 cm, furcate; lower leaves spatulate, slightly crenate, tapering to inconspicuous petiole; cauline leaves sessile, oblong or oblong-lanceolate, obtuse, entire or incised-dentate. Flowers terminal, 1.5 cm long, short-pediceled; teeth of scabrous calyx narrowly lanceolate, acute, declined, with triangular, acute or obtuse appendages, longer than tube; corolla infundibular, pale violet, whitish
191 inside, hairy outside along nerves, 2–2.5 times as long as calyx; capsule pendulous. June–July. (Plate X, Figure 4.)

Pebbly, clayey slopes, taluses and solonetzic sites in submontane belt. – Caucasus: S. Transc. (Nakhichevan ASSR: Ordubad, Dzhul'fa, Aznabyurt, Negram, Karababa; Mikoyan District, Armenian SSR). **Gen. distr.:** E. Anatolia, N. Iran. Described from Iran (Khoi). Type in Leningrad.

Subsection 8. EUCODON (A. DC). Fed. comb. nov. – Sect. Eucodon A. DC. Monogr. (1830) 251, s. str. – Sect. Trachelioideae Charadze in Zam. po sist. i geogr. rast. Tbil. bot. inst. 15 (1949) 28. – Ser. Trachelioideae Boiss. Fl. or. III (1875) 894. – *Drymocodon* Fourr. in Ann. Soc. Linn. Lyon, N. S. XVII (1869) 111. – Calyx without appendages in notches between teeth; flowers in many-flowered racemiform inflorescences, rather short-pediceled; capsule pendulous, dehiscing at base by 3 valves. Stems high, simple. Radical leaves usually petiolate, with cordate base. Perennial herbs with more or less thick roots.

Type species: *C. latifolia* L.

Series 1. **Latifoliae** Charadze in Zam. po sist. i geogr. rast. Tbil. bot. inst. 15 (1949) 28. – Calyx teeth straight, one-third to one-half the length of the corolla. Flowers 3–5 cm long. Plant 1 m high or more, with broad cordate-ovate or ovate, dentate leaves; sometimes 10–12 cm high, else similar.

25. C. latifolia L. Sp. pl. (1753) 233; A. DC. Monogr. 264; idem in DC. Prodr. VII, 2, 882; Ldb. Fl. Ross. II, 2, 882; Boiss. Fl. or. III, 921; Clarke in Hook. Fl. Brit. Ind., III, 439; Shmal'g., Fl. II, 178; Fomin in Mater. Fl. Kavk. IV, 6, 81; Medv. in Tr. Tifl. bot. sada, XVIII, 2, 330; Grossg., Fl. Kavk. IV, ed. 1-e, 73; Maevskii, Fl. ed. 8-e, 548; Kryl., Fl. Zap. Sib. XI, 2636; Rupr. in Bull. Acad. Sc. Pétersb. XI, 196; Korsh. Tent. Fl. ross. or. 272; Somm. et Lev. in Tr. Peterb. bot. sada, XVI, 325; Fedch. and Fler., Fl. Evrop. Ross. 937; Visyulina in Viznachn. rosl. URSR, 509; Kharadze in Fl. Gruzii, VIII, 171; Popov, Rast. i fl. Karpat, 261; Szaf., Kulcz., Pawl. Rosl. Polskie, 643. – *C. macrantha* Fisch. in A. DC. l. c. (1830) 264, pro syn. – *C. eriocarpa* M. B. Fl. taur.-cauc. I (1808) 149; Roem. et Schult. Syst. V, 19. – *C. urticifolia* M. B. l. c. p. p. – *C. latifolia* var. *macrantha* C. A. M. Verzeichn. (1831) 84; Ldb. l. c. – *C. latifolia* var. *eriocarpa* Ldb. l. c. – *C. latifolia* var. *leiocarpa, intermedia, canescens* Trautv. in Tr. Peterb. bot. sada, VI, I (1879) 72, 73. – *C. trachelium* auct. non L.: C. A. M. l. c. – *Dry-*
192 *mocodon latifolium* Fourr. in Ann. Soc. Linn. Lyon, N. S. XVII (1869) 111. – Ic.: Rchb. Ic. Fl. Germ. tab. 238; Syreishch., Ill. fl. Moskv. gub. 224; H. Clifford Crook, Campanulas, 119 (photo). – Exs.: GRF, No. 626; Herb. Fl. Est.-Liv. und Kurland, No. 471; Fl. exs. austro-hung. No. 2985; Fl. stir. exs. No. 1264; Billot. Fl. Gall. et Germ. exs. No. 1032; Fl. Bohem. et Morav. exs. VI, No. 572.

Perennial; root fibrous, with milky juice; stem 1 m and more, straight, simple, glabrous or rather sparsely pubescent, longitudinally striated; leaves glabrous, sometimes scabrous, 7–12 cm long, 3–6 cm wide, pale beneath, irregularly biserrate; radical leaves ovate-oblong, cordate, more distinctly dentate than cauline leaves, with long petioles; lower cauline leaves short-petioled, ovate, acuminate, the upper narrower, sessile, nearly entire. Flowers large, pediceled, solitary in axils of upper leaves, erect, forming narrow, nearly spicate, rather sparse raceme; teeth of pyriform, large glabrous blackish calyx long-acuminate, broadened at base, usually serrate, distinctly shorter than corolla, the latter 6 cm long, blue, sometimes nearly white, infundibular, inflated at middle, glabrous inside but bearded with long soft hairs, shallowly divided into acute somewhat curved lobes; stamens broadened at base, bearded; style slightly shorter than corolla, with 3 (rarely 2) stigmas; capsule triperforate, ovoid, pendulous; seeds ovoid, flattened, pale yellow. June—July.

Mixed deciduous-coniferous forests, riparian woodland and tall, herbaceous alpine vegetation. — European part: Kar.-Lap. (S.), Lad.-Ilm., Dv.-Pech. (S.), V.-Kama, Balt., U. Dnp., U. Dns., M. D., U. V., V.-Don, Transv., Bl., L. Don (S.); Caucasus: all regions; W. Siberia: U. Tob (W.), Alt. **Gen. distr.**: W. Eur., except for N. Scand., Spain and Eur. Turkey; As. Min., W. Himalayas. Described from Eur. (England, Sweden). Type in London.

Note. In its extensive distribution area, this species shows slight variations. Various proposals have been made to separate it into local varieties.

Grown as an ornamental.

26. C. megrelica Mand. et Kuthath. in Zam. po sist. i geogr. rast. Tbil. bot. inst. 18 (1955) 62. — Ic.: Mand. in Kutat., op. cit. 63.

Perennial; plant 10—12 cm high, with reduced rhizome; stem straight, thin, usually solitary, geniculately curved in internodes, indurate below, covered
193 below with small whitish short declined hairs, naked in upper part; radical leaves long-petioled, suborbicular, indistinctly crenate, cordate; cauline leaves short-petioled, sessile above, broadly ovate, with shallow cordate base, cuneately tapering to petiole or with rounded short-cuneate base, twice obtusely dentate, obtuse at apex; all leaves slightly scabrous with small scattered hairs above, sometimes villous beneath. Flower 1, terminal, 3 cm long, violet, sometimes flowers 2—3(6), in short raceme; teeth of gray-haired semispherical calyx lanceolate, straight, darkish, acute, twice as long as tube with ciliate margin, but one-third to one-half as long as the glabrous campanulate dark blue corolla; corolla divided for one-third into acute flat lobes with ciliate margin; calyx appendages inconspicuous; capsule globose, pendulous. August.

Subalpine meadows, on limestone. — Caucasus: W. Transc. (W. Georgia). Endemic. Described from Migariya Mountain in Mingrelia (Gegechkori District). Type in Tbilisi, isotype in Leningrad.

27. C. odontosepala Boiss. Diagn. ser. I, 11 (1859) 66; ej. Fl. or. III, 921; Boiss. et Buhse, Aufzähl. 143; Lipskii, Fl. Kavk. 379; Fomin in Mater. Fl. Kavk. IV, 6, 84; Grossg., Fl. Kavk. IV, 1st ed., 73; Grossg., Opred. rast. Kavk. 423.

Perennial; stems 70—80 cm, glabrous, straight, flexuous, not branching, leafy; leaves thin, glabrous, sometimes very sparsely pubescent beneath along nerves, with ciliate margin, all petiolate, with rounded or slightly cordate base, ovate-oblong, attenuate-acuminate, obtusely or rather acutely toothed with unequal, contiguous, teeth. Flowers axillary, 1—2 on curved pedicels slightly shorter than calyx; calyx teeth linear-lanceolate or linear, spreading or reflexed, small-toothed, with short crisp cilia, twice as long as the glabrous semiglobose tube; corolla 3 cm long, campanulate, divided for one-third, rictus bearded, three times as long as calyx teeth; capsule ovoid, pendulous, opening at base; seeds brown, oblong, bordered. July—August. (Plate XVII, Figure 1.)

Forests of lower mountain belt. — Caucasus: S. Transc., Tal. **Gen. distr.**: N. Iran. Described from Gilyan province. Type in Geneva, topotype in Leningrad.

28. C. trachelium L. Sp. pl. (1753) 235; A. DC. Monogr. 266; idem in DC. Prodr. VII, 2, 469; Ldb. Fl. Ross. II, 2, 882; Boiss. Fl. or. III, 922; Shmal'g., Fl. II, 177; Korsh. Tent. Fl. ross. or. 272; Fedch. in Fler. Fl.
194 Evrop. Ross. 937; O. and B. Fed., Perech. rast. Turk. III, 357; Maevskii, Fl. 8th ed. 548; Szaf., Kulcz., Pawl. Rosl. Polskie, 643; Fomin in Mater. Fl. Kavk. IV, 6, 84, cum dubit.; Grossg., Fl. Kavk. IV, 73, cum dubit.; Kryl., Fl. Zap. Sib. XI, 2673. — C. urticifolia Schmidt, Fl. boem. I (1793–1794) 73; M. B. Fl. taur.-cauc. I, 149, p. p., III, 141. — Ic.: Rchb. Ic. Fl. Germ. tab. 239; Fedch. and Fler., Fl. Evrop. Ross. 935; Syreishch., Ill. fl. Mosk. gub. III, 223. — Exs.: Meinsh. Herb. Fl. Ingr. No. 387; Pl. Finl. exs. No. 1361; Fl. exs. austro-hung. No. 3299.

Perennial; root fibrous, rather thick; stem 1 m or more, straight, simple or branching, angular, reddish, pubescent with white bristles; leaves scabrous, pale beneath, acuminate, strongly crenate-dentate or bicrenate, the lower on long hairy petioles, 8–10 cm long, 6–7 mm wide, cordate-ovate, acuminate, the upper sessile. Flowers 1–4 at tip of branches, short-pediceled, drooping or erect; teeth of calyx pubescent or subglabrous, obconical, blackish, its teeth erect, triangular, acuminate, one-third to one-half the length of the campanulate, blue-violet or nearly white inwardly bearded corolla, divided into shallow declined lobes; stamens with white-ciliate filaments broadened at base; style slightly flexuous, nearly as long as corolla, pubescent; capsule pendulous, ovoid, opening by 3 small holes; seeds ovoid, flattened, yellow. June–July.

Shady mixed forests. — European part: Kar.-Lap. (S.), Balt., Lad.-Ilm., Dv.-Pech. (S.), U. Dns., U. Dnp., M. D., Bes., Bl., L. Don, V.-Don, U. V., V.-Kama, Transv., L. V. (N.), Crim. (mountains); Caucasus: Cisc.? W. Siberia: Alt., U. Tob. (Sverdlovsk). **Gen. distr.**: nearly all of W. Eur., N. Africa, Syria. Described from W. Eur. Type in Leningrad.

Note. In view of the fact that this species grows in Greece, C. athoa Boiss. might be the proper name for this species. Plants from N. Africa and Syria probably represent separate species. C. trachelium, reported for Japan, is really C. punctata.

Series 2. **Rapunculoideae** Charadze in Zam. po sist. i geogr. rast. Tbil. bot. inst. 15 (1949) 28. — Calyx teeth one-fifth to one-fourth as long as corolla, curved or spreading; flowers 2–3 cm long. Herbs 50–70, rarely 100 cm. Leaves cordate-ovate, dentate.

29. C. rapunculoides L. Sp. pl. (1753) 234; A. DC. Monogr. 268, p. p.; idem in DC. Prodr. VII, 2, 469, p. p.; Ldb. Fl. Ross. II, 2, 883, p. p.; Boiss. Fl. or. III, 922, p. min. p.; Shmal'g., Fl. II, 177; Fomin in Mater. Fl. Kavk.
195 IV, 6, 58, p. p.; Trautf. in Tr. Peterb. bot. sada. VI, 1, 75, p. p.; Korsh. Tent. Fl. ross. or. 273; Fedch. and Fler., Fl. Evrop. Ross. 938; O. and B. Fedch., Perech. rast. Turk. III, 357; Pavlov, Fl. Tsentr. Kazakhst. III, 203; Maevskii, Fl. 8th ed. 548; Szaf., Kulcz., Pawl. Rosl. Polskie, 642;

Kharadze in Fl. Gruzii, VIII, 180; Grossg., Fl. Kavk. IV, 73, p. p.; Kryl., Fl. Zap. Sib. XI, 2637. – C. rigida Gilib. Fl. lithuan. I (1792) 51. – C. neglecta Besser, Cat. hort. Cremen. (1816) 28; Rupr. Fl. Ingr. I, 659. – C. ucranica Spreng. ex Roem. et Schult. Syst. V (1819) 121. – C. crenata Link, Enum. hort. Berol. I (1821) 214; Weinm. Enum. stirp. Petropol. 25. – C. rhomboidea Falk, Topogr. Beitr. II (1786) 128. – C. rhomboidalis Gorter, Fl. Ingr. (1764) 34. – C. trachelioides M. B. Fl. taur. taur.-cauc. I (1808) 150 et III (1819) 142. – C. lunariaefolia Rchb. Pl. crit. crit. VI (1832) 750. – C. setosa Fisch. ex A. DC. l. c. (1830) 269, pro syn.– C. rapunculoides var. neglecta Ldb. l. c. – C. rapunculoides var. trachelioides A. DC. l. c. 269; Ldb. l. c. – C. rapunculoides var. glabrata Trautv., ibid. VI, 1, (1879) 75, p. p. – C. rapunculoides var. macrophylla Syr., Ill. fl. Mosk. gub. III (1910) 223, non A. DC. videtur. – Drymocodon rapunculoides Fourr. in Ann. Soc. Linn. Lyon, N. S. XVII (1869) 111. – Cenekia rapunculoides Opiz. Seznam (1852) 36. – Ic.: Rchb. Ic. Fl. Germ. et Helv. tab. 239; Fedch. and Fler., ibid., 936; Syreishch., op. cit. 222; Neimark in Sorn. rast. SSSR, IV, Fig. 441 (hab.) in 440 (semen.); H. Clifford Crook, Campanulas, 170 (photo). – Exs.: Meinsh. Herb. Fl.; Ingr. Cat. III, No. 383; Herb. Fl. Est.-Liv. und Kurland, No. 473; Billot, Fl. Gall. et Germ. exs. No. 2879 et 2879 bis; Fl. ital. exs. No. 670; Callier, Fl. Siles. exs. No. 1251; Rchb. Ic. Fl. Germ. exs. No. 322; Hayek, Fl. stir. exs. No. 9262, 576; Kickxia Belgica, No. 373.

Perennial; root thick, branching, sometimes producing long underground shoots; stem 30–100 cm, usually simple, straight, slightly ribbed, glabrous, often covered like rest of plant with short stiff scabrous hairs; radical and lower cauline leaves long-petioled, cordate-ovate or oblong-cordate-ovate; median cauline leaves ovate, with obtuse base; upper cauline leaves sessile, lanceolate, irregularly and usually acutely toothed. Inflorescence long, racemiform, secund; flowers ca. 2–2.5 cm long, blue-violet, short-pediceled, drooping, solitary; teeth of obconical pubescent calyx linear-lanceolate, curved, many times shorter than infundibular to campanulate corolla, corolla divided for one-third into ovate acute lobes with ciliate margin; calyx appendages absent; style sometimes slightly exserted from corolla; capsule opening at base. June–August.

Forest edges, shrubby formations, fallows, sometimes steep riverbanks and rocks. – European part: all regions except the Extreme North;
196 Caucasus: Cisc.; Centr. Asia: N. Kazakhstan; W. Siberia: Tomsk, Krasnoyarsk Territory. **Gen. distr.**: Centr. Eur., Scand. (S.). Described from W. Eur. Type in London.

Note. This species varies very little within its area of distribution. In the south (Caucasus, Asia Minor) it vicariates with a series of not sharply distinct races, C. grossheimii Charadze, C. cordifolia C. Koch and others. It is very rare in Siberia, where it has only been found near Tomsk, the Krasnoyarsk Territory, near Yenisei. In the north it extends nearly to the forest-tundra belt.

30. C. grossheimii Charadze in Zam. po sist. i geogr. rast. Tbil. bot. inst. 15 (1949) 29; Kharadze in Fl. Gruzii, VIII, 181; Grossg., Opred. rast. Kavk. 423. – C. subalpina Charadze in schedis (teste A. Grossheim, 1947), nom. ined. – C. rapunculoides auct. fl. cauc. p. p. non L.

Perennial; stem 50–100(120) cm, straight or ascending, simple or branching, glabrous or with spreading hairs, leafy; radical leaves triangular-ovate, acuminate, partly irregularly incised-dentate, with truncate or slightly cordate base cuneately tapering to a long, thin petiole; lower cauline and radical leaves large, 10 cm long, 5 cm wide, ovate-oblong to obovate-triangular, with truncate or slightly ovate base, cuneately decurrent along shortened petioles, large-toothed, upper leaves distinctly decreasing in size from ovate-lanceolate to sublinear, sessile; all leaves with appressed hairs above, long spreading hairs below. Flowers 4 cm long, in racemiform inflorescence, on 0.8 cm pedicels, drooping after flowering, the lower sometimes on long axillary branches; teeth of the short obconical calyx linear, acute, curved, like tube more or less covered with stiff white hairs, one-fourth to one-third the length of the broadly infundibular glabrous dark violet corolla; corolla divided for one-third or one-half into triangular acute lobes; style as long as corolla or slightly exserted from it; capsule globose. July–August.

Subalpine meadows to 2,600 m and upper forest belt. – Caucasus: W. and E. Transc. Endemic. Described from Kazbek District in Georgia (banks of the Chkheri River). Type in Tbilisi.

Note. This species is vertically vicarious with C. rapunculoides. In the Caucasus it is confined to the forest belt. It differs from C. rapunculoides by its large, broad leaves, obtuse base, large and broad corolla and globose capsule.

197 **31. C. cordifolia** C. Koch in Linnaea, XIX (1847) 22; Grossg., Fl. Kavk. IV, 74; Kolak., Fl. Abkh. IV, 180; Kharadze in fl. Gruzii, Viii, 181. – C. rapunculoides var. cordata C. Koch in Linnaea, XXIII (1850) 641. – C. rapunculoides f. cordifolia Alb. Prodr. Fl. Colch. (1895) 160. – C. rapunculoides var. cordifolia Fomin in Mater. Fl. Kavk. IV, 6 (1906) 94, p. p. – C. rapunculoides var. glabrata Trautv. in Tr. Peterb. bot. sada, VI, 1 (1897) 75, p. p.

Perennial; stem 1 m, usually lower, straight, slightly branching or simple, sometimes 1 m, covered with short spreading hairs or glabrous; lower and cauline leaves petiolate, cordate, with short ovate or slightly oblong-ovate irregularly large-toothed lamina, bristly-hairy along nerves; upper leaves very small, sessile; inflorescence long, racemiform, secund; flowers 1–2.3 cm long, drooping, bright blue; teeth of subglabrous or very remotely pubescent calyx linear-lanceolate, curved, as long as tube but several times shorter than the infundibular to campanulate corolla, corolla divided into acute ovate lobes. June–September.

Forest mountain belt, forest edges, cliffs above riverbanks. – Caucasus: Cisc. (western part of Main Range up to S. Osetia), W. Transc., E. Transc. (Tedzami ravine in Kartlia). **Gen. distr.**: As. Min. Described from As. Min. Type in Berlin.

Note. This species is closely related to the lowland C. rapunculoides L. and vicariates with it in mountains, primarily in W. Caucasus. C. cordifolia is distinguished by its deeply cordate lower median cauline leaves, calyx and stem subglabrous, flowers rather small, dark blue, without a violet tinge. In the subalpine and alpine belts it vicariates with the closely related C. grossheimii Charadze, with a lower stem, basally obtuse leaves, and to 3 cm long flowers.

32. C. bononiensis L. Sp. pl. (1753) 234; M. B. Fl. taur.-cauc. I, 150, III, 142; Roem. et Schult. Syst. V, 122; A. DC. Monogr. 270; idem in DC. Prodr. VII, 2, 469; C. A. M. Verzeichn. 84; Ldb. Fl. Ross. II, 2, 884; Boiss. Fl. or. III, 923; Shmal'g., Fl. II, 177; Fomin in Mater. Fl. Kavk. IV, 6, 94; Grossg., Fl. Kavk. IV, 73; Kryl., Fl. Zap. Sib. XI, 2638; Rupr. in Bull. Acad. Sc. Pétersb. XI, 197; Trautf. in Tr. Peterb. bot. sada. VI, 2, 76; Korsh. Tent. Fl. ross. or 273; Fedch. and Fler., Fl. Evrop. Ross. 938; O. and B. Fedch., Perech. rast. Turk. III, 357; Pavlov, Fl. Tsentr. Kazakhst. III, 203; Maevskii, Fl. 8th ed. 547; Kharadze in Fl. Gruzii, VIII, 182; Szaf., Kulcz., Pawl. Rosl. Polskie, 643. — C. urticifolia Gilib. Fl. lithuan. I (1792) 521, non al. auct. — C. pyramidalis Gilib. l. c. 52, non L. — C. pyramidata Gilib. Hist. d. pl. II (1798) 209. — C. petraea Habl. Phys. Beschr. d. Taur. (1789) 275, non L. — C. rapunculoides Pall. Phys.-top.
198 Gemälde Taur. (1806) 99, non L. — C. ruthenica M. B. Fl. taur.-cauc. I (1808) 151, III (1819) 142. — C. thaliana Wallr. Sched. crit. (1822) 86. — C. simplex Lam. ex DC. Fl. Franc. III (1790) 730, non Stev. — Drymocodon bononiense Fourr. in Ann. Soc. Linn. Lyon, N. S. XVII (1869) 111. — Cenekia simplex Opiz, Seznam (1852) 36. — Ic.: Rchb. Ic. Fl. Germ. et Helv. Cent. 2, tab. 112; H. Clifford Crook, Campanulas, 47 (photo); Syreishch, Ill. fl. Mosk. gub. III, 221; Fl. yugo-vost. Fig. 664. — Exs.: GRF, No. 466; Fl. exs. austro-hung. No. 2986; Fl. Gall. et Germ. exs. No. 695.

Perennial; root more or less fusiformly thickened; stem 70 cm, straight, scabrous, more or less finely pubescent, simple or slightly branching; leaves ovate, acuminate, scabrous above, usually scabrous and canescent beneath, serrate, acutely toothed, dark green above; radical leaves petiolate, cauline leaves sessile, upper leaves amplexicaul. Flowers 2–2.5 cm long, numerous, in spicate raceme, drooping, very short-pediceled, sometimes inflorescence slightly branching; calyx glabrous or scabrously pubescent, suborbicular, blackish, its teeth acute, more or less spreading, scabrous, narrowly triangular-lanceolate, several times shorter than corolla, corolla infundibular, blue-violet, glabrous outside, bearded inside; filaments membranous, broadened at base, pubescent; style as long as corolla, sparingly pubescent at first, with 3 stigmas; capsule globose, small, pendulous. June–August.

Dry meadows, forest edges, shrubby formations. — European part: nearly all regions except for the north; Caucasus: Cisc., W., E. and S. Transc.; W. Siberia: U. Tob., Irt., Alt.; Centr. Asia: Ar.-Casp., Bal.-As. Min. (W.), Med. (N.). Described from Italy. Type in London.

Note. This species does vary to some extent, but nowhere in its extensive distribution area have distinct varieties been recognized.

Subsection 9. INVOLUCRATAE (Fom.) Fed. comb. nov. — Sect. Involucratae (Fom.) Charadze in Zam. po sist. i geogr. Tbil. bot. inst. 15 (1949) 30. — Ser. Involucratae Fom. in Mater. Fl. Kavk. IV, 6 (1906) 100. — Syncodon Fourr. in Ann. Soc. Linn. Lyon, N. S. XVII (1869) 111. — Weitenwebera Opiz, Seznam (1852) 36. Appendages in notches between calyx teeth absent or inconspicuous; flowers sessile, crowded in heads, whorls or clusters, enveloped by large bracts. Stems straight or ascending, simple or weakly branching in upper part, leafy. Radical leaves petiolate, not forming rosettes; cauline leaves sessile. Plant usually more or less densely pubescent, rarely subglabrous.

199 Type species: C. glomerata L.

Series 1. **Glomeratae** Charadze in Zam. po sist. i geogr. rast. Tbil. bot. inst. 15 (1949) 30. – Inflorescence a single terminal head or head and several whorls or clusters, sometimes branching; flowers always exserted from bracts. Radical leaves with rounded, rarely cordate or subcuneate base.

33. C. cephalotes Nakai in Bull. of the Nat. Sci. Mus. Tokyo, 31 (1952) 111; descr. n. in Addenda XXIII, 462. – C. cephalotes Fisch. ex A. DC. Monogr. (1830) 254, pro syn. – C. cephalantha Fisch. ex A. DC. l. c., pro syn. – C. speciosa Hornem. Hort. Hafn. II (1813–1815) 957, non Gilib. et al. auct. – C. glomerata var. speciosa A. DC. lc. – C. glomerata var. dahurica Fisch. ex A. DC. l. c. 256, in obs. pro syn.; Ohwi, Fl. Jap. 1123. – C. glomerata var. canescens Maxim. Prim. Fl. Amur. (1859) 185. – C. glomerata var. salviifolia Kom. Fl. Man'chzh. III (1907) 555, non Willd. – C. glomerata var. aggregata Kom., op. cit. 555, non Willd. – C. glomerata var. farinosa Kom. in sched. non Roch. nec. C. farinosa Andrz. ex Bess. – C. glomerata var. farinosa Freyn in Österr. bot. Zeitschr. 10 (1902) 399, non Roch. nec C. farinosa Andrz. ex Bess. – C. glomerata Rgl. Tent. Fl. Ussur. (1861) 100, non L. – Ic.: Kom. and Alis., Opred. rast. Dal'nevost. kr. tabl. 299 1 (mediocris); Sims in Bot. Mag. tab. 2649.

Perennial; entire plant densely covered with short gray hairs, making it scabrous or velutinous, rarely subglabrous; rhizome short, with numerous dark brown fibrous roots; stem high, usually 1 m, straight, sturdy, with fine longitudinal striations, leafy, green or reddish, simple or slightly branching; axillary branches thinner, bearing flowers; radical leaves very long-petioled, oblong-ovate, acute, cordate at base, finely and irregularly serrate-dentate, green above, more or less canescent beneath, abruptly tapering to petiole three times as long; lamina 10–15 cm long, 3–7 cm wide; cauline leaves narrower, more or less long-petioled below, sessile in middle and upper parts, semiamplexicaul, lanceolate or oblong-ovate, long-acuminate, serrate-dentate, with unequal obtuse teeth. Inflorescence elongate, interrupted, more or less leafy, sometimes branching and nearly paniculate, clusters of subsessile flowers forming compact terminal head; flowers ca. 2 cm long, violet; teeth of obconical scabrous calyx lanceolate, acute, green, without appendages, pubescent, one-third as long as corolla, corolla
200 tubular-campanulate, pubescent outside; subglabrous inside, divided for one-third into ovate acute lobes; style not exceeding corolla. July–August. (Plate XI, Figure 1.)

Meadows, leafy groves, shrubby formations in mountains and adjacent lowlands. – Far East: Uda, Ze.-Bu., Uss.; E. Siberia: Dau. **Gen. distr.**: NE China, Japan, Korea. Described from Dauria. Type in Copenhagen.

Note. Differs from other species of the order Glomeratae in its very high stems, large, very long-petioled radical leaves, long and rather often branching, nearly paniculate inflorescence, with pubescence either velutinous or scabrous, also developed on outer surface of corolla.

PLATE XI. 1 – Campanula cephalotes Nakai.; 2 – C. trautvetteri Grossh.

34. C. glomerata L. Sp. pl. (1753) 235; Roem. et Schult. Syst. V (1819) 124, p. p.; A. DC. Monogr. 253; p. p.; idem in DC. Prodr. VII, 2, 467, p. p.; Ldb. Fl. Ross. II, 2, 880, p. p.; Trautf. in Tr. Peterb. bot. sada. VI, 1, 68, p. min. p.; Shmal'g., Fl. II, 175, p. p.; Sorn. rast. SSSR. IV, 175, p. p.; Kryl., Fl. Zap. Sib. XI, 2633; Korsh. Tent. Fl. ross. or. 271; Fedch. and Fler., Fl. Evrop. Ross. 936; O. and B. Fedch., Perech. rast. Turk. III, 375; Pavlov, Fl. Tsentr. Kazakhst. III, 201; Maevskii, Fl. 8th ed. 544; Szaf., Kulcz., Pawl. Rosl. Polskie, 642. — C. congesta Roem. et Schult. l. c. 125. — C. conglomerata Güldenst. Reise, II (1791) 343. — C. tubiflora Tausch. in Ldb. l. c. pro syn. — C. betonicaefolia Gilib. Fl. lithuan. I (1792) 53, non Sibth. nec Biehl. — C. salviaefolia Mart. Prodr. Fl. Mosq. ed. 2 (1817) 40, non Wallr. — C. hirsuta Mart. l. c. — C. diffusa Mart. l. c., non Vahl. — C. aggregata Willd. Enum. hort. Berol. Suppl. (1813) 10. — C. glomerata var. typica Trautv. l. c. — C. glomerata var. glabra Korsh l. c., non Bluff, nec C. glabra Mor. — C. glomerata var. glabra Syr. in Syreishch., Ill. fl. Mosk. gub. III (1910) 228, non Bluff. — C. glomerata var. oblongata Trautv. l. c. 69; Maevskii, op. cit. — C. glomerata var. umbrosa Trautv. l. c. 69. — C. glomerata var. vulgata Syr. l. c. 68, non G. B. — C. glomerata var. pusilla Syr. l. c. 228, non DC. nec C. pusilla Haenk. — C. glomerata var. glabra B. Schischk in Kryl., op. cit. 2634, non Bluff. — Marianthemum aggregatum Schrank in Flora, VII (1824) 11. — Syncodon Fourr. in Ann. Soc, Linn. Lyon, N. S. XVII (1869) 111. — Weitenwebera glomerata Opiz, Seznam (1852) 36. — Ic.: Rchb. Ic. Fl. Germ. tab. 235; Fedch. and Fler., Fl. Evrop. Ross. Fig. 934; Syreishch., op. cit., 227; Sorn. rast. SSSR IV, Fig. 440, 2. — Exs.: GRF, No. 676; Herb. Fl. Ingr. Cent. III, No. 384; Annenkov, Fl. Mosq. exs., sine numero; Fl. pol. exs. No. 454; C. Regel, Fl. lithuan. No. 18; Pl. Finl. exs. No. 376a et b.

203 Perennial; plant with thick, partly woody rhizome; stems (15) 30–50 (80) cm, rather thick or relatively thin, straight, simple or sometimes weakly branching, inconspicuously angular-faceted, leafy, often reddish, covered with soft, rarely stiff but not bristly hairs, rarely subglabrous; leaves dense, finely crenate, green above, darker green beneath, pubescent like stem but hairs shorter, rarely subglabrous, sometimes pubescent only along nerves; radical and lower cauline leaves long-petioled, ovate-oblong or ovate-lanceolate, with shallow cordate, rounded or obtuse base, acute or obtuse; upper leaves sessile, narrower and smaller, sometimes semiamplexicaul, nearly flat, not undulant; largest laminas over 10 cm long and 3 cm wide. Inflorescence narrow, interrupted, not branching, comprising compact terminal head and sometimes rather numerous dense axillary clusters of flowers; calyx with not markedly pubescent, green, lanceolate, acuminate teeth, one-fourth to one-third the length of the corolla, without appendages; corolla tubular-infundibular, glabrous or very faintly short-haired, outside, dark violet, 2.5 cm long, divided for one-third into oblong-ovate acuminate or obtuse ciliate-margined lobes; style not exserted. June–August.

Meadows, shrubby formations, forest glades in the taiga belt; in the south encountered in similar habitats, but rare in steppes; in the mountains of Siberia it reaches the alpine belt. — European part: nearly everywhere, with the exception of the Arctic; W. Siberia, E. Siberia: eastwards to Dauria; Centr. Asia (except for the south). **Gen. distr.:** W. Eur., Scand. Described from W. Eur. (England, France, Sweden). Type in London.

Note. It appears that one and the same variety growing from W. Europe to Scandinavia also grows in theEuropean part of the USSR, with the exception of the Caucasus, in Siberia (excluding the Far East and Dauria), and to the south in Kazakhstan. In the south down to the Central Transvolga area, C. *farinosa*, which does not grow in Siberia, is sympatric with C. *glomerata*. C. *cephalotes*, the distinctive vicariant, grows in Dauria, Soviet Far East (Maritime Territory, lower reaches of the Amur, Bureya mountains), Korea and the adjacent parts of China. In the Caucasus and Near Asia there grow some distinct species confined to various altitude zones or to other physical or geographical conditions. The lowland taiga C. *glomerata* s. l. is not quite homogeneous even after the exclusion of the Far Eastern C. *cephalotes*, C. *farinosa* of the SW steppe, and all Caucasian and more southern species. Specimens collected in the subarctic
204 region of Siberia, Altai, and Sayans deviate most strongly from the type, but is insufficient evidence to warrant the recognition of separate species.

35. C. farinosa (Roch.) Andrz. ex Bess. Enum. pl. Volhyn. (1822) 10; Höfft. Cat. Kursk. 17. — C. *glomerata* var. *farinosa* Roch. ex Besser, Cat. hort. Cremen. (1816) 271; Trautf. in Tr. Peterb. bot. sada, VI, 1 69; Korsh. Tent. Fl. ross. or. (1895) 271. — C. *glomerata* var. *farinosa* A. DC. Monogr. (1830) 258, excl. syn.; idem in DC. Prodr. VII, 2, 468, excl. syn.; Fedch. and Fler., Fl. Evrop. Ross. 227. — C. *glomerata* auct. p. min. p. non L.: Ldb. Fl. Ross. II, 2, 880. — Exs.: Fl. Bohem. et Morav. exs. No. 1278; Pabo and Cholovskii, Mogilevsk., gerb. V, 1, sine numero; Novopokr., Gerb. donsk. flory No. 11; Fl. Hung. No. 1044 (Herb. Mus. Nat. Hung.).

Perennial; stem 50—70 cm, simple, straight, sometimes slightly branching, more or less pubescent, reddish at times, angular-faceted, leafy; radical leaves basally cordate or obtuse, oblong or oblong-ovate, rarely broadly lanceolate, long-petioled; lower cauline leaves of similar shape but smaller and narrower short-petioled; upper cauline leaves sessile, semiamplexicaul, markedly decreasing in size, usually slightly crisp-undulant; all leaves unequally finely crenate, gray-tomentose beneath, when young almost indistinguishable from the leaves of C. *bononiensis*. Flowers in terminal heads or dense axillary clusters, enveloped by broad cordate bracts; inflorescence narrow, interrupted, attenuate; calyx and ovary gray-pubescent; corolla as in C. *glomerata*. June—July.

Mainly in steppes, dry forests, sometimes at dry forest edges. — European part: Bes., U. Dnp., M. D., Bl., V.-Don, L. Don, Transv. **Gen. distr.**: Centr. Eur., Bal.-As. Min. (except for Turkey). Described from "Podolia and Volhynia." Type in Kiev, isotype in Leningrad.

Note. Distinguished from typical C. *glomerata* by its predominantly more northern forest habitat in Europe and Siberia, chiefly determined by ecological factors. It is almost exclusively confined to the steppe zone in the south, scarcely touching the more northern steppes, exending from the south to the Middle Volga area; in the north absent in Siberia. Its area of distribution overlaps that of C. *glomerata* but its habitats are different.

C. *farinosa* may appear as merely a form of the common C. *glomerata* with gray-tomentose leaves and other characters reflecting arid
205 conditions, but not being hereditary. Yet Andrzheiovskii (l. c.) proved that its characters did not change under cultivation.

The epithet "farinosa" is poorly chosen; the gray tomentum of the leaves must have looked like meal to Andrzheiovskii.

36. C. subcapitata M. Pop. Rast. i fl. Karpat. (1949) 295 and 259. – Ic.: M. Popov, op. cit. 260.

Perennial; rhizome horizontal, 4–5 cm long, 2–3 mm thick, with short branches, roots sturdy, slightly cordlike, stems 3–50 cm, solitary, straight, thin, strict, attenuate, simple, nearly tetrahedral from base, internodes thin, 2–3 mm thick, subcylindrical, covered with stiff, very acute, not glandular spreading or reflexed hairs, very densely so in inflorescence below stem apex; leaves of sterile shoots (very rarely observed) oblong, acute, with cordate-rounded base, very long-petioled, ; lower cauline leaves oblong or oblong-lanceolate, gradually or nearly abruptly tapering to long petiole; median and upper leaves short-petioled or sessile, oblong-lanceolate, tapering at base, acute; leaves often finely crenate-serrate, covered with slightly appressed, stiff, acute hairs, appressed along nerves beneath, scattered above. Inflorescence terminal, 5–10-flowered, capitulate, small, as big as a hazel nut or slightly bigger, rarely in axils of upper leaves, clusters of 1–3 flowers; bracts of inflorescence ovate, stiff-haired, shorter than head, slightly membranous, subglabrous at base, with inconspicuous nerves; flowers 15–17 mm long, violet; calyx-tube turbinate, glabrous at base, the teeth linear, longer than tube, with curved cilia along margin, exceeding corolla tube; corolla 15 mm long, 12 mm wide, incised to middle into oblong-lanceolate lobes; filaments very short; style completely pubescent; stigmas 3, spirally twisted; capsule unknown. July.

Clearings in the forest mountain zone, usually in clayey soil. – European part: U. Dns. (Carpathians: upper reaches of the Tisza River near Rakhov, Zabroda, Cherna-Gora, Yasinya). Endemic. Described from the upper reaches of the Tisza River near Rakhov. Type in Leningrad, isotype in L'vov.

Note. Popov (op.cit.) seems to regard this species as an endemic confined to a very restricted area, not even extending to the Soviet Carpathians.

37. C. oblongifolia (C. Koch) Charadze in Zam. po sist. i geogr. Tbil. bot. inst. 15 (1949) 30; Grossg., Opred. rast. Kavk. 422; Kharadze in Fl.
206 Gruzii, VIII, 182. – C. glomerata var. oblongifolia C. Koch in Linnaea, XXIII (1850) 639. – C. glomerata var. cervicarioides Fom. in Mater. Fl. Kavk. IV, 6 (1906) 105, non A. DC. nec C. cervicarioides Roem. et Schult.

Perennial; densely grayish pubescent plant; stems 1 m, with arcuate base, straight, simple, reddish, more or less densely covered with crisp short hairs; radical and lower cauline leaves oblong or elliptic, acute, with rounded or slightly cuneate base, tapering to long petiole; median and upper leaves ovate-oblong, sessile, semiamplexicaul, all finely crenate or dentate, more or less densely covered with stiff fine red hairs, making them scabrous. Flowers sessile, terminal or axillary, in few-flowered, compact, interrupted spicate inflorescences, with head of terminal flowers, enclosed by bracts, growing from base of flower clusters, not exceeding inflorescence; calyx teeth triangular-lanceolate, whitish owing to dense, fine, appressed pubescence of calyx, one-third to one-half the length of the tubular to infundibular finely pubescent corolla, corolla 1.5–2.5 cm long, divided into acute lobes; style usually exserted. July–August.

Lower and middle mountain belts, broad-leaved and pine forests, subalpine belt, mountain steppes and forest strip of Quercus macranthera. — Caucasus: S. and E. Transc. (SE). **Gen. distr.**: Near Asia (N). Described from Armenia. Type in Berlin.

Note. This new combination of Kharadze (op. cit.) and the detailed diagnosis refer to specimens from Mount Aragats in the Armenian SSR, but Koch's specimens from Armenia, probably kept in Berlin, should be regarded as the type. In the USSR the distribution area is bounded by the Lesser Caucasus, and extends to Tbilisi in the north. A closely similar form grows in the mountains of Talysh.

38. C. symphytifolia (Alb.) Kolak. Fl. Abkh. IV (1949) 180; Kharadze in Fl. Gruzii, VIII, 184. — C. glomerata var. symphytifolia Alb. Prodr. Fl. Colch. (1895) 161; Fomin in Mater. Fl. Kavk. IV, 6, 105, excl. syn.

Perennial; plant covered with short scabrous hairs; stem 20—60 cm straight, usually simple, radical and lower cauline leaves petiolate, cordate or ovate, other cauline leaves, with the exception of bracts, ovate-oblong or broadly oblong-lanceolate, with cordate or ampelxicaul base, sessile; leaves in inflorescence ovate, cuneately tapering above, longer than flowers. Calyx
207 teeth lanceolate-linear, with ciliate margin, half the length of the infundibular to campanulate bright blue corolla, corolla 2.5 cm long, divided for one-third into ovate-oblong acuminate lobes; anthers twice as long as filaments. July—September.

Alpine meadows of limestone mountain ranges, shrubby formations. Very frequent in Woronovia speciosa and Carex pontica meadows. — Caucasus: W. Transc. (Abkhaziya). Endemic. Described from Mamdzyshkha Mountain in Abkhaziya. Type in Geneva.

39. C. maleevii Fed. sp. nov. in Addenda XXIII, p. 330. — C. glomerata var. hispida Fomin in Mater. Fl. Kavk. IV, 6 (1906) 104.

Perennial; plant stiff-haired, with reflexed hyaline bristles; rhizome oblique, bearing numerous blackish roots; stem ca. 50 cm, straight, thick, firm, longitudinally striated, green or reddish; radical leaves ovate, acute, with subtruncate or cuneate base, abruptly tapering into half as long petioles, irregularly crenulate-dentate; cauline leaves similar in shape and size, short-petioled, semiamplexicaul, 5—7 cm long, 1.5—2.5(3) cm wide. Flowers blue, crowded in dense terminal heads, hidden by upper leaves; teeth of obconical subglabrous calyx lanceolate, straight, green, with white bristles along margins, half the length of the corolla; corolla glabrous outside, slightly bearded inside, tubular-campanulate, divided for one-third into ovate acute lobes; style pubescent, not exserted.

Broadleaved forests in mountain foothills. — Caucasus: W. Transc. (eastwards, not beyond Kutaisi). Endemic. Described from the area between Tuapse and Dzhubga on the Black Sea. Type in Leningrad.

Note. This species is distinguished from its allies of the order Glomeratae by markedly bristly, declined stiff hairs, shorter leaves, which are similar in shape and size from the radical to the upper leaves, upper leaves enclosing the heads and completely glabrous corolla. Also, its appearance is different from that of all related species.

40. C. panjutinii Kolak. in Dokl. AN Arm. SSR, VII, 5 (1947) 223; Kolak., Fl. Abkh. IV, 180; Grossg., Opred. rast. Kavkaza, 422; Kharadze in Fl. Gruzii, VIII, 184.

Perennial; plant densely covered with soft spreading hairs; stem 25–30 cm, simple; lower leaves long-petioled, oblong-lanceolate, gradually tapering to base; median and upper cauline leaves tapering to base, ovate-oblong, irregularly denticulate, 3–5 cm long, 1.2–1.8 cm wide. Flowers in
208 dense capitate inflorescence, enclosed in ovate-oblong, spreading leaves longer than flowers – the inner ovate or linear-oblong, acute or obtuse, incised-dentate, especially in lower part, usually as long as flowers; calyx teeth linear, acute, with short-ciliate margin, half as long as corolla; corolla tubular-campanulate, bright blue, glabrous outside, pubescent inside, divided for one-half into oblong acute lobes; anthers as long as filaments. July–August.

Limestones in alpine belt. – Caucasus: W. Transc. (Abkhaziya). Endemic. Thus far known only from pastures of the Kopshikho Bzyb Range, from where it has been described. Type in Sukhumi, cotype in Leningrad.

41. C. trautvetteri Grossh. sp. nov. in Addenda XXIII, p. 331; Grossg., Opred. rast. Kavk. 422, diagn. brev. ross.; Kharadze in Fl. Gruzii, VIII, 183, descr. georg. – C. sosnowskyi Grossh. in schedis ad Herb. Azerb. – C. glomerata var. caucasica Trautv. in Tr. Peterb. bot. sada. II, 2 (1873) 564; Fomin in Mater. Fl. Kavk. IV, 6, 106. – C. glomerata var. speciosa Trautv., op. cit. 564, p. p. – C. glomerata f. pumila grandiflora Fomin in schedis.

Perennial; plant more or less pubescent or subglabrous, with short oblique rhizome bearing brown fibrous roots; stem 10–20–30 cm, usually low, sometimes elongate, straight, simple, green or reddish, longitudinally striated, sparsely leafy, terminating in solitary few-flowered head; radical leaves petioled, ovate-oblong, acuminate, with slightly toothed margins, tapering abruptly from truncate or cuneate base to thick petiole, as long as lamina; cauline leaves gradually decreasing in size, similar in shape to radical leaves, short-petioled on lower part of stem, remaining leaves sessile, sometimes semiamplexicaul; uppermost leaves acuminate, sometimes nearly subulately acuminate, broadened at base, membranous, incised-dentate or entire, form obconical involucre, not exceeding inflorescence. Flowers 2–3–4 cm long, lilac; teeth of obconical calyx lanceolate, subulate-acuminate-ciliate, one-third the length of the tubular-campanulate calyx, the latter sparingly pubescent outside, subglabrous inside, divided for nearly half into oblong acuminate lobes; style not longer than corolla. June–July. (Plate XI, Figure 2).

Alpine meadows and bluffs. – Caucasus: Cisc. (C. and E. Caucasus), E. Transc. (Lesser Caucasus). Endemic. Described from Armenia (Gezaldar Mountain in Daralagez) from Radde's specimen chosen as the lectotype and preserved in Leningrad.

209 Note. This high-mountain species of the Glomeratae is endemic to the eastern part of the Main Range and the Lesser Caucasus. It is remarkable for its large flowers, "alpine appearance," low stem, shape of terminal leaves with membranous broadening at base, often abruptly tapering into narrow, sometimes filiform ending, not recurved and not exceeding head. More or less close to C. panjutinii Kolak. from the mountains of Abkhaziya but distinctly differing from the characters indicated above.

Grossgeim, the author of this species, did not describe it in Latin nor did he select a type. We have done so in the Addenda to this volume. As lectotype we propose the above-mentioned specimen from Armenia, marked by Trautfetter as C. glomerata var. caucasica Trautv.

Series 2. **Cervicariae** Fed. – Inflorescence long, with small terminal head and several remote whorls or clusters of flowers below it. Radical leaves with attenuate-cuneate base, the lamina decurrent along petiole. Stems coarse, thick, like entire plant covered with spreading bristly hairs.

42. C. macrostachya Waldst. et Kit. ex Willd. Enum. pl. hort. Berol. I (1809) 213. – C. multiflora Waldst. et Kit. Pl. rar. Hung. III (1812) 292; A. DC. Monogr. 260; idem in DC. Prodr. VII, 548; Shmal'g., Fl. II, 175; Fedch. and Fler., Fl. Evrop. Ross. 936; Visyulina in Viznachn. rosl. URSR, 509. – C. cervicaria var. multiflora Rchb. Ic. Fl. Germ. et Helv. VI (1826) 12. – Ic.: Waldst. et Kit. l. c. tab. 263; Rchb. l. c. f. 779. – Exs.: Fl. exs. austro-hung. No. 2983; Fl. Hung. exs. Cent. III, No. 255(1) et 255(2).

Biennnial; root simple, fusiform, as thick as a small finger, brown or whitish; stem 70 cm, usually simple, straight, slightly flexuous, circular in cross section, with longitudinal striae, leafy, bearing many-flowered influorescences, like leaves covered with stiff bristly hairs; leaves sessile, oblong, crenate, obtuse, with prominent nerves; lower leaves broadly lanceolate, subelliptic, the upper oblong-cordate, the uppermost similar to cauline leaves in shape but smaller. Flowers to 100 and more, in interrupted spicate inflorescences, secund or in clustered whorls of 6–7 bracteate flowers, with two broad ovate acuminate bracts at base of whorls; calyx bristly-haired, obovoid, the teeth linear-lanceolate, obtuse, ciliate; corolla tubular, pale violet-blue, glabrous, twice as long as calyx teeth, with recurved ovate acute lobes; filaments with ovate base pubescent, linear, curved; anthers linear; style glabrous, slightly shorter than corolla; stigma trifid,
210 thick, pubescent; capsule obtusely 3-faceted, opening by small pores; seeds subcylindrical, obtuse at both ends, dark brown. July.

Meadows, fallows, steppes. – Eurpean part: Bl., M. D., U. Dns. **Gen. distr.**: Centr. Eur., Bal.-As. Min. (N.). Described from Hungary. Type in Berlin.

43. C. cervicaria L. Sp. pl. (1753) 235; A. DC. Monogr. 258; idem in DC. Prodr. VII, 2, 468; Ldb. Fl, Ross. II, 2, 881; Turcz. Fl. baic.-dahur. II, 478; Shmal'g., Fl. II, 175; Kryl., Fl. Zap. Sib. XI, 2635; Korsh. Tent. Fl. ross. or. 272; O. and B. Fedch., Perech. rast. Turk. III, 357; Fedch. and Fler., Fl. Evrop. Ross. 937; Scaf., Kulcz., Pawl. Rosl. Polskie, 641. – C. cervicaria var. albiflora Syreissch. Ill. fl. Mosk. gub. III, 227. – C. echiifolia Rupr. Fl. Ingr. I (1860) 655. – Syncodon cervicarium Fourr. in Ann. Soc. Linn. Lyon, N. S. XVII (1869) 111. – Weitenwebera cervicaria Opiz, Seznam (1852) 36. – Ic.: Gmel. Fl. Sibir. III, tab. 31 (mediocr.); Syreishch., op. cit. – Exs.: GRF, No. 726; Pl. Finl. exs. No. 1363; Fl. Bohem. et Morav. exs. No. 1085.

Biennial; root 10–15 cm, as thick as a little finger or thicker, cylindrical, simple or branching; stem 70 cm, straight, rather thick, firm, longitudinally striated, somewhat angular, deeply sulcate, usually simple, covered with stiff white spreading hairs; leaves pubescent, crenate or nearly entire; radical leaves obtuse, 10–15 cm long, 2 cm wide or wider, acute, attentuate into short petiole; cauline leaves sessile, erect, 6 cm long, nearly linear-lanceolate, acuminate; upper leaves decreasing in size, semiamplexicaul. Inflorescence nearly capitate at apex, interrupted below, spicate or whorled, with rather small subsessile flowers and bracts; bracts ovate, acute, curved, markedly broadened at base, pubescent, nearly entire, slightly shorter than flowers; teeth of ovate-obconical whitish glabrous or sparingly pubescent calyx oblong, obtuse, erect, pubescent, half the length of the corolla; corolla campanulate, bluish, pubescent along nerves, shallowly divided into ovate acute lobes; filaments broadened below, pubescent; style as long as or longer than corolla, pubescent, with 3 stigmas; seeds obliquely ovoid, slightly rimmed, shiny, yellowish. June–July.

Forest plots, meadows, thinned-out coniferous and mixed forests. – European part: Kar.-Lap. (S.), Dv.-Pech. (S.), Lad.-Ilm., U. V., U. Dnp., U. Dns., M. D., V. -Don, Transv., L. Don (N.), V.-Kama; W. Siberia; Ob (S.), U. Tob., Irt., Alt.; E. Siberia: Yenis. (S.), Ang.-Say. (foothills of Sayan,
211 Tuva). **Gen. distr.**: nearly all of W. Europe, except for the extreme north and south. Described from Europe (Switzerland, Sweden, Germany). Type in London.

N o t e. This species varies very little throughout its area of distribution. In Asia Minor it replaces C. c e r v i c a r i a Nabelek (see: Publ. Fac. Sc. Univ. Masaryk, 70, 5 (1926); Reching. in Symb. Bot. Upsal. XI, 5 (1952) 35).

Series 3. **Macrochlamydeae** Charadze in Zam. po sist. i geogr. rast. Tbil. bot. inst. 15 (1949) 30. – Inflorescence a terminal head of crowded flowers entirely concealed by broad bracts forming an infundibular involucre. Radical leaves with rounded or cordate base. A series of one species.

44. C. macrochlamys Boiss. et Huet in Boiss. Diagn. ser. II, 3 (1853) 111; Boiss. Fl. or. III, 927; Lipskii, Fl. Kav. 379; Fomin in Mater. Fl. Kavk. IV, 6, 100; Grossg., Fl. Kavk. IV, 63; Kharadze in Fl. Gruzii, VIII, 185.

Biennial; root thick, fusiform; stems 20–40 cm, solitary or few, ascending or straight, rather thick, simple or branching from root collar, finely bristly-haired, particularly under inflorescence; leaves appressed-hairy, obtusely crenate-incised; radical leaves long-petioled, broadly ovate, obtuse, with cordate or rounded base; lower cauline leaves oblong, obtuse, decurrent along petiole; median leaves few, sessile, broadly oblong-linear, somewhat constricted at middle, semiamplexicaul. Inflorescence capitate, terminal, solitary, compact, enveloped by oblong obtuse green bracts longer than flowers, forming broad infundibular involucre; calyx teeth long-lanceolate, with long white cilia, one-third the length of the obconical tube, appendages obsolete, triangular, recurved; corolla blue, glabrous, tubular-infundibular or tubular, nearly three times as long as calyx teeth; capsule obovoid; seeds flattened, brown, spongy-verrucose.

Dry rocks, stony slopes, taluses, montane belt. – Caucasus: W. Transc. (Chorokh). **Gen. distr.**: As. Min. (E.). Described from As. Min. Type in Geneva.

Subsection 10. DICTYOCALYX Fed. in Addenda XXIII, 464. — Ser. Spicatae Boiss. Fl. or. III (1875) 894, o. p. — Appendages in incisions between calyx teeth growing markedly after flowering, becoming inflated-broadened, and concealing tube, reticular. Flowers small, sessile, in interrupted spicate inflorescence. Plant scabrous, its bristly hairs sessile on verrucae. Leaves stiff, ovate-oblong.

Type species of subsection: C. stricta L.

212 **45. C. stricta** L. Sp. pl. (1753) 238; Roem. et Schult. Syst. V, 149; A. DC. Monogr. 289; idem in DC. Prodr. VII, 463; Boiss. Fl. or. III, 923; Fomin in Mater. Fl. Kavk. IV, 6, 98; Grossg., Fl. Kavk. IV, 62. — C. stricta var. muricata Trautv. in Tr. Peterb. bot. sada, II, 2 (1873) 563; Lipskii, Fl. Kavk. 379. — Exs.: Kotschy, Iter. cilicico-kurd. No. 334.

Perennial; markedly scabrous with fine short white slightly hamate bristly-prickly hairs, sessile on verrucae; rhizome branching, woody, producing numerous strict slightly virgate stems, 50 cm high, 1–2 mm across, leafy over entire length, simple or slightly branching; rosettes none; stem leaves stiff, elongate; lower leaves ovate-oblong, acute, gradually attenuate into petioles, entire or with somewhat contiguous teeth, with petiole 2–4 cm long; upper leaves sessile, sometimes slightly amplexicaul, gradually decreasing in size at stem apex, lanceolate or linear-lanceolate, acuminate. Flowers blue, sessile, in axil of small bracts, forming a nearly spicate inflorescence, sometimes flower terminal, single; teeth of tomentose calyx triangular-oblong or sublinear, obtuse, one-third to one-half the length of the corolla; appendages markedly accrescent, reticular, ovate, obtuse, completely concealing calyx, corolla tubular-infundibular, pubescent or glabrous, 2–2.5 cm long. June–July.

Central mountain belt, apparently in dry places. — Caucasus: possibly near the Turkish border. **Gen. distr.:** Bal.-As. Min. (As. Min.). Described from Turkish Armenia. Type in Paris.

Note. The records of this species nearest to the USSR are for Turkey (Kaziki-Paran, Palanteken Range, Malyi Bogaz and Gasan-Kuli mountains near Erzerum). It is sharply distinguished from all common species of Campanula in the flora of the USSR by verrucose-hamate hairs on the leaves, recalling those of Onosma gracile. The color of the corolla is mentioned only by Roemer and Schultes (l. c.). Linnaeus (l. c.) described the lower leaves as cordate, but this is not so. Possibly, his plants were not from Asia Minor or else he made a mistake; none of the authors following Linnaeus ever indicated cordate leaves, neither for Asia Minor nor for the Lebanon and Syria. Roemer and Shultes (l. c.) pointed out the incompatability of Linnaeus' diagnosis. A. De Candolle (in Prodr.) separated the plants from the Lebanon and Syria as C. libanotica DC, but if Linnaeus' description refers to plants from these countries as C. stricta then C. libanotica should be rejected in favor of a new name for the plants from Asia Minor.

213 Subsection 11. CORDIFOLIAE (Fom.) Fed. comb. nov. — Sect. Cordifoliae Charadze in Zam. po sist. i geogr. Tbil. bot. inst. 15 (1949) 21, p. p. — Ser. Cordifoliae Fom. in Mater. Fl. Kavk., IV, 6 (1904) 32, p. p. — Ser. Alliariaefoliae Kem.-Nath. in Zam. po sist. i geogr. rast. Tbil. bot. inst. 15 (1949) 10. — Appendages in notches between calyx teeth as long as tube,

usually longer; flowers large or medium-sized, in spicate-racemiform or racemiform inflorescences. Radical leaves long-petioled, with broad, usually deeply cordate triangular or slightly oblong lamina. Stems high. Pubescence tomentose or setulose. Perennial plants, with well-developed rhizome or indurate thick root.

Type species of subsection: C. alliarifolia Willd.

This subsection comprises a genetic series of forms which does not warrant its separation into species. Kharadze (op. cit.) recognized it as a section, but it is distinctly heterogenous. We are selecting two new subsections Latilimbus Fed. and Trigonophyllon Fed. — on the basis of a completely different structure of the calyx.

46. C. sclerotricha Boiss. in Kotschy, Pl. Pers. bor. ed. Hohenacker (1846) No. 502. diagn. brev.; Boiss. Diagn. ser. I, 2 66; ej. Fl. or. III, 901; Grossg., Fl. Kavk. IV, 63. — Exs.: Kotschy, l. c. (isotypus).

Biennial or perennial; scabrous with short setulose hairs; stem to 70 cm, straight, rather thick, 0.6—0.8 cm across in lower part, slightly sulcate, reddish, simple or short branching; lower leaves broadly ovate, acute, shallowly cordate, strongly incised- or lobate-dentate, long-petioled, with 10 cm long and 8 cm wide lamina; median cauline leaves short-petioled, acute, acutely dentate; bracts in inflorescence lanceolate, as long as flowers or longer. Flowers few, axillary, short-pediceled, declinate; calyx lobes broadly- and long-lanceolate, becoming larger, pectinate-bristly along margins and nerves, nearly as long as corolla; appendages broadly ovate, obtuse, as long as calyx-tube, corolla setulose-hairy, narrowly tubular-infundibular. July.

Shady rocks, also gardens. — Caucasus: S. Transc. (Nakhichevan ASSR near Ordubad). **Gen. distr.:** Iran (N.). Described from Elburz Range near Derbent, Iran. Type in Geneva, isotype in Leningrad (exs.).

47. C. brotheri Somm. et Lev. in Nuov. Giorn. Bot. Ital. IV, 2 (1897) 199, non C. brotherorum Feer (1890); Lipskii, Fl. Kavk. 377; Somm. et Lev. Enum. pl. Cauc. 315.

214 Apparently perennial; plant pale green, subglabrous; stem ca. 30 cm, straight, remotely leafy, simple, glabrous, 1.5 mm thick at base; radical and lower cauline leaves triangular-ovate, the laminas more than 2 cm long and as wide, acute, cordate, acutely and firmly crenate-dentate, petioles 9 cm, more than four times as long as lamina; median cauline leaves gradually decreasing in size, with shorter petioles; upper leaves ovate-lanceolate and lanceolate, the uppermost sessile, shorter than flowers. Inflorescence a secund, few-flowered loose raceme, ca. 10 cm long, flowers drooping when young, pedicels 1-flowered, ca. 1 cm or shorter, bracteoles of upper pedicels shorter than calyx; bracteoles linear, nearly as long as or longer than pedicels; teeth of glabrous calyx 7 mm, lanceolate, acuminate, with bristly ciliate margin, appendages ovate-oblong, obtuse, also with bristly-ciliate margin, longer than tube, as long as or only half the length of the calyx teeth; corolla blue, longer than calyx, divided into short lobes with pubescent margin. June.

Probably growing in the central mountain belt. — Caucasus: E. Transc. Endemic? Described from Kartlia near Akhaltsikhe. Type in Florence, isotype in Helsinki.

Note. Little is known about this species. It was described from Brotero's collections at the above-mentioned locality (No. 590 Brotero herbarium), but was never found again in the Caucasus. Judging by the description Sommier and Levier (l. c. 200) consider this species close to C. sclerotricha Boiss. from which it differs by the nakedness of all parts, the remote pedicels longer than the leaves, narrower and shorter calyx teeth, the calyx appendages exceeding the tube before anthesis, and the corolla without small bristles, not tubular. There seems to be no ground for identifying it with C. raddeana.

48. C. dolomitica E. Busch in Tr. Bot. muz. AN SSSR, XXII (1930) 215; Grossg., Fl. Kavk. IV, 63. – Ic.: E. Busch, op. cit. 216, f. optima.

Perennial; rhizome oblique, rather thin; stem 30–45 cm, ascending at base, straight higher up, usually branching from base, sometimes simple, rather thick, 2–3 mm across, with small spreading hairs, obsoletely ribbed or rounded, leafy; leaves densely tomentose beneath at first, becoming sparingly grayish-tomentose; radical leaves with very long slender petioles
215 several times as long as lamina, with petioles 25 cm long, reniform, orbicular-cordate or acuminate-cordate, irregularly crenate-dentate, together with lamina 2.2–5.5 cm long, 3.5–4 cm wide; cauline leaves long-petioled below, gradually reduced upwards, acuminate, cordate, crenate-dentate. Flowers very large, white, becoming yellow when dry, 4 cm long, long-pediceled in axils of leaves, on 4–20 cm pedicels; teeth of broad obconical sparingly pubescent calyx acuminate, ovate-lanceolate, one-sixth to one-fourth the length of the corolla; appendages ovate-sagittate, reflexed, concealing calyx-tube; corolla infundibular or campanulate, glabrous, with long-ciliate margin; divided for one-third into ovate slightly acuminate lobes, style not exserted; stamens with filaments broadened at base, fimbriate below. July. (Plate XII, Figure 2.)

Dolomites in subalpine meadows. – Caucasus: Cisc. (C. Caucasus). Endemic. Described from "Skalistyi" Range (Digoriya, Tetors, Digor-Tyrs). Type in Leningrad.

Note. A very beautiful species, close to C. alliarifolia and its relatives from which it differs by very large flowers on very long pedicels, and very long-petioled leaves. Formally, the type specimen is a very poorly developed plant from Agashtan (ascent to Dumagur Mountain), but its description refers to other specimens from Digor-Tyrs, one of which is shown in a drawing (E. Busch, op. cit., 116). Hence the plant is the true type. The plant was later collected on the "Skalistyi" Range near Upper Mizur, along the Ardon River, and at the Khyzny mountain pass.

49. C. alliariifolia Willd. Sp. pl. I (1797) 910; A. DC. Monogr. (1830) 241; idem in DC. Prodr. VII, 464; Boiss. Fl. or. III, 902; Shmal'g., Fl. II, 178; Fomin in Mater. Fl. Kavk. VI, 6, 32; Medv. in Tr. Tifl. bot. sada, XVIII, 2, 332; Grossg., Fl. Kavk. IV, 63; Kharadze in Fl. Gruzii, VIII, 141. – C. lamiifolia Ad. in Web. u. Mohr. Beitr. I (1805) 48; M. B. Fl. taur.-cauc. I, 154, 423, III, 144; Ldb. Fl. Ross. II, 878; Rupr. in Bull. Acad. Sc. Pétersb. XI, 194. – C. lamiifolia var. albotomentosa Rupr. l. c. 195. – C. macrophylla Sims in Bot. Mag. tab. 912. – C. alliariifolia var. macrophylla Trautv. in Tr. Peterb. bot. sada, VI (1879) 67. – Ic.: Buxb.

Cent. V, tab. 18 (C. foliis lamii etc.); Sims, l. c. tab. 912; H. Clifford Crook, Campanulas, 26 (photo). – Exs.: Herb. Fl. Cauc. No. 392; Sintenis, Iter or. No. 7082.

Perennial; rather densely pubescent, especially on lower surface of leaves; root thick; stems straight, or ascending, simple or branching, 70 cm
216 high, 3–4 mm thick; radical leaves triangular-cordate or cordate, sometimes nearly hastate, obtuse or acute, long-petioled, lamina 5–8 cm long and as wide, the petioles sometimes up to 20 cm; cauline leaves usually sharply decreasing in size, short-petioled; uppermost leaves subsessile, all leaves rather densely pubescent, often white-tomentose beneath, sparingly pubescent or subglabrous above, irregularly, sometimes very strongly, acutely toothed. Flowers large, in long secund raceme, on short pedicels, straw-colored, sometimes markedly yellowing when dry; teeth of tomentose calyx broadly lanceolate, acute, with ciliate margin, one-fifth to one-fourth the length of the infundibular to campanulate corolla, the latter pubescent outside, bearded at rictus; appendages of calyx lanceolate, as long as tube, curved, concealing tube; style hidden in corolla. July–August.

Rocks, mostly limestones, in forest mountain belt. – Caucasus: W. and E. Cisc., W., E. and S. Transc. **Gen. distr.**: Bal.-As. Min. (E.). Described from As. Min. Type in Berlin.

50. C. ochroleuca Kem.-Nath. in Zam. po sist. i geogr. rast. Tbil. bot. inst. 15 (1949) 10; Kharadze in Fl. Gruzii, VIII, 142. – C. alliariifolia ssp. ochroleuca Kem.-Nath. et var. rupestris, silvatica, alpestris Kem.-Nath. in Tr. Tifl. bot. inst. II (1938) 136. – C. alliariifolia var. cordata Trautv. in Tr. Peterb. bot. sada., VI, 1 (1879) 67. – Ic.: Kharadze, op. cit., Plate 372 (mediocr.).

Perennial; rhizome oblique; stem 80 cm, rather thick and firm, simple and straight, sparsely leafy, from middle passing over to loose, spicate-racemiform inflorescence, with irregularly arranged, more or less pubescent, longitudinally striated flowers; radical leaves triangular-cordate, 5–12 cm long, 3.5–10 cm wide, acute, crenate, abruptly tapering to petiole as long as or one and a half times as long as lamina, rather sparingly pubescent, especially beneath; cauline leaves strongly decreasing in size, very small in inflorescence, short-petioled or subsessile, the uppermost lanceolate, subentire. Flowers large, pale, ochre-yellow; teeth of short calyx lanceolate, hairy, with appendages one-third the length of the narrowly campanulate corolla, which is divided for one-third to one-fourth into ovate lobes, acute, erect; style slightly shorter than corolla. June–July.

Rocks in upper forest and subalpine mountain belts. – Caucasus: Cisc., W. and E. Transc. (only in middle part of Main Range). Endemic. Described from W. Georgia (Verkhnyaya Racha). Type in Tbilisi.

217 Note. Differs from the allied C. alliariifolia and from C. letschchumensis, C. leskovii and C. makaschvilii by the yellow flowers, from C. dolomitica by subsessile flowers and relatively short petioles, and from C. kirpicznikovii by the triangular-cordate, not oblong, radical leaves.

51. C. leskovii Fed. in Bot. mat. Gerb. Bot. inst. AN SSSR, XV (1953) 378. – Ic.: Fed., op. cit. 375, fig. 2.

Perennial; plant more or less covered with short opaque hairs broadened at base or subglabrous; stem straight, ascending, in nature possibly nearly drooping, simple — sometimes not branching — thin, weak (not firm and thick), leafy; radical and cauline leaves of similar shape, gradually decreasing in size toward stem apex, long-petioled in lower part of stem; uppermost leaves short-petioled or subsessile, all leaves broadly ovate or nearly reniform, with cordate or truncate base, acute or obtuse, rarely with rounded apex, obsoletely obtusely crenate, green, subglabrous above, dull beneath, covered with opaque appressed, very short hairs (never tomentose, or profusely pubescent), without petioles (1.5) 3–4 cm long, (1) 3–4 cm wide. Flowers medium-sized, short-pediceled, solitary, axillary; teeth of slightly tomentose obconical calyx broadly lanceolate or oblong, acute, with finely ciliate margin, half the length of the whitish (when alive, possibly bluish) infundibular corolla, bearded at rictus; calyx appendages ovate, obtuse, reflexed, enveloping tube to middle; style as long as corolla; capsule not known. August–September.

Limestones in forest mountain belt. – Caucasus: W. Transc. (Sochi, Caucasian game reservation, near Kulak glacier in C. Caucasus). – Endemic. Described from limestones above the Sochi River, near the watershed of the Mzymta and Shakhe rivers. Type in Leningrad.

Note. Although close to C. alliariifolia Willd., this species has a unique appearance, recalling not Alliaria but rather Glechoma hederacea. It is also marked by the smallness of the leaves, their obtuse crenation, slender stems, sparing pubescence and axillary flowers.

52. C. kirpicznikovii Fed. in Bot. mat. Gerb. Bot. inst. AN SSSR, XV (1953) 379.

Perennial; plant nearly entirely covered with appressed opaque and
218 short, often scabrous hairs, broadened at base; stems 2 (always?), leafy, 10–15 cm, covered with reflexed hairs; radical leaves large, ovate-oblong, with small basal notch, rarely slightly cordate, acute, irregularly obtusely crenate-dentate, without petioles 3–5–10 cm long, 2–3–4 cm wide, abruptly tapering into slender petiole, as long as lamina, leaves longer than stem; cauline leaves markedly decreasing in size, short-petioled, the uppermost small, subsessile. Flowers in racemes, axillary, medium-sized, ca. 1.5 cm long, drooping, whitish-yellow, on short twisted pubescent pedicels; teeth of obconical subglabrous calyx oblong, broadly lanceolate, acute, spreading or reflexed, slightly pubescent, with finely ciliate margin, half the length of the infundibular corolla, the latter subglabrous on both sides, calyx appendages triangular, recurved, sparingly pubescent; style strongly branching; stigmas 3. August.

Rocks, ravines. – Caucasus: Cisc. (C. Caucasus). Endemic. Described from rocks in Chegem ravine (Kabardian SSR). Type in Leningrad.

Note. Though described from one specimen, collected by Kirpichnikov in 1951, this is certainly a distinct species, clearly distinguished from the somewhat related C. alliariifolia Willd. by the oblong leaves, downturned calyx teeth and subglabrous corolla.

53. C. makaschvilii E. Busch in Bot. mat. Gerb. Bot. inst. AN SSSR, VII, 6 (1938) 128; Grossg., Opred. rast. Kavk. 416; Kharadze in Fl. Gruzii, VIII, 146. – Ic.: E. Busch, op. cit. 129, fig. bona.

(219)

PLATE XII. 1 – Campanula collina M.B.; 2 – C. dolomitica E. Busch; 3 – C. bayerniana Rupr.; 4 – C. autraniana Alb.

Perennial; root radishlike, thick, usually producing several stems; stems straight, 25–50 cm, with fine, sparse, spreading hairs, sometimes branching from middle, leafy; leaves gray-green, pubescent, not usually tomentose beneath, with short sparse hairs above; radical leaves long petioled, laminas acuminate, deeply cordate, 3–6 cm long, 3–7 cm wide, petioles 8–15 cm; petioles of leaves gradually shortening, laminas gradually decreasing in size and constricting, from the radical through the cauline to the uppermost terminal leaves, in addition to these characters also with truncate, not cordate, base, all leaves crenate-dentate. Inflorescence racemiform secund;
221 flowers drooping, short-pediceled, pink or nearly white; lobes of broadly conical calyx pubescent, lanceolate, half the length of the finely pubescent, infundibular, 2.5 cm long corolla; corolla divided for one-third into acute lobes, calyx appendages oblong, recurved, one-third the length of the calyx lobes, as long as tube; style not exserted. July.

Shrubby formations in lower belt, extending up to tea plantations. – Caucasus: W. Transc. (Adzhariya). Endemic. Described from Ochkhaluri (Batum District). Type in Leningrad.

Note. Very close to C. alliariifolia, but differing by its pink flowers.

54. C. letschchumensis Kem.-Nath. in Zam. po sist. i geogr. rast. Tbil. bot. inst. 15 (1949) 11; Kharadze in Fl. Gruzii, VIII, 146.

Perennial; plant tomentose, with thick root and ascending or straight, simple or branching stems, 60–120 cm high; radical leaves long-petioled; 18 cm long, 12 cm wide, triangular cordate, green above, with scattered short hairs, white-tomentose beneath, crenate or nearly entire; lower cauline leaves similar in shape to the radical, also long-petioled, upper leaves strongly decreasing in size, short-petioled. Flowers rather large, pink-violet or wine-red, drooping; teeth of tomentose calyx lanceolate, appendages ovate, as long as or shorter than tube, markedly shorter than campanulate corolla, style exserted from corolla. June–August.

Lower forest belt, shady ravines in limestone rocks. – Caucasus: W. Transc. (Kutaisi District, Ladzhanur River gorge, Rion gorge near Achariskhidi). Endemic. Described from Ladzhanur River gorge near Orpiri (in Lechkhume). Type in Tbilisi.

Note. This species differs from C. makaschvilii E. Busch, the closest ally of C. alliariifolia (and like it pink-fowered), by the apically green, subglabrous leaves, and by the shape of the leaves and the corolla.

Subsection 12. LATILIMBUS Fed. subs. nov. in Addenda XXIII, p. 332. – Appendages in notches between teeth of calyx very small, inconspicuous, sometimes filiform, or nearly absent; teeth not accrescent after flowering, usually straight. Radical leaves with cordate, truncate or broadly cuneate base, oblong-ovate, abruptly tapering to petioles, dentate or crenate; stems simple, or more or less branching; flowers in oblique raceme or arranged
222 in panicle on branching stems, drooping before blossoming. Capsule pendulous.

Type species: C. collina M. B.

Series 1. **Collinae** Kolak. s. str. – Appendages in notches between calyx teeth absent or inconspicuous. Lamina of radical leaves usually oblong-ovate or elliptic.

55. C. collina M. B. Fl. taur.-cauc. I (1808) 152; Roem. et Schult. Syst. V, 128; A. DC. Monogr. 272; idem in DC. Prodr. VII, 470; Rupr. in Bull. Acad. Sc. Pétersb. XI, 195; Boiss. Fl. or. III, 203; Shmal'g., Fl. II, 177; Alb. Prodr. Fl. Colch. 155, p. p.; Lipskii Fl. Kavk. 377; Somm. et Lev. in Tr. Peterb. bot. sada, XVI, 317; Fomin in Mater. Fl. Kavk. IV, 6, 38, p. p.; Grossg., Fl. Kavk. IV, 64, p. p.; Kharadze in Fl. Gruzii, 153. – C. collina var. major M. B. l. c. – C. collina var. pumila M. B. l. c. – C. collina var. eriocalyx Trautv. in Tr. Peterb. bot. sada, VI (1879) 65; Lipskii, op. cit. 377. – C. collina var. leiocalyx Trautv. op. cit. – Ic.: M. B. Cent. pl. rar. ross. I, tab. 42; H. Clifford Crook, Campanulas, 60.

Perennial; rhizome slightly creeping; stem straight, 30–40(50) cm, simple or arcuate at base, few-flowered, cylindrical, sparingly pubescent; leaves broadly lanceolate or ovate-oblong, slightly pubescent or subglabrous, regularly dentate; radical and median cauline leaves with long subglabrous petioles; lamina ca. 5–10 cm long, 2–3 cm wide, obtuse or acute base slightly cordate or truncate; uppermost leaves gradually passing into narrow, linear, sessile bracts. Flowers usually unilaterally declinate, remote, with pedicels of different length forming secund inflorescence, violet-blue, 3 cm long; teeth of subglabrous or else bristly-haired, semiglobose or short-conical calyx lanceolate, obtuse or acute, slightly pubescent or glabrous, one-third to one-half the length of the infundibular corolla; corolla glabrous outside, pubescent inside, divided for one-third into deflexed acuminate lobes; style as long as or slightly longer than corolla; seeds small, brown, not rimmed. June–July. (Plate XII, Figure 1.)

Subalpine and alpine meadows, not in rocks and snow. – Caucasus: Cisc., Dag., W., E. and S. Transc. Absent in mountains of S. Armenia and Talysh. **Gen. distr.**: Bal.-As. Min. (E. Anatolia). Described from Kislovodsk ("circa acidulam Narzana"). Type and paratype in Leningrad.

Note. A rather variable species, locally forming very poorly circumscribed races.

56. C. annae Kolak. in Zam. po sist. i geogr. rast. Tbil. bot. inst. 16 (1951) 55; Kharadze in Fl. Gruzii, VIII, 154.

223 Perennial; stems 30–45 cm, much branched, remotely bristly; radical leaves 10 cm long, the petioles long, narrowly winged, the lamina oblong or oblong-lanceolate, 4–6 cm long, 3.6–3.9 cm wide, short-acuminate, with cordate or truncate base, dentate; lower cauline leaves broadly lanceolate, the cuneate base decurrent to winged petiole; median and upper leaves sessile, lanceolate, attenuate-acuminate. Inflorescence racemiform, secund, with long arcuate pedicels, sometimes branching, the lower branches 3-flowered, the upper 1-flowered; teeth of pyriform calyx triangular-lanceolate, remotely setulose, nearly one-third the length of the broadly infundibular-campanulate blue-violet corolla; corolla 2.5 cm long, with ciliate margin; capsule oblong, tapering at base. July–August.

Gravels and subalpine meadows. – Caucasus: Cisc., E. Transc. (C. Caucasus). Endemic. Described from Kazbek District, Georgian SSR (Tskhal-Gori, taluses). Type in Tbilisi, cotype in Leningrad.

Note. Differs very little from the type of C. collina M. B.; its much larger flowers to 3.5 cm long, provide a distinguishing character.

57. C. schistosa Kolak. in Zam. po sist. i geogr. rast. Tbil. bot. inst. 16 (1951) 58; idem Fl. Abkh. IV (1949) 172; Kharadze in Fl. Gruzii, VIII, 155. – C. schistosa var. uniflora Kolak. op. cit. (1951) 58; idem, op. cit. (1949) 172. – Ic.: Kolak., op. cit. (1949) Plate XVI (fig. dextra).

Perennial; rhizome branching with rosettes of leaves producing reduced shoots, diverging below rosette and laterally, stems 20–30(40) cm arcuately curved at base, straight above, slightly leafy, usually bearing few flowers; radical leaves stiffly setulose, broadly ovate-oblong, with truncate or slightly cuneate base; crenate-dentate at margin, obtuse or acuminate apex, petioles nearly as long as lamina or slightly longer; cauline leaves broadly lanceolate, petiolate, the upper sessile, narrower. Bracts narrowly linear, 0.5–2 cm long, 1–2 mm wide; pedicels 5 mm; calyx semiglobose, densely pubescent with reflexed white bristles, teeth lanceolate-linear, 8–11 mm long, distinctly declined, appendages small; corolla 3.5 cm long, broadly campanulate, blue-violet, with ciliate margins. August.

Gravelly taluses in alpine belt (2,300 m). – Caucasus: W. Transc. (Abkhaziya). Endemic. Described from Malaya Dzyshra Mountain in Bzyb Range. Type in Leningrad.

224 Note. Differs markedly from the rather closely related C. collina M. B. by the short lamina, larger flowers and deflexed calyx teeth.

58. C. kluchorica Kolak. in Zam. po sist. i geogr. rast. Tbil. bot. inst. 16 (1951) 56; idem, Fl. Abkh. IV (1949) 175, nom. et diagn. ross.; Kharadze in Fl. Gruzii, VIII, 156. – C. microflora Kolak. in schedis (nomen hybridum et improprium).

Perennial; stem 20–30 cm, glabrous, often geniculately bent; radical leaves elongate, 3.5–5 cm long, 1.3–1.6 cm wide, ovate-lanceolate, attenuate-acuminate, with truncate or subcordate base, irregularly large-toothed, with long narrowly-winged petioles broadened below, slightly amplexicaul with ciliate margin; lower cauline leaves 4 cm, the uppermost 0.5–2 cm long; teeth of glabrous semiglobose calyx appressed to corolla, hamately curved at apex, one-third to one-fourth the length of the infundibular blue-violet corolla; corolla 15–20 mm long, divided into narrow acute lobes; style exserted, arcuately curved after flowering. July.

Gravels in subalpine and upper forest belts. – Caucasus: W. Transc. (Abkhaziya). Described at the junction of the Klukhori River gorge with the Nakhar River. Endemic. Type in Leningrad.

Note. The smallest-flowered species of the order Collinae Kolak., with weak, slender stem and small leaves.

The type specimen bears the label of the original, unpublished and incorrect description by Kolakovskii under the new name of "C. microflora;" not to mention the ungrammatical mixture of Greek and Latin words.

59. C. albovii Kolak. in Zam. po sist. i geogr. rast. Tbil. bot. inst. 16 (1951) 54; idem, Fl. Abkh. IV (1949) 175, descr. ross.; Kharadze in Fl. Gruzii, VIII, 155.

Perennial; stems slender, long, 30–60 cm, slightly leafy, somewhat flexuous, arcuately ascending in lower part, above nearly straight, spreading-haired below, subglabrous above; radical leaves usually long-petioled, with lamina oblong or slightly ovate-oblong, sometimes broadly lanceolate, (4)7–8(10) cm long, (1)1.5–3 cm wide, obtuse or acute, with cordate or broadly cuneate base, crenate-dentate. Inflorescence few-flowered, sometimes branching; teeth of glabrous or setulose semiglobose calyx lanceolate, spreading, 4–5 mm, one-third to one-half the length of the campanulate-infundibular corolla, calyx appendages inconspicuous; corolla 1.5–2 cm
225 long, violet-blue, divided for one-third into acute lobes; style as long as corolla. July–August.

Subalpine meadows. – Caucasus: W. Transc. (W. Georgia). Endemic. Described from the pastures of Afshtrakh in the upper reaches of Bzyb River. Type in Tbilisi, autotype [?] in Leningrad.

Note. This species has in common with C. kluchorica Kolak. the weak and slender stem and the large long radical leaves. Inflorescence few-flowered, drooping.

60. **C. sphaerocarpa** Kolak. in Soobshch. AN Gruz. SSR; VIII, 4 (1947) 238; idem, Fl. Abkh. IV (1949) 168; Kharadze in Fl. Gruzii, VIII, 154. – C. sphaerocarpa var. grandiflora Kolak. op. cit. (1947) 239. – C. sphaerocarpa var. multiflora Kolak. op. cit. (1949) 171. – C. collina var. abchasica Alb. in Bull. Herb. Boiss. II (1894) 118; Prodr. Fl. Colch. 156; Fomin in Mater. Fl. Kavk. IV, 6, 44; Grossg., Fl. Kavk. IV, 64. – Ic.: Kolak., op. cit (1949) Plate XVII (fig. manca).

Perennial; rhizome thin, producing reduced shoots with rosettes of leaves, the stems diverging laterally from under rosette, stem and rosette base covered with brown squamous remnants of petioles; radical leaves rounded-ovate to ovate-oblong, with cuneate or cordate base and attenuate acute or obtuse apex, irregularly serrate-dentate or subcrenate, stiff-haired, with short or long petioles covered with scattered hairs; cauline leaves short-petioled, decreasing in size, similar to radical leaves but narrower, uppermost leaves subsessile. Flowers solitary, slightly drooping; calyx usually violet, obconical, with scattered hairs, the hairs slightly bristly, teeth lanceolate-linear, strict, 3-nerved, appendages small, inconspicuous; corolla infundibular-campanulate, blue, 2–3(3.5) cm long, glabrous outside and at rictus with pubescent margin; capsule globose; seeds ovoid with narrow marginal rim. July–September.

Limestones in alpine mountain belt (2,000–2,500 m). – Caucasus: W. Transc. (Abkhaziya). Endemic. Described from Arbik Mountain in Gagry Range. Type in Leningrad.

Note. Kolakovskii's (op. cit.) var. grandiflora differs from the type and may be a distinct species.

61. **C. fonderwisii** Alb. in Bull. Herb. Boiss. II (1894) 117; Kharadze in Zam. po sist. i geogr. rast. Tbil. bot. inst. 17 (1953) 108; idem, Fl. Gruzii, VIII, 151. – M. migarica B. Schischk. in Tr. Azerbaidzh. otdel.
226 Zakavk. fil. AN SSSR, sekt. bot. 1 (1933) 49; Grossg., Fl. Kavk. IV, 70. – C. collina var. fonderwisii (Alb.) Lipsky, Fl. kavk. (1899) 377. – Ic.: Kharadze, op. cit. (1953) Plate 3 (mediocris).

Perennial; plant covered with short thin hairs; root woody, producing rosette of leaves and numerous thin arcuately ascending, 10–20 cm long stems, densely stiff-haired below, glabrous above, few-branched, with thin, leafless, elongate, 1-flowered branches forming few secund paniculate inflorescences; rosette leaves ovate or ovate-elliptic, rounded or acute with slightly cordate or attenuate base and long petioles; lamina 2–2.5 cm long, 1–1.5 cm wide, crenate, with short sparse stiff-hairs above and beneath; lamina of cauline leaves decreasing in size, ovate-lanceolate, acute; calyx teeth narrowly lanceolate, 5 mm long, 1 mm wide, acute, glabrous, one-fourth the length of the corolla, appendages recurved, with hamate mucro, shorter than or as long as calyx-tube; corolla blue, 2 cm long, campanulate, incised for more than one-third into acute, apically ciliate lobes; style shorter than corolla, barely reaching base of its lobes; stigmas 3. June.

Limestones in subalpine belt. – Caucasus: W. Transc. Endemic. Described from Mingrelia Mountain in W. Georgia. Type in Geneva, cotype in Tbilisi, topotype in Leningrad.

Note. This species, described by Al'bov (l. c.), was reduced by Lipskii (op. cit.) to a variety of C. collina. Fomin (Mater. Fl. Kavk. IV, 6 (1904)) would not even consider it as such; to him it was merely "a weakly developed shade form." Forgotten until Shishkin again described it from Mingrelia Mountain in 1933, this species now came to be called C. migarica. As such it was immediately accepted until Kharadze in 1952 discovered in Tbilisi one of Al'bov's specimens which served to establish the identity of C. fondervisii with C. migarica; new collections of this plant from its "locus classicus" cleared all doubts as to its merits as a distinct endemic species of limestone rocks in Mingrelia.

Not having seen the type (preserved in Geneva) associated with the original name, we rather doubt the validity of the name C. fondervisii Alb. The Tbilisi sheet mentioned by Kharadze was collected later (op. cit.), and is moreover mixed of two different species. If the type specimen also proves to be mixed, the C. fondervisii Alb. will become a nomen confusum, and preference must be given to C. migarica Schischk., all the more so since Al'bov's data are not quite accurate.

227 **62. C. irinae** Kuthath. in Zam. po sist. i geogr. rast. Tbil. bot. inst. 17 (1953) 96; Fl. Gruzii (1952) VIII, 152, descr. georgica. – Ic.: Kutat., op. cit. Plate 1 (bona).

Perennial; rhizome straight, woody, producing thin slightly angular shoots, covered with gray squamous remnants of leaves, partly disintegrated into short fibers; stems few, (6)10–15(20) cm, thin, arcuately curved, thin-sulcate, straight or ascending, in upper part with some thin elongate leafy branches, like leaves densely covered with microscopic scabrous hairs with remnants of leaves below; in radical leaves, long subfiliform petioles 6–10 cm, abruptly broadened to broadly oblong-ovate or suborbicular lamina, 3–5 cm long, 2–2.5 cm wide, half the length of the petiole, obtuse or rounded, rarely acute, with truncate or cordate base; bicrenate-serrate, densely covered above and beneath, especially beneath, with very short, stiff appressed gray hairs; cauline leaves sharply diminishing in size, 1–1.5 cm long, 0.5–1 cm wide, subsessile or sessile, obovate, with cuneate base, crenate-serrate, the teeth more acute than in radical leaves. Flowers 2.5 cm long, blue-violet, with filiform pedicels bearing small bracts, declinate or slightly drooping; calyx obconical, shortly and finely ciliate along ribs, teeth narrowly lanceolate or sublinear, acute, arcuately curved, finely ciliate, several times shorter than the narrow, infundibular glabrous corolla, divided for one-third or more into rather narrow lobes, appendages nearly as long as tube, one-third to one-half the length of the calyx lobes, finely ciliate, reflexed; style not exserted; stigmas 3; capsule pendulous, broadly obconical. August–October.

Apparently grows in rock fissures in forest mountain belt. – Caucasus: W. Transc. (W. Georgia). Endemic. Described from Tskhra-Dzhvari Mountain of Nakeral Range (Imeretia, Tkibuli District). Type in Tbilisi, autotype in Leningrad.

Note. Close to C. fonderwisii, for which it differs in the shape of the leaves, longer calyx teeth, different pubescence, and isolated distribution area.

Series 2. **Sarmaticae** Charadze s. smpl. – Appendages in notches between calyx teeth distinct, but short, sometimes dentiform, or elongate, subulate, hamately curved. Lamina of radical leaves usually triangular-elongate-ovate, with acute attentuate apex.

228 **63. C. calcarea** (Alb.) Charadze in Zam. po sist. i geogr. rast. Tbil. bot. inst. 13 (1947) 44; Grossg., Opred. rast. Kavk. IV, 417. – C. calcarea Alb. in schedis. – C. sarmatica var. calcarea Alb. Prodr. Fl. Colch. (1895) 155; Fomin in Mater. Fl. Kavk. IV, 64, 37; Grossg., Fl. Kavk. IV, 64.

Perennial; rhizome rather thick, covered with brown squamous leaf-remnants of previous year; stems 80 cm, robust, thick, arcuately curved below, straight above, simple, with long white hairs, inflorescences many-flowered, racemiform-paniculate, flowers on obsolete pedicels, crowded in clusters along rachis, radical and lower cauline leaves ovate or elliptic, cordate, cuneately decurrent along petiole or with suberect base, acute or obtuse, irregularly dentate with slightly crisp margin, grayish beneath, green above, both sides densely covered with rather long, partly soft, partly bristly hairs; cauline leaves smaller, sessile, lanceolate or linear-lanceolate; calyx white-haired, bristly, the teeth triangular-lanceolate, the appendages linear, shorter than tube; corolla blue, 3.5 cm long, 3–4 times as long as calyx teeth; capsule semiglobose, pendulous. August–September.

Alpine meadows, limestones and taluses (2,000–2,300 m). – Caucasus: W. Transc. (Abkhaziya: Mamdzyshkha, Akhavgash, Kukudzynshta, Atsetuka, Chugzyrkhv, Adzituka, Arbika, Ashkhabashkha mountains, Atskhun pasture). Endemic. Described from Kopeim'e Mountain in Gagry Range. Type in Tbilisi, cotype in Leningrad.

Note. The pubescent, thick stems recall C. cervicaria. The corolla is blue – not white as claimed by Kharadze (op. cit.) – as is evident from specimens collected in Gagry Range by Kolakovskii.

64. C. sarmatica Ker.-Gawl. in Bot. Reg. (1817) tab. 237; A. DC. Monogr. 240; idem in DC. Prodr. VII, 2, 464; Ldb. Fl. Ross. II, 2, 878; Rupr. in

Bull. Acad. Sc. Pétersb. XI, 195; Boiss. Fl. or. III, 902; Shmal'g., Fl. II, 179, p.p.; Fomin in Mater. Fl. Kavk. IV, 35, excl. var.; Grossg., Fl. Kavk. IV, 64, excl. var. nonnull.; Kharadze in Fl. Gruzii, VIII, 149. – C. betonicaefolia Biehler, Pl. nov. herb. Spreng. (1807) 14; M. B. Fl. taur.-cauc. I (1808) 153 et III (1819) 144, non Sibth. et Smith. (1806). – C. gummifera Willd. ex Roem. et Schult. Syst. V (1819) 144. – C. commutata Roem. et Schult. l. c. 143. – C. albiflora C. Koch in Linnaea, XXIII (1850) 634. – C. brotherorum Feer in Journ. of Bot. XXVIII (1890) 273. – C. sarmatica var. subtomentosa Trautv. in Tr. Peterb. bot. sada. VI (1879) 67. – Marianthemum betonicaefolium Sthrank in Flora, VII (1824)
229 11. – Ic.: Ker-Gawl. l. c. tab. 237; Kharadze, op. cit., Plate 373 (bona); H. Clifford Crook, Campanulas, 191. – Exs.: Brotherus, Pl. Cauc. No. 591 et 592 (C. brotherorum).

Perennial; root thick, slightly woody; stems 50 cm thick, sometimes longer, straight, simple or more or less branching, pubescent; radical leaves with cordate base, oblong or ovate-oblong, lamina 10 cm long 3, rarely 5 cm wide, obtuse, abruptly tapering to winged petiole, sometimes edge of wings lobed; cauline leaves ovate-oblong, shorter, with cordate or cuneate base and short petioles; uppermost leaves reduced, sessile, lanceolate, all leaves above and beneath, especially beneath, densely tomentose, crisp-crenate or dentate. Flowers 2.5–3 cm long, in loose slightly secund racemes, the lower on more or less long pedicels, the upper on short pedicels or subsessile; calyx teeth triangular-lanceolate, acute, one-third to one-half the length of the campanulate pale blue, sometimes nearly white corolla, appendages usually small, dentiform or attenuate from dentiform base, nearly one-fourth the length of the semiglobose, usually densely covered white-stiff-haired calyx-tube; corolla glabrous or with scattered stiff hairs along nerves, bearded at rictus, divided for one-fourth into parabolic acute lobes; capsule pendulous; style not exserted; seeds pale brown, ovoid, flattened, rimmed. June–August.

Rocky and stony habitats in mountains, from forest to subalpine belt. – Caucasus: Cisc., W. and E. Transc. (throughout the Main Range, but mainly on its northern slopes). Endemic. Described from the basin of the Kuban River. Type in Berlin.

Note. Recently, C. calcarea, C. sommieri and several others were removed as separate species from the allied C. sarmatica. C. brotherorum, earlier described, is a distinct form of the common C. sarmatica, long ago established by Fomin (op. cit.) and corroborated by new material from Osetia upon comparison with the isotype of C. brotherorum (No. 592, Brotero herbarium).

Biehler (l. c.) in the original diagnosis of C. betonicaefolia adds "E Cuban. Redowsky," which points to the basin of the Kuban River as the type locality. However, the reference to Redowsky as the collector is more than strange, as he never was in the Caucasus, but traveled and collected only in E. Siberia. Biehler's diagnosis, too, is strange, for it mentions
230 "capsula quinqueloculari calycis villosissimi sinubus reflexis testa," though in fact the capsule of C. sarmatica is always trilocular and the calyx appendages are short and do not conceal the tube. Hence, there is no reason to believe that C. betonicaefolia Biehler is synonymous with C. betonicaefolia M. B. In fact, the only reason for identifying the plants

described by Biehler and by Bieberstein is the latter's reference to Biehler. A. de Candolla (l. c.),and like him many other botanists, did not see Biehler's work.

65. C. woronovii Charadze in Zam. po sist. i geogr. rast. Tbil. bot. inst. 15 (1949) 22; Grossg., Opred. rast. Kavk. 417. — C. sarmatica f. tenuicalis Somm. et Lev. in Tr. Peterb. bot. sada. XVI (1900) 317. — C. sarmatica var. gracilis Fom. in Mat. Fl. Kavk. IV, 6 (1904) 37;p. p.

Perennial; plant covered with fine stiff hairs; stems 40 cm, rather thin, ascending, numerous, simple or slightly branching above, slightly leafy, bearing racemiform secund inflorescences; radical leaves long-petioled, elongate-triangular-lanceolate, with cuneate or slightly cordate base, biserrate-dentate, tapering to petioles 2–3 times as long as the 6–7 cm long, 2–2.5 cm wide laminas; cauline leaves smaller and narrower, the upper sessile. Bracts sessile, dentate; bracteoles small, narrow; calyx-tube narrow, rather long, white-lanate, the teeth linear-lanceolate, acute; corolla infundibular, bluish with ciliate margin, capsule white-lanate, narrowly cylindrical. September.

Alpine meadows. — Caucasus: W. Cisc. Endemic. Described from Fisht Mountain. Type in Leningrad.

Note. Distinguished from C. sarmatica Ker-Gawl. mainly by the very long-petioled leaves and narrow, long capsule.

66. C. siegizmundii Fed. sp. nov. in Addenda XXIII, p. 332. — C. sarmatica var. glabra A. DC. Monogr. (1830) 240; Fomin in Mater. Fl. Kavk. IV, 516.

Perennial; plant nearly quite glabrous or very sparingly pubescent; rhizome thick, branching above and producing shoots, in upper part bearing numerous setulose remnants of petioles; stems 20–40 cm ascending or straight, simple, rather thin, glabrous, slightly leafy, with racemiform elongate inflorescences; radical leaves 1.5–2 cm wide, with very long, thin petioles, with petioles 10–15 cm long, oblong-ovate or narrowly triangular-
231 ovate, attenuate, acute, with obliquely one-sided cordate base or truncate, acutely biserrate, glabrous above and beneath or only above, sometimes finely pubescent beneath, petioles filiform, glabrous, 3 times as long as lamina; lower cauline leaves petiolate, all others sessile, small, lanceolate. Pedicels nearly as long as calyx; calyx lobes lanceolate, acute, erect, subglabrous or barely pubescent, half the length of the corolla, appendages short, small, subulate, spreading; corolla glabrous; capsule spherical to short-obconical, prominently nerved, more or less covered with short hairs, opening at base by 3 small valves. Fr. August.

Rock crevices in forest mountain belt. — Caucasus: Cisc. (C. Caucasus). Endemic. Described from Sovetskii District in the Kabardian ASSR (Goluboe Lake near Babugent). Type in Leningrad, isotype in Kiev.

Note. The branching rhizome, with branches and shoots densely covered with dry, gray, squamous, bristly remnants of petioles, is very typical for this species, and distinguishes it from all other species allied to Sarmaticae. It is a typical chasmophyte.

67. C. sommieri Charadze in Zam. po sist. i geogr. rast. Tbil. bot. inst. 13 (1947) 46; Grossg., Opred. rast. Kavk. 417; Kharadze in Fl. Gruzii, VIII, 150. – C. sarmatica var. ramosissima Somm. et Lev. in Tr. Peterb. bot. sada. XVI (1900) 315. – C. sarmatica var. gracilis Fomin f. ramosissima (Somm. et Lev.) Fomin in Mater. Fl. Kavk. IV, 6 (1904) 37; Grossg., Fl. Kavk. IV, 64. – Ic.: Kharadze, op. cit. (1947) 57.

Perennial; appressed pubescent plant; stems 15–40 cm, branching from base, thin, ascending; radical leaves 2–4(7) cm long, 1.5–3 cm wide, ovate or oblong-ovate, cordate, acute, irregularly serrate-dentate, long-petioled, dark green, appressed-hairy above, grayish, densely covered with fine white hairs beneath, prominently nerved; cauline leaves numerous, 0.5–2.5 cm long, distinctly reduced at stem apex, cuneate, petiolate; uppermost leaves small, lanceolate, very short-petioled. Flowers terminal, on elongate branches, solitary or 2–3 on long pedicels; calyx-tube with fine white hairs, teeth narrowly lanceolate, acute, finely pubescent, appendages small; corolla 1.6 cm long, narrowly infundibular, twice as long as calyx teeth, glabrous outside, with ciliate margin, violet; style slightly exserted; receptacle hairy;
232 capsule obovate-conical, elongate, pendulous; seeds ovoid, flattened, brown, rimmed. July–September.

Caucasus: W. Cisc. (upper reaches of Kuban River, Klukhori). Endemic. Described from cliffs between Indysh and the Khumar fortress. Type in Tbilisi, cotype in Leningrad.

68. C. calcarata Somm. et Lev. in Nouv. Giorn. Bot. Ital. (1895) 96; Tr. Peterb. bot. sada, XVI (1900) 324; Lipskii, Fl. Kavk. 379 and Suppl. I, 69; Fomin in Mater. Fl. Kavk. IV, 6, 45; Grossg., Fl. Kavk. IV, 70; Kharadze in Fl. Gruzii, VIII, 157. – Ic.: Somm. et Lev., op. cit. (1900) Plate 32.

Perennial; sparsely pubescent or subglabrous plant; rhizome creeping, covered with squamous remnants of petioles; stems 15–30 cm, ascending, thin, leafy, 1- or few-branched, sometimes branching below middle, leaves of sterile shoots and radical leaves ovate-cordate or oblong-cordate, bicrenate-dentate, long-petioled; cauline leaves oblong-lanceolate, crenate-dentate, short-petioled; upper leaves sessile. Flowers 1.5–2 cm long, long-pediceled, bracteolate, forming nearly racemiform lateral inflorescence; teeth of glabrous calyx triangular-lanceolate, one third to one half the length of the blue campanulate corolla, appendages subulate-acuminate, recurved, sometimes with hamately curved tip, longer than tube; style nearly as long as corolla or slightly exserted; capsule pendulous; seeds small, yellowish, not rimmed. July (September).

Rocks in forest mountain belt. – Caucasus: Cisc. (upper reaches of the Kuban River). Endemic. Described from cliffs between Madniskhevi (Uchkulan) and Indysh. Type in Leningrad, isotype in Florence.

Note. A rare plant. Besides Sommier and Levier's type the only specimens available in herbaria are those collected by Busch and Klopotov near Kart-Dzhiyurta (along the Kuban River).

Subsection 13. TRIGONOPHYLLON Fed. subs. nov. in Addenda XXIII, p. 333. — Calyx teeth nearly without appendages or with small ones, accrescent after flowering, longer or even many times longer than tube; flowers drooping before anthesis; capsule pendulous. Radical leaves and leaves of sterile shoots usually with thin petioles, lamina long or very long, triangular, with cordate truncate or cuneate base. Stems 1-flowered, simple or slightly branching, with branches terminating by flowers. Rhizome branching, with sterile shoots or without shoots.

Type species: C. dzyschirica Kolak.

233 Series 1. **Dzyschricae** Ser. Collinae Kolak. p. min. p. — Stem simple, 1-flowered or slightly branching with few flowers, slightly leafy. Cauline leaves small, sharply set off from large, long-petioled triangular-ovate radical leaves, usually sessile or subsessile, broadly lanceolate.

69. C. dzyschrica Kolak. in Zam. po sist. i geogr. rast. Tbil. bot. inst. 16 (1951) 57; idem, Fl. Abkh. IV (1949) 172, nomen et descr. ross.; Kharadze in Fl. Gruzii, VIII, 157. — Ic.: Kolak., op. cit. (1949) table XVIII (fig. manca).

Perennial; entire plant grayish, scabrous, with small dense bristly hairs; rhizome branching, rather thin, with creeping shoots and lateral, ascending stems longer than leaves, slightly leafy, terminating by 1–2 drooping flowers; radical leaves long-petioled, ovate-triangular, cordate, with undulant margins and irregular large, acute teeth; laminas 3–5 cm long, 2–4 cm wide; cauline leaves decreasing in size, short-petioled, cuneate at base. Flowers 1 cm long with 1–4 small bracteoles, unequally pediceled; calyx conical, semiglobose in fruit, bristly, the teeth triangular-lanceolate, declined, 1 cm and longer, appendages inconspicuous; corolla pale blue-violet, campanulate, 2–2.5(3) cm long, incised for one-third, with ciliate margin. July–August.

Alpine belt, limestones. — Caucasus: W. Transc. (Abkhaziya). Endemic. Described from Malaya Dzyshra Mountain in Bzyb Range. Type in Leningrad.

Note. This unique species is readily distinguished from its allies by its dense, short-bristly pubescence. It is most closely related to C. jadvigae Kolak, less closely to C. autraniana Alb.

70. C. jadvigae Kolak. in Bot. mat. Gerb. Bot. inst. AN SSSR, XV (1953) 22; idem, Fl. Abkh. IV (1949), 171, nomen et descr. ross.; Kharadze in Fl. Gruzii, VIII (1952) 156.

Perennial; plant subglabrous; rhizome branching, with numerous sterile shoots and 1–2-flowered stems; stems straight, rather thin, arcuately curved at base, diverging from reduced sterile shoots, bearing 2–3 small, leaves; radical leaves as long as or longer than stems, 10–12 cm, with long subfiliform petioles and broadly ovate lamina, with cuneate or truncate base, strong obtuse teeth, subglabrous, with short scattered cilia along margin and nerves; cauline leaves similar to the radical, but much smaller and narrower,
234 the lower leaves petiolate, the upper sessile, the uppermost close to base of slightly drooping flowers. Calyx tube semiglobose, with white bristles,

PLATE XIII. 1 – Campanula kachetica Kantsch.; 2 – C. radula Fisch. ex Tchihatch.; 3 – C. incanescens Boiss.

the teeth linear-lanceolate, 6–8 mm long, 2 mm wide, without appendages, declined, acute; corolla blue-violet, broadly campanulate, incised for one-third, lobes acute, ovate with long scattered cilia along margin. June–August.

Crevices of limestone rocks in cirques. — Caucasus: W. Transc. (Abkhaziya). Endemic. Described from Arabik Mountain in the Gagry Range. Type in Leningrad.

Note. This remarkable species is somewhat reminiscent of C. autraniana Alb., from which it is distinguished by the short, few-flowered stems — as long as the leaves — the obtuse crenation of the leaves, the different shape of the ciliate calyx and its habit. Buried in rock crevices, the rhizome produces obsolete shoots, bearing bundles of leaves, in the axils of which there grow leafy branches, lateral to the shoot axis. C. autraniana produces stems from the heart of the rosette.

71. C. kolakovskyi Charadze in Zam. po sist. i geogr. rast. Tbil. bot. inst. 13 (1947) 108; Kolak., Fl. Abkh. IV, 168; Kharadze in Fl. Gruzii, VIII, 152. — Ic.: Kolak., op. cit. tabl. XVI (fig. manca).

Perennial; plant covered with scattered stiff hairs; rhizome rather thin, branching in upper part; stems 25–35(40) cm, subglabrous, with few leaves and 1–2 drooping flowers; radical leaves elongate-ovate, cordate, acute, from cordate base cuneately decurrent along long thin petiole, irregularly and obtusely toothed or slightly incised, the teeth turned forward; cauline leaves much smaller, short-petioled, narrower, the uppermost lanceolate, sessile. Bracts linear-lanceolate, small; teeth of stiff haired white calyx triangular-lanceolate, acuminate, glabrous, appendages filiform, shorter than tube, white-haired; corolla 3.5 cm long, $2\frac{1}{2}$ times as long as calyx teeth, dark violet, infundibular, incised for one-third, glabrous, with long hairs along margin. June–August.

237 Crevices of limestone rocks in central (forest) mountain belt. — Caucasus: W. Transc. (Abkhaziya). Endemic. Described from Gegi River gorge (tributary of the Bzyb River). Type in Tbilisi, topotype in Leningrad.

Note. Kharadze has compared this species with C. sphaerocarpa Kolak. and C. collina M. B. However, but for the filiform appendages of the calyx it is in all characters closest to C. dzyschrica Kolak.

Series 2. **Antiquae** Kolak. s. str. — Stem branching in upper part, forming loose, densely leafy inflorescence with few flowers. Cauline leaves similar in shape to the radical, gradually decreasing in size, like the lower leaves elongate-triangular-ovate, attenuate, petiolate beneath, sessile in upper part of stem.

72. C. suanetica Rupr. in Bull. Acad. Sc. Pétersb. XI (1867) 190; Boiss. Fl. or. III, 910; Somm. et Lev. in Tr. Peterb. bot. sada, XVI, 323; Fomin in Mater. Fl. Kavk. IV, 6, 74; Grossg., Fl. Kavk. IV, 71; Kharadze in Fl. Gruzii, VIII, 150. — C. suanetica var. appendiculata Somm. et Lev., op. cit. tabl. XXXI.

Perennial; completely glabrous plant; rhizome rather thin, sometimes thickened, covered with squamous or fibrous remnants of petioles; stems ascending, leafy, branching at apex, branches emerging from axils of leaves 1–3-flowered; radical leaves and leaves of sterile shoots reniform- or ovate-cordate, very long-petioled; cauline leaves ovate or oblong-ovate, more or less deeply cordate, attenuate, short-petioled, all leaves twice or irregularly deeply dentate. Flowers medium-sized, on long, slender, erect pedicels; teeth of glabrous calyx strongly declined, linear-lanceolate, small-toothed, one-third to one-half the length of the corolla; calyx appendages finely dentiform, nearly as long as tube; corolla blue infundibular-campanulate, glabrous, incised for one-third into acute oblong lobes, bearded at rictus; style slightly exserted; receptacle glabrous; ripe capsule semiglobose, erect; seeds yellow-brown, shiny, small, flattened, rather broadly rimmed. July–August.

Rock crevices in forest mountain belt. – Caucasus: W. Transc. (Lentekhi, Muri in Svaniya). Endemic. Described from rocks along the Tskhenis-Tskhale River near Muri. Type in Leningrad.

73. C. autraniana Alb. in Bull. Herb. Boiss. II (1894) 115; ej. Prodr. Fl. Colch. 157; Fomin in Mater. Fl. Kavk. IV, 6, 73; Grossg., Fl. Kavk. IV, 71. – Ic.: H. Clifford Crook, Campanulas, 37.

Perennial; completely glabrous plant; rhizome thin, branching; stems thin, weak, often decumbent, flexuous, slightly branching, branches 1-flowered;
238 leaves semicoriaceous; radical leaves ovate-cordate, ovate or ovate-oblong, tapering to acute tip, sometimes with cuneate rounded or one-sided cordate base, irregularly biserrate-dentate, very long-petioled, petioles sometimes with small lateral subulate teeth; cauline leaves oblong, with cuneate base and short petioles; upper leaves oblong-lanceolate or lanceolate, subsessile. Flowers medium-sized, drooping after flowering; teeth of glabrous calyx linear, one-fifth to one-quarter the length of the campanulate corolla; appendages in incisions between calyx teeth inconspicuous; style in corolla; receptacle glabrous; capsule obconical; seeds brown, small, ovoid, not rimmed. August–September. (Plate XII, Figure 4.)

Limestones in subalpine belt, near upper timberland. – Caucasus: W. Transc. Endemic. Described from Fisht Mountain. Type in Leningrad, isotype in Tbilisi and Geneva.

74. C. engurensis Charadze in Zam. po sist. i geogr. rast. Tbil. bot. inst. 1 (1938) 7, 15 (1949) 17; idem, Fl. Gruzii, VIII, 135. – Ic.: Kharadze, op. cit. (1938) Fig. 1 (bona).

Perennial; plant grayish-green, covered with fine transparent hairs; stems numerous; simple, weak, 16–25 cm, sparsely pubescent, borne on thin creeping rhizome; leaves glaucescent beneath, green above, densely and finely pubescent, irregularly largely twice incised-dentate, teeth acute; radical leaves often asymmetrical, long-petioled, ovate-lanceolate, rarely ovate, with notched, sometimes slightly cordate or truncate base, acuminate, 3–8 cm long, 2–4 cm wide; cauline leaves ovate, ca. 4 cm long, with broadly cuneate base, short-petioled; upper leaves decreasing in size, subsessile. Inflorescence 3–7-flowered, loosely corymbiform; bracts lanceolate or linear-lanceolate, dentate or subentire; pedicels erect, shorter than calyx; calyx densely pubescent, broad, short, its teeth triangular-lanceolate, with very short appendages, one-quarter to one-third as long as the length of the corolla. Corolla tubular-campanulate, 2–2.7 cm long, with fine scattered hairs outside, dark blue, violet-blue when dry, incised for one-fifth into ovate acute lobes; anthers long, free, sometimes connate; style not exserted. July.

Rocks in woody ravines.— Caucasus: W. Transc. (Verkhnyaya Svaniya). Endemic. Described from Khaishi (Tkheishi ravine). Type in Tbilisi.

Subsection 14. SYMPHYANDRIFORMES (Fom.) Fed. comb. nov. — Sect. Symphyandriformes (Fomin) Charadze in Zam. po sist. i geogr. rast. Tbil.
239 bot. inst. 15 (1949) 17, excl. spec. nonnull. — Ser. Symphyandraeformes Fom. in Mater. Fl. Kavk. IV, 6 (1907) 894, p. p. — Calyx teeth divergent, limb of calyx often broader than corolla, with more or less developed appendages in incisions between teeth; anthers nearly always adnate to tube when young, later partly divergent. Radical leaves ovate-cordate, long-petioled, dentate. Plant with many stems and branching inflorescence. Rhizome with shoots, covered with squamous remnants of petioles.

Type species of subsection: C. kolenatiana C. A. M.

Series 1. **Finitimae** Fed. — Calyx teeth glabrous or pubescent but without hyaline marginal cilia. Plant more or less pubescent. Flowers white, reddish or violet. Anthers partly connate. Lamina elongate, ovate, shallowly cordate.

75. C. finitima Fom. in Vestn. Tifl. bot. sada, 1 (1905) 15; Kharadze in Zam. po sist. i geogr. rast. Tbil. bot. inst. 15, 17. — Symphyandra finitima Fom. in Mater. Fl. Kavk. IV, 6, rev. and enl. (1907) 151; Grossg., Fl. Kavk. IV, 76.

Perennial; plant with crisp grayish fine hairs; rhizome woody, thick, usually branching; stems rather low, ca. 20 cm thin, ascending, flexuous, leafy, short-spreading to corymbiform, branching in upper part with 1-flowered branches; leaves stiff, small, with 4 cm long lamina; radical leaves rather long-petioled, broadly ovate, with slightly cordate or broadly cuneate base, obtuse, crenate-dentate; lower cauline leaves oblong-ovate, near base obtuse, subulate, short-petioled, crenate-serrate; upper leaves decreasing in size, short-petioled or subsessile, acute. Pedicels erect, much shorter than calyx, with linear-lanceolate bracts; flowers 4 cm long, white, slightly reddish; teeth of broad obconical calyx broadly triangular, acute, one-third the length of the corolla; calyx appendages small, reddish, much shorter than tube; style not exserted; corolla pubescent, poculiform, constricted at base, bearded inside, divided for one-quarter into ovate acute lobes; anthers adnate to tube or free. July.

Rock crevices in central mountain belt. — Caucasus: possibly near the Turkish border. **Gen. distr.**: As. Min. (border with the Caucasus). Described from Tamrut near Ol'tin District. Type in Tbilisi, topotype in Leningrad.

Note. This species, found only by Koenig, is also known from the village of Zekhdzor, not far from the "locus classicus." It was introduced in the
240 Tbilisi Botanical Garden in 1903 by Fomin, who studied it in great detail.

76. C. ossetica M. B. Fl. taur.-cauc. III (1819) 145; Kharadze in Zam. po sist. i geogr. rast. Tbil. bot. inst. 15 (1949) 17; Grossg., Opred. rast. Kavk. 417; Kharadze in Fl. Gruzii, VIII, 139. – Symphyandra ossetica (M. B.) A. DC. Monogr. (1830) 368; idem in DC. Prodr. VII, 2, 495; Ldb. Fl. Ross. II, 2, 896; Boiss. Fl. or. 889; Rupr. in Bull. Acad. Pétersb. XI, 221; Trautf. in Tr. Peterb. bot. sada, VI, 94; Fomin in Mater. Fl. Kavk. IV, 6, 121; Grossg., Fl. Kavk. IV, 76. – Ic.: Fl. Gruzii, VIII, tabl. 139.

Perennial; plant sparsely white-lanate (white-bristly), becoming subglabrous; rhizome as thick as a finger, short-branching, blackish-brown, grumous, woody, in upper part covered with some scabrous remnants of petioles and numerous leaf traces [sic!]; stems thick, curved, declined, densely leafy, elongate and racemiformly branching, 20–40(50) cm, usually reddish; leaves deeply bidentate or partly incised-dentate, with acute glandular teeth, with white bristles along margin; radical leaves long-petioled, cordate, to 10 cm long and more, 3–5 cm wide, ovate or ovate-oblong, acuminate; median leaves sessile, the upper also sessile, narrow. Flowers violet, with filiform slender pedicels, bracteate, sometimes flowers in two's; teeth of glabrous obconical calyx very broadly lanceolate, acuminate, strongly accrescent in fruit, their margin with scattered teeth and cilia, much longer than tube, half the length of the corolla; calyx appendages narrowly lanceolate, very acuminate, nearly as long as tube; corolla tubular-campanulate, shallowly divided into acute lobes; style not exserted; stamens usually not connate; capsule obconical; seeds brown, oblong, acute. June–July.

Rocks in central mountain zone, ravines. – Caucasus: Cisc. (Osetiya, Digoriya). Endemic. Described from Mtiuleti (Kaishaur). Type in Leningrad.

Series 2. **Raddeanae** Fed. – Calyx teeth with transparent bristly cilia broadened at base. Plant glabrous or subglabrous. Flowers violet or bluish; stamens free. Leaf elongate, more or less attenuate.

77. C. raddeana Trautv. in Bull. Acad. Sc. Pétersb. X (1852) 395; Boiss. Fl. or. III, 910; Fomin in Mater. Fl. Kavk. IV, 6, 75; Grossg., Fl.
241 Kavk. IV, 70; Kharadze in Fl. Gruzii, VIII, 140. – Ic.: H. Clifford Crook, Campanulas, 168. – Exs.: A. et V. Brotherus, Pl. Cauc. No. 590.

Perennial; glabrous plant; stems rather thin, erect, simple or slightly shortly and paniculately branching, borne on rather thin, slightly creeping rhizome covered with squamous remnants of petioles; leaves acute-dentate; radical leaves and leaves of sterile shoots long-petioled, deeply cordate-ovate or triangular; cauline leaves short-petioled, ovate or triangular, the upper decreasing in size, subsessile, acuminate. Flowers medium-sized, declined, solitary in axils of upper leaves, pediceled, bracteate and longer than calyx; teeth of glabrous calyx triangular, acute, covered with transparent bristles on broadened base; calyx appendages triangular-lanceolate, bristly, recurved, longer than calyx tube; corolla violet, campanulate, bearded inside, twice as long as calyx teeth; anthers orange; style exserted; receptacle glabrous. June–July.

Central mountain belt. – Caucasus: E. Transc. (W.), S, Transc. (N.). Endemic. Described from Borzhomi. Type in Leningrad.

Note. There seems to be no reason to identify this species with C. brotheri, later described by Sommier et Levier.

78. C. betulifolia C. Koch in Linnaea, XXII (1849) 635; Boiss. Fl. or. III, 909; Fomin in Mater. Fl. Kavk. IV, 6, 147; Grossg., Fl. Kavk. IV, 71. – C. denticulata Boiss. et Huet in Boiss. Diagn. ser. II, 3 (1856) 107. – Ic.: H. Clifford Crook, Campanulas, 47. – Exs.: Herb. Fl. Cauc. No. 291.

Perennial; glabrous or sparingly finely pubescent glaucous-green plant; stems 10–15(20) cm, often numerous, straight or slightly decumbent, brittle, leafy, flexuous, nearly corymbiform, with spreadingly branches in upper half, branches 1–3-flowered. Leaves indurate, slightly fleshy; radical and lower cauline leaves ovate-cuneate, large-toothed, long-petioled; upper leaves subsessile, small, oblong-lanceolate. Flowers erect, shorter than calyx; tube of pubescent calyx very short, with very acute lanceolate teeth and appendages; corolla narrowly campanulate, pubescent, three times as long as calyx teeth. May–June.

Rock crevices in ravines of mountain rivers and slopes. – **Gen. distr.**: As. Min. So far found only at border of Caucasus, near Artvin and Ardanuch. May also occur along the Chorokh River. Described from the Chabants valley, in eastern Asia Minor. Type in Berlin.

242 **79. C. kolenatiana** C. A. M. ex Rupr. in Bull. Acad. Sc. Pétersb. XI (1867) 216; Boiss. Fl. or. III, 910; Fomin in Mater. Fl. Kavk. IV, 6, 74; Grossg., Fl. Kavk. IV, 70 (calami lapsu – "Colenatiana"); Kharadze in Fl. Gruzii, VIII, 136. – Ic.: H. Clifford Crook, Campanulas, 109.

Perennial; subglabrous plant; rhizome rather thick, multicipital; stems numerous, straight or ascending, sometimes flexuous, usually slightly branching, branches uniflorous or bearing few-flowered, nearly one-sided racemes; radical leaves cordate-ovate, obtuse, incised-crenate, with long denticulate petioles; cauline (median) leaves ovate, usually short-petioled; upper leaves oblong, sessile. Pedicels thin, short, bracteate; flowers large, usually drooping; calyx glabrous, its teeth and appendages triangular-lanceolate, acute, with small spreading teeth and bristles on broadened base, appendages one-third as long as teeth and reflexed, longer than calyx; corolla broadly campanulate, violet, glabrous, more or less bearded inside, three times as long as calyx; receptacle glabrous; style slightly exserted. June–August.

Rocks in deep and moist ravines, forest and subalpine belts. – Caucasus: E. Transc., Dag. Endemic. Described from Salvat Mountain near Nukhi. Type in Leningrad.

Note. This species of Meyer was described by Ruprecht (op. cit.) after specimens grown in 1847 from seeds obtained by Kolenati from Salvat Mountain near Nukhi, supplemented by a specimen from rocks along the Bezhit River near Geled, in Dagestan.

80. C. achverdovii Charadze in Zam. po sist. i geogr. rast. Tbil. bot. inst. 15 (1949) 18; Grossg., Opred. rast. Kavk. 421.

Perennial; rhizome multicipital, robust, branching, producing numerous ascending leafy stems 30–45 cm, branching in upper part, covered with

scattered fine hairs in lower half; radical leaves decurrent along long thin petioles, ovate or triangularly ovate, with cuneately tapering, slightly cordate base, acute, acutely dentate-serrate at margins, sometimes incised-dentate beneath, covered above and beneath with fine scattered stiff hairs, 3–3.5 cm wide, with petiole 4–5 cm long; lower cauline leaves ovate, the upper ovate-lanceolate, petiolate, the uppermost sessile; bracts small, lanceolate. Flowers in paniculate inflorescence, solitary, at tips of branches and in axils; calyx glaucous, basally long- and white-haired, its teeth triangular-lanceolate, tapering above, long-acuminate, one-third the length of the corolla; calyx appendages long, lanceolate, appendages and teeth
243 glabrous; corolla violet, tubular-infundibular, glabrous outside, hairy inside, 2–3.5 cm long; receptacle glabrous, concave; capsule pendulous. August.

Dry stony slopes in central mountain belt. – Caucasus: Dag. (upper reaches of Samur River). Endemic. Described from Luchek in Rutul District. Type in Tbilisi.

Note. Very close to C. kolenatiana, from which it differs only by the glaucous calyx with long attenuate teeth and the narrowly infundibular corolla.

81. C. kemulariae Fom. in Tr. Bot. inst. AN SSSR, ser. I, No. 3. (1936) 289; Grossg., Opred. rast. Kavk. 421; Kharadze in Fl. Gruzii, VIII, 139. – Ic.: Fomin, op. cit. 290, Fig. opt.; H. Clifford Crook, Campanulas, 108.

Perennial; sparingly pubescent plant; the thick creeping rhizome bears straight, branching, 19–20 cm, stems, radical and lower cauline leaves ovate, often oval, with deep cordate incision, attenuately acuminate, twice large-toothed, petioles much longer than lamina; median cauline leaves of similar shape but petioles much shorter, as long as or shorter than lamina; upper leaves oval, incised at base, acuminate, short-petioled or subsessile. Flowers medium-sized, in more or less leafy paniculate inflorescence, with clusters of few flowers on terminal branches; bracts narrowly lanceolate; calyx glabrous, with triangular, 10–12 mm long acuminate teeth with transparent marginal cilia, appendages triangular-lanceolate, with ciliate margin, twice as long as tube; corolla campanulate, blue, with bearded margin, nearly twice as long as calyx; style conspicuously exserted from corolla, with 3 spirally curved stigmas. June.

Limestones in the forest belt. – Caucasus: W. Transc. Endemic. Described from W. Georgia (near Chiaturi). Type in Leningrad, cotype in Tbilisi.

Note. More or less closely related to C. kolenatiana and C. raddeana from which it is distinguished by biserrate, deeply cordate but not incised-crenate leaves and multiflorous inflorescence, and by the shape of the leaves and multiflorous inflorescence, with yellow (not orange) anthers respectively.

Series 3. **Bayernianae** Fed.– Calyx teeth without bristly cilia, sometimes with crisp hairs; flowers blue; anthers free. Leaf lamina short, cordate-ovate, sometimes subreniform.

244 **82. C. bayerniana** Rupr. in Bull. Acad. Sc. Pétersb. XI (1867) 188; Boiss. Fl. or. III, 914; Fomin in Mater. Fl. Kavk. IV, 6 (1905) 76, excl. var. Trautvetteri Fom.; Grossg., Fl. Kavk. IV, 69. – C. bayerni Rupr. in schedis.

Perennial; nearly entirely glabrous plant; rhizome branching, producing small shoots with rather thin, 1 mm flexuous, few-flowered stems, 15–20 cm; leaves rather coriaceous; radical leaves and leaves of sterile shoots long-petioled, reniform or ovate-cordate, 1–2(2.5) cm wide, irregularly large-toothed, with hamate declinate teeth and crisp-haired margin; cauline leaves decreasing in size, ovate-orbicular or ovate, large-toothed, short-petioled. Flowers 1.5(2) cm, the bracteate pedicels twice as long as calyx; calyx teeth triangular, long-acuminate, with crisp hairs along margin; appendages very short, tube with hemispherical or conical base; corolla narrowly campanulate, blue, glabrous outside, bearded inside, 4–5 times as long as calyx; style not exserted; broadened part of filaments usually bearded; receptacle glabrous. June–July. (Plate XII, Figure 3.)

Central and subalpine rocks. – Caucasus: S. Transc. (Armenia, Nakhichevan ASSR, Karabakh). **Gen. distr.**: N. Iran. Described from Ordubad, near Shikhaus). Type in Leningrad.

83. C. choziatowskyi Fom. in Tr. Tifl. bot. sada, VI, 3 (1904) 38 and in Mater. Fl. Kavk. IV, 6, 78; Grossg., Fl. Kavk. IV, 69. – Ic.: H. Clifford Crook, Campanulas, 60.

Perennial; glabrous or sparingly pubescent plant; rhizome thick, woody; stems ascending, flexuous, slightly branching, many-flowered; leaves thick, coriaceous when dry; lower leaves long-petioled, cordate-ovate, with large, acute teeth, the margins slightly reflexed, sometimes with scattered cilia; cauline leaves ovate or subcordate or rounded, short-petioled; upper leaves decreasing in size, ovate-lanceolate, subsessile; lamina 1–2.5 cm long, slightly shorter than broad. Flowers medium-sized, bluish, the pedicels filiform, slightly shorter than flowers, bracteate; bracts small, lanceolate to linear-lanceolate; teeth of glabrous calyx subsperhical, triangular, its teeth acute, markedly declined, tube broadly obconical, appendages very short; corolla tubular-infundibular, constricted at base, 1 cm long, glabrous outside, bearded inside; style not exserted; receptacle glabrous; ripe capsule obconical; seeds small, ovoid, brown, shiny. July–August.
245 Crevices in basalt and andesite in central and subalpine belts. – Caucasus: S. Transc. (Armenia). Endemic. Described from the banks of the Zanga River near Arzakend, in the region of Sevan Lake (Gokcha). Type in Leningrad, topotype in Tbilisi.

Note. The type was not available when Fomin wrote on the Caucasian Campanulaceae (op. cit.), but later Busch found in the Leningrad Herbarium a duplicate of the type among plants collected in Armenia in 1896 and 1897 by Khotsytovskii. Even though the label is written by Busch and not Fomin this should be regarded as the type.

C. choziatowskyi is widespread in Armenia, where it appears to be confined to igneous rocks. In habit and shape of leaves it closely resembles Symphyandra armena.

84. C. takhtadzhianii Fed. in Bot. mat. Gerb. Bot. inst. AN SSSR, XV (1953) 374. — C. *bayerniana* var. *andina* Trautv. in Tr. Peterb. bot. sada, II, 2 (1880) 563, non. C. *andina* Rupr. — C. *andina* Trautv. in schedis, non Rupr. — C. *bayerniana* var. *trautvetteri* Fom. in Tr. Tifl. bot. sada, IV, 2 (1902) 7 and in Mater. Fl. Kavk. IV, 6, 77; Grossg., Fl. Kavk. IV, 69. — Ic.: Fed., op. cit.

Perennial; all parts with stiff dense hairs; rhizome branching, woody; stems thin, low; leaves subcoriaceous; radical leaves with long slender petioles, 5—7 cm; lamina 1—1.5(2) cm wide and as long, rounded, broadly ovate, shallowly cordate, large-toothed, the teeth acute, irregularly and broadly triangular; cauline leaves decreasing in size, short-petioled or subsessile. Flowers drooping or erect, 1—1.3 cm long, on thin bracteate pedicels twice as long as calyx; teeth of stiff-haired calyx triangular-lanceolate, acute or nearly subulate, twice as long as corolla; calyx appendages linear-triangular, recurved, twice as long as basally semiglobose tube; corolla blue, narrowly campanulate, stiff-hairy outside, bearded inside; style not exserted. July.

Mostly in limestones of central mountain belt. — Caucasus: S. Transc. (Daralagez in Armenia, Nakhichevan ASSR). Endemic. Described from ravines near the Arpa River in Mikoyan District, Armenia. Type in Leningrad.

Note. This species, until recently regarded as a variety of C. *bayerniana*, was confused by Ruprecht (op. cit.) with C. *andina* Rupr., endemic to the eastern part of the Main Range of the Caucasus. It differs from the more or less closely allied C. *bayerniana* Rupr. by the stiff hairs on the outside of corolla and calyx and on leaves and stems as well as by the fact that it grows in limestones while C. *bayerniana* is most often found
246 on basalts, andesites and granodiorites, in the subalpine mountain zone and not in the central mountain zone.

85. C. elegantissima Grossh. in Tr. Azerbaidzh. otdel. Zakavk. fil. AN SSSR, sekt. bot. 1 (1933) 56; Grossg., Opred. rast. Kavk. 421.

Perennial; rhizome thickened, multicipital, in upper part with squamous remnants of petioles, leaves and stipules; radical leaves subglabrous above, with short stiff hairs beneath, long-petioled, cordate-ovate, 2.5 cm long, large-toothed at margin, teeth declinate, unequal; petioles 2—3 times as long as lamina; stems weak, nearly prostrate, branching, 10—12 cm, multiflorous, more or less leafy; cauline leaves similar to radical, gradually decreasing in size toward stem apex, short-petioled; uppermost leaves subsessile, with cuneate base. Flowers 2.5 cm long, apparently blue in nature, calyx obconical brown, its teeth triangular-ovate, acuminate or with mucro, one-third the length of the corolla; calyx appendages reflexed, nearly twice as long as tube; corolla infundibular, covered outside with small papillae, pubescent inside, ciliate along nerves; style not exserted; capsule unknown. August.

Slopes of central mountain belt. — Caucasus: S. Transc. (Armenia). Endemic. Described from Daralagez near Kodukh-Vany, Soganli Mountain. Type in Baku.

Note. This species is most closely related to C. *takhtadzhianii*, from which it is distinguished by the absence of dense stiff hairs which in the latter are even found on the corolla. Superficially, the two species are very similar.

Subsection 15. OREOCODON Fed. subsect. nov. in Addenda XXIII, p. 333. – Sect. Saxicolae Charadze in Zam. po sist. i geogr. rast. Tbil. bot. inst. 15 (1949) 20, p. p. non Ser. Saxicolae Boiss. – Ser. Rupestres Boiss. Fl. or. III (1875) 894, p. p. – Notches between calyx teeth with or without appendages; flowers in corymbiform panicles or in racemes, drooping, pediceled, sometimes cleistogamic. Cauline leaves all short-petioled or subsessile, rounded or broadly ovate; lower leaves distinctly petioled, never in dense rosettes. Pubescence dense, lanate or velutinous, gray- or white-tomentose, sometimes scabrous. Stems developing from thick branching grumous rhizome numerous, brittle, ascending or arcuate, sometimes weak, subfiliform.

Type species of subsection: C. incanescens Boiss.

247 Note. We feel that Kharadze (op. cit.) erroneously formed section Saxicolae with C. kachetica as type – a species very close to C. incanescens and similar species. However, in Boissier (l. c.) C. incanescens is not in the related series Saxicolae Boiss. (basonym for Saxicolae), but in Rupestres Boiss. According to Kharadze, section Saxicolae is thus based on a type which evidently belongs to quite another affinity of species, most notably to Boissier's Rupestres. Although in part identifying it with Saxicolae Boiss., Kharadze did not include in it any of the species included in this series by Boissier whose series is not at all analogous to the section described by Kharadze. In order to obviate misunderstandings we have referred the bellflowers close to C. incanescens and C. kachetica to a subsection with a new name and description.

Series 1. **Kantschaveliae** Fed. – Calyx with well-developed appendages nearly as long as tube, margins of teeth and appendages with bristly cilia. Pubescence of plant villous, spreading.

86. **C. kantschavelii** Zagareli in Bot. zhurn. SSSR XXI, 3 (1936) 285; Kharadze in Fl. Gruzii, VIII, 136. – Ic.: Zagareli, op. cit. 187, fig. mediocr.

Perennial; rhizome thick, densely short-branching, grumous, producing few stems, with remnants of stems and leaves in the shape of short scales; stems 25–30 cm, thick, arcuate below, straight or flexuous above, like entire plant covered with stiff spreading hairs, leafy, corymbiformly branching in upper part, multiflorous; leaves irregularly incised-dentate, sometimes slightly lobate-dentate, with acute or rounded teeth, densely short-haired margin and nerves with long hairs; lower cauline and radical leaves broadly and shortly ovate or subrounded, with long remote lobes in upper part; petioles winged; median leaves rounded-oval or rounded-rhombic, with winged, sometimes lobed petioles; upper leaves oval, sessile or broadly cuneately attenuate into short petiole. Inflorescence about 25-flowered, rarely few-flowered; flowers 2 cm long, short-pediceled, violet; calyx teeth lanceolate, acute, bristly-ciliate, half the length of the corolla; corolla narrowly infundibular, glabrous outside, white-haired inside; calyx appendages oblong, obtuse, ciliate, reflexed, as long as or longer than tube; style not exserted; capsule trihedral. June, Fr. October.

248 Rock crevices of dry limestones. – Caucasus: E. Transc. (Georgian SSR). Endemic. Described from Kvareli District, Kakhetia. Type in Leningrad.

Note. Zagareli (op. cit.) and Kharadze (op. cit.) erroneously related this species to C. kolenatiana. It is related to C. incanescens, being the largest species in this group with the largest leaves. Grossgeim did not include it in his Key to Plants of the Caucasus (1949).

Series 2. **Coriaceae** D. Sosn. in Tr. Azerbaidzh. otdel. Zakavk. fil. AN SSSR, sekt. bot., 1 (1933) 56. – Calyx teeth triangular-lanceolate, the appendages one-quarter the length of the teeth but distinct. Plant pubescence very stiff, acutely scabrous.

87. C. radula Fisch. ex Tchihatch. Asie Mineure. III, 2 (1860) 395. – C. radula Fisch. in mssr. ad specim. Kotschyanum e Kurdistania (1848) No. 620; Boiss. Fl. or. III, 909; Grossg., Opred. rast. Kavk. 416; Sosnovsk. in Tr. Azerb. fil. AN SSSR, 1 (1938) 45; Kharadze in Zap. po sist. i geogr. rast. Tbil. bot. inst. 15, 20. – C. garensis Fisch. in sched. ad specim. typ. – C. coriacea Boiss. et Kotschy in Kotschy, Iter cilic.-kurd. (1859) No. 472, nomen. – C. radula var. minor Boiss. l. c.; Grossg., Fl. Kavk. IV, 63. – Exs.: Kotschy, Iter cilic.-kurd. No. 472.

Perennial; rhizome thick, multicipital; entire plant scabrous with short stiff hairs, glaucescent-green; stems 10–30(45) cm, numerous, thick, brittle, straight or ascending, densely leafy, paniculately branching in upper part; leaves rather large-toothed, indurate, slightly fleshy but dry and very brittle, ovate, cuneately attenuate to short petiole, margins crisp-undulate; cauline leaves gradually decreasing in size, sessile above, slightly amplexicaul. Branches of inflorescence 1–3-flowered, scabrous, with small leaves; pedicels erect, slightly longer than calyx; calyx markedly scabrous, its teeth triangular-lanceolate, spreading, twice as long as short obconical tube, appendages one-quarter the length of the teeth, rounded-obtuse, recurved; corolla to 1–2 cm long, blue or whitish, narrowly campanulate, four times as long as calyx teeth. August–September. (Plate XIII, Figure 2.).

Calcareous rocks in river gorges of central mountain belt. – Caucasus: S. Transc. (Arpa River gorge in Daralagez, Armenia). **Gen. distr.**: Arm.-Kurd. Described from Gara Mountain, Kurdistan. Type in Leningrad.

Note. The description of this species first appeared in Tchihatcheff (l.c.), having been prepared at his request by Fenzl, after a specimen
249 collected by Kotschy. Tchihatcheff probably never saw the specimens or the plant. In any case, C. radula Fisch. ex Tchihatch., recently proposed by Sosnovskii (l. c.), is preferable to C. radula (Fisch.) Tchihatch. A detailed diagnosis (in ms.) of C. radula by Fischer is available based on a specimen (No. 623) collected by Kotschy on Gara Mountain in Kurdistan, to which this diagnosis and the name C. radula Fisch. was attached. Fischer had earlier attached to this specimen a temporary label with C. garensis Fisch. Fenzl made his description from three herbarium specimens: 1) from the vicinity of Van Lake, 2) from the vicinity of Amadi, north of Mosul Mountain, 3) from rocks in the upper part of Gara Mountain, northeast of Mosul Mountain. If the plant from the vicinity of Van Lake, mentioned first in Tchihatcheff's list of specimens, is accepted as the type, then C. coriacea Boiss., subsequently published, though not described from this locality, should be called C. radula, irrespective of whether the Van and

Kurdistan plants are distinct species. If, however, it is proven that the plants from these two localities really belong to different species, then the plant from Kurdistan will have to be renamed. Fortunately C. radula sensu Fischer seems indistinguishable from C. coriacea Boiss. et Ky., and it is therefore preferable to consider the plant from Gara Mountain as the type of C. radula, the more so since Fenzl's description closely agrees with this plant, rather than with the formal (Van) type.

Observations in Daralagez, Armenia, have shown both forms to grow nearly side by side, as reflecting differences in moisture conditions. The type form grows in shady and more moist places whereas the form sometimes marked as C. coriacea or C. radula var. minor Boiss. grows in dry, barren soil.

Campanula amana Rech. fil., recently described from Amanus Mountain in Syria by Rechinger (Rechinger fil. in Ann. Naturhist. Mus. Wien, 57 (1949–1950) 85), appears to be far more isolated. Its stems are pubescent with longer hairs, the cauline leaves are never basally cordate or amplexicaul, but attenuate or rounded; the flowers form dense clusters, crowded in an apparently spicate inflorescence, etc.

Series 3. **Canescentes** Fed. — Calyx with short, inconspicuous appen-
250 dages or appendages lacking calyx teeth, triangular. Plant with short scabrous hairs or densely tomentose.

88. C. kachetica Kantsch. in Vest. Tifl. bot. sada, nov. ser., Nos. 4–5 (1931) 2; Grossg., Fl. Kavk. IV, 70; Kharadze in Fl. Gruzii, VIII, 141. — Ic.: H. Clifford Crook, Campanulas, 107.

Perennial; plant and leaves scabrous, with very short stiff hairs; rhizome thick, branching, its branches 1.2 cm across, densely covered with short, somewhat scarious white remnants of petioles, producing numerous stems; stems ascending, partly arcuately curved, simple or slightly branching, with uniflorous branches, stiff-haired, slightly sulcate, densely leafy; leaves indurate, coriaceous, glaucous; radical leaves oblong-triangular-ovate, with cordate base, large-crenate or large-toothed, sometimes with small lobes, the petioles narrowly winged, as long as blades; cauline leaves broadly rounded-ovate or ovate, irregularly large-toothed, with petioles 0.8–2.5 cm long and nearly as wide, slightly decreasing in size toward apex, often larger toward tip of stem than at middle. Flowers terminal or axillary, whitish-pink or white-pink-bluish; lobes of the rounded-obconical calyx narrowly ovate-triangular, sparsely large-toothed, densely stiff-haired, one-third the length of the corolla; calyx appendages very short, slightly recurved, obtuse; corolla narrowly campanulate, hairy outside, bearded inside, divided for one-third into ovate-triangular lobes; style not exserted; capsule dehiscing at base by valves. July–August. (Plate XIII, Figure 1.)

Rock crevices in central mountain belt, sometimes also in crevices of old stone structures. — Caucasus: E. Transc. Endemic. Described from Kakhetia (Shiraki, near Krasnye Kolodtsy, next to the ruins of "Tamara Castle"). Type in Tbilisi, topotype in Leningrad.

(251)

PLATE XIV. 1 – *Campanula dzaaku* Alb.; 2 – *C. ciliata* Stev.; 3 – *C. fominii* Grossh.; 4 – *C. ardonensis* Rupr.

Note. Kantschaveli (op. cit.) comments on the similarity of this species with Symphyandra, though it really is related to C. incanescens Boiss. and belongs to subsection Oreocodon.

89. C. incanescens Boiss. Diagn. ser. 1, 7 (1846) 16; Fl. or. III, 912;
O. and B. Fedch., Perech. rast. Turk. III, 356, p. p.; Grigor'ev, Opred. rast.
253 Stalinabada, 258; Nikitin in Fl. Turkm. VI, 374. — O. fedtschenkoana
Trautv. in Tr. Peterb. bot. sada, VI, 1, (1879) 77. — C. incanescens var.
typica Korsh. in Izv. Akad. Nauk., IX, 5 (1898) 226; O. and B. Fedch.,
op. cit. — Ic.: Korzh., op. cit. Fig. 1 (optima).

Perennial; rhizome deep in rock crevice, branching in upper part, producing stems in dense tufted fascicles; entire plant canescent, pubescence on stem and branches slightly spreading, leaves with short thin matted appressed tomentum; stems 10—20 cm, weak, thin, brittle, rather densely leafy, with thin nearly capilliform branches sometimes arcuately curved but more or less straight in upper part of stem, forming loosely corymbiform inflorescence; leaves 0.5—0.7 long and as wide, gray, broadly ovate, with few large teeth or slightly crenate; lower leaves short-petioled, the upper larger, chasmogamous, to 1 cm long, usually proterandrous, those on lateral and particularly on lower branches, cleistogamous, very small teeth of pubescent calyx of the chasmogamous flowers lanceolate-triangular at base, acuminate, one-quarter the length of the corolla; corolla pubescent, narrowly campanulate, pink or bluish, divided rather deeply, sometimes to middle, into oblong scute lobes; calyx appendages very short or none; cleistogamous flowers borne on very short capilliform branches, ca. 2 mm long, spherical, with poorly developed corolla, nearly without tube, the stamens strongly reduced, the style with entire stigma; capsule opening at base by small holes; seeds abundant, very small, June—July. (Plate XIII, Figure 3.)

Rocks in central mountain zone. — Centr. Asia: Mtn. Turkm., Pam.-Al. (excluding Shugnan), T. Sh. (excluding northwest). **Gen. distr.**: Iran., Arm.-Kurd. Described from Shiraz Mountain in S. Iran. Type in Geneva, isotype in Leningrad.

Note. A very variable species; within the USSR, the form which agrees
with the isotype grows in Kopet Dagh. In the mountains of Tadzhikistan,
with the exception of Shugnan, there is found a form with larger leaves, also
collected on Fergana Range. C. evolvulacea Royle, described from the
western part of the Himalayas, is encountered in Shugnan and further north
in Talass Ala-Tau. Korzhinskii (op. cit.) accepted this as a variety of
C. incanescens var. mollis et var. holosericea, treating it as
transitional between the latter and C. incanescens. He remarked that
254 var. mollis, for example, "could be accepted as a distinct species."
Judging by available herbarium specimens, if "transitional" forms occur in
Central Asia, then they are absent in the south. In Iran, for example, there
grow no plants even remotely resembling C. evolvulacea, nor have the
Himalayas yielded anything resembling C. incanescens. Incidentally,
the species most closely related to C. evolvulacea is the Himalayan
C. cashmiriana Royle, not C. incanescens.

90. C. evolvulacea Royle, Illustr. Bot. Himal. I (1830) 253, nom. nud.; A. DC. in DC. Prodr. VII, 2, 473, diagn. — C. incanescens, auct. non Boiss.:

Korzh. in Izv. Akad. Nauk. IV, 5 (1898) 425, p. p.; O. and B. Fedch., Perech. rast. Turk. III, 356, p. p. — C. incanescens var. holosericea Korsh. op. cit. 433; O. and B. Fedch., op. cit. — C. incanescens var. mollis Korsh. op. cit. 426; O. and B. Fedch., op. cit. — C. cashmiriana var. evolvulacea (Royle) Clarke in Hook. Fl. Brit. Ind. III (1882) 441; Boiss. Fl. or. Suppl. I, 333. — Ic.: Korzh., op. cit. 2 (optima).

Perennial; rhizome buried in rock crevices, producing loose simple or branching tufts of almost capilliform stems; entire plant with thin gray appressed tomentose pubescence; stems numerous, ascending, thin, flexuous, reddish, 0.5—0.7 mm thick, cylindrical, densely covered with gray strongly appressed, partly matted short hairs, visible only under magnification; leaves densely fine-tomentose, nearly entire or with few inconspicuous teeth; lower leaves smaller, obovate, rarely subrounded; median leaves large, to 0.7 cm long, 0.5—0.7 cm wide, usually subrhombic, very short-petioled or sessile; upper leaves ovate, sessile, smaller than the median, very small on branches of inflorescence. Upper branches of inflorescence often furcate, filiform, spirally twisted when young; flowers small, on long or short capilliform peduncles, rarely dimorphous, chasmogamous or cleistogamous; in chasmogamous flowers teeth of the gray broadly turbinate calyx narrowly ovate or lanceolate, sometimes lanceolate-triangular, thinly tomentose, acute, entire, one-third to one-half the length of the corolla; corolla velutinous, nearly without tubes, deeply divided into obtuse lobes; style not exserted; cleistogamous flowers much smaller, on very thin, rather long capilliform branches, arcuately curved beneath with obsolete stamens and style. May—June.

Rock crevices. — Centr. Asia: Pam.-Al. (Shugnan), T. Sh. (Talas Ala-Tau). **Gen. distr.**: Ind.-Him. (from Kashmir to Kumaon, also probably in Karakorum). Described from N. India. Type in London, isotype in Geneva.

255 Note. This species, described from the Himalayas, was first introduced by Vvedenskii in the flora of Central Asia (in schedis ad Herb.). It is distributed from Shugnan to the northwest and after a break, to Talass Ala-Tau. In the mountains of the Pamir-Alai system (Zeravshan and Gissar ranges) it is replaced by C. incanescens Boiss., which is also common to Kopet Dagh and Fergana. (See Note to description of that species.)

There are no data on the exsiccates of this species. We have considered the widely distributed collections from the Himalayas, and from some other herbaria, which fully agree with material from Central Asia, for example: Herb. Chitral Relief Exped. No. 1895; Duthie, Fl. of Northwest India, 606a; Aitchison, Herb. of Kurrum Valley, Afghanistan, sine num.

91. C. massalskyi Fom. in Mater. Fl. Kavk. IV, 6 (1905) 78; Grossg., Fl. Kavk. IV, 69; idem, Opred. rast. Kavk. 420..

Perennial; rhizome thick, branching, multicipital; stems 10 cm, thin, subfiliform, flexuous, brittle, with short, 1—3-flowered branches with sparse spreading hairs; leaves 0.5—0.7 cm long and as wide, thick, the lower ovate-rounded, attenuate into petiole; cauline leaves rounded- or ovate-cordate, short-petioled, all with irregularly dentate-undulate margin, bright green above, pubescence velutinous, gray-tomentose-velutinous beneath. Flowers small, axillary, or terminal, long-pediceled; teeth of greenish velutinous calyx broadly triangular, acute, the notches between them with inconspicuous recurved appendages; tube semiglobose; corolla tomentose, narrowly tubular-infundibular, 3–4 times as long as calyx; style not exserted; receptacle glabrous. July.

Limestone in central mountain zone. – Caucasus: S. Transc. (Nakhichevan ASSR, Negram ravine). **Gen. distr.**: former Kars region in Turkey. Described from near Kagyzman (along the Digor-Chai and Besh-Kilisa rivers). Type in Leningrad.

92. C. ketzkhovelii Sosn. in Grossg., Fl. Kavk. ed. 1-e, IV (1934) 69, ross.; Sosn. ex Charadze in Zam. po sist. i geogr. rast. Tbil. bot. inst. 15, 20.

Perennial; small semishrub; root thick, multicipital, woody; stems 10 cm, numerous, whitish, thin, flexuous, very small, covered with short, rather dense appressed hairs, shortly branching in upper part, leaves small, thick, the lower ovate-rounded, petiolate, the median and upper rounded or
256 nearly ovate-cordate, 0.6–0.7 cm long, 0.5–0.6 cm wide, all with irregularly dentate, slightly undulate margin, green with dense crisp hairs above, canescent with more dense hairs, nearly tomentose beneath, gradually diminishing toward stem apex. Flowers small, terminal or axillary, few, short-pediceled; calyx semiglobose, very small, more or less densely pubescent with acute ciliate margin, the teeth nearly half the length of the corolla; appendages short, reflexed, not concealing tube; corolla narrowly tubular-infundibular, 1–1.2 cm long, sparingly pubescent; style not exserted. August.

Andesites in central mountain zone. – Caucasus: S. Transc. (Armenia). Endemic. Described from the extinguished volcano Areg (Bogutlu), near Pirmalak, at the southern spurs of Mount Aragats (Alagez). Type in Tbilisi.

Note. Very closely related to the previously described C. massalskyi Fomin from the Turkish border, from which it is distinguished by the short, not particularly acute calyx teeth.

Subsection 16. SCAPIFLORAE (Boiss.) Fed. comb. nov. – Sect. Scapiflorae (Boiss.) Charadze in Zam. po sist. i geogr. rast. Tbil. bot. inst. 15 (1949) 25. – Ser. Scapiflorae Boiss. Fl. or. III (1875) 894. – Notches between calyx teeth with more or less developed appendages, usually as long as tube, rarely shorter; flowers erect but capsule pendulous; corolla campanulate, usually with globose base; capsule dehiscing at base by 3 valves. Stems uniflorous, slightly leafy, usually short. Leaves in rosettes at stem base, sometimes at apices of sterile shoots. Rhizome often thickened, sometimes columelliform, multicipital.

Type species of subsection: C. ciliata Stev.

Series 1. **Ciliatae** Charadze in Zam. po sist. i geogr. rast. Tbil. bot. inst. 15 (1949) 23. – Receptacle glabrous; corolla glabrous; calyx appendages as long as or much shorter than tube. Leaves with long spreading cilia or curved prickles at margins.

93. C. ciliata Stev. in Mém. Soc. Nat. Mosc. (1812) 256; A. DC.
Monogr. 229; idem in DC. Prodr. VII, 461; M. B. Fl. taur.-cauc. III, 147;
Ldb. Fl. Ross. II, 876; Boiss. Fl. or. III, 903; Fomin in Mater. Fl. Kavk.
IV, 6, 45; Grossg., Fl. Kavk. IV, 68; Kharadze in Fl. Gruzii, VIII, 164. –
C. speciosa M. B. in schedis. – Medium saxifragum Fisch. ex
A. DC. l. c., pro syn. – C. tridentata var. ciliata Trautv. in Tr.
Peterb. bot. sada, IV, 1 (1879) 163. – C. ciliata var. typica Rupr. in
257 Bull. Acad. Sc. Pétersb. XI (1861) 174. – C. ciliata var. longifolia
Rupr. l. c. – C. ciliata var. pontica Alb. in Bull. Herb. Boiss. II
(1894) 118; ej. Prodr. Fl. Colch. 156.

Perennial; rhizome vertical, rather thick; stems 7–10(15) cm, ascending, 1-flowered, slightly leafy; radical leaves forming dense rosettes, subsessile, linear-lanceolate or oblong-lanceolate, obtuse or acute, glandular crenate-dentate, ciliate, glabrous above and beneath; cauline leaves narrowly linear, with ciliate margin. Calyx glabrous, its teeth white-ciliate, linear-lanceolate or linear, appendages ciliate, triangular-ovate, reflexed, enveloping tube; corolla blue, obconical or campanulate, glabrous, sometimes ciliate along nerves, twice as long as calyx teeth. June–July. (Plate XIV, Figure 2.)

Rocks, taluses, sometimes moraines in alpine mountain zone. – Caucasus: Cisc., Dag., W. and E. Transc. Endemic. Described from Azerbaidzhan near Khinalyg. Type in Helsinki.

94. C. dzaaku Alb. in Bull. Herb. Boiss. II (1894) 114; ej. Prodr. Fl. Colch. 156; Fomin in Mater. Fl. Kavk. IV, 6, 47; Grossg., Fl. Kavk. IV, 68; Kolak., Fl. Abkh. IV, 176; Kharadze in Fl. Gruzii, VIII, 167.

Perennial; rhizome rather thin, creeping, producing tufts of thin, brittle, sometimes subfiliform, erect or ascending, leafy, uniflorous, glabrous stems and sterile shoots; radical leaves and leaves of sterile shoots in dense rosettes, sessile, coriaceous, lanceolate or oblong-lanceolate, glabrous and shiny, obsoletely denticulate or nearly entire from middle, with curved margins and recurved cilia; cauline leaves sessile, small, lanceolate, multiserrate; teeth of glabrous and shiny calyx linear, one-sixth to one-fifth the length of the corolla; appendages dentiform or sometimes none, tube shortly ovate-obconical; corolla violet, glabrous when dry, receptacle glabrous; capsule ovoid. July–August. (Plate XIV, Figure 1.)

Limestones in alpine zone. – Caucasus: W. Transc., Abkhaziya, Mingrelia. Endemic. Described from limestones, Okhachku Range in Abkhaziya (also Migariya, Dzhavri and Kvira mountains). Type in Geneva.

Note. The odd name was chosen by Al'bov to name this plant after a young Abkhazian women by the name of Dzaku, whom he met while traveling through the Caucasus.

Series 2. **Tridentatae** Charadze in Zam. po sist. i geogr. rast. Tbil.
bot. inst. 15 (1949) 23. – Receptacle and corolla glabrous; calyx appendages
258 as long as tube. Leaves sparingly pubescent; radical spatulate, short-
petioled, leaves always with 3 distinctly expressed teeth at obtuse apex.

95. C. tridentata Schreb. Ic. et. descr. pl. (1766) 3; L. Mantissa (1767) 44; Roem. et Schult. Syst. V, 147; A. DC. Monogr. 227; Boiss. Fl. or. III, 904, p. p.; Fomin in Mater. Fl. Kavk, IV, 6, 47, p. p.; Grossg. Fl. Kavk. 1. IV, 67, p. p.; Kharadze in Zam. po sist. i geogr. rast. Tbil. bot. inst. 15, 25; idem, Fl. Gruzii, VIII, 167. – C. tridens var. ciliata et var. araratica Rupr. in Bull. Acad. Sc. Pétersb. IX (1867) 175. – C. tridentata var. gracilis Fomin, op. cit. 51; Medv. in Tr. Tifl. bot sada, XVIII, 2, 340. – Ic.: Schrb. l. c. tab. 3, manca; H. Clifford Crook, Campanulas, 253.

Perennial; rhizome branching or simple, with few shoots; plant not cespitose, often forming large compact cushions on alpine carpets; stems 2–7(15) cm erect or arcuate at base, 1-flowered, slightly leafy, glabrous or very sparingly pubescent, thin, about 0.5 mm across; rosetted leaves oblanceolate, spatulate, gradually tapering to petioles, with petioles 2–3 cm long, glabrous or with few rather long marginal hairs, entire but for 3-toothed at apex; cauline leaves sublinear, sometimes also 3-toothed at apex. Flowers medium-sized, 2 cm long, blue, with basally globose corolla; calyx teeth linear-lanceolate, acute, like narrowly lanceolate appendages, dark, not violet with ciliate margin; corolla and receptacle glabrous. June–August.

Alpine carpets in glacial cirques or lava plateaus adjacent to perpetual snow and pereletoks in mountains. – Caucasus: S. and E. (NW) Transc. **Gen. distr.**: Bal.-As. Min. (E. Anatolia). Described from the "East." It is not clear whether the type exists; Schreber's drawing is poor and does not adequately represent it. As lectotype any specimen, for example from the Alagez Mountain in Armenia, could be chosen.

Note. This species was so badly described from Asia Minor that its main characters are not clear. Schreber's drawing shows a 5-locular capsule, obviously an anomaly, bearing little resemblance to all known herbarium specimens from Asia Minor (see notations from Ruprecht's work). In fact, no plants agreeing with Schreber's drawing were subsequently collected in Asia Minor and the description is obviously erroneous. The plants which most closely agree with the description grow in the alpine zones of
259 mountains in the eastern part of Asia Minor and the Lesser Caucasus. Like the related, endemic Caucasian species, C. biebersteiniana they possess such distinctive characters as a tridentate leaf apex. Because of this very character it is impossible to confuse these two species with others and notwithstanding the faulty initial diagnoses and the absence of a type, the Caucasian tridentate bellflower can confidently be regarded as C. tridentata Schreb. Bieberstein, who saw one of the Tournefort specimens, concluded that it agreed fully with his description of the Caucasian C. rupestris (i. e. C. biebersteiniana) and these species were therefore accepted as one. But Ruprecht (l. c.) correctly pointed out that to Schreber this specimen was but a variety of C. tridentata. Indeed, its similarity with the type of the Caucasian C. biebersteiniana is accidental (see Note to C. biebersteiniana) and the tendency to unite C. biebersteiniana with C. tridentata, cannot be justified. A third species, C. bythinica DC., of the series Tridentatae was described from Asia Minor, but its leaves were not apically tridentate and generally were entire.

96. C. biebersteiniana Roem. et Schult. Syst. V (1819) 147; A. DC. Mongr. 460; Ldb. Fl. Ross. II, 876, p. p. – C. rupestris M. B. Fl. taur.-cauc. I (1808) 154, non Sibth. – C. tridens Rupr. in Bull. Acad. Sc. Pétersb.

XI (1867) 175; Kharadze in Zam. po sist. i geogr. rast. Tbil. bot. inst. 15, 27; idem, Fl. Gruzii, VIII, 168. — C. tridentata auct. non Schreb.: Boiss. Fl. or III (1879) 904, p. p.; Lipskii, Fl. Kavk. (1899) 378; p. p.; Somm. et Lev. in Tr. Peterb. bot. sada, XVI, 317, p. p.; Grossg., Fl. Kavk. IV, 67, p. p. — C. biebersteiniana var. macrantha Ldb. l. c. 282. — C. tridentata var. rupestris Trautv. in Tr. Peterb. bot. sada, IV, 1 (1879) 163, p. p. — C. tridentata var. barbata Fomin in Mater. Fl. Kavk. IV, 6, 50; Medv. in Tr. Tifl. bot. sada, XVIII, 2, 339. — C. tridens var. barbata (Fomin) Charadze, op. cit. — Ic.: A. DC. l. c. tab. 10, f. manca. — Exs.: Baenitz, Herb. Europ., sine numero; Pl. or. exs. No. 349.

Perennial; forming carpets, not densely cespitose; rhizome thick, brown, with few remnants of leaves at tip; stems 7–10 cm, slightly leafy, 1-flowered, erect or ascending, glabrous or sparingly pubescent; leaves of rosettes nearly spatulate, gradually tapering to petioles, with petioles 3–4(5) cm long, 0.5–0.7(1) cm wide, glabrous or subglabrous, entire, apex rounded, distinctly obtusely or acutely toothed; cauline leaves narrow, oblanceolate or sub-
260 linear, the uppermost subsessile. Calyx canescent at base with long soft tangled hairs, the teeth violet, linear-lanceolate, rather broad, long, obtuse, half the length of the corolla, appendages ovate, obtuse, lanate, reflexed, concealing tube; corolla dark blue, campanulate, 3–4 cm long, glabrous, divided for one-quarter into rounded abruptly acuminate lobes; flowers erect, until anthesis, drooping in fruit; receptacle glabrous; seeds elongate-ovoid, flattened, with slightly thickened margins, pale, dull. June–July. (Plate XV, Figure 3.)

Alpine carpets, on platforms of glacial cirques, very rarely in rocks. — Caucasus: Cisc., Dag., W. Transc. Endemic. Described from "rocks in high mountains of the Caucasus." Type in Leningrad.

Note. The taxonomic history of this species is somewhat confusing. Bieberstein described it as C. rupestris M. B. from collections which Ruprecht thought were gathered partly from near Mtiuleti (Kaishaura) (2 specimens) and partly from another locality (1 specimen). A further specimen belonged to another species, not found since, which Ruprecht believed to be his C. meyeriana, The type of C. rupestris was confused by Bieberstein, whose description is therefore unintelligible. The drawing based on Alphonse de Candolle's data (l. c.) does not at all resemble C. rupestris. Also, the epithet was misleading as C. biebersteiniana never grows in rocks. Roemer and Schultes (l. c.) had renamed it C. biebersteiniana using an earlier homonym of the Soviet bellflower and giving it characters not generally typical. Accepting the related C. tridentata Schreber, they confused it with C. biebersteiniana, ascribing to the latter small flowers ("flore minore") and to the former C. tridentata large flowers ("flore magno"), whereas in fact the reverse is true. The type C. biebersteiniana (2 specimens specified by Ruprecht) is the same plant, though in a bad state of preservation, which Ruprecht had accurately described as C. tridens based on a series of new collections from different localities. Later authors lumped C. biebersteiniana and C. tridentata together, but when these are accepted as distinct,
261 Kharadze, Grossgeim and Kolakovskii should be followed in preferring Roemer and Shultes' name over Ruprecht's name, because they correctly describe C. biebersteiniana. If C. biebersteiniana appears

inapplicable because of the mixed type specimens, then C. tridens too is unsuitable, as it is based on the same type. All considered we propose that C. biebersteiniana be given preference over C. tridens, if only because of priority.

C. biebersteiniana is distributed over the whole of the Main Range and in the Adzhar mountains, and C. tridentata in the Lesser Caucasus and adjacent parts of Asia Minor, from where it was first described (see Note to C. tridentata).

In his "Flora," Bieberstein refers to the habitat "in rupestribus alpium caucasicarum," but the type is labeled: "Georgia." Ruprecht, as shown above, gave the correct locality (Mtiuleti).

Series 3. **Aucherianae** Charadze in Zam. po sist. i geogr. rast. Tbil. bot. inst. 15 (1949) 23, s. str. ("Aucheri"). – Receptacle pubescent; corolla pubescent; calyx appendages as long as tube. Leaves pubescent or subglabrous; radical leaves toothed or entire, spatulate, narrowly lanceolate or linear-lanceolate, long- or short-petioled.

97. C. aucheri A. DC. in DC. Prodr. VII (1839) 460; Boiss. Fl. or. III, 905, p. p.; Fomin in Mater. Fl. Kavk. IV, 6, 569; p. p.; Medv. in Tr. Tifl. bot. sada, XVIII, 2, 338, p. p.; Grossg., Fl. Kavk. IV, 66; Kharadze in Fl. Gruzii, VIII, 168; Grossg., Opred. rast. Kavk. 419. – C. pallidiflora Rupr. in Bull. Acad. Sc. Pétersb. XI (1867) 179. – C. szovitsiana Rupr. in schedis. – C. pubiflora var. szovitsiana Rupr. in schedis. – C. saxifraga var. transcaucasica Rupr. l. c. 184.

Perennial; more or less pubescent plant, canescent with rather dense soft hairs; rhizome cespitose, multicipital, brown, thick, producing small cushions of entangled branches, bearing rosettes or leaves and numerous stems; stems 10–15 cm, pubescent, ascending, leafy, uniflorous, radical leaves lanceolate, sometimes narrowly lanceolate, with petioles 3–5(7) cm long, 0.3–0.5(0.7) cm wide, densely covered with short stiff hairs, toothed mostly at apex, entire beneath, very gradually attenuate into petioles; cauline
262 leaves sessile, narrowly lanceolate or linear, few. Flowers erect, large, 3–4 cm long, blue-violet; teeth of pubescent calyx lanceolate, acute, broadened at base, one-third to one-half the length of the corolla, appendages triangular, acuminate, one-third the length of the teeth, exceeding tube; corolla campanulate, pubescent outside; style not exserted; receptacle pubescent; capsule pendulous; seeds straw-yellow, ellipsoid, flattened, bordered. May–July.

Alpine zone, often on fixed taluses and gravel platforms on mountain slopes. – Caucasus: Cisc. (C. Caucasus), S. and E. Transc. **Gen. distr.**: Bal.-As. Min., Arm.-Kurd. Described from Turkish Armenia. Type in Geneva.

Note. The recently described Rechinger C. froedinii Rech. f. (in Anzeiger d. math. naturw. Kl. Acad. Wiss. Wien, IX (1950) 8; Symb. Bot. Upsaliensis, XI (1952) 36, Fig. 18) seems to be identical with C. aucheri. Rechinger (l. c.) relates it to C. saxifraga M. B., but it is probably a form of C. aucheri, since C. froedinii has more in common with C. aucheri than with C. saxifraga. Unfortunately, we have not seen the type or the diagnosis of species and our conclusion is based on a good drawing and a short description of the characters. In comparing his new species with C. *saxifraga*, Rechinger writes that the "stem is larger, tufts denser, leaves thicker and more indurate, pubescent and not only ciliate, calyx teeth pubescent over entire surface." These characters also apply to C. *aucheri*. Again, judging from the drawing of C. *froedinii* it agrees very closely with C. *aucheri*, and also somewhat resembles C. *ruprechtii* Boiss. (= C. *gilanica* Rupr.). Unfortunately Rechinger does not say whether the corolla of C. *froedinii* is pubescent or not. If it is, then it is close to C. *aucheri*, if not it is related to C. *ruprechtii*, but not to C. *saxifraga*. The latter is characterized by the absence of dense tufts and by all the characters which Rechinger himself uses to distinguish his new species from C. *saxifraga*.

98. C. radčhensis Charadze in Zam. po sist. i geogr. rast. Tbil. bot. inst. 13 (1947) 53, 15 (1949) 25 ("C. *ratschensis*"); Fl. Gruzii, VIII, 170.

Perennial; rhizome sometimes branching, producing numerous shoots and ascending, 1-flowered, thick, stems, 10 cm tufted at base, more or less
263 white-tomentose with tangled long soft hairs, with clustered silvery remnants of old leaves; rosetted leaves spatulate, attenuate into short petioles, entire or slightly crenate in upper part, densely pubescent above and beneath, with soft tangled hairs along margin; cauline leaves numerous, spatulate, short-petioled, entire, 3–5-toothed at apex, densely pubescent at base. Calyx densely and softly pubescent, the teeth oblong, obtuse, nearly half the length of the corolla, appendages triangular-lanceolate, reflexed, exceeding tube, white-tomentose along margin; corolla violet-blue, broadly infundibular, densely and softly pubescent, only 1.5 cm long; receptacle hairy; capsule pendulous. July–August.

Limestones in central mountain belt. – Caucasus: W. Transc. (Racha) Endemic. Described from Satsalike and Potskhrevi mountains in Racha Range. Type and paratype in Tbilisi, isotype in Leningrad.

Note. In addition to the localities indicated above this bellflower has been collected only twice in W. Georgia. According to Kharadze (op. cit.) it is distinguished from C. *aucheri*, C. *alpigena* and other related species by the basally woody stems, their tomentose pubescence, the silvery remnants of dead leaves, and some further specific differences.

99. C. armazica Charadze in Zam. po sist. i geogr. rast. Tbil. bot. inst. 13 (1947) 51, 15 (1949) 25; idem, Fl. Gruzii, VIII, 169. – C. *aucheri* var. *compacta* Fomin in Mater. Fl. Kavk. IV, 6 (1905) 61.

Perennial; rhizome thin, transversely rugose from leaf traces, branching, producing dense tufts with numerous stems and sterile shoots, covered with imbricate grayish remnants of leaves, hidden by rosettes of live leaves; stems only 5–7 cm, 1-flowered, with scattered tangled hairs, leafy; rosetted leaves broadly oblanceolate or spatulate, rather abruptly attenuate into petioles, glabrous and shiny above, dentate, with appressed hairs; lower cauline leaves oblong-lanceolate, sessile, the upper smaller, lanceolate, also sessile. Teeth of the densely pubescent white calyx ovate-oblong, tapering above, obtuse, one-third the length of the corolla, appendages lanceolate, reflexed, longer than tube; corolla dark blue, tubular-campanulate, densely covered outside with fine hairs, only 1.5 cm long; receptacle hairy; capsule pendulous. June.

Rocks in upper forest belt. – Caucasus: E. Transc. (Mtskheta, Manglisi). Endemic. Described from Armaza near Mtskheta. Type in Tbilisi, topotype in Leningrad.

264 Note. Known only from the two localities mentioned, this species differs from others of the series Aucherianae by obsolete stems, shiny leaves and compactly pulvinate growth. The species epithet "armazica" refers to Harmozica, the ancient Roman name for Mtskheta (armazica).

100. S. alpigena C. Koch in Linnaea, XXIII (1850) 638; Kharadze in Zam. po sist. i geogr. rast. Tbil. bot. inst. 13, 49, descr. emend, 15, 25; idem, Fl. Gruzii, VIII, 169. – C. aucheri auct. non DC.; Boiss. Fl. or. III, 905, p. p.; Fomin in Mater. Fl. Kavk. IV, 6, 57, p. p.

Perennial; rhizome branching, with numerous sterile shoots and stems, cespitose, brown, in upper part covered with remnants of leaves; stems 7–10(15) cm slightly leafy, 1-flowered, ascending, nearly straight from arcuate base, sparsely soft-haired; leaves or rosettes and lower cauline leaves oblong-spatulate, rather abruptly attenuate into reduced petioles, with petioles 2–3(5) cm long, 0.5–0.7 cm wide, irregularly serrate-dentate or slightly incised-dentate, appressed-hairy above and beneath; median and upper cauline leaves lanceolate-oblong, obtuse. Flowers rather large, 2.5–3 cm long, violet; calyx densely appressed-hairy, the teeth broadly lanceolate-oblong, obtuse, one-third the length of the corolla, appendages triangular-lanceolate, acute, nearly twice as long as calyx-tube, concealing tube; corolla broadly campanulate, hairy outside; receptacle hairy; capsule pendulous. June–July.

Gravels of alpine carpets, sometimes in rocks. – Caucasus: W. Transc. Adzhar-Imeretian Range, Askhi plateau). **Gen. distr.:** Bal.-As. Min. (Pontiiskii and Arsianskii ranges). Described from Pontiiskii Range. Type in Berlin.

Note. This neglected species of Koch was recently revived by Kharadze (op. cit.). It differs from C. aucheri by obsolete, spatulate leaves and smaller flowers. The shape of the leaves recalls C. argunensis, but taxonomically it is farther from it than from C. aucheri.

267 Series 4. **Saxifragiformes** Fed. – Receptacle pubescent; corolla glabrous; calyx appendages as long as tube, acutely triangular or attenuate into filiform ending. Leaves linear or lanceolate-linear, gradually tapering into rather short petioles.

101. C. saxifraga M. B. Fl. taur.-cauc. I (1808) 155, III (1819) 147; Roem. et Schult. Syst. V, 147; A. DC, Monogr. (1830) 228; idem in DC. Prodr. II (1839) 460; Rizenkamf in Byull. Mosk. obshch. ispyt. prir. 2, 31; Ldb. Fl. Ross. II, 876; Rupr. in Bull. Acad. Sc. Pétersb. XI, 183; Boiss. Fl. or. III, 904 et Suppl. 231; Shmal'g., Fl. II, 179; Fomin in Mater. Fl. Kavk. IV, 6, 54; Medv. in Tr. Tifl. bot. sada. XVIII, 2, 338; Grossg., Fl. Kavk. IV, 66; Kharadze in Fl. Gruzii, VIII, 170. – C. tridentata var. saxifraga Trautv. in Tr. Peterb. bot. sada, IV, 2 (1876) 387. – C. saxifraga var. leptorrhiza Somm. et Lev. in Tr. Peterb. bot. sada, XVI (1900) 318; Medv., op. cit. – Ic.: A. DC. l. c. (1830) tab. 10.

(265)

PLATE XV. 1 – Campanula dasyantha M.B.; 2 – C. chamissonis Fed.; 3 – C. biebersteiniana Roem. et Schult.

Perennial; rhizome 0.5 cm thick, blackish, producing — especially on taluses — numerous branching uniflorous stems and sterile shoots with rosettes of leaves; stems leafy, 10 cm or longer; radical leaves and leaves of sterile shoots glabrous, but for crisp margins, linear or linear-lanceolate, obsoletely crenate or entire from middle up, lamina 3—5 cm long, as long as petiole, 0.3—0.5(0.6) cm wide, cuneately tapering to petioles; cauline leaves smaller, narrowly campanulate, the uppermost linear, sessile. Flowers large, somewhat declined, violet-blue, 3 cm; calyx teeth oblong-lanceolate, usually obtuse, one-third to one-half the length of the glabrous or subglabrous campanulate corolla; appendages triangular, acute, recurved and exceeding tube; corolla lobes somewhat acute along sides; style not exserted; receptacle stiff-haired; seeds straw-yellow, flattened, narrowly white-scarious. June—July.

Alpine zone, rocks and taluses, rarely grass plots. — Caucasus: Cisc. (C. Caucasus mostly). Endemic. Described from the top of Beshtau Mountain. Type in Leningrad.

Note. This species is more or less closely related to C. aucheri, from which it is sharply distinguished by the always glabrous corolla (see Note to C. aucheri).

102. C. ruprechtii Boiss. Fl. or. III (1875) 905; Grossg. in Tr. Azerb. fil. AN SSSR, I, 57; idem, Fl. Kavk. IV, 72. — C. gilanica Rupr. in Bull. Acad. Sc. Pétersb. XI (1867) 182, non Roem. et Schult. — C. tridentata Hablitzl. in schedis, non Schreb. — C. hablitzlii Rupr. in schedis. — C. gilanica var. obovata Rupr. l. c. 183.

Perennial; rhizome thick, branching, multicipital, with few whitish or brown remnants of petioles; stems rather short, not exceeding 7—10 cm,
268 ascending, thin, slightly flexuous, uniflorous; leaves pubescent, narrowly linear-oblong or spatulate, tapering to ciliate petioles, entire or 2—3-toothed at apex; cauline leaves narrowly linear. Calyx glabrous, the teeth oblong-lanceolate, 1—1.3 cm long, half the length of the corolla, appendages long and acutely attenuate, nearly filiform at ending, with ciliate margin. Corolla glabrous. June—July.

Rocks and gravels in Alpine zone. — Caucasus: S. Transc. (Daralagez, Nakhichevan ASSR). **Gen. distr.**: N. Iran. Described from the alpine part of Gilyan. Type (holotype) in Leningrad.

Note. First found in Gilyan, this species was later collected in the USSR by Karyagin, in Daralagez and found to be identical with the type, preserved in Leningrad, on which both Ruprecht and Boissier based their descriptions. Also found in Salvarty, Yaglu, Ak-Yurt, Kapudzhikh mountains in the Zangezur Range, in mountains near the village of Alagez in Daralagez, and on Kyapaz Mountain in the Gyandzhin Range.

Series 5. **Argunenses** Fed. — Receptacle pubescent; calyx appendages acuminate or acute, longer than or as long as tube; corolla pubescent. Leaves abruptly or gradually tapering into long or short petioles; lamina of radical leaves rounded or nearly spatulate, serrate or crenate.

103. C. argunensis Rupr. in Bull. Acad. Sc. Pétersb. XI (1867) 181; Fomin in Mater. Fl. Kavk. IV, 6, 61; Medv. in Tr. Tifl. bot. sada, XVIII, 2, 339; Grossg., Fl. Kavk. IV, 67; Fl. Gruzii, VIII, 172. — C. tridentata var. argunensis Trautv. in Tr. Peterb. bot sada, IV, 2 (1876) 388. — C. tridentata var. pubiflora Trautv. op. cit. V, 2 (1878) 457 ex p. — C. tridentata var. petrophila Trautv. op. cit. X (1890) 120 et in schedis, ex p. — C. aucheri Boiss. Fl. or. III (1875) 905, ex p. — C. aucheri var. foliis parvis Boiss. in sched. ad specim. typ. — C. aucheri ssp. argunensis (Rupr.) Fom. in schedis. — C. glabrescens Rupr. in schedis. — C. canopubescens Rupr. in schedis. — C. pubiflora var. pubifolia Rupr. in schedis. — C. glabrescens var. grandifolia Rupr. in schedis. — C. aucheri var. pubiflora (Rupr.) Boiss. in schedis. — C. pubiflora var. subrotundifolia Rupr. in schedis. — C. subrotundifolia Rupr. in schedis.

Perennial; rhizome thick, woody; stems 10—12 cm, ascending, leafy; radical leaves rounded or ovate-rounded, long-petioled, serrate; cauline leaves of similar shape or spatulate, distinctly serrate; all leaves more or less canescent, often subglabrous. Calyx teeth triangular, one-third to one-half the length of the corolla, appendages triangular, acute, much longer than
269 tube; corolla broadly campanulate, violet, densely pubescent; style not exserted; receptacle stiff-haired; capsule pendulous, semiglobose; seeds ellipsoid, pale brown, bordered. July—August.

Rocks in alpine zone. — Caucasus: Cisc., Dag., E. Transc. Endemic. Described from limestones near Evdokimov fortress, Argun River. Type in Leningrad.

Note. C. argunensis is common to the eastern half of the Main Range of the Caucasus. Grossgeim (op. cit.) reported it for the western Caucasus, but all specimens from there determined as C. argunensis probably are C. circassica Fom. Judging from the labels Ruprecht originally distinguished two independent species, C. canopubescens and C. glabrescens, which he subsequently united. The type belongs to the latter form. The most eastern forms (Dagestan) are the most pubescent, and if Ruprecht's forms prove to belong to a separate species, the Dagestan specimens will have to be given a new name.

104. C. meyeriana Rupr. in Bull. Acad. Sc. Pétersb. XI (1867) 178. — C. aucheri Boiss. Fl. or III (1875) 905, p. p. non DC. — C. biebersteiniana C. A. M. Verzeichn. (1831) 83, p. p. non Roem. et Schult. — C. anomala Fom. in Mater. Fl. Kavk. IV, 6 (1905) 53, p. p. — C. glareosa Fom. in schedis. — C. saxifraga C. A. M. in schedis, non M. B.

Perennial; rhizome thick, distally covered with remnants of petioles and calcareous exudations; stems few, 10 cm, arcuately curved, ascending or nearly prostrate; radical leaves obovate or oval-lobate, with short or rather long petioles, crenate or serrate, especially at apex, slightly pubescent or subglabrous, lamina 2—2.5 cm long, 7—10 mm wide. Calyx colored, usually grayish pubescent at base, the teeth obtuse, linear, with triangular base, appendages ovate-lanceolate, acute; corolla 3 cm long; pubescent outside, broadly campanulate; stigmas straight at anthesis. July.

Limestones in alpine mountain belt. — Caucasus: Cisc., Dag., E. Transc. Endemic. Described from Shakh-Dag Mountain. Type in Leningrad.

Note. Ruprecht's description of this species was not accepted by Boissier and later authors who identified it with C. aucheri. Describing from the western Caucasus the new species C. anomala Fom., Fomin applied this name to the type of C. meyeriana Rupr. Like Ruprecht he doubted Meyer's labels and decided that Ruprecht's type of establishing
270 C. meyeriana was not collected on Shakh-Dag, as stated on the label, but in the western Caucasus. However, Elenevskii has recently collected on Shakh-Dag a bellflower identical with Ruprecht's autotype. Excellent specimens of this species were also collected by Kharkevich in Gunib District. Thus, C. meyeriana obviously grows in the eastern part of the Main Range and C. anomala is not identical with it, but rather a well-expressed endemic race allied to C. argunensis. C. meyeriana is distinguished from C. anomala Fom. by the pubescent corolla, few leaves, nearly prostrate stem, oblong-spatulate, apically pubescent radical leaves with large, 1 cm wide lamina. C. meyeriana may be regarded as the high mountain race of C. argumensis Rupr.

105. C. doluchanovii Charadze in Zam. po sist. i geogr. rast. Tbil. bot. inst. 13 (1947) 54, 15 (1949) 25; idem, Fl. Gruzii, VIII (1952) 175.— Ic.: Kharadze, op. cit. (1952) tabl. 375, bona.

Perennial; rhizome branching, producing loose tufts of sterile elongate columelliform shoots, distally densely covered with whitish remnants of petioles, partly concealing the young leaves; stems 20–25 cm, leafy, arcuate, reddish, sparsely covered with fine long hairs; leaves of sterile shoots and radical leaves in loose rosettes, elliptic or obovate-elliptic, 3.5–6 cm long, 1.5–2.5 cm wide, with petioles to 25 cm long [sic!], tapering petioles 2–3 times as long or much longer than lamina, irregularly rounded-crenate, covered above and beneath with short scattered, somewhat appressed hairs, long-haired along margin and midrib; lower cauline leaves similar in shape but poorly developed, petiolate; upper leaves sessile, oblong-lanceolate or sublinear, crenate, soft-haired. Calyx considerably accrescent becoming membranous after flowering, finely pubescent, the teeth oblong, obtuse, appendages acuminate, rounded-ovate, longer than tube; corolla broadly infundibular, pale lilac, twice as long as calyx teeth, densely and finely pubescent; receptacle pubescent, subglabrous in fruit; capsule pendulous; seeds ovoid-ellipsoid, flattened, straw-yellow, with broadly membranous margins.

Crevices of laminated rocks in subalpine open woodlands, at ca. 2,000 m.—
271 Caucasus: E. Transc. (Lagodekhi District). Endemic. Described from Khochal-Dag Mountain. Type in Tbilisi, cotype with different date in Leningrad.

Note. This species is very similar to C. argunensis Rupr. with which it is undoubtedly related. It differs in all parts being much larger, shoots very long, large leaves, obtuse calyx lobes, and bordered seeds. Its distribution area is included in that of C. argunensis. Ecologically, C. argunensis is a typical alpine plant whereas C. doluchanovii grows in rocks, in shady, subalpine, open woodland.

Series 6. **Bellidifoliae** Fed.— Receptacle pubescent; corolla glabrous; calyx appendages lanceolate, as long as or nearly as long as tube. Radical leaves long-petioled, spatulate or broadly spatulate, abruptly or gradually tapering to petioles, acute and incised or crenate-dentate.

106. C. fominii Grossh. in Tr. Azerb. fil. AN SSSR, 1 (1933) 56; idem, Fl. Kavk. IV, 66; Kharadze in Zam. po sist. i geogr. rast. Tbil. bot. inst. 15, 25. – Ic.: Grossg., op. cit. (1933) tab. separata, sine numero, fig. manca.

Perennial; pulvinate cespitose plant with reduced sterile columelliform shoots and branching rhizome, producing stems; stems 10–15 cm, 1-flowered, arcuately ascending, leafy, sometimes covered with long transparent hairs, with semi-decayed remnants of rosetted leaves; radical leaves and leaves of sterile shoots obovate-cuneate or broadly oblanceolate, nearly spatulate, gradually tapering to petioles, nearly as long as or longer than lamina, incised-dentate mostly toward apex, at apex more deeply incised, nearly lobate, sparsely and shortly pubescent or subglabrous, ciliate along petioles, pale green, 5–10(15) cm long, 1–1.5 cm wide, with acutely toothed lobes; cauline leaves reduced, similar to the radical but less dentate; uppermost leaves subsessile, lanceolate. Flowers 2.5 cm long, drooping, pale lilac; teeth of broadly obconical subglabrous calyx broadly lanceolate, acute, sparingly pubescent, 3–4 times as long as tube, declined after flowering, appendages lanceolate, reflexed, longer than and concealing calyx tube; corolla three times as long as calyx, glabrous inside and outside or very sparingly pubescent, divided to nearly one-third into acute or obtuse lobes; style not exserted; receptacle stiff-haired. June–July. (Plate XIV, Figure 3.)

272 Rock crevices in central mountain belt. – Caucasus: E. Transc. (E.). Endemic. Described from Kuba District, Azerbaidzhan. Type in Tbilisi, topotype in Leningrad.

Note. C. fominii is unique in the order Bellidifoliae by its incised-dentate, nearly lobate leaves. A typical chasmophyte growing in horizontal crevices of limestone rocks.

107. C. bellidifolia Ad. in Weber and Mohr, Beitr. z. Naturk. I (1805) 47; Rupr. in Bull. Acad. Sc. Pétersb. XI, 181; Boiss. Fl. or. III, 906; Shmal'g., Fl. II, 179; Fomin in Mater. Fl. Kavk. IV, 6, 64; Medv. in Tr. Tifl. bot. sada, XVIII, 2, 339; Grossg., Fl. Kavk. IV, 67; Kharadze in Fl. Gruzii, VIII, 176. – A. adami M. B. Fl. taur.-cauc. I (1808) 155; Stev. in Mém. Soc. Nat. Mosc. III, 356; A. DC. Monogr. (1830) 226; idem in DC. Prodr. VII, 460; Ldb. Fl. Ross. II, 875. – C. rupestris Ad. in schedis, non M. B. – C. tridentata var. bellidifolia Trautv. in Tr. Peterb. bot sada, IV, 2 (1876) 388. – C. bellidifolia var. longisepala Fomin, op. cit. 65; Medv., op. cit. – Ic.: M. B. Cent. pl. ross. I, tab. 16; A. DC. l. c. (1830) tab. 11, fig. 1.

Perennial; cespitose plant with branching root, with scaly remnants of petioles at base; stems rather low, 10–12(15) cm, weak, uniflorous; leaves subglabrous or sparsely pubescent; radical leaves long-petioled, small, ovate-rounded, crenate-dentate; cauline leaves like radical but shorter, short-petioled and more sparsely dentate. Teeth of glabrous or subglabrous calyx oblong, obtuse, one-fifth to one-quarter the length of the campanulate lilac corolla; appendages lanceolate, obtuse or acute, longer than tube; receptacle pubescent; capsule nearly rounded; seeds brown, not bordered. June–July.

Rock crevices in central and alpine zones, in ravines. — Caucasus: Cisc. (C. Caucasus). Endemic. Described from Osetia near Larsi. Type in Berlin, isotype (under different name) in Leningrad.

Note. No specimens referred by Adams to C. bellidifolia are preserved at Leningrad herbaria. Only one specimen, certainly a C. bellidifolia Ad., bears a label of this: "Campanula rupestris mihi, habitat in rupium fissuris ad Lars." Obviously what Adams first called C. rupestris he later changed to C. bellidifolia, without making the necessary alteration on the label. Hence, that specimen is the authentic type of C. bellidifolia Ad., with Larsi in Osetia as the classical habitat. This would accord with Bieberstein's report on C. adami: "circa portas
273 caucasicas frequens" (i. e. at the entrance to Dar'yal Gorge, where Larsi is located). Bieberstein's description of C. adami was based on the plant given him by Adams.

108. C. sosnowskyi Charadze in Zam. po sist. i geogr. rast. Tbil. bot. inst. 13 (1947) 109, 15 (1949) 25; Grossg., Opred. rast. Kavk., 419; Kharadze in Fl. Gruzii, VIII (1952) 175. — Ic.: Kharadze, op. cit. (1952) tabl. 376, opt. delin.

Perennial; rhizome densely cespitose, branching, with few columelliform shoots and stems, very densely covered at base with canescent leaf-remnants; stems 10—15(20) cm, uniflorous, straight or slightly arcuate at base, leafy, covered with scattered hairs; radical leaves and leaves of sterile shoots with rounded or rounded-spatulate lamina, (0.5)1.5—2.5 cm long, 0.5—1.7 cm wide, abruptly tapering to thin petioles, 3—4 times as long as lamina, rounded-obtuse or truncate, serrate or dentate-crenate especially above middle, with scattered hairs above, glabrous beneath or pubescent only along margin; cauline leaves reduced, rounded-lobed, gradually tapering to reduced winged petioles, crenate-dentate at apex; uppermost leaves oblanceolate, cuneately tapering at base, sessile. Calyx sparingly pubescent, with oblong-ovate-acute teeth and lanceolate, thinly acuminate appendages, covered with tangled hairs, not elongating in fruit; flowers 1.5 cm long, corolla glabrous, twice as long as calyx teeth, incised to half its length, violet; receptacle pubescent at first, becoming glabrous; capsule semiglobose above, conical below. June—July.

Limestone rocks in subalpine zone. — Caucasus: Cisc. (C. Caucasus). Endemic. Described from Skalistyi Range, near Chmi in Kazbek District.

Note. C. sosnowskyi is distinguished from C. bellidifolia Ad. by broader acutely toothed radical leaves with longer petioles, abruptly tapering to subrounded lamina, cauline leaves very small, flowers small, with acutely toothed calyx. It is also more or less closely related to C. fominii Grossh. and C. kryophila Rupr. In addition to morphological differences it is also distinguished from all the above species by its limited distribution area. C. fominii Grossh. grows in limestones in the Kuba District, C. bellidifolia Ad. in the upper reaches of the Terek River and Gizel-Don, C. kryophila Rupr. in the upper reaches of the Ardon River, whereas C. sosnovskyi is confined to limestones in the Skalistyi Range.

274 Series 7. **Anomalae** Fed. — Receptacle and corolla glabrous; calyx appendages ovate, as long as calyx tube and concealing it. Radical leaves rather short-petioled, spatulate or oblanceolate, serrate-dentate at apex.

109. C. anomala Fom. in Grinevetsk., Result. dvukh botan. putesh. na Kavkaz (1903) 126; nomen; Fomin in Mater. Fl. Kavk. IV, 6, 53, descr., excl. syn. nonnul.; Medv. in Tr. Tifl. bot. sada, XVII, 2, 341; Grossg., Fl. Kavk. IV, 67; Kharadze in Fl. Gruzii, VIII, 171.

Perennial; rhizome cespitose, thick, angular, the upper part densely covered with leaf-remnants; stems 10–15 cm, rather thin, sometimes slightly flexuous, leafy reddish, uniflorous; rosetted leaves lanceolate-spatulate or oblong-lanceolate, obtuse, cuneately tapering to petioles as long as lamina; margin pubescent, dentate from middle or only at apex (teeth small, declinate); cauline leaves lanceolate, the uppermost sublinear, all usually with tridentate apex and pubescent margin. Flowers large, 4 cm and longer; corolla broadly campanulate, violet, glabrous; calyx teeth triangular-lanceolate, obtuse, one-third the length of the corolla, appendages ovate, obtuse, longer than tube, with more or less lanate margin; receptacle quite glabrous; style sometimes slightly exserted from corolla or nearly as long. June–July.

Rocks in alpine zone. — Caucasus: Cisc. (W. and C. Caucasus). Endemic. Described from Fisht Mountain; type at the herbarium of the former Yur'evskii University.

Note. This species is very closely related to C. meyeriana Rupr., from which it differs by glabrous leaves and consistently reddish stems. It also resembles C. circassica Fom. and C. tridentata Schreb., differing from the former by larger growth and thick rhizome, with abundant leaf-remnants, and from the latter by a broadly gaping violet, not blue, corolla. In the western part of the Caucasus this species seems to be rare; it does not grow in Abkhaziya, though it occurs to the west of Abkhaziya (Fisht Mountain, Pseashkha and others) and in the northeast, from Elbrus to Digoria). It is not mentioned in Kolakovskii's "Flora Abkhazii."

110. C. circassica Fom. in Grinevetsk., Result. dvukh botan. putesh. na Kavkaz (1903) 126, nomen; Fomin in Mater. Fl. Kavk. IV, 6 (1905) 52,
275 descr.; Medv. in Tr. Tifl. bot. sada, XVIII, 2, 342; Grossg., Fl. Kavk. IV, 64; Fl. Gruzii, VIII, 172; Kolak., Fl. Abkh. IV, 179. — C. saxifraga var. tridentata Rupr. in Bull. Acad. Sc. Pétersb. XI, 184, non C. tridentata Schreb. — C. tridentata var. rupestris Trautv. in schedis, non C. tridentata Schreb. nec C. rupestris M. B. — C. saxifraga auct. p.p. non M. B.: Alb. Prodr. Fl. Colch. (1895) 121. — C. aucheri auct. p.p. non DC.: Alb. l. c. 157.

Perennial; rhizome cespitose, with numerous thin glabrous sterile shoots; stems 10 cm; uniflorous, few-leaved, thin arcuately ascending, straight in upper part, glabrous; rosetted leaves of sterile shoots and at base of stem lanceolate-spatulate, faintly acutely toothed at apex, nearly edentate, cuneately tapering to petioles twice as long as leaves, the leaves 3–4 cm long, 0.5 cm wide, glabrous or sparsely pubescent; cauline leaves lanceolate, short-petioled, few-toothed at apex. Teeth of subglabrous calyx lanceolate, acute, very sparingly pubescent or subglabrous, one-fifth to one-quarter the length of the corolla; appendages ovate, as long as tube; corolla broadly campanulate, violet when dry, glabrous; style not exserted; receptacle glabrous. June–July.

Alpine plots, sometimes in taluses near melting snow. – Caucasus: Cisc., W. Transc. (from Pseashkha to Nakhar mountain pass). Endemic. Described from alpine meadows of Pseashkha Mountain. Type in the herbarium of the former Yur'evskii University, paratype in Leningrad.

Note. A very rare species, the only available specimens being those of Fomin, from the Radde collections at the Nakhar mountain pass, of Voronov from the Irtsikh Range (Main Range from Marukhskii mountain pass to Klukhorskii), of Al'bov from Chedymskii Range (Abkhaziya) and some others. It most nearly resembles C. tridentata Schreb. and C. biebersteiniana Roem. et Schult., from which it differs by several acute teeth at the leaf apex – as against 3 – the presence of thin glabrous underground shoots, the absence of tomentose pubescence on the calyx and the broady campanulate corolla.

Series 8. **Ardonenses** Fed. – Receptacle and corolla glabrous. Calyx appendages lanceolate, acute, longer than tube. Radical leaves long-petioled, with linear or ovate-rounded lamina. Calyx teeth thin, filiform.

111. C. ardonensis Rupr. in Bull. Acad. Sc. Pétersb. XI (1867) 185; Boiss. Fl. or. III, 906; Fomin in Mater. Fl. Kavk. IV, 6, 67; Medv. in Tr.
276 Tifl. bot. sada, XVIII, 2, 340; Grossg., Fl. Kavk. IV, 67. – C. ardonensis var. latifolia Fom., op. cit.; Medv. op. cit. – C. tridentata var. ardonensis Trautv. in Tr. Peterb. bot sada, VI (1879) 58.

Perennial; subglabrous, generally quite glabrous cespitose plant; rhizome thick, branching, multicipital; stem base densely covered with remnants of petioles; leaves of sterile shoots and radical leaves narrowly linear, serrate, tapering to filiform petioles; cauline leaves very narrowly linear; stems rather long, 10–20(25) cm, thin, subfiliform, uniflorous, longer than radical leaves. Calyx glabrous, its teeth narrowly linear, nearly subulate, very acute, half the length of the corolla, appendages lanceolate, acute, slightly longer than tube; corolla 1.5–2 cm long, dark blue, glabrous, narrowly campanulate, divided for one-half into lobes; style not exserted; receptacle quite glabrous; capsule semiglobose to conical; seeds pale brown, oblong, not bordered. May–July. (Plate XIV, Figure 4.)

Herbaceous rocky habitats in river gorges in the upper part forest, part subalpine mountain zone. – Caucasus: Cisc. (C. Caucasus). Endemic. Described from Osetia (Ardon River below Misurtsy). Type in Leningrad.

Note. Related to C. kryophila, this species always differs from it by its narrow leaves. In C. kryophila the lamina of the radical leaves is often rounded; in both species the lamina tapers to long, filiform petioles. In all other characters they are very similar.

112. C. kryophila Rupr. in Bull. Acad. Sc. Pétersb. XI (1867) 184; Fom. in Mater. Fl. Kavk. IV, 6, 66; Medv. in Tr. Tifl. bot. sada, XVIII, 2, 341; Grossg., Fl. Kavk. I V, 67; Kharadze in Fl. Gruzii, VIII, 179. – C. ardonensis var. kryophila (Rupr.) Boiss. Fl. or. III (1875) 906. –

C. tridentata var. saxifraga Trautv. in Tr. Peterb. bot sada, IV (1876) 58, ex p. — C. saxifraga var. leptorhiza Somm. et Lev. in Tr. Peterb. bot. sada, XVI (1900) 318, ex p. — C. ardonensis ssp. kryophila (Rupr.) Fom. in sched. ad specim. typ. — C. petrophila Lipsky in schedis, non Rupr.

Perennial; subglabrous cespitose plant covered at base with numerous remnants of petioles; rhizome multicipital, branching, squamous, producing 10–15(20) cm, uniflorous, leafy stems; leaves of sterile shoots and radical leaves long- and thin-petioled, lamina ca. 1 cm long, ovate-rounded or ovate-oblong, obscurely crenate-dentate in upper half, entire below; cauline leaves lanceolate, tapering to rather short petioles. Teeth of calyx acute, narrowly lanceolate or linear; appendages lanceolate, acute, longer than calyx; corolla narrowly campanulate, glabrous, divided for one-third into lobes twice as
277 long as calyx teeth; receptacle quite glabrous; capsule broadly ovoid; seeds brown, oblong-ellipsoid, not bordered. June–August (September).

In mountains on rocks at edges of glaciers. — Caucasus: Cisc. (C. Caucasus). Endemic. Described from rocks near Tsei glacier (along the Voenno-Osetia highway). Type in Leningrad.

Note. C. kryophila is closely related to C. ardonensis (see Note to this species) as well as to C. bellidifolia, with which it is easily confused. Reliable distinctions are the quite glabrous receptacle and linear (not spatulate) cauline leaves in C. kryophila.

Series 9. **Besenginicae** Fed. — Receptacle and corolla glabrous; calyx appendages small, shorter than tube. Radical leaves narrowly lanceolate, glabrous beneath, with retrorse appressed hairs above. Calyx teeth glabrous outside, white-tomentose inside. Of this series only one species is known.

113. C. besenginica Fom. in Tr. Tifl. bot. sada, VI, 2 (1902) 8; idem, Mater. Fl. Kavk. IV, 6, 66; Medv. in Tr. Tifl. bot. sada, XVIII, 2, 341; Grossg., Fl. Kavk. IV, 68.

Perennial; cespitose plant with numerous remnants of petioles at stem base; stems 15–10(25) cm thin, elongate, straight or ascending, uniflorous; leaves entire, covered above with appressed retrorse hairs, quite glabrous and shiny beneath; radical leaves and leaves of sterile shoots narrowly lanceolate, long-petioled; cauline leaves linear. Calyx teeth triangular-lanceolate, obtuse, quite glabrous, shiny above, densely white-tomentose inside, one-third the length of the calyx, appendages dentiform, very short, nearly inconspicuous; corolla blue, campanulate, glabrous; receptacle glabrous. July.

Moraines, rocks, rarely in subalpine meadows, near glaciers. — Caucasus: Cisc. (C. Caucasus). Endemic. Described from rocks near Bezengi glacier. Type in Leningrad.

Note. This endemic species appears to occupy a very limited area in the C. Caucasus. It was found only near the Bezengi and Mizhirgi glaciers.

Series 10. **Dasyanthae** Fed. — Receptacle glabrous; corolla glabrous outside, ciliate-hairy along margins, sometimes also in lower part; calyx appendages shorter than tube. Radical leaves with lanceolate or ovate glabrous or pubescent lamina.

114. C. dasyantha M. B. Fl. taur.-cauc. III (1819) 147. – C. pallasiana Vest in Roem. et Schult. Syst. V (1819) 138; D. C. Prodr. VII, 461; Kryl.,
278 Fl. Zap. Sib. XI, 2631. – C. pilosa Pall. ex Roem. et Schult. l. c. 148; Ldb. Fl. Ross. II, 2, 877; DC. l. c. 460; Turcz. Fl. baic.-dahur. II, 475; Kryl., Fl. Alt. III, 773. – C. altaica A. DC. Monogr. (1830) 229, non Ldb.; idem in DC. l. c. 461, p. p. – C. redowskyi Fisch. in Tr. Peterb. bot. sada, I (1873) 288, non C. redowskiana Cham. – Ic.: Ldb. Ic. pl. Fl. Ross. I, 7, tab. 209; A. DC. l. c. (1830) tab. 10, f. 3.

Perennial; rhizome branching, producing solitary or few stems, stems 5–15 cm, simple, sometimes with 1 branch, glabrous or long-haired in upper part; rosetted leaves long-elliptic, lanceolate or linear-lanceolate, 2–12 cm long, 3–8 mm wide, small-toothed, with contiguous teeth terminating in yellowish mucro tapering to petiole; cauline leaves reduced, lanceolate-linear, sessile; all leaves more or less long-haired. Flowers solitary (rarely 2), declined or straight; teeth of pubescent calyx broadly lanceolate, one-third to one-half the length of the corolla, appendages much shorter than teeth, reflexed; corolla pale blue, 2–2.5 cm long, incised for one-third into ovate acute lobes, long-ciliate at margins, pubescent inside in lower part. June–August. (Plate XV, Figure 1.)

Gravel-moss montane tundra, also rocks on treeless mountain tops (goltsy – alpine tundra belt above the timberline). – W. Siberia: Alt.; E. Siberia: Ang.-Say., Dau., Lena-Kol. (S.); Far East: Okh. (Ayan), Kamch. **Gen. distr.:** Mong. (Mong. Alt.), N. Am. (Canada). Described from E. Siberia. Type in Leningrad.

Note. Fischer (op. cit.) described specimens from the shores of the Sea of Okhotsk as C. redowskyi. These are distinguished by thin stems and narrower leaves.

C. dasyantha M. B. was believed to grow mostly on Kamchatka, the Kurile Islands, Sakhalin and Japan and C. pilosa Pall. ex Roem. et Schult. in the mountains of Siberia. This proved not to be so. Bieberstein's type of C. dasyantha was collected in Siberia by Sievers who is known to have traveled east through the Siberian mountains, reaching Dauria. Not only had he never been on the islands mentioned, or on Kamchatka, but he never even reached the shores of the Sea of Okhotsk. C. dasyantha M. B. was collected in the Sayan Mountains or Transbaikalia, probably in mountains near Baikal itself, where Sievers, too, collected it. The plants from Kamchatka, Japan and the other islands which passed for C. dasyantha since the time of de Candolle, Chamisso and Ledebour turned out to be a distinct species, C. chamissonis. For the Siberian plant C. dasyantha is preferable to the two other names which were published simultaneously,
279 C. pilosa and C. pallasiana. For the "Flora of the USSR," it is relevant that C. dasyantha M. B. was preserved in Leningrad and that Bieberstein's description of this species (Fl. taur.-cauc.) was published in Russia simultaneously with the work of Roemer and Schultes (see also Note to C. chamissonis and C. lasiocarpa).

115. C. chamissonis Fed. nom. nov. – C. dasyantha Cham. in Linnaea, IV (1829) 37, non M. B. – C. dasyantha auct. mult. p. p. non M. B.: A. DC. Monogr. (1830) 230; idem in DC. Prodr. II, 2, 461; Ldb. Fl. Ross. II, 877; Miyabe et Miyake, Fl. Saghal. 292; Kom. Fl. Kamch. III, 115; Hulten, Fl.

Aleut. Isl. 311; ej. Fl. of Kamtch. IV (1930) 149; ej. Il. of Alaska a. Yukon, IX, 1456; Sugawara, Ill. Fl. of Saghal. IV, 1745. — *C. pilosa* β *dasyantha* Herder in Tr. Peterb. bot. sada, I, 2 (1872) 288. — Ic.: A. DC. l. c. (1830) tab. 10, f. 4 (mediocr.); Hula. l. c. (1930) tab. 5, f. a, b; Sugawara, l. c. tab. 800 (mediocr.); H. Clifford Crook, Campanulas, No. 156.

Perennial; root cylindrical, simple, whitish, thick above, bearing rosette of leaves; stems not exceeding 10 cm, uniflorous, rather thin, sparingly pubescent or glabrous, sometimes reddish, slightly leafy, straight or arcuately curved, sometimes stems few; rosetted leaves 2–3 cm long, 1 cm wide, obovate, acute, crenate, subglabrous, tapering to petioles as long as lamina; lower cauline leaves obtuse, nearly entire, the upper sessile, lanceolate, entire, pubescent, only 3–5(10) mm long. Flowers nearly straight 3–3.5 cm long, very large in relation to the plant; calyx pubescent, tube obconical, small, the teeth ovate-lanceolate, broad, straight, appendages very small, acute, barely visible through the hairs; corolla infundibular, blue, with darker nerves, subglabrous outside, bearded inside and along margins, with ovate acute lobes; stamen short, with broadened membranous, ciliate filaments; style as long as corolla, with 3 stigmas. July. (Plate XV, Figure 2.)

Alpine mountain zone, stony places. — Far East: Kamch., Sakh., Kurile Islands, Moneron Island. **Gen. distr.:** Ber. (Aleutians), Jap. (Honshu and Hokkaido islands). Described from Unalaska Island. Type in Leningrad.

Note. It was necessary to introduce the new name *C. chamissonis* Fed. because the closely related *C. dasyantha* M. B. could not really be identified with it. The type of *C. dasyantha* in Leningrad comprises specimens of the two simultaneously published species, *C. pilosa* Pall. ex Roem. et Schult. and *C. pallasiana*, which have no relation with *C. da-*
280 *syantha* M. B. from the Aleutians, Kamchatka, Kurile Islands and Sakhalin, as accepted and described in detail by Chamisso, which is accepted in the newest flora (for example, Hulten, Fl. of Alaska a. Yukon), even though the type was collected by Sievers in Siberia. This is evident from the label ("ex Sibiria") and as corroborated by the characters of the plant itself — certainly the distribution area of the Altai-Sayan-Transbaikalia species, which was described by Roemer and Shultes and by Bieberstein. Moreover, it is known that Sievers never visited either Kamchatka or the islands of the Sea of Okhotsk or Bering Sea. This confusion goes back to the time of Chamisso, A. de Candolle and Ledebour. Chamisso probably never saw Bieberstein's type and recognized the Unalaska Island as *C. dasyantha* M. B. In A. de Candolle's monograph the above name refers to specimens collected by Chamisso and Choris; consideration is also given to Gmelin's description based on the Steller collections which belong to another species — *C. lasiocarpa* Cham. Ledebour "*C. dasyantha*" includes plants from the Chamisso and Eschscholtz collections, as well as Gmelin's description *C. lasiocarpa*. He erroneously extended the area of *C. dasyantha* to the Bering Strait, where only *C. lasiocarpa* grows, an error perpetuated by Hulten (Fl. of Kamtch.; Fl. Aleut. Isl.). Hulten reported *C. dasyantha* for the Commander Islands, Later (Fl. of Alaska a. Yukon) he corrected the area, but failed to correct the name. He appropriately notes that Choris, mentioned by Ledebour, collected the plant only on Unalaska Island, and that there was no reason to claim it for the Bering Strait. All the available herbarium specimens definitely confirm this view.

116. C. aldanensis Fed. et Karav. Sp. nov. in Addenda XXIII, p. 334.

Perennial; rhizome short, branching, producing loose tufts of grayish thin shoots bearing stiff transparent leaf remnants; stems weak, 10–12 cm, sometimes subfiliform, arcuately ascending from base, straight above, twice to three times as long as rosettes of leaves, uniflorous, sparsely, finely pubescent; radical leaves and leaves of shoots lanceolate, 5–10 cm long, 0.3–0.5 cm wide, gradually tapering to filiform petioles, nearly as long as
281 lamina, obtuse, irregularly and obscurely incised-denticulate; teeth obtuse, antrorse; lamina glabrous above and beneath, pale green; cauline leaves reduced, linear, subsessile. Flowers 1.5 cm long, pale blue; teeth of calyx stiff-haired, lanceolate, with arcuately curved tips, nearly half the length of the corolla, appendages short, curved, triangular, stiff-haired; corolla infundibular, sparingly pubescent outside, divided for one-third into crisply pubescent lobes; style not exserted. July–August.

Limestones, riverbanks in the taiga. – E. Siberia: Lena-Kol. (Aldan, Lena, Olekma rivers and their tributaries in Yakut ASSR). Endemic. Described from the rocky banks of the Aldan River at its confluence with the Mili. Type a specimen collected by Korzhavin in Leningrad.

Note. Karavaev has repeatedly collected this plant in the basin of the Lena and Aldan rivers in the Yakut ASSR.

Taxonomically, *C. aldanensis* is rather close to *C. dasyantha* M. B., which may be regarded as its vicariant as *C. aldanensis* is confined to limestones in the forest belt whereas *C. dasyantha* grows in the high mountains.

Series 11. **Ledebourianae** Fed. – Receptacle and corolla glabrous; calyx appendages as long as or longer than tube. Rosetted leaves cuneate-spatulate, dentate at apex or obscurely crenate. Rhizome marked by columelliform branching, crowned by rosettes of leaves, forming dense pillows.

117. C. ledebouriana Trautv. in Tr. Peterb. bot. sada, II, 2 (1873) 477; Boiss. Fl. or. III, 905; Fomin in Mater. Fl. Kavk. IV, 6, 71; Medv. in Tr. Tifl. bot. sada, XVIII, 2, 336. – *C. ledebourii* Trautv., op. cit. 562 et in schedis ad sp. typ.; Grossg., Fl. Kavk. IV, 72; Fomin in schedis.

Perennial; cespitose plant with creeping nodose branching rhizome; stems 8–10(15) cm, numerous, 1–4-flowered, covered with sparse antrorse appressed hairs; leaves of sterile shoots and radical leaves narrowly spatulate to oblong-lanceolate, often linear-lanceolate, tapering to more or less long petioles; cauline leaves linear-lanceolate, small, sessile, glandular-denticulate at margin, glabrous beneath, with sparse appressed lanate
282 reflexed hairs above. Calyx glabrous, subcoriaceous, its teeth broadly lanceolate, obtuse, sparsely and shortly ciliate, one-fourth to one-third the length of the corolla, appendages lanceolate, acute, longer than tube; corolla glabrous, blue, campanulate, 1.5–2 cm long; style slightly exserted; receptacle glabrous. June.

Rocks in alpine zone. – **Gen. distr.:** Montane border region between the USSR and Turkey. Described from Lake Kugel in Great Ararat. Type in Leningrad, isotype in Tbilisi.

(283)

PLATE XVI. 1 – Campanula andina Rupr.; 2 – C. karakuschensis Grossh.; 3 – C. petrophila Rupr.; 4 – C. lehmanniana Bge.

Note. This species, never found in the USSR, may well grow in S. Armenia.

118. C. minsteriana Grossh. in Tr. Bot. inst. AN Azerbaidzh. SSR, II (1936) 256; Grossg., Opred. rast. Kavk. (1949) 415.

Perennial; very small glabrous plant, not more than 2–5 cm high; rhizome producing columelliform branches clustered in dense pulvinate tufts crowned by rosettes of leaves and stems; stems ascending, arcuately curved, leafy; rosetted and cauline leaves of nearly the same shape, with petioles 1–1.5 cm long, obovate-cuneate, acute and rather largely 3–5-toothed at apex, sometimes with small teeth along margins, tapering to poorly distinguishable petioles. Flowers ca. 0.5 cm long, pale blue, terminal, short-pediceled; lobes quadripartite (sometimes with obsolete linear fifth lobe); calyx triangular, obscurely toothed, barely as long as campanulate corolla, appendages linear-oblong, acute, as long as tube; corolla glabrous, stamens 4, with short triangularly broadening filaments and long anthers. May.

Limestones in deep gorges of mountain rivers. – Caucasus: S. Transc. (Nakhichevan ASSR). Endemic. Described from Negram gorge, near Negram railway station. Type in Baku, isotype in Leningrad.

Note. No further specimens, besides the type, are known to have been collected. As the capsule and styles are lacking, further study is required. Grossgeim included it in Rupestres, but the well-developed calyx appendages indicate its inclusion in subsection Scapiflorae. Unfortunately, though, it is impossible to determine its affinity because of the lack of material.

Subsection 17. RUPESTRES (Boiss.) Fed. comb. nov. – Sect. Rupestres
285 (Boiss.) Charadze in Zam. po sist. i geogr. rast. Tbil. bot. inst. 15 (1949) 24, p. p. – Ser. Rupestres Boiss. Fl. or. III (1875) 906, p. p. – Ser. Saxicolae Boiss. l. c. 894. p. p. – Sect. Saxicolae Charadze, op. cit. 20, p. p. – Notches between calyx teeth with small appendages or none, teeth lanceolate or linear-lanceolate; flowers erect up to blossoming, pediceled; capsule pendulous. Stems few-flowered, sometimes uniflorous. Leaves forming dense rosettes at base of stem and apex of sterile shoots, rosetted and cauline leaves spatulate, sometimes lanceolate or linear, petiolate or sessile, very rarely cordate, ovate or reniform; radical leaves distinctly broader and longer than the cauline. Rhizome usually thick, multicipital, sometimes columelliform, cespitose, sometimes creeping, covered with remnants of petioles. Pubescent or subglabrous herbs. Type species of subsection: C. petrophila Rupr.

Note. This subsection, which Beissier considered as a series based on C. petrophila, comprises species related to C. incanescens, as well as to C. lehmanniana (referred to the series Saxicolae). The latter species is retained though removed from section Saxicolae Charadze, for at the time of its description none of the species included by Boissier in Saxicolae – the basonym for this section – had been taken into consideration (see also Note to subsection Oreocodon).

Series 1. **Petrophilae** Fed. – Calyx appendages short, dentiform. Rosetted leaves ovate, abruptly tapering to petioles; cauline leaves spatulate or reniform, distinctly petiolate. Rhizome loosely cespitose, rather thin, with shoots prostrate in part.

119. C. petrophila Rupr. in Bull. Acad. Sc. Pétersb. XI (1867) 186; Boiss. Fl. or. III, 906; Fomin in Mater. Fl. Kavk. IV, 6, 69; Medv. in Tr. Tifl. bot. sada, XVIII, 2, 336; Grossg., Fl. Kavk. IV, 71; Kharadze in Fl. Gruzii, VIII, 158. — C. tridentata var. petrophila Trautv. in Tr. Peterb. bot. sada, VI (1879) 58, p. p. — C. petrophila var. linoides Rupr. l. c. — C. petrophila var. barbalensis Rupr. l. c. — C. adami auct. p. p. non M. B.: C. A. M. Verzeichn. (1831) 83. — C. petrophila var. longiflora Rupr. l. c.; Fomin, op. cit. 70; Medv. op. cit. — C. petrophila var. exappendiculata Somm. et Lev. in Tr. Peterb. bot sada, XVI (1900) 320 and in Fomin, op. cit. 71; Medv., op. cit. — C. petrophila var. angustiflora Fom. op. cit. 71. — C. petrophila var. laxa Rupr. in schedis ad specim. typ. — C. petrophila var. glabrescens Rupr. in schedis. — C. petrophila f. parviflora Rupr. in schedis. — C. petrophila var. longelobulosa Rupr. in schedis. — C. linoides Rupr. in schedis. — C. barbalensis Rupr. in schedis. — C. petrophila f. um-
286 brosa Fom. in schedis. — C. rupicola Rupr. in schedis. — C. rupicola f. glabrescens Rupr. in schedis. — Ic.: H. Clifford Crook, Campanulas, 155.

Perennial; loosely cespitose, densely pubescent or subglabrous plant, with multicipital apically branching rhizome producing numerous thin leafy slightly prostrate 1- or 2—3-flowered stems and sterile shoots; radical leaves and leaves of sterile shoots small, long-petioled, ovate or obovate, rounded, crenate-dentate at apex; cauline leaves short-petioled, obovate or elliptic. Calyx pubescent, its teeth triangular-lanceolate, obtuse, one-fourth to one-third the length of the corolla, appendages very short, dentiform; corolla blue, glabrous or pubescent along nerves, campanulate or nearly tubular-campanulate, bearded inside; receptacle glabrous; capsule semiglobose; seeds oblong, flattened with narrow border. July—August. (Plate XVI, Figure 3.)

Caucasus: Cisc. (C. Caucasus), Dag., E. Transc. Endemic. Described from E. Caucasus. Type and numerous cotypes in Leningrad.

Note. C. petrophila is a very variable species. Initially Ruprecht intended to recognize a few species, such as C. petrophila, C. barbalensis, C. linoides and C. rupicola, but he later lumped them in one species with three varieties. His specimens of C. petrophila were collected at Bogos Range, in the interior of Dagestan, and in the upper reaches of Samur River. Bogos' specimen should be regarded as the type.

120. C. andina Rupr. in Bull. Acad. Sc. Pétersb. XI (1867) 189; Grossg., Opred. rast. Kavk. 421; Kharadze in Fl. Gruzii, VIII, 159. — C. andia Rupr. ex Lipsky in Tr. Peterb. bot. sada, XIV (1898) 312; Fomin in Tr. Tifl. bot. sada, IV, 2 (1902) 4; idem, Mater. Fl. Kavk. IV, 6, 72; Medv. in Tr. Tifl. bot. sada, XVIII, 2, 337; Grossg., Fl. Kavk. IV, 69. — C. gumbetica Boiss. Fl. or. III (1875) 914. — C. andia var. alexeenkoi Fom., op. cit. (1902) 5.

Perennial; plant canescent with dense pubescence; rhizome thin, producing subfiliform, flexuous stems not more than 10 cm, 1—2-flowered, slightly prostrate and ascending, leafy; leaves small, indurate, with prominent transparent nerves, irregularly and obtusely serrate, 0.5—0.7(1) cm long and as wide; radical leaves ovate-cordate, the petioles as long as or shorter than lamina; cauline leaves rounded-reniform, short-petioled, the upper reduced,

oblong, subsessile. Flowers medium-sized, rather large in relation to leaves, 1.5–2 cm long, the pedicels with bracteoles as long as calyx; calyx canescent, its teeth narrowly lanceolate, acute, one-fourth to one-third the length of the corolla, appendages very short, much shorter than tube; corolla
287 violet, infundibular-campanulate, bearded inside, pubescent outside; style much shorter than corolla; receptacle glabrous. July. (Plate XVI, Figure 1.)

Limestones in central mountain belt. – Caucasus: Dag., E, Transc. Endemic. Described from Gumbet near Danukh in Dagistan (Andi Range). Type in Leningrad.

Note. Presumably a very rare species, as yet collected only in Dagestan and near Balta, on the E. Georgian highway, and Stolovaya Mountain in the C. Caucasus. Fomin described variety var. Alexeenkoi Fom. from the Alekseenko collections, but this may represent a species, as the only available specimen of the type of C. andina, collected by Ruprecht, is in very bad condition.

The proper specific epithet is "andina," not "andia" as arbitrarily proposed by Lipsky, by reference to the type locality – the Andi Range in Dagestan. Boissier's earlier proposal of C. gumbetica probably reflects a confusion with plants of the S. American Andes. Ruprecht's C. andia he considered absurd, not being aware of the existence, outside the Andes, of the Andi Range. Ruprecht's "in Andibus caucasicis" does indeed sound strange, and it is interesting that he himself followed Boissier, and labeled the type: C. gumbetica

Series. **Saxicolae** Boiss. Fl. or. III (1875) 894. – Calyx appendages short, dentiform or none. Rosetted leaves spatulate-cuneate or lanceolate, narrowly lanceolate, short- or long-petioled. Rhizome not developing prostrate shoots; plant often forming dense cushions.

121. C. karakuschensis Grossh. in Izv. Azerb. fil. AN SSSR, 1–2 (1939) 118; Grossg., Opred. rast. Kavk. 416. – C. prilipkoana Grossh. in schedis ad specim. isotyp.

Perennial; glaucescent plant, scabrous, covered with very short stiff-bristly hairs clearly visible under magnification; rhizome producing numerous columelliform branches crowned by rosettes of leaves and stems, covered below by short squamous remnants of leaves, forming dense pulvinate tufts; stems 5–10 cm, leafy, ascending, arcuately curved, corymbiformly branching above, rosetted leaves spatulate-obovate, acutely toothed
288 or crenate, cuneately tapering to short petioles; cauline leaves reduced, sessile, the uppermost nearly lanceolate, dentate. Flowers dark blue, 1–1.5 cm long, short-pediceled; calyx with spreading stiff hairs, the teeth ovate-lanceolate, acute, nearly one-third the length of the corolla, appendages linear, shorter than tube; corolla infundibular-tubular, with scattered hairs on the outside; style slightly exserted. May–July. (Plate XVI, Figure 2.)

Dry limestones in central mountain belt. – Caucasus: S. Transc. Endemic. Described from Kara-Kush Mountain near Aznabyurt. Type in Baku, isotype in Leningrad.

Note. Grossgeim's description (op. cit.) is faulty: the corolla is dark blue, not pale, not as long, but twice as long as the calyx. Outside the type locality, it has also been found on limestones between Khachik and Gnishik, in Armenia, at not more than 50 km from Kara-Kush Mountain.

122. **C. lehmanniana** Bge. Reliq. Lehm. (1851) 211; Boiss. Fl. or III, 917; Trautf. in Tr. bot. sada, VI, 1, 81; O. and B. Fedch., Perech. rast. Turk. III, 357. — C. lehmanniana var. pubiflora et var. nudiflora Zakirov in herb.

Perennial; rhizome indurate, woody, branching, the branches densely covered with grayish-brown scarious remnants of petioles, rugose from leaf traces, producing rosettes of sterile leaves and rosettes with single stems; stems short, usually 10 cm with inflorescence, sometimes to 20 cm, simple, leafy, few-flowered, straight, pubescent at first, becoming naked; leaves pubescent at first, becoming glabrous; leaves of sterile rosettes with petioles, usually 6 cm, oblong, acuminate, acutely toothed, tapering to a rather long petiole, slightly shorter than lamina; leaves at base of stem obovate-elliptic, acute, serrate-dentate, short-petioled, with petiole 2.5 cm long; cauline leaves much smaller, sessile, elongate at both ends, sometimes linear-oblong, acute and entire, sometimes oblong, acute and serrate-dentate; leaves of inflorescence smaller than the cauline, linear-lanceolate, acuminate, entire. Flowers straight, 1.7 cm long, singly in axils of terminal leaves and at stem apices, pediceled, declined, as long as or longer than leaves, and in racemiform very short or slightly corymbiform, few-flowered, terminal, inflorescence, ebracteate or with 1—3 small, linear-subulate, entire bracts; calyx short, subglobose, pubescent, the teeth linear-subulate, entire,
289 acute, about as long as corolla at first, becoming nearly as long, about twice the length of the ripe capsule; corolla tubular-campanulate, very finely pubescent, divided into ovate acute lobes; style with 3 stigmas not exerted; capsule ovoid or subglobose, thinly and finely pubescent, faceted, pendulous, opening by 3 pores above middle. July. (Plate XVI, Figure 4.)

Rocks, usually limestone, in gorges. — Centr. Asia: Pam.-Al. (Zeravshan, Gissăr ranges), T. Sh. (W.). Endemic. Described from Kara-Tau Mountain in Zeravshan Range. Type in Paris.

Note. This species was described by Bunge (l. c.) after an incomplete specimen from Kăra-Tau, without roots or flowers. The first detailed description is due to Trautfetter, who had at his disposal plants from Zeravshan and Kokan, though without having seen the type. Though no specimens from the type locality have been seen, both descriptions certainly treat one and the same species. (Kara-Tau Mountain is in the Zeravshan Range; today though Kara-Tau is associated with a mountain in the Tien Shan system.)

123. **C. capusii** (Franch.) Fed. comb. nov. — C. lehmanniana var. capusii Franch. Pl. du Turkest. (1883) 115 et in Ann. Sc. Nat. VII, ser. XVII (1884) 208; O. and B. Fedch., Perech. rast. Turk. III, 357. — Ic.: Franch. l. c. tab. 14, f. A.

Perennial; rhizome branching, cespitose, multicipital, with indurate grayish-brown slightly nodose branches, upper part covered with dried gray remnants of petioles, bearing flowering stems and sterile rosettes of leaves; stems low, 5—10 cm, simple, usually 1-flowered, slightly leafy, rather thin; radical leaves and leaves of sterile rosettes narrowly lanceolate ca. 2—3(4) mm wide, with petiole 3—4 cm long, grayish, somewhat scarious, indistinctly and remotely dentate, acute, long-attenuate at apex, very gradually tapering to rather short petiole; cauline leaves 2–3(4), lanceolate or linear, acuminate, sessile, much smaller than the radical. Calyx short, conical, very sparingly pubescent becoming globose, its teeth linear-subulate, markedly shorter than corolla; straight at first, becoming recurved; corolla shallowly lobed, glabrous, blue, 0.5–1.8 cm long; capsule globose, membranous, slightly ribbed, opening by small, round holes at sides of lower half. June–July.

290 Stony slopes and bluffs, sometimes in Savin juniper open woodland, in central mountain zone. — Centr. Asia, Pam.-Al. (Alai Range, Zeravshan Range?), T. Sh. (Talas Ala-Tau). Endemic. Described from "Shivata." Type in Paris.

Note. Drobor recently collected on the northern slope of Alai Range (Shakimardan River basin) a specimen which perfectly fits the description and drawing in Franchet's work (l.c.). This species is related through transitional forms with C. lehmanniana. In the eastern part of its distribution area there occur, in addition to C. lehmanniana, only the narrow-leaved, few-flowered forms, which may belong to C. capusii. C. lehmanniana, with broad acutely toothed leaves and few flowers, is characteristic of the Zeravshan mountains, and is not found elsewhere. C. capusii may therefore represent a geographically, rather distinct species, rather than a variety. Incidentally, " Shivata," the locality listed by Franchet (l. c.) following Capus, who collected C. capusii somewhere in Central Asia, has not been located. Lipskii (Fl. Sr. Azii, III), too, deciphering the mysterious localities contained in Capus' labels and cited by Franchet, failed in this. In July 1881 Capus collected plants in Zeravshan and Shivata probably lies in that area.

124. C. eugeniae Fed. Sp. nov. in Addenda XXIII, 446. — C. alberti var. stenosepala Ik.-Gal. in schedis, non C. alberti Trautv. (1879).

Perennial; low subglabrous plant, with cespitose multicipital branching rhizome producing numerous reduced gray shoots covered with short scaly remnants of petioles, bearing loose rosettes of leaves, or else flowering stems; stems 10–15 cm, simple, thin, filiform, slightly longer than radical leaves and leaves of sterile shoots, leafy, uniflorous, rarely 2-flowered, lower leaves rosetted, (5)8–10 cm long, with elongate thin petioles, the laminas narrowly lanceolate, tapering at both ends, acute, nearly half the length of the filiform petioles, glabrous, thin, nearly entire or obscurely toothed; cauline leaves linear-lanceolate or alternate, sessile. Flowers erect or drooping, solitary, 1.5 cm long; teeth of calyx obconical subglabrous, its teeth narrowly linear, slightly shorter than corolla but several times longer than tube, appendages none; corolla narrowly campanulate, blue, shallowly divided into acute triangular lobes; style not exserted;
291 capsule (unripe) obconical, short, becoming rounded, opening at base by small holes. August.

Crevices of marble bluffs, in subalpine and central mountain zones. — Centr. Asia: T. Sh. (Fergana Range). Endemic. Described from Baubashata Mountain. Type in Leningrad.

Note. This species is closely related to C. lehmanniana Bge. from which it is readily distinguished by the filiform stems, thin, very long petioles of the radical leaves and nearly always 1-flowered stems. The same distinctions apply to C. capusii (Franch.) Fed. with the exception of the flowers which are usually solitary in both species. In fact, C. capusii is much closer to C. lehmanniana than C. eugeniae is to either of these species. All three species form a natural series of the alliance.

C. eugeniae used to be confused with C. alberti Trautv., to which it bears some superficial resemblance. The narrow corolla and, even more, the short, obconical, later rounded, not ribbed capsule opening at the base by pores, indicate its inclusion in section Medium, whereas C. alberti belongs to section Rapunculus. Yet, even Ikonnikov-Galitskii erroneously referred the Fergana Range C. eugeniae to C. alberti var. stenosepala Ik.-Gal. (in schedis ad specimen e Kara-Ungur). He may have contemplated a promotion of var. stenosepala to the rank of a species, as the label reads: "pro specie."

Subsection 18. HYPOPOLION Fed. subs. nov. in Addenda XXIII, p. 335. – Notches between calyx teeth with triangular recurved appendages, as long as the short semiglobose tube; calyx teeth lanceolate or sublinear, much longer than tube, arcuately curved; corolla campanulate, glabrous outside, nearly three times as long as calyx. Flower in loose corymbs, straight at first, later drooping; capsule nearly erect. Stems flexuous, weak, branching, developing from nearly creeping, thinly branching rhizome. All leaves cauline, linear, often obscurely small-toothed in upper part, tomentose beneath.

Type species of subsection: C. hypopolia Trautv.

125. C. hypopolia Trautv. in Tr. Peterb. bot. sada, IV (1874) 61; Boiss. Fl. or. Suppl. 331; Fomin in Mater. Fl. Kavk. IV, 6, 68; Medv. in Tr. Tifl. bot. sada, XVIII, 2, (191) 335; Grossg., Fl. Kavk. IV, 71; Fl. Gruzii, VIII (1952) 159. – Ic.: H. Clifford Crook, Campanulas, 97. – Exs.: GRF, No. 72; Herb. Fl. Cauc. No. 196.

292 Perennial; loosely cespitose plant, 10–15(25) cm; underground shoots very thin, elongate; stems thin, leafy, producing a cluster of branches at apex, all reaching the same length; radical leaves and leaves of sterile shoots narrowly linear, long attenuate at base, subglabrous or sparingly pubescent above, densely gray-tomentose beneath, small-toothed, usually with convolute margin; cauline leaves very narrowly linear, sessile. Flowers straight at first, later drooping, medium-sized, 2 cm long; calyx pubescent, its teeth linear-lanceolate, obtuse, markedly recurved, one-fourth to one-third the length of the corolla, appendages broadly lanceolate, acute, nearly as long as tube; corolla campanulate, blue, pubescent outside; style slightly or conspicuously exserted; receptacle glabrous; capsule nearly straight; seeds ellipsoid, brown, bordered. July–August. (Plate XVII, Figure 2.)

Rocks in subalpine and alpine zones. – Caucasus: Cisc. (E.), E. Transc. (W.). Endemic. Described from Main Range (Gudaur). Type in Leningrad.

Note. Restricted to the C. Caucasus (Gudaur, Ermani, Kobi, Lisri, Ksanka, Kabustinskoe Gorge, Zoromag, Sbitskoe Gorge).

Subsection 19. HETEROPHYLLA (Nym.) Fed. comb. nov. – Heterophylla Nym. Consp. Fl. Europ. (1878–1882) 479, nomen, pro ser.;

Witasek in Abh. Zool.-bot. Ges. Wien, I, 3 (1902) 8, pro cycl. sine diagn. lat. — Calyx without appendages, the teeth linear-lanceolate, acute, more or less declined or appressed to corolla; flowers drooping; capsule opening by valves at base. Rosetted leaves and leaves of sterile shoots usually rounded or broadly ovate, petiolate, very different from lanceolate, sessile, cauline leaves, sometimes all leaves equal (but then radical leaves undeveloped). Usually subglabrous perennial plant, with rhizome or more or less thickened roots.

Type species of subsection: C. rotundifolia L.

Series 1. **Vulgares** Witasek in Abh. Zool.-bot. Ges. Wien, I, 3 (1902) 8.— Rosetted leaves (if developed) rounded or broadly ovate, petiolate; cauline leaves linear or linear-lanceolate, sessile. Plant 1-flowered or with paniculate inflorescences. Rhizome with shoots.

126. C. langsdorffiana Fisch. ex Trautv. et Mey. Florula Ochotens. (1856) 60; Kom. Fl. Man'chzh. III, 2, 556; Kom. and Alis., Opred. rast. Dal'nevost. kr. II, 993.—C. linifolia auct. non Scop.: Ldb. Fl. Ross. II, 2, 888; Turcz. Fl. baic.-
293 dahur. II, 479, excl. syn. nonnull. — C. rotundifolia auct. p.p. non L.: Ldb. Fl. alt. I (1829) 238.— C. rotundifolia var. arctica Miyabe et Miyake, Fl. Saghal. 293; Sugawara, Ill. Fl. of Saghal. IV, 1947, non Lange. — C. rotundifolia var. albiflora Sugawara, Pl. Saghal. (1937) 292. — C. linifolia δ langsdorffiana A. DC. Monogr. (1830) 279; idem in DC. Prodr. VII, 471, quoad pl. ex Sibiria. — C. linifolia γ heterodoxa Ldb. Fl. Ross. II, 2 (1845—1846) 888. — C. rotundifolia var. linifolia Kryl. Fl. Alt. III (1904) 790; idem, Fl. Zap. Sib. XI, 2639. — C. heterodoxa auct. p. p. non Vest.: Witasek in Abh. Zool.-bot. Ges. Wien, I, 3, 53. — Ic.: Sugawara, l. c. tab. 801, f. A (bona).

Perennial; stems usually few, 30—40 cm, slightly branching or simple, rather thin, sometimes nearly filiform, glabrous like entire plant; radical leaves cordate, petiolate, not always developed; cauline leaves linear, 7 cm long, 2—3 mm wide, usually narrower and often shorter. Inflorescence few-flowered, but mostly flowers solitary, slightly drooping; calyx-tube obconical, slightly ribbed with prominent nerves, teeth long, recurved-linear or linear-subulate, sometimes nearly as long as corolla; corolla blue, campanulate, glabrous; anthers usually longer than filaments. June—August.

Most frequent in wooded bluffs, extending to treeless tundra of the Arctic; also in alpine mountain belt. — W. Siberia: Alt.; E. Siberia: Ang.-Say., Dau., Yenis., Lena-Kol.; Far East: Uda, Ze.-Bu., Okh., Sakh. **Gen. distr.:** NE China. Described from "between Okhotsk and Yakutsk." Type in Leningrad.

Note. This species differs very little from C. rotundifolia. The type separated by Fischer is very similar to the common C. rotundifolia. A. de Candolle combined it with collections from the Aleutians and NW America under C. linifolia δ langsdorffiana, although the latter are quite different, with nearly lanceolate even serrate cauline leaves and very large flowers. One might think that the earlier described C. heterodoxa Vest is identical with the Siberian C. langsdorffiana of Fischer. But in spite of the fact that C. heterodoxa is reported "in Siberia," the identity of this species is dubious since Vest's diagnosis agrees poorly with C. langsdorffiana. Moreover, Turchaninov (l. c.) claims that true C. heterodoxa is an American plant which does not grow in Siberia. Nor can we use C. linifolia Lam. or C. linifolia Scop., a name originally given to the W. European species but no longer in use.

Witasek's (l.c.) confidence in identifying C. heterodoxa Vest with
C. langsdorffiana must be due to the fact that she probably never saw
294 Fischer's type, or the Siberian plants and was not aware that these plants
have linear cauline leaves, never lanceolate, as in C. heterodoxa. In
America, plants with lanceolate leaves predominate.

127. C. rotundifolia L. Sp. pl. (1753) 163; Roem. et Schult. Syst. V, 94; A. DC. Monogr. 281; idem in DC. Prodr. VII, 2, 474; Ldb. Fl. Ross. II, 2, 888; Shmal'g., Fl. II, 176; Witasek in Abh. Zool.-bot. Ges. Wien, I, 3, 10; Fomin in Mater. Fl. Kavk. IV, 6, 79; Kryl., Fl. Zap. Sib. XI, 2639; Korsh. Tent. Fl. ross. or. 274; Szaf. Kulcz., Pawl. Rosl, Polskie, 646. — C. uniflora Gorter, Fl. Ingr. (1761) 33, non L. — C. filiformis Gilib. Fl. lithuan. I (1781) 53. — C. tenuifolia Mart. Prodr. Fl. Mosq. (1817) 38. — C. rotundifolia var. reflexa Syreishch., Ill. Fl. Mosk. gub. II, 221. — C. rotundifolia var. linifolia Korsh. Tent. Fl. Ross. or. (1898) 274, non Wahlenb. — Ic.: Rchb. Ic. Fl. Germ. IX, tab. 1603; Maevskii, Fl. (1954) 205; Syreishch., op. cit. III, 221; Fedch. and Fler., Fl. Evrop. Ross. fig. 937 (mala); H. Clifford Crook, Campanulas, 181. — Exs.: GRF, No. 370.

Perennial; rhizome thin, branching, procumbent, sparsely cespitose; stems 15—40(65) cm, numerous, sometimes solitary, more or less densely leafy, glabrous, sometimes shortly and sparsely hairy especially in lower part; radical leaves and leaves of reduced sterile shoots long-petioled, reniform, cordate-rounded or cordate-ovate, large-toothed, rarely nearly entire, often wilted at flowering; cauline leaves narrowly lanceolate, linear-lanceolate, linear or narrowly linear, at stem apex tapering, usually entire, with slightly curved margins, rarely remotely toothed. Flowers in spreading paniculate inflorescence, sometimes numerous, long-pediceled, drooping or declinate; calyx glabrous short obconical, the teeth linear-subulate, usually shorter than corolla, appressed to corolla, sometimes declinate; corolla blue or pale lilac, 12—20 mm long, shallowly cleft into short acuminate lobes; capsule pendulous, opening at base by small apertures; seeds oblong, shiny. July—September.

Meadows, forest edges, shrubby formations. — European part: all regions except for high-latitudes in the Arctic, Crimea and southern steppe belt; W. Siberia: all regions; E. Siberia: all regions except for the Yakut ASSR; there, as in the Far East, replaced by the vicariant C. langsdorffiana. **Gen. distr.**: Scand., Centr. and Atl. Eur. Described from Europe. Type in London.

Note. All reports notwithstanding, this species does not grow in the
Caucasus. Trautfetter described var. hirta (Tr. Peterb. bot. sada, VI)
from specimens supposedly collected in Daryal Gorge by Lagovskii. As
295 already observed by Witasek, this plant is very similar to one of the forms
found in W. Europe. So once again we are involved with the botanical
mysteries so typical of Lagovskii; in all probability the so-called var.
hirta was really collected in the mountains of Switzerland or Italy.

128. C. giesekiana Vest. ex Roem. et Schult. Syst. V (1819) 89; Witasek in Abh. Zool.-bot. Ges. Wien. I, 3, 50; Mishkin, Fl. Khibinsk, gor., 66. — C. rotundifolia var. linifolia Wahlenb. Fl. lapp. (1812) 64, non C. linifolia Scop. et al. auct.; Fedch. and Fler., Fl. Evrop. Ross. 238. — C. rotundifolia var. arctica Lge. Fl. Danica, XVI (1867) tab. 724. — C. uniflora Gieseke in sched. (sec. Vest). — C. linifolia β scheuchzeri Ldb. Fl. Ross. II, 2, 888, non C. scheuchzeri Vill. — C. rotundifolia β linifolia Rupr. Fl. Samojed. Cisural. (1845) 185, non C. linifolia Scop. et al. auct. — C. scheuchzeri auct. non Vill.: Stankov and Taliev, Opred. rast. (1949) 747, probabiliter p. p.

Perennial; rhizome thin, branching, brownish, with slightly prostrate shoots, stems 5—10(20) cm, few, arcuate at base, ascending, straight in upper part, glabrous, usually reddish, thin, 1-flowered or few-flowered; radical leaves rounded-ovate, long-petioled, small, 1 cm wide, often absent or drying up early; cauline leaves confined to lower half of stem, 2—3(5) cm long, 2 mm wide, lanceolate-linear or linear, broadened toward apex, thus slightly spatulate, sometimes furcately curved, glabrous like entire plant, with slightly convolute margins. Flowers large, 2—3 cm long, drooping or declinate; calyx reddish-black, semiglobose or obtusely obconical, glabrous, the teeth subulate or linear, declinate or appressed to corolla, nearly as long as tube but much shorter than corolla; corolla bluish-lilac, glabrous, campanulate, divided into broad acuminate lobes; style nearly as long as corolla; capsule hidden by calyx, becoming ribbed with prominent nerves. July. (Plate XVII, Figure 3.)

Stony meadows, sandy habitats in tundra of Arctic plain, seashores, mountains in alpine zone. — Arctic: Arc. Eur. (Kola Peninsula, Kildin Island, Bear Islands, Olenii Island, Kolguyev Island, Novaya Zemlya); European part: Kar.-Lap. (Khibiny). **Gen. distr.**: Arctic (Greenland, Labrador). Described from Greenland.

Note. Probably a very young Quaternary species evolved under the influence of an Arctic environment at the northwestern limit of the distribution area of C. rotundifolia, thus representing an Arctic descendant of the taiga flora. Very typical are some characters common to many of the
296 Arctic plants: reddish-black pigmentation of the calyx, reddish color of the upper part of the stems, comparatively large flowers, low growth, etc. Specimens known from the Polar Urals, Kara tundra, Dudinka and the lower reaches of the Yenisei are very similar, but because of their very long green calyx teeth, relatively small flowers, longer leaves, etc. they are not included in C. giesekiana. All considered, the Siberian arctic form has developed quite independently of the Lapland, Novaya Zemlya and Greenland forms, presumably in the broad Siberian distribution area of C. langsdorffiana, not in the area of typical C. rotundifolia. In fact, the Siberian Arctic form is unquestionably closer to the original C. langsdorffiana than C. giesekiana is to C. rotundifolia, and it is for this reason that we accept only the European Arctic C. giesekiana, which also grows in Greenland and Labrador, as a distinct species. The Siberian form we are inclined to regard as a deviation from the taiga C. langsdorffiana, which has not achieved specific status.

C. giesekiana is sometimes erroneously identified with C. uniflora L., but as already shown by Witasek Vest's name C. giesekiana as well as his diagnosis refer to the Arctic representative, of the allied C. rotundifolia, which we have retained. Stankov may have reported C. giesekiana for the Arctic region of the western part of the USSR as C. scheuchzeri but it is more likely that he was referring to C. uniflora L., which he considers a synonym for true C. scheuchzeri Vill. Else, like some other authors, Stankov may have lumped C. uniflora L. with C. giesekiana, in spite of the fact that these species belong to different sections: in fact, reporting C. scheuchzeri for the Arctic, Stanker does not mention such a typical species as C. uniflora L., long since known to all (see Note to C. uniflora).

129. K. kladniana (Schur) Witasek in Abh. Zool.-bot. Ges. Wien I, 3 (1902) 39, s. str.; idem in Magyar Bot. Lapok, V, 237; M. Popov, Rast. i fl. Karpat, 261. — C. scheuchzeri var. kladniana Schur, Enim. pl. Transs. (1866) 443.

Perennial; rhizome thin, creeping; stems 10—15 cm, weak, thin, straight or nutant, usually numerous; radical leaves thin-petioled, cordate, acute, finely
297 or obscurely serrate, lamina 1 cm wide; cauline leaves declinate, glabrous, the lower ovate, petiolate, the others lanceolate or linear, entire, obtuse or acute, 2—3(5) cm long. Flowers 1 or 2, sometimes few; receptacle glabrous, small, elongating after flowering; calyx teeth narrowly linear, elongate, declinate or recurved below; corolla campanulate, 2(2.5) cm long, incised for one-third into acute lobes; anthers slightly longer than filaments; style pubescent for two-thirds; capsule glabrous, not sulcate, pendulous. July—September.

Meadows of alpine, subalpine and upper mountain forest zone. — European part: U. Dns. (Carpathians: Cherna Gora, Petrosh, Pop Ivan and Goverla mountains). **Gen. distr.:** Centr. Eur. (Transylvanian Alps, Sudeten). Described from Banat, Hungary. Type in Vienna.

Note. Witasek's two studies quoted above survey many of the small species of C. rotundifolia L. s. l., of which this is one. Most of her species are so-called "small" species; hardly distinguishable. C. kladniana, at first accepted in a rather broad sense, is later broken down into smaller units, one of which is our C. polymorpha.

130. C. polymorpha Witasek in Magyar Bot. Lapok, V (1906) 239; Szaf., Kulcz., Pawl. Rosl. Polskie, 646; Visyulina in Vizn. rosl. URSR, 509. — C. scheuchzeri auct. non Vill. nec Lodigg: Sag et Schn. Fl. Zentralkarp. II (1891) 369. — C. kladniana (Schur) Witasek in Abh. Zool.-bot. Ges. Wien, I, 3 (1902) 39, p. p.

Perennial; rhizome thin, sometimes branching, whitish in upper part; stems 10—15 cm, sometimes few, ascending, thin but not weak and not nutant, glabrous, sometimes rather densely leafy; radical leaves reniform, at times cordate, crenate-serrate, usually wilting early or obsolete; cauline leaves ovate-lanceolate, lanceolate or linear, obtuse, acute, sessile or short-petioled, usually all entire, very rarely some ovate serrate and long-petioled. Flowers 1 or few, usually 2—5, sometimes many, large, to 1.7 cm long; receptacle glabrous, distinctly but not considerably accrescent after flowering; corolla glabrous, campanulate, shallowly divided into nearly triangular lobes; calyx teeth of variable length, usually one-fourth to two-thirds the length of the corolla, linear, appressed to corolla; tube obconical, slightly globose, dark green or blackish; style as long as corolla. July.

298 Stony places in alpine zone. – European part: U. Dns. (Carpathians). **Gen. distr.:** Centr. Eur. (Tatras). Described from the Tatras (above Tatranská Lomnica). Type in Budapest (Degen herbarium).

Note. A "small" species of *C. kladniana* Schur, separated by Witasek while working on the Degen herbarium, and considered by her as a subspecies (l. c. 237) though formally given the rank of a species. She is inclined to separate 4 varieties (α, β, γ, δ), calling attention to the marked variability of *C. polymorpha* which is most pronounced in the form of the leaves. Thus, in variety α the leaves are thick and short, in β thin and delicate (these varieties are more similar to *C. kladniana* s. str.). Of the other two varieties – γ and δ – the first is a rather high, densely leafy plant, with leafy branches produced in the axil of the narrow leaves; the second is a small, 1-flowered plant.

Our specimens of *C. polymorpha* were collected inside the USSR and may be referred to variety δ. They are all uniflorous and narrow-leaved (see Igoshina's collections from the Carpathians: Rakhov District, Goverla Mountain, Bliznitsa Mountain).

Series 2. **Lanceolatae** Witasek in Abh. Zool.-bot. Ges. Wien. I, 3 (1902) 9. – Rosetted leaves none; cauline leaves lanceolate, sessile, mostly about middle of stem. Inflorescence a one-sided raceme, roots often more or less thickened.

131. C. napuligera Schur, Enum. pl. Transs. (1866) 444; Szaf., Kulcz., Pawl. Rosl. Polskie, 644; Visyulina in Viznacnh. rosl. URSR, 510. – *C. pseudolanceolata* auct. non Pantocsek: Witasek in Abh. Zool.-bot. Ges. Wien, I, 3 (1902) 78, p. p.; M. Popov, Rast. i fl. Karpat, 261.

Perennial; root oblong, simple or slightly branching, with turniplike thickening, fleshy (tart taste recalling radish); stems 30, sometimes 50 cm, straight, densely leafy, thickish, slightly flexuous; radical leaves drying early; lower cauline leaves 4–5 cm long, 0.7–0.8 cm wide, elliptic-oblong
301 to lanceolate, the upper sublinear, the uppermost narrowly linear, becoming squamous; all leaves often nearly horizontal, like stems completely glabrous, obscurely dentate or nearly entire. Flowers 1–2 in terminal secund raceme, 1.3–1.8 cm long, bluish; calyx semiglobose, small and glabrous, the teeth three times as long as the tube, nearly subulate-linear, keeled at back, declinate, one-third to one-half the length of the corolla; corolla infundibular-campanulate, glabrous, divided for one-third into acute lobes; style slightly exserted. July–August.

Sunny, herbaceous places in mountains. – European part: U. Dns. (Carpathians). **Gen. distr.:** Centr. Eur. (Transylvanian Alps, Tatras). Described from Poyana in Transylvania. Type in Vienna.

Note. The Carpathians plant apparently is *C. napuligera*, not *C. pseudolanceolata*, as reported by some authors. We have the exsiccates of the type of *C. pseudolanceolata* Pantocsek, collected by the author (Fl. exs. austro-hung., No. 3303), but these do not agree with

(299)

PLATE XVII. 1 – Campanula odontosepala Boiss.; 2 – C. hypopolia Trautv.; 3 – C. giesekiana Vest. ex Roem. et Schult.

plants of the Soviet Carpathians. In Vitasek's interpretation of C. pseudolanceolata there is some confusion due to her ignoring earlier descriptions, including that of C. napuligera Schur., a species close to C. hostii Baumg. and C. pseudolanceolata, as well as to the common C. rotundifolia and other small species of the series Rotundifoliae. P. napuligera was collected in the Soviet Carpathians by Popov (Pop Ivan Mountain) and by Madal'skii (Cherna Gora).

Section 2. RAPUNCULUS (Fourr.) Boiss. Fl. or. III (1875) 895. — Rapunculus Fourr. in Ann. Soc. Linn. Lyon, N. S. SVII (1869) 111, pro gen. — Sect. Eucodon A. DC. Monogr. 214, p. p. — Capsule opening at apex or at middle by pores or valves; calyx without appendages and without folds (reduced appendages). Other characters variable.

Type species of section: C. rapunculus L.

Subsection 1. CAMPANULASTRUM (Small) Fed. comb. nov. — Gen. Rapunculus Fourr. in Ann. Soc. Linn. Lyon, N. S. XVII (1869) 111, p. p. — Sect. Eucodon A. DC. Monogr. (1830) 214, p. p. — Gen. Čenekia Opiz, l. c. (1852) 36. — Campanulastrum Small, Fl. South.-East. U. St. (1903) 1141, p. p. — Calyx usually without appendages; corolla infundibular or campanulate; capsule opening by small holes at apex or sometimes about middle, not at base. Stems 1-flowered or few-flowered, sometimes flowers
302 in paniculate inflorescences. Leaves gradually tapering to petioles or sessile. Plant with creeping, sometimes branching rhizome or with simple, usually thickened roots.

Type species of subsection: C. rapunculus L.

Series 1. **Lactiflorae** Fed. — Calyx teeth ovate-lanceolate or lanceolate, sometimes serrate; flowers in more or less branching, many- or few-flowered inflorescences. Cauline leaves oblong-ovate, dentate or serrate. Plant with rhizomes.

132. C. lactiflora M. B. Fl. taur.-cauc. I (1808) 153; A. DC. Monogr. (1830) 311; idem in DC. Prodr. VII, 477; Ldb. Fl. Ross. II, 890; Boiss. Fl. or. III, 935; Shmal'g., Fl. II, 176; Fomin in Mater. Fl. Kavk. IV, 6, 109; Grossg., Fl. Kavk. IV, 74; Kharadze in Fl. Gruzii, VIII, 186; Poletiko in Tr. Bot. inst. AN SSSR, ser. VI, 2, 273—286. — C. hispida M. B. l. c. III (1819) 134, nomen. — C. celtidifolia Boiss. Diagn. ser. II, 3 (1843) 111. — C. biserrata C. Koch in Linnaea, XIX (1850) 29. — C. aspera M. B. in schedulis ad spec. typ. (nom. provis.). — C. lactiflora var. pilosa Trautv. in Tr. Peterb. Bot. sada, V, 2, (1878) 454. — C. lactiflora var. glabra Trautv., op. cit. IV, 2 (1876) 390. — Ic.: M. B. Cent. tab. 10; A. DC. l. c. (1830) tab. 7; Curtis, Bot. Mag. tab. 1973; Poletiko, op. cit. fig. 3 (flowers), 4 (root), 6 (seeds); H. Clifford Crook, Campanulas, 101. — Exs.: Pl. or. exs. No. 350.

Perennial; pubescent or subglabrous plant; stems to 150 cm and more, straight, thickish, leafy; radical leaves ovate or ovate-oblong, acute, short-petioled; cauline leaves ovate-oblong or oblong, sessile; all leaves biserrate, teeth glandular. Flowers 2.5–3 cm long, in paniculate inflorescences; calyx teeth ovate-lanceolate, small-toothed, glandular along margin, nearly half as long as corolla, tube obconical, glabrous or covered with white bristly hairs; corolla white or pale blue, broadly campanulate, divided to middle. June–July.

Upper forest and subalpine mountain zones, penetrating the formations of tall herbaceous vegetation. – Caucasus: nearly all regions. **Gen. distr.:** As. Min. Described from Mtiuleti (Kaishaur) and Ordzhonikidze (Vladikavkaz) in Osetia. Type in Leningrad, cotype in Geneva and Berlin.

Note. The type is a mixture of specimens collected "in meadows near Vladikavkaz and on a mountain (?) in Kaishaur."

Economic importance. An exquisite garden plant, which has long been in cultivation.

133. C. hieracioides Kolak. in Soobshch. AN Gruz. SSR VIII, 4 (1947) 236; idem, Fl. Abkhaz. IV, 181; Kharadze in Fl. Gruz., VIII, 186. – Ic.: Kolak., op. cit. table XIX (fig. manca).

303 Perennial; plant covered with stiff short bristles; stems 20–30 cm, few, straight or ascending, flexuous-geniculate; rosetted leaves none; basal leaves squamiform, 1–2 cm long; cauline leaves sessile, prominently nerved grayish beneath, dark green above, ovate-lanceolate or broadly lanceolate, 3–6 cm long, 1–2(2.5) cm wide, acute, irregularly and obtusely serrate-dentate. Flowers in short apical raceme or corymb; pedicels glabrous, rather short; calyx 2–3 mm long, densely covered with spreading bristles, the teeth long, narrow, curved, lanceolate-subulate, 15–20 mm long, with bristly pubescence, distinctly nerved; corolla blue-violet, infundibular-campanulate, 25–30 mm long, glabrous, bearded inside, with pubescent margin, divided for one-third into ovate-triangular lobes; style not exserted; capsule trilocular, opening by apical holes. June–July.

Limestones in upper forest belt (800–1,000 m above sea level). – Caucasus: W. Transc. (Abkhaziya). Endemic. Described from the Gegi River Gorge (tributary of the Bzyb). Type in Sukhumi, topotype in Leningrad.

Note. A very unique species sharply differentiated by the geniculate stems, the very stiff, whitish, bristly pubescence. In habit it is reminiscent of Hieracium umbellatum.

134. C. pontica Alb. in Bull. Herb. Boiss. II (1894) 116; Prodr. Fl. Colch. 162; Fomin in Mater. Fl. Kavk. IV, 6, 112; Grossg., Fl. Kavk. IV, 74; Kharadze in Fl. Gruzii, VIII, 185.

Perennial; completely glabrous plant; root oblique, distally covered with scale-like remnants of leaves; stems 50–70 cm, rather thick, striated with longitudinal furrows between decurrent petioles; radical leaves ovate-oblong, abruptly tapering to narrowly winged petiole as long as or slightly longer than lamina, crisp-crenate, with glandular teeth, with few small spiny hamately curved bristles; upper leaves oblong-lanceolate, sessile, amplexicaul. Inflorescence paniculate, spreading, with long axillary bracteolate pedicels; bracteoles linear-lanceolate, dentate; calyx teeth ovate-lanceolate, acute, hamately small-toothed, glandular, half the length of the corolla; the tube shortly obconical, covered with white papillae and verrucae; corolla broadly campanulate, 2–3 cm long, blue, divided for one-third; style cleft to middle. May–June.

Dry glades in xerophylous open woodland, mountain slopes, sometimes
304 bluffs and rocky slopes. — Caucasus: W. Transc. (Batum District). **Gen. distr.:** As. Min. (Artvin District). Described from Ponti Range. Type in Geneva.

Economic importance. An ornamental plant.

Series 2. **Silenifoliae** Fed. — Calyx teeth linear or linear-lanceolate; corolla narrowly infundibular-campanulate with ciliate margin; flowers solitary or 2–5. Leaves lanceolate or linear. Stem covered at base with fibrous remnants of leaves. Only one species known.

135. C. turczaninovii Fed. nom. nov. — C. silenifolia Fisch. ex. A. DC. Monogr. (1830) 320, non Host (1827); idem in DC. Prodr. VII, 479; Ldb. Fl. Ross. II, 886; Turcz. Fl. baic.-dahur. II, 480; Kryl., Fl. Zap. Sib. XI, 2640. — C. uniflora Georgi, Beschr. Russ. Reichs, III, 4 (1800) 772, p. p. non L. — C. ciliata Patrin in A. DC. l. c. (1830) pro syn., non Stev. — C. baicalensis Pall. ex A. DC. l. c. (1830) pro syn. — C. simplex var. dasycarpa Trautv. in schedis.

Perennial; root 2–5 mm across, slightly branching, fusiform; stems 10–40 cm and higher, one or few, straight, simple, glabrous or sparingly pubescent, base covered with fibrous dried petioles; radical and lower cauline leaves oblong-elliptic, lanceolate, sometimes sublinear, acuminate or obtuse, usually entire or denticulate, the petioles nearly equal to lamina, with lamina 5–15 cm or slightly longer, 0.2–1.2 cm wide, sometimes wider; median and upper leaves sessile, lanceolate-linear, acuminate, with ciliate margin, sparingly pubescent or glabrous above and beneath, entire. Flowers solitary or 2–5 in racemiform inflorescences; pedicels 1–3(5.5) cm; bracts ovate, ovate-lanceolate or sublinear, acute; calyx more or less pubescent or subglabrous, its teeth linear or linear-lanceolate, one-third the length of the corolla; corolla large, 3–3.5 cm long, blue, narrowly infundibular, campanulate, glabrous outside, divided for one-third into ovate acuminate lobes with slightly ciliate margin. June–July. (Plate XVIII, Figure 4.)

Mainly in alpine tundra above the timberline (Siberia), sometimes in forest belt. — W. Siberia: Alt.; E. Siberia: Ang.-Say., Dau., Lena-Kol.; Far East: Ze.-bu. **Gen. distr.:** N. Mong. Described from Dauria. Type in Geneva, autotype in Leningrad.

Note. Ever since Ledebour it has been unclear whether C. rigescens Pall. ex Roem. et Schult. and C. infundibulum Vest ex Roem. et Schult. are identical with C. turczaninovii. Ledebour believed that C. infundibulum did not belong here but yet he lists C. rigescens with a
305 question mark as a synonym of C. silenifolia (see list of doubtful species).

C. rigescens and C. infundibulum are surely synonyms of two species now known by another name. C. rigescens described from Baikal ("in frigidis and lacum Baical") is probably C. turczaninovii, but Roemer and Schultes' description is too poor to be reliable. In fact, the description and the relevant note could refer to any Campanula, even C. glomerata. Since (excluding C. langsdorffiana, C. rotundifolia and C. dasyantha which are absolutely inappropriate) only these species occur in Baikal,

C. rigescens may be synonymous with two very differing species, i. e. C. turczaninovii, and, strange as it may seem, C. glomerata. As to C. infundibulum, this might prove to be a synonym of C. altaica Ldb. The description by Roemer and Schultes is so bad that one would flounder without end in trying to select synonyms for C. turczaninovii and C. altaica. A decisive conclusion might be reached only if the specimens from the Pallas herbarium, to which Roemer and Schultes referred, could be examined; yet, it is possible that they no longer exist or Vest's notes have been lost.

The narrow-leaved form prevails in the Yakut ASSR and is the only one encountered along the Aldan River. The broad-leaved form is found most often near Lake Baikal and in the Sayan Mountains. However, these characters are not correlated with the distribution area and therefore the forms cannot be regarded as separate species, despite the fact that their extreme populations are rather different from each other.

Series 3. **Rapunculiformes** Fed. — Calyx teeth narrowly lanceolate or nearly subulate, reflexed; inflorescence usually branching, paniculate. Cauline leaves narrowly lanceolate, lanceolate or sublinear, obscurely crenate to nearly entire. Plants with more or less thickened roots, rarely with rhizomes.

136. C. rapunculus L. Sp. pl. (1753) 232; A. DC. Monogr. (1830) 325; idem in DC. Prodr. VII, 2, 480; M. B. Fl. taurc.-cauc. III, 138; Ldb. Fl. Ross. II, 2, 887, p. p.; Shmal'g., Fl. II, 176; Korsh. Tent. Fl. ross. or. 274; Fomin in Mater. Fl. Kavk. IV, 6, 107, p. p.; Maevskii, Fl., 8th ed. 546; Szaf., Kulcz., Pawl. Rosl. Polskie, 643; Visyulina in Viznachn. rosl. URSR, 510. — C. fastigiata Gmel. Reise, I (1774) 153, non Dufour. — C. coarctata Gilib. Fl. lithuan. I (1781) 48. — C. rapunculus γ calycina
306 A. DC. l. c. (1830) 326. — Rapunculus verus Fourr. in Ann. Soc. Linn. Lyon, N. S. XVII (1869) 111. — Ic.: Rchb. Ic. Fl. Germ. tab. 252; H. Clifford Crook, Campanulas, 179.

Biennial; root thick, often thickened-napiform, white, with milky juice; stems straight, simple, 70—100 cm, sometimes branching in upper part, longitudinally striate, glabrous or covered especially below with spreading white hairs; radical leaves obovate, obtuse or acuminate, attenuate into petioles, subentire, glabrous or sparingly pubescent; cauline leaves sessile, linear-lanceolate, entire, the uppermost reduced. Racemes in long spicate flowers, sometimes branching to form slightly paniculate inflorescence; flowers subsessile or on short thin erect pedicels; calyx obconical, glabrous or bristly in strips, teeth straight, narrowly long-acuminate, sometimes setaceous, entire or slightly dentate, glabrous, shorter (sometimes by half) than corolla; corolla whitish-blue or nearly white, campanulate, glabrous, divided for one-fourth or less into ovate acute lobes; filaments dilated, pubescent; style hidden in corolla, pubescent for nearly entire length, with 3 stigmas; capsule erect, obconical, elongate, striated with prominent nerves, trilocular, opened by 3 terminal pores; seeds very small, flattened, ovate, shiny, with blackish groove. June—July.

Forest edges, meadows, edges of fields, roadsides. — European part: U. Dnp., Bes., M. D., Bl., L. Don, V.-Don (west only), Crim.; Caucasus: Cisc. **Gen. distr.:** Centr. and S. Eur. Described from Europe (Switzerland, England, France). Type in London.

Note. Korzhinskii (l. c.) disproved the earlier reports of this species from Kazan.

Economic importance. The young leaves and especially the roots are eaten as salad.

137. C. lambertiana A. DC. Monogr. (1830) 327; idem in DC Prodr. III, 2, 480; Kharadze in Fl. Gruzii, VIII, 189. — C. rapunculus var. lambertiana Boiss. Fl. or. III (1875) 940; Lomak. in Tr. Tifl. bot. sada, I, 60 (1898) 58. — C. rapunculus auct. mult. p.p. non L.: Ldb. Fl. Ross. II, 2 (1847) 887, p.p.; Shmal'g., Fl. II, 176; Fomin in Mater. Fl. Kavk. IV, 6, 107; Grossg., Fl. Kavk. IV, 75. — C. rapunculus var. spiciformis Boiss. l. c.; Fomin, op. cit. 109.

Biennial; root usually not napiform-thickened, thin-fusiform, whitish; stems simple, or slightly branching, 70 cm, longitudinally striate, covered for the entire length with spreading white recurved hairs; radical leaves
307 drying early, broadly elliptic or obovate, obtuse, abruptly attenuate into slightly winged petioles; cauline leaves remote, sessile or short-petioled, broadly oblanceolate or oblong, acute, ca. 5 cm, pubescent, irregularly dentate or crenate. Flowers in spicate racemes or slightly paniculate long-branched inflorescence, short-pediceled; calyx glabrous, obconical, teeth long-acuminate, nearly filiform or setaceous at apex, declined, shorter than corolla; corolla infundibular, usually glabrous, shallowly lobed; style as long as corolla. July.

Dry meadows, forest edges, shrubby formations. — Caucasus: Cisc., Dag., W. and E. Transc., Tal. **Gen. distr.:** Iran (N.), Bal.-As. Min. Described from Gilyan Province, in Gmelin's collections. Type in London, isotype in Leningrad.

Note. This species is closely related to C. rapunculus, but obviously different by its rather thin, narrowly fusiform, not napiform-thickened root. Most of the specimens from the Caucasus belong to C. lambertiana, but those from Ciscaucasia belong to C. rapunculus. Grossgeim (op. cit.) described thickened roots for the Caucasian plants, but the herbarium material is not as such. Only the specimens from Ciscaucasia, which should be referred to C. rapunculus, have thickened roots. The roots of the Iranian specimens known to us (isotype and topotype) are not thickened.

138. C. persicifolia L. Sp. pl. (1753) 164; A. DC. Monogr. 322; idem in DC. Prodr. VIII, 2, 470; Ldb. Fl. Ross. II, 2, 885; Boiss. Fl. or. III, 935; Shmal'g., Fl. II, 176; Korsh. Tent. Fl. ross. or. 278; Fomin in Mater. Fl. Kavk. IV, 6, 112; Syreishch., Ill. fl. Mosk. gub. III, 226; Grossg., Fl. Kavk. IV, 74; Kryl., Fl. Zap. Sib. XI, 2643; Szaf. Kulcz., Pawl. Rosl. Polskie, 644. — C. speciosa Gilib. Fl. lithuan. I (1781) 50. — C. dasycarpa Kit. apud Schult. Oesterr. Fl. Ed. 2 (1814) 900. — C. persicifolia α eriocarpa Syreisch. op. cit. 226; Fedch. and Fler., Fl. Evrop. Ross. 939. — C. persicifolia laevicaulis et var. lasiocarpa Korsh. l. c.

273.— Rapunculus persicifolius Fourr. in Ann. Soc. Linn. Lyon, N. S. XVII (1869) 111.— Ic.: Rchb. Ic. Fl. Germ. tab. 252; H. Clifford Crook, Campanulas, 146; Syreishch., op. cit. 226.— Exs.: GRF, No. 884; Tarachkov and Poganko, Gerb. Orl. gub. No. 215; Fl. exs. austro-hung. No. 2992; Billot, Fl. Gall. et Germ. exs. No. 1269; Callier, Pl. Hercegov. exs. No. 205; Hayek, Fl. stir. exs. No. 1265; Petrak, Fl. Bohem. et Morav. exs. No. 1157.

Perennial; root fibrous, not thickened; stems straight, 50—70 cm, completely glabrous, somewhat longitudinally striate, simple or slightly branching,
308 leafy, with milky juice; leaves glabrous, shiny, indurate, pale beneath, crenate-dentate; radical leaves oblong or obovate, acute or obtuse, attenuate into 7—12 cm long petioles; cauline leaves regularly remote, sessile, linear-lanceolate, 7—10 cm, acuminate; upper leaves linear, narrow. Flowers few, terminal or axillary, solitary, pediceled, slightly drooping, not longer than pedicels, large; calyx obconical, glabrous or rough-pubescent, grooved, teeth glabrous, acuminate, entire, nearly erect, half as long as corolla; corolla broadly campanulate, slightly inflated at middle, blue, 4—5 cm long, divided into broadly ovate short erect lobes; stamens as long as calyx teeth, filaments triangularly dilated, ciliate at base; style as long as corolla, divided nearly to middle into 3 pubescent stigmas; capsule erect, ovoid, with 10 nerves and 3 grooves; seeds ovate, flattened, shiny. June—July.

Meadows, forest clearings. — European part: U. Dns., U. Dnp., Balt., Lad.-Ilm., Dv.-Pech. (S.), V.-Kama, U. V., V.-Don, Transv., L. Don (W.), Crim.; Caucasus: Cisc. (Stavropol'); W. Siberia: U. Tob. (Sverdlovsk). **Gen. distr.**: W. Eur., except for Scandinavia, Denmark, England and the south. Described from Centr. Eur. Type in London.

Note. Actually absent in the Caucasus (known only from the vicinity of Stavropol'), and in Siberia (encountered only near Sverdlovsk at the foothills of the Ural Mountains). The corresponding records from Asia Minor and Greece apparently refer to C. latiloba A. DC. or to an independent species.

139. C. patula L. Sp. pl. (1753) 163; A. DC. Monogr. 329; idem in DC. Prodr. VII, 2, 480; Ldb. Fl. Ross. II, 2, 887; Shmal'g., Fl. II, 176; Korsh. Tent. Fl. ross. or. 274; Fomin in Mater. Fl. Kavk. IV, 6, 115; Syreishch., Ill. Fl. Mosk, gub. III, 225; Grossg., Fl. Kavk. IV, 75; Neimark in Sorn. rast. SSSR, IV, 179; Kryl., Fl. Zap. Sib. XI, 2643; Szaf., Kulcz., Pawl. Rosl. Polskie, 643.— C. decurrens L. Sp. pl. (1753) 164.— C. patens Güldenst. Reise, II (1791) 171.— C. neglecta Roem. et Schult. Syst. V (1819) 104.— C. patula α flaccida Syreishch. op. cit.— C. patula β albiflora Syreishch. op. cit.— Rapunculus patulus Fourr. in Ann. Soc. Linn. Lyon, N. S. XVII (1869) 111.— Ic.: Rchb, Ic. Fl. Germ. tab. 253; Libesh. and Trin., Fl. Sankt-Peterb. in Mosk. (1818) Plate 23 (color); Syreishch., op. cit. fig. 206; Neimark, op. cit. fig. 440, 4, c; H. Clifford Crook, Campanulas, 145.— Exs.: Fl. lithuan. exs. No. 79; Tarachkov and Poganko, Gerb. Orl. gub. No. 213; Fl. exs. austro-hung. No. 2994; Hayek, Fl. stir. exs. No. 891; Schultz. Herb. norm. No. 905 et 905 bis; Tausch, Herb. Bohem., sine numero; Petrak. Fl. Bohem. et Morav. exs. No. 1156; Billot,
309 Fl. Gall. et Germ. exs. Nos. 1034 et 1034 bis; Pl. Finl. exs. No. 374; Woloszczak, Fl. pol. exs. No. 457; Meinsgauzen, Herb. Fl. Ingr. Cent. III, No. 390.

Biennial; root small, thin, whitish, branching; stems 50–70 cm, longitudinally striate, slightly angular, glabrous or roughly pubescent, branching, branches spreading, nearly leafless; radical leaves obovate, petiolate, ca. 3–6 cm, crenate, glabrous; cauline leaves few, lanceolate-linear, sessile, entire or slightly serrate, shorter than the radical. Flowers in loose paniculate inflorescences, terminal or axillary, long-petioled, up to 3 cm long; calyx elongate obconical, glabrous or roughly pubescent, teeth long-acuminate, sometimes finely toothed at base, usually declinate, nearly twice as long as corolla; corolla infundibular, violet-blue, sometimes nearly white, with distinctly colored nerves, glabrous, divided for one-third into ovate acute recurved lobes; stamens with membranous pubescent filaments, subcapillary at the rounded base; style as long as corolla, bluish, pubescent with 3 stigmas; capsule erect, ovoid-cylindrical, 10-nerved, glabrous or sparingly pubescent, opened by three small pores; seeds ovate, slightly flattened, very small, shiny. May–July.

Meadows, forest edges. — European part: nearly all regions except for the Crimea, Caucasus, Arctic and most of Dvina-Pechora; W. Siberia: Ob (near Tyumen). **Gen. distr.**: nearly all Europe except for the extreme North and South. Described from England and Sweden. Type in London.

Note. This species has been often (doubtfully) reported from the Caucasus (Grossg., op. cit.; Kryl., op. cit.), but definitely it is not found there. The report on Asia Minor as another locality of C. patula (Kryl., op. cit.) is also erroneous.

140. C. abietina Griseb. et Schenk in Wiegeman's Archiv, XV, 1 (1852) 333; Fuss. Fl. Transs. exc. 419; M. Popov, Rast. i fl. Karpat, 261; Szaf., Kulcz., Pawl. Rosl. Polskie, 644. — C. patula var. pauciflora Rochel, Pl. Banat. rar. (1826) 42. — Ic.: Rochel, l. c. tab. 6; H. Clifford Crook, Campanulas, 25. — Exs.: Fl. exs. austro-hung. No. 2995; Fl. exs. reip. Boh.-Slov., No. 1272; Woloszczak, Fl. pol. exs. No. 970.

Biennial or perennial (?); root whitish, producing thin sterile shoots and usually one stem; stem straight, thin, 1–1.5 mm thick, 50 cm high, usually few-flowered, sometimes 1-flowered, smooth and glabrous like leaves; radical leaves and leaves of sterile shoots ovate-orbicular, crenate,
310 cuneately attenuate into petioles as long as lamina, cauline leaves elliptic-lanceolate or lanceolate, obtuse, the upper linear sessile; all leaves thin, brittle, semitransparent when dry. Inflorescence loose, spreading, corymbiform-paniculate, 5-flowered; pedicels elongate, thin, 8–10 cm; teeth lanceolate-linear, declinate, entire, more than half as long as corolla and slightly longer than calyx-tube; corolla with narrowly infundibular tube; lobes ovate, acute; capsule obovate, clavate, nearly as long as calyx. July. (Plate XVIII, Figure 3.)

Shady places in coniferous forests of the upper mountain belt, amid carpets of moss. — European part: U. Dns. (Carpathians, especially Marmaresh Mountain). **Gen. distr.:** Centr. Eur. (Carpathians, Transylvanian Alps, some parts of the Balkans). Described from Banat. Type in Berlin.

Note. This species is more or less related to C. patula, the latter differing from it by its pubescence in lower part of the stem, the lanceolate and short-hairy leaves, large number of inflorescences, smaller flowers and dentate calyx lobes. C. abietina is also related to C. spathulata, a species not found in the USSR, and to a slighter degree to species of the series Stevenianae.

141. C. hemschinica C. Koch in Linnaea, XXIII (1850) 644; Boiss. Fl. or. III, 940; Fomin in Mater. Fl. Kavk. IV, 116; Medv. in Tr. Tifl. bot. sada, XVIII, 2, 331; Grossg., Fl. Kavk. IV, 75; Kharadze in Fl. Gruzii, VIII, 187. — C. patula var. confertiflora Trautv. in Tr. Peterb. bot sada, IV, 1 (1876) 165. — Ic.: H. Clifford Crook, Campanulas, 86. — Exs.: Herb. Fl. Cauc. No. 143; Pl. or. exs. No. 295.

Biennial and perennial; root short, fibrous, oblique or vertical; stems 1—3, thick, 30—45(50) cm, acutely angular at bases of decurrent petioles, straight, with paniculate or subcorymbiform inflorescence above middle; lower leaves ovate-oblong or oblong, attenuate into short petioles; cauline leaves oblong-lanceolate, sessile, rigid, slightly amplexicaul at base, acute; all leaves obtusely crenate at margin. Flowers small, approximate, terminal or divergent, with erectly elongated pedicels; calyx teeth narrowly lanceolate, erect, one-fourth to one-third as long as corolla, tube obconical, sometimes covered with papillae; corolla blue, 2.5 cm long, narrowly conical-infundibular, divided for one-third into acute lobes; capsule obconical; seeds small, ovate-oblong, shiny. June—July.

Forests, the subalpine mountain zones. — Caucasus: W. Transc. (Adzharistan, Guriya). **Gen. distr.:** As. Min. (E.), N. Iran. Described from Khemshin in As. Min. Type in Berlin.

311 **142. C. vaidae** Pénzes in Bot. Közlemények. XXXIX, 1—2 (1942) 93. — Ic.: Penzes, l. c. fig. 1, mala.

Rhizome short, creeping; stems straight, 8—10 cm, glabrous, simple, 1-flowered; with sterile leafy shoots at base; leaves lanceolate, glabrous above, spreading-ciliate beneath, crenate, attenuate into petioles, 30 mm long, 8 mm wide. Flowers long-pediceled, erect; calyx sithout appendages between teeth, teeth broadly lanceolate, 12—14 mm long, 4 mm wide; corolla 12—20 mm long, infundibular, violet, 5-partite; stigmas 3. July.

Gravelly places in thickets of Alnus viridis in mountains ca. 1,600 m above sea level. — European part: U. Dns. (E. Carpathians). Endemic. Described from the southwestern slope of Sesul Mountain near Tiszaborkut in Marmaresh (Carpathians). Type in Hungary.

Note. Since we did not see specimens of this species we present the author's description according to which he related C. vaidae to C. abietina but noted that the new species differs from C. abietina by its 1-flowered, low stems and the shape of the calyx lobes. From the drawing it can be seen that the calyx teeth in V. vaidae are very broad and the species bears very little resemblance to C. abietina in stature. We have placed C. vaidae after C. abietina though it probably pertains to quite another group.

Series 4. **Stevenianae** Fed. — Calyx teeth lanceolate or narrowly lanceolate, rarely nearly subulate, appressed to corolla; flowers solitary or in few-flowered or few-branched inflorescence. Cauline leaves lanceolate or narrowly lanceolate, obscurely crenate or very sparingly toothed. Plants with creeping or branching rhizomes.

143. C. alberti Trautv. in Tr. Peterb. bot sada, VI, 1 (1879) 83; O. and B. Fedch., Perech. rast. Turk. III, 358; Vvedensk. in Sched. ad Herb. Fl. As. Med., VII, 26. — Phyteuma alberti Rgl. in schedis. — Exs.: HFAM, No. 175.

Perennial; rhizome indurate, branching in upper part where densely covered with long coarse bristly grayish-brown remnants of dead stems and petioles; plant completely glabrous, many stemmed, 20–30(40) cm; stems straight, usually in dense tufts, rather thin, stiff after flowering, brittle, with spreading leaves; lower leaves linear-oblong, ca. 5 cm, acute, attenuate into petioles, with declinate, acute, spreading-glandular teeth; median leaves long, subentire, sessile; upper leaves linear, mucronate, reduced. Calyx obconical-cylindrical, teeth linear-lanceolate or linear-
312 subulate, without appendages, acute, as long as tube but shorter when capsule ripe, nearly 3 times as long as corolla; corolla deeply 5-lobed, glabrous, violet; capsule oblong-turbinate, ca. 1 cm long, erect, opening at apex by 3 small pores, very finely verrucous-dotted outside, 10-nerved, hence slightly ribbed, crowned by erect and brittle calyx teeth. July. (Plate XVIII, Figure 2.)

Rock crevices in the subalpine zone. — Centr. Asia: T. Sh. (Chotkal, Zailiiskii Ala-Tau, Kirghiz Ala-Tau, Talass Ala-Tau, Fergana ranges). Endemic. Described from mountains along the Chotkal river and also from Alexander Range (now Kirghiz Ala-Tau) and from Zailiiskii Ala-Tau. Type in Leningrad.

Note. The combination of authors Campanula alberti (Rgl.) Trautv., used by Lipskii (in sched.), is incorrect since Regel noted on the label only the names C. alberti — Phyteuma alberti Rgl. without any valid publication.

The small specimens of C. alberti resemble superficially two bell-flowers, C. eugeniae Fed. and C. capusii (Franch.) Fed. of another group (the section Medium). These species were earlier misidentified as C. alberti; but the latter distinctly differs from them by its narrowly obconical calyx and the 10-ribbed capsule opening by small pores at the apex. The other two species have short calyxes, short capsules nearly globular, with their pores at the base. In appearance it also resembles Cylindrocarpa sewerzowii Rgl.

144. C. altaica Ldb. Ind. sem. hort. Dorpat. (1824) 2, non DC.; Fedch. in Maevskii. Fl. ed. 7-e, 87, quoad specim. e Ross. centr.; Kryl., Fl. Zap. Sib. XI, 2641 (et var. macranthera Kryl.) — C. infundibulum Ldb. Fl. alt. I (1829) 239, non Vest. — C. stevenii β sibirica A. DC. Monogr. (1830) 321; idem in DC. Prodr. VII, 479. — Stevenii var. sibirica Fom. in Mater. Fl. Kavk. IV, 6 (1906) 115, p. p.; Voron and Shelk. Schaed. ad Herb. fl. cauc. No. 145, in adnot. (diagn. brev.): — C. stevenii auct. non M. B. Ldb. Fl. Ross. II (1847) 886, quoad specim e Sibir.; Kryl., Fl. Alt. III, 778; Sukach. in Tr. Bot. sada Yur'evsk. univ. IV, 113, p. p. — C. stevenii β altaica C. A. M. in schedis. — C. simlex O. et B. Fedtsch. Perech. rast. Turk. III (1909) 358, p. p. non Stev.; Fedch. and Fler., Fl. Evrop. Ross. 939. — C. uralensis Nevski ex B. Fedtsch. in Maevskii, Fl.
315 ed. 6-e (1933) 643, p. p. in nota, nom. provis. ac. confusum. — C. uralensis Vissul. in Viznacnh. rosl. URSR (1950) 510. — Exs. GRF No. 1171.

(313)

PLATE XVIII. 1 – Campanula altaica Ldb.; 2 – C. alberti Trautv., plant and capsule; 3 – C. abietina Griseb. et Schenk.; 4 – C. turczaninovii Fed., plant and capsule.

Perennial; rhizome oblique, short-ascending or nearly horizontal, with many brownish roots; stems solitary or few, simple or slightly branching toward inflorescence, straight or arcuately curved at base, 40–50(60) cm high, 1.5–2 mm wide, orbicular or slightly angular at cross section, glabrous; rosetted leaves few, broadly lanceolate or oblanceolate, obtuse or short-acuminate, narrowly cuneate at base then attenuate into long petioles; petioles 1–2 cm wide, 15 cm long, lamina as long as petiole or slightly shorter, crenate or obscurely toothed at margin; cauline leaves narrower, subsessile, the apical usually sublinear. Inflorescence 2–3–(5)-flowered, rarely more, flowers rather long-pediceled; corolla broadly infundibular, 4(5) cm across, 2.5–5 cm wide [?], pale blue (not violet!), cleft for one-half into few acuminate lobes; calyx conical, glabrous, teeth 1.5 mm wide, oblong-lanceolate, obtuse, carinate at apex, concave inside, herbaceous, as long as or slightly longer than tube. June–July. (Plate XVIII, Figure 1.)

Meadows, forest glades, meadow-steppes, foothills, lowlands sometimes in the subalpine mountain zone. – European part: U. V., V.-Don (W.), M. D. (W.), U. Dns. (S.), Bl.; W. Siberia: Alt. **Gen. distr.**: Mong. (N.). Described from Altai. Type in Leningrad.

Note. This species is found in Siberia in Altai and Kuznetskii steppe but does not extend beyond Yenisei being restricted to the forest-steppe area. There is an interval in its distribution in the European part of the USSR, then it is again encountered in the Central Russian Upland and in the Ukraine (plains). The intermediate zone, i. e. the Volga and Transvolga areas, Ural area and the eastern part of W. Siberia, is the distribution area of C. wolgensis extending into Kazakhstan. Vissulin (op. cit.) reported the Altai bellflower from the Ukraine under an incorrect name.

Roemer and Shultes described C. seminuda Vest ex Roem. et Schult. (Syst. veg. V (1819) 91) from Siberia. Ledebour placed this name among the synonyms of C. stevenii s. l., but with a question mark. Judging from the original diagnosis Ledebour's suggestion is very probable, since in Siberia there is only one bellflower, the so-far called C. altaica, which might be referred to C. stevenii; when this is confirmed, the name C. seminuda should replace the later name C. altaica. (See also dubious species).

316 **145. C. beauverdiana** Fom. in Vestn. Tifl. bot. sada, 1 (1905) 12; Fomin in Mater. Fl. Kavk. IV, 6, 149; Medv. in Tr. Tifl. bot. sada, XVIII, 2, 331; Grossg., Fl. Kavk. IV, 74; Kharadze in Fl. Gruzii, VIII, 188. – C. beauverdiana var. aissori Tamamsch. in Fedde, Report. XXXVIII (1935) 171. – Ic.: Fomin, op. cit. (1905) Plate 1 (optima).

Perennial; glabrous or hardly pubescent plant; rhizome thin, creeping, with underground shoots; stems thin, straight or ascending, 15–20(40) cm high, 1-flowered or bearing 2–3 flowers; lower leaves broadly ovate or oblong-ovate, attenuate into thin petioles, obtusely crenate at margin; cauline leaves lanceolate or often linear-lanceolate, acute, small-toothed at margin (teeth glandular); upper leaves linear, acute. Flowers rather long-pediceled; calyx teeth narrowly lanceolate, acute, nearly twice as long as the narrowly cylindrical tube, tube covered with inflated (spongy) white cells; corolla sky-blue, glabrous, campanulate, 3 cm long, more than twice as long as calyx teeth, divided for one-third into lobes; style blue, bipartite, hidden in calyx. June–July.

Mountains, dry meadows, stony slopes. — Caucasus: E., S. and W. Transc. **Gen. distr.**: As. Min. Described from Armenia and from mountains near Mtskhet in Georgia. Type probably in Tbilisi, only the "autotype" is in Leningrad.

Note. This species is vicarious to C. stevenii in the southern and eastern Caucasus; in general it is taxonomically more related to the Near Asian race C. phytidocalyx Boiss. It was collected in Armenia and the Nakhichevan ASSR but there are also specimens from the eastern part of the Main Range (summits of Tfandag and Shakhdag). The rest of its distribution area is the eastern part of Asia Minor and probably also in N. Iran.

146. C. stevenii M. B. Fl. taur.-cauc. III (1819) 138; A. DC. Monogr. 321; idem in DC. Prodr. VII, 479; Ldb. Fl. Ross. II. 886, p. p.; Boiss. Fl. or. III, 936, pro max. parte; Sukach, in Tr. Bot. sada Yur'evsk. univ. IV, 113, pp.; Fomin in Mater. Fl. Kavk. IV, 6, 113 excl. var. — C. simplex Stev. in Mém. Soc. Nat. Mosc. III (1812) 225, non DC.; Rupr. in Bull. Acad. Sc. Pétersb. XI, 197; Grossg. Fl. Kavk. IV, 74; Kharadze in Fl. Gruzii, VIII, 187. — C. steveniana Roem. et Schult. Syst. V, 91. — C. vitinghoffiana Roem. et Schult. l. c. 102. — Ic.: H. Clifford Crook, Campanulas, 203. — Exs.: Herb. Fl. Cauc. No. 145.

Perennial; rhizome thin, partly creeping, with shoots; stem straight, simple or slightly branching toward inflorescence, rounded, leafy, 1-flowered, or 2–3–(4)-flowered, 20–50 cm long; radical leaves oblong-spatulate, 1 cm wide, acute or obtuse, attenuate into petioles about as long as lamina; lower
317 cauline leaves oblong-lanceolate, long-petioled; upper leaves subsessile, linear-lanceolate or linear; all leaves membranous, obtusely serrate at margin, teeth glandular. Flowers in corymblike raceme; calyx teeth lanceolate-linear, as long as or nearly as long as the elongate cylindrical tube; corolla purple-violet, broadly infundibular, nearly rotate, twice as long as calyx teeth, partite into ovate acute lobes; stamens with short filaments and longer anthers; style blue, 3-partite into 3 stigmas; capsule elongate-oblong. May–July.

Subalpine and alpine mountain zones, dry meadows and mountain steppes, also on turfed cliffs, sometimes in open pine forests (Bakuriani). — Caucasus: Cisc., Dag., S. and E. Transc., Tal. **Gen. distr.**: Bal.-As. Min. (only As. Min.), Iran. Described from Azerbaidzhan (the subalpine Khinalyg Mountain). Type in Helsinki, isotype in Moscow.

Note. This species is not found in the W. Caucasus since it is affined to the mountain-steppe vegetation not common to the Colchis. Taxonomically, it is more related to C. wolgensis than to C. altaica which is distinguished by its broad, apically keeled sepals.

147. C. wolgensis P. Smirn. in Byull. Mosk. obshch. ispyt. prir. nov. ser. LII, 3 (1947) 57; Kryl., Fl. Zap. Sib. XI, 2642. — C. uralensis Nevski in sched. — C. uralensis Nevski ex B. Fedtsch. in Maevskii, Fl. ed. 6-e (1933) 643, in nota, nomen provis. ac confusum; Stankov and Taliev, Opred. rast. 749, p. p.; Maevskii, Fl. ed. 8-e, 547. — C. stevenii var. sibirica Fom. in Mater. Fl. Kavk. IV, 6 (1906) 115, p. p. — C. stevenii auct. non M. B.: Ldb. Fl. Ross. II, 886, quoad specim. e Ross. or.: Korsh. Tent. Fl. ross. or. 274; Sukach. in Tr. Bot. sada Yur'evsk. univ. IV, 113, p. p.,

Nekras. in Fl. yugo-vost. VI, 288. — C. altaica auct. non Ldb.: Pavlov, Fl. Tsentr. Kazakhst. III (1938) 202. — C. simplex auct. non Stev., nec A. DC.: O. and B. Fedch., Perech. rast. Turk. III (1909) 358, p. p. — Ic.: Fl. yugo-vost. VI, 288.

Perennial; rhizome ascending or nearly horizontal, 4—6 cm long and even longer and 2—3 mm thick, lignified, more or less angular with remnants of stems and petioles, many-rooted; stems usually 2—3, rarely more (sometimes solitary), 30—40(50) cm, straight or arcuate at base and usually slightly flexuous, angular at cross section, glabrous or sometimes covered below with curved hairs; leaves grayish-green, stiff, glabrous, sparsely ciliate at base of petioles; radical and rosetted leaves long-petioled, petioles nearly
318 as long as lamina or one and a half times as long; lamina ovate-lanceolate or oblanceolate, long-cuneate tapering at base, distinctly crenate; upper leaves sessile, linear, obscurely serrate or entire. Inflorescence paniculate, 3—4-flowered, sometimes reduced with a single terminal flower; pedicels 5—10 cm, more or less flexuous, with 2 short lanceolate-linear, approximate or remote bracteoles; calyx-tube glabrous, obconical, 4—7 mm, the teeth lanceolate or often linear-lanceolate, (6)7—9(10) mm, gradually acuminate, straight or concave inside, flat at apex, glabrous, obscurely (discernible under illumination) nerved, without appendages; corolla infundibular, violet, glabrous, tube 7 mm long, lobes ovate, ca. 10 mm long; stigma violet, nearly 3 times longer than style; capsule oblong, with 10 longitudinal nerves; seeds elliptic, slightly flattened, ca. 1 mm long, shiny. June—July.

Forest-steppe zone but not in the feather-grass steppe or in forest outliers, preferring the meadow-steppe habitat. — European part: V.-Don, (E.); W. Siberia: U. Tob. (SW), Irt. Endemic. Described from the Volga area. Type in Moscow.

Note. This species was described in great detail by Smirnov under the name C. wolgensis. The name C. uralensis, which had been given to this plant by Nevskii (in schedis) and recorded by Fedchenko (1933) in a footnote at the above-indicated page of the 6th edition of "Flora Srednei Rossii" by Maevskii, cannot be adopted because it is only a preliminary name, used in doubt in case someone would definitely prove the specific status of that taxon. In other words, it is a "nomen provisorium" which would be rejected according to the nomenclatural rules in effect now. In addition, the name C. uralensis Nevski ex B. Fedtsch. maintains elements which are regarded as typical for nomina confusa, thus being once more against the rules. Fedchenko understood incorrectly the range of C. uralensis, having in view plants from Siberia and Central Russia of quite another species, namely C. altaica Ldb. Fedchenko combined two different species under the name C. uralensis and attributed his erroneously interpreted C. uralensis to Nevskii. What Nevskii left behind is a herbarium specimen with the inscription "C. uralensis Nevsky," collected at the Argayash District of Bashkiria (near Kuyanba), and which can be undoubtedly identified as C. wolgensis P. Smirn. Since
319 Nevskii did not describe the noted species and later even discarded his first idea to describe it, and since Fedchenko offered as C. uralensis a mixture of two species, the name C. uralensis must be rejected.

C. wolgensis differs from C. altaica by very small but remarkably consistent characters, the most differentiating being the narrow, nearly filiform calyx teeth in the Volga bellflower and the oblong-lanceolate calyx teeth in the Altai population. The distribution range of the Volga type is in the Volga area, the Ural area, W. Siberia up to Altai and in Kazakhstan. It separates two disjunctive areas of another species, C. altaica, distributed in Altai and in the west European part of the USSR. This is, as Smirnov notes (op.cit.), due to historical and floristic causes.

Subsection 2. ROTULA Fed. subs. nov. in Addenda XXIII, p. 335.— Notches between calyx teeth without appendages, teeth triangular at base, lanceolate, acute, often with small lateral denticles, declinate after flowering. Corolla broadly infundibular, nearly rotate. Capsule erect, cylindrical-oval, longitudinally striate with 10 nerves, opening at apex by 3 small pores. Stems branching, leafy. Rosetted leaves absent; lower cauline leaves ovate-cordate, abruptly tapering into more or less long thin petioles, irregularly crenate-dentate.

Type species of subsection: C. carpatica Jacq.

148. C. carpatica Jacq. Hort. bot. Vindobon. I (1770) 22; Roem. et Schult. Syst. V, 93; A. DC. Monogr. 319; idem in DC. Prodr. VII, 479; Sagorski et Schneid. Fl. Carpat. Centr. II, 371; Fedch. and Fler., Fl. Evrop. Ross. 939; Szaf., Kulcz., Pawl. Rosl. Polskie, 644.— Ic.: Jacq. l. c. tab. 57; Curtis Bot. Mag. tab. 117; H. Clifford Crook, Campanulas, 49.— Exs.: Fl. exs. austro-hung., No. 2993; Fl. pol. exs. No. 363 et 363a; Domin et Krajina, Fl. Cechoslv. exs. No. 193.

Perennial; root fibrous, whitish; stems 15—50 cm, erect, branching, leafy, brittle, cylindrical, orbicular at cross section, glabrous;flowering branches and pedicels elongate, glabrous; leaves glabrous, the radical long-petioled, blade ovate-orbicular, cordate, irregularly crenate-dentate, up to 5 cm long, petioles the same length or longer; cauline leaves petioled, ovate, acute, nearly cordate, sinuate or crenate-dentate, rapidly reduced toward apex, the uppermost narrow, subsessile. Pedicels glabrous 10—15 cm long, erect;
320 flowers large; calyx glabrous, tube obconical, short, ca. 3 mm, 10-nerved, the teeth nearly twice as long as tube, triangular at base then lanceolate, acute, entire or denticulate at margin, erect, but declinate at flowering; corolla infundibular-campanulate, broad, glabrous, 3 cm long and as wide, with lobes ovate, acute declinate but not reflexed; stamens membranous, dilated and ciliate at base, filiform above, with yellow anthers as long as filaments; style bipartite; capsule erect, oval-cylindrical, 10-nerved, glabrous; seeds flattened, pale yellow, ovate. (Plate XIX, Figure 1.)

Limestone sliffs in the upper mountain zone.— European part: U. Dns. (Carpathians). **Gen. distr.:** Centr. Eur. (Carpathians, Tatras, Transylvanian Alps). Described from the Carpathians. Type in Vienna.

Note. From the USSR area we had only specimens collected near Trebuzan (Marmarosh Mountains in the Carpathians). All were glabrous.

Economic importance. A very beatiful plant long grown in gardens.

Subsection 3. MELANOCALYX Fed. subs. nov. in Addenda XXIII, 467.— Notches between calyx teeth without appendages, teeth nearly as long as corolla, erect, tube inflated at apex subanthesis and especially after, elongate, gradually tapering toward base, blackish, finely pubescent, striate with prominent nerves; corolla deeply incised into obtuse, nearly erect lanceolate lobes; capsule inflated at summit, nearly clavate, crowned by the persistent erect calyx teeth, opened by three apical valves. Stems 1-flowered, more or less leafy, developing from the partly prostrate branching rhizome. Leaves crowded mostly at base of stem as if in a rosette, lanceolate or sometimes nearly obovate, obscurely crenate or subentire. Plants glabrous, except for calyx.

Type species of subsection: C. uniflora L.

149. C. uniflora L. Sp. pl. (1753) 231; Roem. et Schult. Syst. V, 89; A. DC. Monogr. 329; idem in DC. Prodr. VII, 2, 482; Ldb. Fl, Ross. II, 890; Fedch. and Fler. Fl. Evrop. Ross. 938; Kom., Fl. Kamch. III, 116 (in Kamtsch. desideratur); Hult. Fl. Aleut. Isl. 313; ej. Fl. of Alasca a. Yukon, IX, 1463. — C. scheuchzeri Stank. in Stankov and Taliev, Opred. rast. (1949) 747, non C. scheuchzeri Vill. et auct. aliorum omnium. — Ic.: Wahl, Fl. Dan. tab. 1512 (bona); Fedch. and Fler., op. cit. fig. 938 (mediocr.). — Exs.: Dorfler, Herb. norm. No. 4897.

Perennial; root simple, blackish, furrowed, sometimes thickened; stems
321 solitary or few, simple, 1-flowered, usually not more than 10—15 cm, glabrous, erect, leafy; leaves glabrous, entire or crenulate; lower leaves obovate, obtuse, ca. 2 cm long, attenuate into short petioles; median leaves obovate-lanceolate, the upper reduced, linear-lanceolate, acute. Flowers drooping; calyx sparingly pubescent, teeth erect, acute, subglabrous, short, nearly as long as corolla (rarely shorter), the tube elongate, pyriform, slightly inflated above; corolla blue, infundibular; stamens nearly as long as calyx lobes, with slightly broadened filaments; style pubescent; capsule erect, long, somewhat pyriform or nearly cylindrical, narrow, ribbed, opening at apex by three pores; seeds small, ovate, yellowish, dull. July—August. (Plate XIX, Figure 2.)

Stony tundras in the Polar region and in similar places in the alpine mountain zone. — Arctic: Arc. Eur. (Novaya Zemlya, Vaigach, Polyarnyi Urals), Arc. Sib., Chuk. (Wrangel Island), An. **Gen. distr.:** Arctic (Iceland, Spitsbergen, Greenland, Scand., particularly Lapland), N. Am. (Aleutians, Alaska and Yukon, Baffin Island, Labrador to Newfoundland). There are data recording this species from the American mountains, 39°NL. Described from Lapland. Type in London.

Note. A widespread species, varying very little. Komarov (op. cit.) thought that the plant might very likely be found in the northern mountainous part of Kamchatka. C. uniflora is reported by de Candolle (l. c.) from St. Petersburg province (prov. Ingrica), but this is obviously a misidentification.

Stankov's (op. cit.) C. scheuchzeri is an ambiguous species. This name was often incorrectly applied to the Arctic race of the related C. rotundifolia sl., i. e. C. giesekiana Vest. Stankov's comment, that this species has "fleshy leaves," leads us to suspect that in this case it was referred to C. uniflora L. A bellflower growing in the Alps and only in the Tatra Mountains adjacent to the USSR had been described under the name C. scheuchzeri Vill. It therefore appears that Stankov accidentally described C. uniflora L. again giving it a name that had been given many times to other bellflowers (but always from the same one group, i. e. from the subsection Heterophylla). It is even more probable that Stankov combined two quite different species — C. uniflora and C. giesekiana Vest (see Note to the latter species).

Subsection 4. ODONTOCALYX Fedt. subsect. nov. in Addenda XXIII, p. 336. — Notches between calyx teeth without appendages; calyx teeth linear-
322 lanceolate, with acute bilateral dentiform lobes; tube pubescent; corolla glabrous; capsule erect, cylindrical, opening at apex by three small pores. Radical and cauline leaves oblanceolate or lanceolate-linear, with acute or nearly mucronate teeth at margins. Rhizome loosely cespitose, producing 1-flowered stems and sterile shoots.

Type species of subsection: C. lasiocarpa Cham.

150. C. lasiocarpa Cham. in Linnaea, IV (1829) 39; A. DC. in DC. Prodr. VII, 2, 482; Hook. Fl. Bor. Amer. II, 29; Ldb. Fl. Ross. II, 890; Fedtsch. Fl. d. Il du Command. 80; Kom., Fl. Kamch. III, 113; Hult. Fl. Aleut Isl. (1937) 318; Fl. of Kamtch. IV (1930) 150; Sugawara, Ill. Fl. of. Saghal. IV, 1747; Hult. Fl. of Alaska a. Yukon, IX, 1458; Tolmach, in Bot. zhurn. SSSR, XXXV, 4, 351. — C. algida Fisch. ex A. DC. Monogr. (1830) 338. — C. dasyantha A. DC. Monogr. (1830) non M. B. quoad pl. ex ins. Beringii. — C. dasyantha Hult. l. c. (1930) 149, non M. B. quoad pl. ex Command. ins. — C. stelleri Steph. in schedis. — Ic.: A. DC. Monogr. tab. 11. f. B (mediocr.); Sugawara, l. c. tab. 801, f. dextra (mediocr.); H. Clifford Crook, Campanulas, 119.

Perennial; rhizome thin, whitish, horizontally creeping, with short decumbent branches, thickened at ends and bearing rosettes of leaves on sterile shoots and at base of fertile stems; stems leafy, 1-flowered, straight, 3—10 cm, covered with sparse hairs or glabrous; radical leaves and rosetted leaves of sterile shoots long-petioled, oblanceolate to narrowly obovate, acute, usually large-toothed, sometimes nearly aristate-dentate; cauline leaves sessile, the higher the narrower, uppermost leaves linear, acute, subentire. Flowers large, pale blue-violet, sometimes drooping, 3—3.5 cm; calyx densely lanate-pubescent, teeth lanceolate, mucronate, usually lobed-serrate, one-third to one-half as long as corolla, appendages absent; corolla glabrous, campanulate-infundibular, with broadly ovate mucronate lobes; capsule cylindrical, opening at apex. July—August. (Plate XIX, Figure 3.)

Alpine mountain zone, cliffs, lichen tundra of high mountain, sometimes riparian gravels. — Arctic: An.; Far East: Kamch. (and Commander Islands), Sakh. (S. and Kurile Islands). **Gen. distr.**: Jap., N. Am. (Aleutians, from Yukon and Alaska to British Columbia). Described from Unalaska (Aleutians). Type in Leningrad, cotype in Berlin.

Note. This species was at first confused with C. chamissonis which in turn was mixed up with C. dasyantha M. B. For example, A. de Candolle did not describe Bieberstein's C. dasyantha but quite another plant
323 collected on Unalaska Island by Chamisso together with another species described by Chamisso himself as C. lasiocarpa Cham. At the same time de Candolle noted that C. dasyantha was first discovered by Steller on Bering Island and therefore referred to Gmelin (Fl. sib.) who had

described Steller's plant. This double error by de Candolle was due to misunderstanding of Bieberstein's C. dasyantha M. B. which had been described from Siberia. This plant has never grown on Bering Island and is a synonym of the simultaneously published C. pilosa Pall. ex Roem. et Schult. and C. pallasiana Vest ex Roem. et Schult. Moreover, Bieberstein himself was in doubt whether to include Steller's plant described by Gmelin in C. dasyantha, but nevertheless he recorded the island in the Bering Sea as another locality of his species. Thus, de Candolle assumed from Chamisso's inaccurate classification that C. dasyantha M. B. grew in the Aleutian Islands and from Gmelin's diagnosis (Fl. sib.) also on the Commander Islands. Ledebour repeated this error, aggravating the situation by denoting the Bering Strait as a habitat of "C. dasyantha." It has now been definitely proven (Fedchenko, 1906, Hulten, 1949) that only C. lasiocarpa Cham. grows on the Commander Islands and that this is the plant described by Gmelin. Its area extends to the Bering Strait and is much broader than that of the other species which Chamisso and de Candolle accepted as C. dasyantha M. B. and which has now been renamed C. chamissonis.

Komarov (op. cit.) distinguished two forms in Kamchatka: f. caulescens Kom., with developed stems profusely foliate, and f. longifolia Kom., with long upper leaves overtopping the flowers. Hulten (l. c. 1937) described var. latisepala Hult., with broad, triangular calyx lobes, but considered it to be a hybrid of C. lasiocarpa × C. dasyantha, growing side by side in the Aleutians. However, this assumption is not particularly convincing since these two species belong to different and taxonomically remote sections.

Dubious and excluded species

C. adscendens Vest ex Roem. et Schult. Syst. V (1819) 90; A. DC. Monogr. 342; idem in DC. Prodr. VII, 2, 848; Ldb. Fl. Ross. II, 889.

Roemer and Schultes (l. c.) and later Ledebour (l. c.) found resemblance between the specimens preserved under this name at the Pallas herbarium
324 in Siberia and C. uniflora L.; in fact, C. adscendens is not related to C. uniflora because, as both Roemer and Schultes pointed out, the capsule opens at the base in C. adscendens but not in C. uniflora, Hence, it is most likely that C. adscends, as Pallas himself believed, is related to the group of C. rotundifolia and is possibly none other than C. langsdorffiana. If this is proven then C. langsdorffiana will have to be renamed C. adscendens.

C. calycina Boeber ex Roem. et Schult. Syst. V (1819) 102; Spreng. Syst. I, 731; Ldb. Fl. Ross. II, 888 in obs. — C. persicifolia var. calycina A. DC. Monogr. (1830) 323; idem in DC. Prodr. VII, 479; p. p. non Rchb.

Alphonse de Candolle pointed out that a deformed form of C. persicifolia had been described under the name C. calycina. The truth is that de Candolle had based his findings on a specimen grown "e regione populi Kirguis" and not in the Crimea from which C. calycina was described. Therefore, it is possible that de Candolle's C. persicifolia var.

calycina is not identical to the plant which Beber had collected in the Crimea. Ledebour (l. c.) saw the authentic specimen at the Willdenow herbarium (under label No. 3809). According to him, "one specimen with leaves is actually the middle part of the stem; the floriferous branch is emerging from the axil of a middle leaf on this part of the stem whereas the axils of the other leaves, both the upper and the lower, are sterile. Leaves are ovate, acuminate, slightly cordate and amplexicaul at base, sparingly pubescent, like the stem. Another specimen is a floriferous branch, branching, bearing linear and inconspicuously serrate leaves subtending the flowers. The calyx teeth in both specimens are very long-mucronate, more or less serrate below middle, nearly twice as long as the corolla, rarely equal to it."

It can be seen from this description that what is called C. calycina is probably a deformed form not of C. persicifolia but more likely of C. rapunculus. This is supported by Ledebour's description of the shape of the flowers, in particular the calyx lobes. Since the flora of the Crimea has been adequately investigated and nothing resembling C. calycina has evern been found, C. calycina should not be considered as an independent species.

C. camtschatica Pall. ex Roem. et Schult. Syst. V (1819) 152; Spreng. Syst. I, 725; A. DC. Monogr. 343; DC. Prodr. VII, 2, 484; Ldb. Fl. Ross. II, 889 ("kamtschatica"); Trautf. in Tr. Peterb. bot. sada, VI, 1, p. 92 ("kamtschatica"); Kom., Fl. Kamch. III, 117, cum dubit.; Hult. Fl. of Kamtch. IV, 149, cum dubit.

325 Roemer and Schultes (l. c.), based their description of this species upon a specimen from the Pallas herbarium in Siberia which had been preserved in the Willdenow herbarium; they pointed in their diagnosis to the following characters of the plant: ". . . radical leaves cordate, dentate; cauline leaves oblong-lanceolate, entire. Stem filiform, usually 1-flowered. Calyx teeth recurved, linear, as long as corolla." They also noted that: "Root filiform, strongly branching; radical leaves orbicular, slightly angular-dentate, membranous, glabrous, smooth, with long filiform petioles; cauline leaves attenuate into petioles, obtuse. Stem filiform, weak, with one flower at apex; calyx lobes equal, subulate acuminate, recurved. Perennial. On Kamchatka."

Ledebour (l. c.) saw the authentic specimen (Willdenow herbarium, No. 3813) and emended the diagnosis by pointing out that some of the radical leaves were reniform, some of the cauline leaves irregularly dentate stem few-flowered, some of the calyx lobes exceeded the corolla. Further he denoted: "A delicate plant. Root filiform, prostrate undergound. Stem of the specimen from the Willdenow herbarium with only one flower, and a bifurcate middle branch. Calyx lobes recurved, unequal, 3 lobes, as long as corolla, the 4th slightly shorter but equal to the three in length and width, the 5th lobe wider and longer and slightly broadening at apex."

Hulten (l. c.) indicated that ". . . this specimen is composed of a flower and a few leaves in a way which makes it impossible to determine the relationship of the plant to this flower." Hence, he regards C. camtschatica as a doubtful species, outside of the Kamchatka flora.

This opinion is shared by Komarov (op. cit.) who added that the species is a synonym of C. uniflora L., which might occur in Kamchatka. He also pointed out that C. camtschatica is "possibly . . . one of the forms of C. simplex Stev., known from the Kolyma River and widespread in N. Siberia." He doubted the accuracy of the report of a plant from Kamchatka at the Pallas herbarium, adding "perhaps not on Kamchatka but in the Yakut Republic."

On reviewing the facts we draw the conclusion that two different species have been preserved under the name C. camtschatica in the Willdenow herbarium. The fragment of the stem with flower should be referred to C. langsdorffiana. In this species the calyx lobes are often as long as the corolla but strongly declinate or recurved. Due to this feature of the calyx teeth it is definitely impossible to refer this part of the plant to C. uniflora, since in the latter the calyx lobes are always directed upward
326 although as long as the corolla. The leaves and root of that specimen (or perhaps rhizome with shoots) unquestionably belong to another species. This specimen probably does not refer to C. simplex (or C. altaica), the corolla of which always is longer than the calyx and the lobes are obtuse and keeled at apex. It is almost certain that the plant is not from Kamchatka but is a stray among the Kamchatka plants in the Pallas herbarium. C. camtschatica is not only a dubious species (species dubia) but also a nomen confusum. Hence, the name must be rejected in any case.

C. elegans Roem. et Schult. Syst. V (1819) 105; Spreng. Syst. I, 729; A. DC. Monogr. 343; idem in DC. Prodr. VII, 484; Ldb. Fl. Ross. II, 891, sp. dub. – C. speciosa Willd. ex Roem. et Schult. l. c. pro syn. non Hornem. nec Gilib.

Neither de Candolle (l. c.) nor Ledebour (l. c.) knew whether to possibly accept as an independent species the so-called C. elegans and where to refer it. This species had been described by Roemer and Schultes (l. c.) according to a specimen in the Pallas herbarium. Since then the name has been listed among the dubious species. What the plant is and where it was collected is unknown to this day. We present here a translation of the original diagnosis of Roemer and Schultes (l. c.): "Leaves linear-lanceolate, completely entire, sessile, the lower wider, shortly petiolate, attenuate into petiole, smooth. Stem pubescent; flowers axillary, upward in spicate inflorescences; calyx teeth as long as corolla, hirsute, spreading, nearly subulate acute." And further: "Stem 1 foot high and even taller, simple, orbicular at cross section, weak; leaves alternate, the lower lanceolate, with few small teeth. Flowers very short-pediceled, disposed in axils of upper leaves and forming an interrupted, spicate inflorescence."

C. heterodoxa Vest ex Roem. et Schult. Syst. V (1819) 98.

There are reasons to believe that the plant used as the authentic specimen for the description of this species was not from Siberia as reported in the original diagnosis, but from N. America. For this reason it should be excluded from the flora of the USSR (see Note on C. langsdorffiana Fisch.).

329 **C. infundibulum** Vest ex Roem. et Schult. Syst. V (1819) 106. – C. steveni auct. p. p. non M. B.: A. DC. Monogr. (1830) 321, cum dubit.; Ldb. Fl. Ross. II, 886. – C. steveni var. sibirica A. DC. in DC. Prodr. VII (1839) 479, cum dubit.

(327)

PLATE XIX. 1 – Campanula carpatica Jacq., plant and capsule; 2 – C. uniflora L., plant and capsule; 3 – C. lasiocarpa Cham.

It is quite likely that C. infundibulum Vest and C. seminuda Vest are the earlier and simultaneously published names for the later established C. altaica Ldb. C. infundibulum Vest was regarded at first as a synonym of C. stevenii, described much earlier when it was not known that the related species C. altaica grows in Siberia. The names of these bellflowers should now be compared and re-examined. There are no grounds yet for giving C. infundibulum Vest the priority over C. altaica Ldb. C. infundibulum Vest was considered at first a synonym of C. stevenii and of C. altaica, but with doubt. Both de Candolle (l. c.) and Ledebour (l. c.) put a question mark after C. infundibulum as a synonym of C. stevenii. It might be explained by the fact that Reichenbach (Ic. pl. crit. I, Plate 75, Figure 158) apparently presented as C. infundibulum the plant which Fisher had described as C. silenifolia and which owing to homonymy received the new name C. turczaninovii. The absence of the authentic specimen, not seen even by de Candolle, leaves some doubt as to who understood C. infundibulum correctly – de Candolle or Reichenbach. Added to this situation is the recent discovery near the southern Siberian frontiers of C. wolgensis P. Smirn., which besides C. altaica is affined to C. stevenii s. l. It is now more difficult to decide what C. infundibulum represents and if it is reallly identical with C. altaica Ldb. Although the probability of the correctness of identifying these names is high, it is much higher than the possibility of identifying C. altaica Ldb. and C. seminuda Vest (see Note to C. altaica Ldb.).

C. limariaefolia Roem. et Schult. Syst. V (1819) 92. – C. lunariaefolia A. DC. Monogr. (1830) 343, non Rchb.

Alphonse de Candolle apparently regarded the species epithet "limariaefolia" as an error and changed it into "lunariaefolia." He regarded the species described under this name by Roemer and Schultes (l. c.) as ambiguous and said that it could only be C. alliariifolia Willd. He also noted that in Reichenbach (Pl. crit.) C. rapunculoides L. was actually put under the name C. lunariifolia. C. limariaefolia Roem. et Schult. could be considered now as a synonym of C. alliariifolia Willd. because several species related to it are now known from the Caucasus from where Roemer and Shultes described their plant (Adams herbarium).
330 Therefore, there is no way to accurately identify the names. The problem might be solved after examining the authentic C. limariaefolia in the Willdenow herbarium in Berlin.

C. polyantha Roem. et Schult. Syst. V (1819) 130; A. DC. Prodr. VII, 484; Ldb. Fl. Ross. II, 891, sp. dub. – C. multiflora Willd. ex Roem. et Schult. l. c. pro syn., non Waldst. et Kit.

Thus far it is not known to which of the Siberian bellflowers the above name refers. The diagnosis made by Roemer and Schultes (l. c.) does not resemble any of the Siberian plants. The translation of the diagnosis is as follows: "Hirsute plant; leaves oblong-lanceolate, serrate; flowering raceme paniculate-glomerate (paniculatoglomerati); calyx sharply (argute) serrate." The description is based on a specimen in the Pallas herbarium in Siberia. No botanist has ever seen the authentic material.

C. rigescens Pall. ex Roem. et Schult. Syst. V (1819) 102; A. DC. Monogr. 342; idem in DC. Prodr. VII, 483.

It is possible that C. rigescens is the earliest name for C. turczaninovii Fed. or C. silenifolia Fisch., but there are substantial doubts on the priority of C. rigescens (see Note to C. turczaninovii).

C. scandens Pall. ex Roem. et Schult. Syst. (1819) 120; A. DC. in DC. Prodr. VII, 485, sp. excl.; Ldb. Fl. Ross. I, 891, sp. dub.

Alphonse de Candolle (l. c.) considered that the plant described from the Pallas herbarium by Roemer and Schultes (l. c.) under the name C. scandens belonged to some other genus. A probable candidate might be the genus Codonopsis but some characters (crenation of calyx lobes) preclude referring C. scandens to this genus. Not one of the Siberian Campanulaceae can be related to C. scandens, and this taxon is still a mystery.

C. seminuda Vest ex Roem. et Schult. Syst. V (1819) 91.

Alphonse de Candolle (Monogr. Campan. 321) and later Ledebour (Fl. Ross. II, 886) submitted that the species described by Roemer and Schultes (l. c.) under the name C. seminuda was a synonym of C. stevenii s. l. Since C. seminuda was not described from the Caucasus but from Siberia, where only C. altaica Ldb., related to C. stevenii, is found, it might be that C. seminuda is the earlier name for C. altaica. This might be proven on finding the authentic C. seminuda in the Pallas herbarium,
331 and then the name C. seminuda would have priority over C. altaica (see also Note to C. altaica Ldb.). After more relevant investigation the prior name may be proven to be C. infundibulum Vest ex Roem. et Schult. (see Note to C. infundibulum).

C. violaefolia Lam. Encycl. Meth. I. (1783) 587; Roem. et Schult. Syst. V, 145; A. DC. Monogr. 242; idem in DC. Prodr. VII, 464. — C. violae Pers. Enchir. I (1805) 192.

Simultaneously with C. punctata, Lamarck (l. c.) described under the name C. violaefolia a garden specimen of bellflower which was very like the former species and had been grown in Paris in 1765 from seeds probably obtained from Siberia ("e Sibiria creditur"). The dried plant was preserved in the Jussieu herbarium. Alphonse de Candolle, who studied the specimen, decided that it resembled C. punctata. The corolla of C. violaefolia was only one and a half times as long as the calyx teeth (and not 2–3 times as in C. punctata) and spots inside the corolla, characteristic of C. punctata, could not be observed. It is quite probable that C. violaefolia is indeed a synonym of C. punctata. Obliterating any final determination is the ambiguity of the origin of C. violaefolia; Fischer pointed out, in denial of de Condolle, that nothing resembling it has evern been seen among the Siberian plants ("Clarissimus Fischer scripsit hanc speciem nunquam inter plantas sibiricas videsse").

C. volubilis Pall. ex Roem. et Schult. Syst. V (1819) 98.

Alphonse de Candolle (Monogr. Campan. 311) regarded this taxon as a synonym of the Caucasian plant C. lactiflora M. B., and Ledebour

(Fl. Ross. III, 890) treated it likewise. Roemer and Schultes (l. c.) indicated, however, that C. volubilis was described according to a specimen in the Pallas herbarium in Siberia. It is conceivable that de Candolle did not see the authentic material since he referred to Steven who had examined the specimen and found it resembling the later-described C. lactiflora from the Caucasus. There is no way of verifying this and therefore we are enumerating C. volubilis among the dubious species, particularly since the label of the authentic specimen was mixed up.

Genus 1432. **SYMPHYANDRA*** A. DC.

A. DC. Monogr. (1830) 365.

Calyx 5-partite, rather detached from corolla, with folds or sometimes appendages between teeth. Corolla 5-lobed, campanulate or infundibular,
332 pubescent or glabrous, white or blue. Stamens 5, with free membranous ciliate filaments; anthers united into a tube. Style pubescent, with 3 stigmas. Capsule trilocular, opening at sides of the lower part by small valves. Seeds ovate, more or less flattened, shiny. Perennial plants, usually chasmophytes, with indurate, often woody rhizome and few-leaved stems. Leaves alternate, petiolate, often cordate, the lower very long-petioled. Flowers in spherical or partly paniculate inflorescence, axillary and terminal, pediceled.

Type genus: Symphyandra cretica A. DC. (from Crete).

This genus is very close to Campanula and sympatric with it in the Mediterranean area, Caucasus and N. Iran. Alphonse de Candolle, who established the genus, pointed the characters differentiating it from Campanula (concrescence of anthers to form a tube); but with the discovery of a large number of seemingly intermediate species, these characters became less significant. Recently Kharadze (l. c.) has proposed that only a few species in which the anthers are always united should be maintained as Symphyandra. However, nearly all the species which Kharadze excludes from Symphyandra are so close to the type of this genus in all other characters that their transfer from it to Campanula is admittedly artificial. The preservation of Symphyandra is justified only in tribute to the memory of de Candolle; it would be more accurate to regard the species of this genus as being species of Campanula.

Another argument in support of preserving Symphyandra is that, apart from the species which are usually referred to this genus but in general are close to Campanula, there have been established two species which are, on the one hand, related to Campanula and, on the other, to Symphyandra as is usually understood. One of these species was recently described from Iran by the British botanist P. H. Davis (Davis in Hooker's icones plantarum (1950) tab. 3497) under the name Zeugandra, the other described long ago from Korea by the Japanese botanist Nakai (Nakai in Bot. Mag. Tokyo, XXXV (1921) 147) under the name Hanabusaya. It is noteworthy that Zeugandra because of its characters appears to be intermediate not between two but three related genera of the family

* From the Greek symphysis – growing together and aner – male (in this case meaning the concrescence of the anthers).

(333)

PLATE XX. 1 – Symphyandra lezgina Alexeenko ex Lipsky; 2 – S. zangezura Lipsky; 3 – S. pendula (M.B.) A. DC., plant and flowers (cross section).

Campanulaceae — Campanula, Symphyandra and Adenophora. In taxonomic characters Hanabusaya is close to Symphyandra but 335 in general appearance is vividly reminiscent of the Far Eastern species Campanula punctata.

When compared with other genera making up the Campanulaceae, these groups of bellflowers appear transitional but still have an independent combination of characters owing to which they could be ranked as separate genera. This has its advantage in an understanding of the phylogenesis within this family.

With respect to Parageranion and Petrocodonia, the new sections established by us, they too could subsequently be ranked as two genera.

According to Grossheim (Rastitel'nye bogatstva Kavkaza, 2nd ed. (1952)), the leaves of some species of Symphyandra contain rubber. The leaves of S. pendula from Ciscaucasia contain 1.62% rubber.

The species of Symphyandra are ornamental plants grown in rocky soil in parks and gardens.

1. Leaves only cauline, the lower oblong-lanceolate, dentate, cuneate at base, short-petioled, the upper sessile; stems straight, ascending, thin; corolla broadly campanulate, deeply (more than half) dissected into lobes . . . . 7. **S. lezgina** Alexeenko.
\+ Leaves cauline and radical, the lower ovate-cordate, orbicular or pentagonal, long-petioled . . . . 2.
2. Leaves palmately 5-partite, with rhombic lobes, reminiscent of the leaves of geranium, cordate; stems numerous, thin, developing from thick slightly cespitose rhizome; corolla broadly campanulate, nearly rotate . . . . 6. **C. zangezura** Lipsky.
\+ Leaves entire or dentate but not lobate; corolla campanulate or infundibular, never nearly rotate . . . . 3.
3. Notches between calyx teeth without appendages; leaves cordate or ovate, bidentate; pedicels very long; stems pendulous; corolla tubular-campanulate, pubescent . . . . 1. **S. lazica** Boiss. et Bal.
\+ Notches between calyx teeth with appendages . . . . 4.
4. Leaf blade not more than 2—3 cm across, orbicular or short-cordate, large-toothed; flowers white or bluish, 1.5—2 cm long . . . . 6.
\+ Blades 7—10 cm long, 8 cm wide, oblong, ovate or elliptic; flowers white or yellowish, ca. 3 cm long . . . . 5.
5. Canescent, soft-hairy plants; leaves with blades often more than 10 cm long, 8 cm wide; inflorescence dense, multiflowered . . . . 3. **S. transcaucasica** (Somm. et Lev.) Grossh.

336 + Plant velutinous at first, later sparingly pubescent or often glabrous; laminas up to 7 cm long and not more than 4—5 cm wide; inflorescence loose . . . . 2. **S. pendula** (M.B.) A. DC.
6. Calyx-tube grayish-pubescent; upper leaves sessile . . . . 4. **S. armena** (Stev.) A. DC.
\+ Calyx-tube glabrous. All leaves petiolate . . . . 5. **S. daralaghezica** Grossh.

Section 1. ANOTOCALYX A. DC. in DC. Prodr. VII, 2 (1839) 494. – Calyx without appendages. Leaves simple, cordate or ovate, the lower long-petioled. Corolla tubular-campanulate; anthers united.

Type species of section: the same as the type of the genus S. cretica A. DC.

1. S. lazica Boiss. et Bal. in Boiss. Fl. or. III (1875) 888; Fomin in Mater. Fl. Kavk. IV, 6, 119; Grossg. Fl. Kavk. IV, 75. – Campanula lazica (Boiss. et Bal.) Charadze in Zam. po sist. i geogr. rast. Tbil. bot. inst. 15 (1949) 18.

Perennial; plant pubescent; stems developing from thick rhizome, elongate, thin, drooping, forming spreading few-flowered paniculate inflorescence at apex; leaves cordate, ovate or oblong, nearly bidentate, the radical long-petioled; the cauline short-petioled. Pedicels very long, with 1–2 elliptic dentate bracteoles; calyx short-hairy, teeth partly spreading, broadly triangular-oblong, acuminate, entire, three-fourths as long as tubular-campanulate corolla.

Rocks and slopes, the central mountain zone. – Caucasus: may be found in Adzharistan. **Gen. distr.**: As. Min. (Pontic mountain range.) Described from Lazistan near Kabakhar. Type in Geneva.

Note. Both Fomin (op. cit.) and Grossheim (op. cit.) thought it quite possible for this species to grow in the Caucasus.

Section 2. OTOCALYX A. DC. in DC. Prodr. VII, 2 (1839) 494. – Calyx with appendages. Leaves simple, ovate-cordate or orbicular-cordate, large-toothed at margin, lower leaves long-petioled. Corolla campanulate or infundibular; anthers concrescent but later usually partly free.

Type species of section: S. pendula (M. B.) A. DC.

Series 1. **Pendulae** Fed. – Blades ovate, elongate, velutinous. Flowers 3 cm long.

2. S. pendula (M. B.) A. DC. Monogr. (1830) 366; idem in DC. Prodr. VII, 2, 494; Ldb. Fl. Ross. II, 2, 896; Rupr. in Bull. Acad. Sc. Pétersb. XI, 221; Boiss. Fl. or. III, 886; Medv. in Tr. Tifl. bot sada, XVIII, 2, 327;
337 Fomin in Mater. Fl. Kavk. IV, 6 (1906) 119, excl. var.; Grossg., Fl. Kavk. IV, 76, excl. var.; Kharadze in Fl. Gruzii, VIII, 190, excl. var. – Campanula pendula M. B. Fl. taur.-cauc. I (1808) 154, III (1819) 144; Roem. et Schult. Syst. V, 144; Shmal'g., Fl. II, 179. – C. alliariifolia Rchb. Pl. crit. III, 345, non Willd. – Ic.: Sweet. Flow. Gard. Ser. II, tab. 66. – Exs.: GRF, No. 467.

Perennial; rhizome thick, covered with short brown remnants of petioles, buried in rock crevices; stems slightly drooping, numerous, glabrous, or velutinous, 20–40 cm, woody at base, leafy, loosely and somewhat paniculately branching; leaves glabrous or slightly velutinous, markedly crenate-dentate or bicrenate; lower leaves with rather long thin petioles, slightly cordate, ovate, acute, 7 cm; upper cauline leaves sessile, similar in shape to the lower but angustate-cuneate at base, numerous. Flowers whitish, sulphur-yellow when dry, 1–3 in paniculate inflorescence, with short velutinous pedicels; calyx pubescent, obconical, teeth lanceolate, erect, much longer than tube, distinctly accrescent in fruit, persistent, half as long as corolla, calyx appendages recurved, ovate, acuminate, shorter than tube; corolla campanulate, tomentose outside, barbate inside, 3 cm long, divided into ovate acute lobes; filaments profusely pubescent, membranous-rufous at base, filiform above, bent inside; anthers always connate; style as long as corolla, sometimes longer or shorter, filiform, pubescent from base nearly to apex, with 3 filiform stigmas, erect in aestivation but later curved outside; capsule pendulous, ovoid, angular; seeds many, ovate, shiny. July—August. (Plate XX, Figure 3.)

Rock crevices in the central mountain zone. — Caucasus: Cisc. (Mineral'nye Vody, Mount Elbrus and further to the east), Dag. Endemic. Described from Konstantinogorsk (Zheleznovodsk). Type in Leningrad.

3. S. transcaucasica (Somm. et Lev.) Grossh. Opred. rast. Kavk. (1949) 424; Kolak., Fl. Abkh. IV, 182 (erron. nom. auct. "Charadze"). — S. pendula var. transcaucasica Somm. et Lev. Enum. pl. Cauc. (1900) 313; Fomin in Mater. Fl. Kavk. IV, 6, 121; Grossg., Fl. Kavk. IV, 76; Kharadze in Fl. Gruzii, VIII, 190.

Perennial; rhizome thick, brown, covered with squamous remnants of dried petioles; plant glaucescent, soft-hairy; stems arcuately curved, drooping, thick, 4—5 mm across, 60 cm long, short-branching from base, leafy, multiflorous, lower branches 25 cm long, arcuately flexuous; lower
338 leaves coriaceous, ca. 16 cm long, 8 cm wide, ovate-oblong or broadly ovate-lanceolate, deeply cordate, with 16 cm long petioles; cauline leaves broadly ovate, acute, the lower cordate at base, the upper nearly cuneate, leaves of branches broadly ovate-lanceolate. Flowers like in S. pendula. June—July.

Rock crevices in mountain river gorges. — Caucasus: W. Transc. (Abkhaziya, Imeretiya). Endemic. Described from the Rion River between Mekven and Alpan. Type in Leningrad.

Note. This species is closely related to S. pendula but differs from it by the soft grayish pubescence on all parts of the plant, the thicker stems, the very large leaves, and the larger number of flowers in a denser inflorescence. In establishing this species Grossgeim designated the specimen from Abkhaziya (Askhi Plateau) comb. nova, but the real (nomenclatural) type is the original specimen var. transcaucasica Somm. et Lev. from the above-indicated locality which is now in Leningrad, the isotype being in Florence. The specimen noted by Grossgeim is comparatively few-leaved and is more reminiscent of C. pendula than the type of the variety described by Sommier and Levier.

Series 2. **Armenae** Fed. — Blades short, suborbicular or shortly ovate-cordate, large-toothed. Flowers 2 cm long.

4. S.. armena (Stev.) A. DC. Monogr. (1830) 367; idem in DC. Prodr. VII, 2, 495; Ldb. Fl. Ross. II, 2, 896; Boiss. Fl. or. III, 890; Fomin in Mater. Fl. Kavk. IV, 6, 122; O. and B. Fedch., Perech. rast. Turk. III, 358; Medv. in Tr. Tifl. bot. sada, XVIII, 2, 327; Grossg., Fl. Kavk. IV, 76. — Campanula armena Stev. in Mém. Soc. Nat. Mosc. III (1812) 256; M. B. Fl. taur.-cauc. III, 145. — Exs.: Fl. Cauc. exs. No. 48.

Perennial; plant more or less densely pubescent, sometimes subglabrous; rhizome thick, branching, 20—40(60) cm, grayish-brown, covered in upper part with rather long intertwined or spreading dry gray remnants of petioles and old stems; stems straight or drooping, thin, flexuous, branching in the upper part or from the middle, branches strict, uniflorous or with 3—5-flowered corymbs forming nearly paniculate inflorescence; leaves acutely large-toothed, 2—3—4 cm across; lower leaves long-petioled, cordate-ovate, the upper subsessile, reduced. Flowers medium-sized, 2 cm long, with thin filiform pedicels provided with small bracteoles; calyx
339 pubescent-canescent, teeth declinate, triangular, acute, one-fourth as long as corolla, appendages very short, acute, corolla white, finely velutinous, campanulate, short-lobate; style exserted; capsule obovoid-turbinate; seeds brown, oblong, acute. June—August.

Rocks in the upper forest, the subalpine, sometimes the alpine zones. — Caucasus: E. Transc. (Lesser Caucasia), S. Transc., Tal.; Centr. Asia: Mtn. Turkm. (Kopet Dagh?). **Gen. distr.:** N. Iran. Described from rocky place near Gandzhi (now Kirovabad). Type in Helsinki.

Note. In spite of old reports, this species is not listed in the "Flora Turkmenii," VI (1954), thus there is a doubt as to its occurrence in Turkmenia. On the other hand, it is possible that excluding this species, as well as the genus, from the list was a result of an oversight.

5. S. daralaghezica Grossh. in Tr. Azerb. otd. Zakavk. fil. AN SSSR, sekt. bot. 1 (1933) 57; idem, Fl. Kavk. IV, 76. — S. armena var. serratosepala Fom. in Mater. Fl. Kavk. IV, 6 (1906) 152; Grossg., op. cit. (1934) 76.

Perennial; rhizome thick, branching, densely covered with short squamous brownish remnants of dried petioles; stems and lower leaves sparsely pubescent, leaves rather dense pubescent; stems numerous, straight or curved, 40—60 cm, usually strongly branching, forming a many-flowered paniculate inflorescence, subglabrous or sometimes short-hairy, gradually becoming denser toward upper branches; lower leaves petiolate, ovate, acutely large-toothed, subglabrous, blade 2 cm long; upper leaves reduced, short-petioled, densely short-pubescent. Flowers white, 2 cm long, long-pediceled, drooping; calyx glabrous, teeth thinly and finely pubescent at margins, spreading, broadly triangular, acute, entire or 1—2-toothed, half as long as corolla, elongating after flowering, with hardly developed short acute appendages; corolla tubular-campanulate, finely pubescent inside and out, partite for one third into acute lobes; anthers connate, sometimes partly separate; style distinctly exserted. July—August.

Rocks in the central mountain belt. — Caucasus: S. Transc. (Daralagez in Armenia, and S. Zangezur). Endemic. Described from Daralagez, near Alagez. Type in Leningrad, isotype in Baku.

Note. Differs from the closely related S. armena (Stev.) A. DC. by its long calyx considerably accrescent in fruit, the always white flowers, and the pubescent, petiolate upper leaves.

Section 3. PARAGERANION Fed. sect. nov. in Addenda XXIII, p. 336. — Leaves palmatipartite, pentagonal, with sinuate-dentate lobes, the lower long-
340 petioled. Corolla broadly campanulate, nearly rotate; anthers connate.

Type species of section: S. zangezura Lipsky.

6. S. zangezura Lipsky in Tr. Peterb. bot sada, XIII, 2 (1889–1893) 317; Fomin Mater. Fl. Kavk. IV, 6, 123; Medv. in Tr. Tifl. bot sada, XVIII, 2, 327; Grossg. Fl. Kavk. IV, 77.

Perennial; rhizome thick, brown, woody, branching, usually with abundant quantity of rather long coarsely setaceous grayish strict remnants of dried petioles in upper part; stems often many, suberect or declinate, rather thin, weakly branching, rarely simple, finely hairy at slower part, 20–30(40) cm, with declinate, 1-flowered branches; lower leaves long-petioled, reminiscent in shape of geranium leaves, deeply cordate, pentagonal, dissected into nearly rhombic oblong-triangular or oblong segments, large-incised-dentate; upper leaves gradually becoming smaller toward apex, subsessile, the uppermost nearly lanceolate, subentire. Flowers on thin, sometimes subfiliform pedicels markedly longer than calyx, drooping, rather large, 3 cm across, violet-blue; calyx slightly hirsute or tomentose, teeth curved, long-lanceolate, nearly half as long as corolla, appendages short, acute, with hamate mucro at tip; corolla glabrous outside, slightly ciliate inside, broadly open and nearly rotate, divided to middle into lobes; style long-hairy, with 3 long bent stigmas. June–August. (Plate XX, Figure 2.)

Rock crevices, stony taluses, gorges and cliffs, the forest and subalpine zones. — Caucasus: S. Transc. (Zangezur, Karabakh). Endemic. Described from Pirdaudan at the foothills of Kapudzhikh along the Okhchi River. Type in Leningrad.

Note. A rare plant that at times is found in masses. It is known to grow in Kapudzhikh Mountain (northern and southwestern slopes), near Urmis, Megri, Budakur, Lishk, Okhchi, Gedzhalan, Khustup Mountain, upper reaches of the Paraga-Chai River. It differs sharply from the other species of Symphyandra by its leaves, recalling those of the geranium, and by the nearly rotate corolla. Grown as an ornamental in "rocky knolls" in parks and gardens.

Section 4. PETROCODONIA Fed. sect. nov. in Addenda XXIII, p. 336. — Leaves simple, oblong-lanceolate, cuneate, short-petioled or sessile.
341 Corolla broadly campanulate, nearly rotate; anthers connate.

Type species of section: S. lezgina Alexeenko ex Lipsky.

7. S. lezgina Alexeenko ex Lipsky, Fl. Kavk. Suppl. I (1902) 68; Fomin in Mater. Fl. Kavk. IV, 6, 124; Grossg. Fl. Kavk. IV, 77. — Exs.: Fl. Cauc. exs. No. 49 (isotypus!).

Perennial; sparsely pubescent or subglabrous and glaucescent plant; stems many, developing from thick woody root, ascending or slightly drooping, leafy, simple or rarely branching in upper part where sparsely and finely crisp-pubescent, 30–40(50) cm, 1- or often few-flowered on small branches, floral branches sometimes branching and leafy; rosetted leaves absent; cauline leaves ovate-oblong or elliptic-oblong, cuneate into short petioles, sometimes nearly spatulate, the uppermost and sometimes lower leaves broad, nearly rhombic or even orbicular, reduced, all leaves crenate-dentate at margin, with declinate teeth, sparingly ciliate, (0.5)1–2 cm long, 0.7(1) cm wide. Flowers 2 and sometimes 3 cm long, blue, violet, short-pediceled; calyx glabrous, teeth linear, broadly lanceolate at base, acute, remotely denticulate, longer than tube, crisp-ciliate at margin; appendages broadly triangular, acute, recurved; corolla nearly twice as long as calyx teeth, broadly campanulate, glabrous, incised deeper than middle into oblong-lanceolate, acute lobes barbate at margin; style exserted, with 3 filiform falcately curved stigmas. July–September. (Plate XX, Figure 1.)

Rock crevices in rocky ravines of the central mountain zone. – Caucasus: E. Transc. (NE Azerbaidzhan), Dag. Endemic. Described from the mountains of Azerbaidzhan near Myudzhi. Type in Leningrad, isotypes in herbaria of other cities.

Note. This plant is very rare and common only to the easternmost part of the Main Range in the Caucasus. In addition to Myudzhi, it is found only in Azerbaidzhan near its frontier with Dagestan, near Budug, Chichi, Dzhul'yan, Adur, Aput, Sumakhally, Teng'ya Gorge, the source of Bandam-Chai River, Yatukh-Chai River.

Genus 1433. **BRACHYCODON*** FED., gen. nov.

Fed. in Addenda XXIII, 468.– Campanula, sect. Eucodon A. DC. Monogr. (1930) 214, p.p.; sect. Rapunculus Boiss. Fl. or. III (1875) 895, p.p.

Calyx dissected nearly to base, teeth without appendages, linear-lanceolate, elongate, nearly concealing corolla. Corolla small, obconical, 5-lobed for two-thirds of its length, half as long as calyx; stamens longer;
342 style half as long as corolla, partite into 3 short stigmas. Ovary trilocular. Capsule obconical, pentagonal, 5-ribbed, opening at apex by valves. Biennial, small plant with thin root. Stems furcately branching with spreading-ascending branches reaching the same height. Leaves alternate or nearly opposite, small, rosetted leaves absent. Flowers axillary, very short-pediceled, subsessile, inconspicuous.

This species is intermediate in character between Campanula and Legouzia and should unquestionably occupy a position between the two genera (within the phylogenetic scheme).

A monotypic genus whose only species is distributed in the southern regions of the Mediterranean countries, from Italy and North Africa, through Asia Minor up to the Caucasus and also in Central Asia (Gissar Range).

Type species of genus: B. fastigiatus (Duf.) Fed.

1. B. fastigiatus (Duf.) Fed. comb. nov. – Campanula fastigiata Duf. ex A. DC. Monogr. (1830) 340; DC. Prodr. VII, 483; Ldb. Fl. Ross. II 891; Boiss. Fl. or. III, 944; Fomin in Mater. Fl. Kavk. IV, 6, 107; M. Popov in Tr. Turkest. nauchn. obshch., I, 18; Grossg., Fl. Kavk. IV, 62. – C. galae M. Pop. in schedis. – Ic.: A. DC. l. c. tab. 12. – Exs.: F. Schultz, Herb. norm. No. 2761.

Annual or biennial; plant slightly fleshy, grayish, scabrous, small, 3–5(8) cm high; stems furcately branching from tip of the thin root, branches antrorse and usually reaching same length; leaves small, pubescent or glabrous; lower leaves simple, short-petioled, ovate-cuneate, 3–5-toothed;

* From the Greek brachys – short and codon – bellflower.

uppermost leaves crowded in upper part of stem; calyx teeth linear-lanceolate, obtuse, erect, longer than tube or nearly as long, distinctly longer than corolla after flowering; corolla small, light blue, obconical, incised for two-thirds into lobes, concealed in calyx-tube; capsule long-obconical, 5-ribbed, trilocular; seeds ovate. May—June. (Plate XXI, Figure 3.)

Dry solonetzic places in the lower mountain belt. — Caucasus: E. Transc. (near Kirovobad); Centr. Asia: Pam.-Al. (Gissar Range). **Gen. distr.:** Spain, N. Africa, As. Min. Described from Spain. Type in Geneva, topotype in Leningrad.

Note. A plant very rare in the Caucasus: according to Fomin (op. cit.) it is nearly a weed and therefore might have been introduced (from Asia Minor). It was found by Hohenacker in great quantities near the Elenendorf
343 colony (now Khanlar) and never collected after that. In the Leningrad herbarium there is a specimen with Lagovskii's label reading "near Nukhi above Kit-Chai River where it meets the Agry-Chai." However, in spite of the seeming accuracy of the label, it is probably false as is always the case with this collector.

According to Popov (op. cit.) this short bellflower of Central Asia is related to those species "which are characterized as having derived strains in Central Bukhara" (Tadzhikstan). It was found near Derbent in Baisun District at the foot of Gissar Range and hence it may not be an introduced weed, its presence here is another indication of the ancient Mediterranean floristic influence on the flora of Central Asia and Transcaucasia.

Genus 1434. **ADENOPHORA*** FISCH.

Fisch. in Mém. Soc. Nat. Mosc. IV (1823) 165; idem in Ldb. Ind. sem. hort. Dorpat. (1822) 1, nom. nud.; A. DC. Monogr. (1830) 354; Korsh. Mém. Acad. Sc. Pétersb. 7 sér. XLII, 2 (1894) 26.— Floerkea Spreng. Anleit. II (1818) 523, non Willd. (1801).— Floerkea Spreng. Syst. I (1825) 735, pro sect. gen. Campanulae.

Calyx quinquepartite. Corolla shallowly divided into 5 lobes, campanulate, tubular-campanulate or infundibular. Stamens free but adjacent, filaments erect, membranous, strongly ciliate; anthers elongate. Style encircled at base by hollow cylindrical (tubiform or cup-shaped) or annular disk (nectary), usually distinctly exserted or as long as corolla, when young finely pubescent above with hairs in 10 longitudinal rows, later glabrous. Stigma tripartite. Ovary trilocular. Capsule opening at lower part by 3 valves or pores. Seeds ovate, more or less flattened. Perennial herbs, with thickened roots, erect stems, alternate or whorled leaves and usually racemiform or paniculate inflorescences. Floral buds clavate.

Eurasian Palearctic genus with many species in E. Asia (China, Korea, parts of Japan and India, Himalayas).

Most of the species of this genus in the USSR are found in Transbaikalia, in particular in the Far East. One species grows in Sakhalin and the Kurile Islands.

There are over 50 species in Adenophora.

Type species of genus: A. stenanthina (Ldb.) Kitagawa.

* From the Greek aden — gland and phoreo — carry.

344 Note. Fischer (l. c.) based the genus Adenophora on the type A. marsupiiflora Fisch. Later it became clear that the earlier described Campanula stenanthina Ldb. by Ledebour was the same plant as A. marsupiiflora.

Adenophora was established in 1816 by Fischer on account of the presence of a disk (nectary). He informed Sprengel of this in a letter and also sent him the plant. The latter was quick to publish the new genus under the new name Floerkea (1818). Thus, Fischer's priority would have been lost were it not for the earlier honomym Floerkea Willd. (1801).

The genus is taxonomically regarded as a difficult one, mainly because of its extensive and complicated synonymy. Actually, its species are rather sharply distinguished and much easier to recognize in nature and herbaria than one would expect. A great deal of work for an understanding of the taxonomy of this genus was accomplished by Korzhinskii (op, cit.) in his monograph on ladybells endemic to the USSR. Although he reduced all the taxa of Adenophora to five basic ancestral species with many conjectured hybrids between them, he clearly described the sharply defined morphological and geographical differences of these forms and hybrids, even if he did not separate them into specific geographical races. This is apparent, not only from his monograph, but also from the notes he made at different herbaria. His work considerably facilitated our determination of the taxonomic composition of Adenophora. We proceeded not on Korzhinskii's basis seeking for the origin of this or that form or species, sometimes quite successfully, but in order to establish concrete taxa of the genus, regardless of their origin, whose areas of distribution were independent and included the USSR. We endeavored to establish binary specific names for these taxa to which Korzhinskii had given a complex nomenclature in his attempt to present hybrid combinations.

In addition, the recent investigations of Sukachev (see reference in A. taurica) indicate that although hybridization between the species of Adenophora is possible it is very difficult and often accompanied by a break-down of progeny, or, more important, by the formation of apomictic forms. The presence of apomictic, and hence completely constant, forms among the species of Adenophora makes it more justifiable for the botanist to establish concrete, taxonomic units than to adopt Korzhinskii's
345 hypothesis in its original form. Accordingly, we have re-introduced many of the species names rejected by Korzhinskii and applied them to clearly defined geographical races; to many races the earlier epithets have been given, in accordance with the acting nomenclatural rules.

There is no up-dated and complete study of Adenophora. Following the first survey of the genus by Fischer (op. cit.) under the title "Adumbratio generis Adenophorae," Alphonse de Candolle's general monograph (l. c.) and the above-noted work by Korzhinskii, the only works on the subject, have been a few articles or treatments in different "floras," including new data on taxonomy of the genus, and contributed by such authors as Komarov, Freyn, Nannfeldt, Nakai, Kitagawa, Reverdatto, Borbás, and others. From the nomenclatural aspect, the most important of these works is Kitagawa's "Lineamenta Florae Manshuriae" (1949). Borbás work (Magyar Bot. Lapok, No. 6/7 (1904)) appears to be a critical study, but in many ways it is a step backward as far as the taxonomical perception of Adenophora is concerned. The confusion and inaccuracies in the names, the failure to establish sections and other taxonomic subdivisions, the useless descriptions of "new species," and the erroneous conception of Adenophora as a subgenus of Campanula – all this combined considerably lessen the value of his endeavors.

In spite of these shortcomings, we are using some of Borbás' taxonomic subdivisions in our new system of Adenophora, basing our choice on the principle of priority and conception of the type. The first outline of a correct system for this genus was drawn up much earlier by Korzhinskii, who classified the species of Adenophora in three natural groups ("series") according to the shape of the disk (nectary). Unfortunately, in describing these subdivisions he neglected to name them. We are preserving these subdivisions in essence, ranking them as sections and supplying names which we have adopted from Borbás epithets for more or less similar species groups. Thus, Korzhinskii's "Series III" now becomes sect. Microdiscus Fed., preserving its original volume; "Series II" becomes sect. Thyrsanthae (Borbás) Fed., but splitting from it the section Platyphyllae (Borbás) Fed. Korzhinskii's original "Series I" is preserved intact under the name sect. Pachydiscus Fed. Within some of these sections we present new series with their corresponding names.

346 Some of the species of Adenophora, notably A. liliifolia, and probably those similar to it like A. lamarckii, A. taurica and A. tricuspidata, may be of secondary importance as food plants. The fleshy root of Adenophora is sweet and has a biscuit taste. During the Second World War the roots were tried as food-resource in Kazan.

All the species of Adenophora are suitable as ornamental plants in gardens and parks.

1. Cauline leaves in whorls . . . . 2.

\+ All leaves alternate . . . . 3.

2. Floral branches in regular whorls; corolla small, narrowly campanulate, ca. 1 cm long; flowers with markedly exserted style . . . . 9. **A. tetraphylla** (Thunb.) Fisch.

\+ Branches of inflorescence not whorled, paniculate; corolla rather large, 1.5–2 cm long, broadly campanulate; style hardly exserted or shorter than corolla . . . . 10. **A. pereskiifolia** (Fisch.) G. Don.

3. All cauline leaves with well-developed petioles, thin, oblong-ovate, cordate, long-mucronate; flowers with broad corolla . . . . 8. **A. trachelioides** Maxim.

\+ Cauline leaves sessile or short-petioled or partly short-petioled . . 4.

4. Disk (nectary) thick, short-cylindrical, ca. 4 mm in diameter and long lanate-pubescent. Glaucescent glabrous plant, 15–40 cm. Lower leaves ovate-lanceolate, petiolate, remotely dentate, the rest linear-lanceolate. Flowers solitary or few, 2–2.5 cm long . . . . 16. **A. himalayana** Feer.

\+ Disk (nectary) narrowly tubular or short but then not more than 1 mm in diameter and long, glabrous or subglabrous . . . . 5.

5. Cauline leaves linear, entire or subentire . . . . 6.

\+ Cauline leaves lanceolate, ovate, oblong, always serrate or dentate . . . . 7.

6. Corolla narrow, tubular-campanulate, constricted below lobes; style considerably exserted; inflorescence usually with drooping flowers . . . . 11. **A. stenanthina** (Ldb.) Kitagawa.

\+ Corolla broad, campanulate, not constricted below lobes; style as long as corolla; inflorescence compressed, often racemiform . . . . 13. **A. gmelinii** (Spreng.) Fisch .

347 7. Cauline leaves lanceolate, oblong- or elongate-lanceolate, sometimes sublinear, remotely runcinate-dentate; plant 70 cm and taller. Flowers 2 cm long, with broadly campanulate corolla concealing the style; sepals lanceolate, rather broad . . . . 14. **A. coronopifolia** Fisch.

\+ Cauline leaves ovate, obovate, rhombic, oblong, if narrowly lanceolate then plant up to 15–20(30) cm . . . . 8.

8. Leaves narrowly lanceolate, arcuately curved, folded along midrib, acutely toothed, attenuate at apex; plant 15–20(30) cm, multicaulescent; flowers campanulate, with exerted style . . . . 15. **A. rupestris** Reverd.

\+ Leaves different; plant 70–150 cm . . . . 9.

9. Leaves markedly crisp, ca. 2–3 cm long, with acute, even aristate teeth, caudate, attenuate and acute at apex, sessile; stems ca. 70 cm; inflorescence with thin branches; flowers ca. 1 cm, with narrow corolla and exserted style . . . . 12. **A. crispata** (Korsch.) Kitagawa.

\+ Leaves not crisp, usually longer than 2–3 cm . . . . 10.

10. Style considerably or distinctly exserted . . . . 11.

\+ Style as long as or shorter than corolla or hardly exserted . . . . 14.

11. Leaves dense, imbricate, irregularly toothed, sessile, 2–3–4 cm long, 1–2 cm wide; inflorescence usually 5–10-flowered, rarely with more; corolla tubular-infundibular; stem 50 cm . . . . 2. **A. taurica** (Sukacz.) Juz.

\+ Leaves less dense, not imbricate; plant 70–150 cm . . . . 12.

12. Calyx teeth lanceolate, rather wide, erect; cauline leaves ovate-oblong, elongate, sessile, serrate-dentate; inflorescence large, branching, paniculate, multiflorous . . . . 1. **A. liliifolia** (L.) Bess.

\+ Calyx teeth narrowly linear-subulate, long, declinate or recurved . . . . 13.

13. Leaves very large-toothed, teeth triangular, 1–1.5–2 cm long, with curved callous mucro, 7–8(9) at each side . . . . . . 7. **A. jacutica** Fed.

\+ Leaves contiguously bidentate, teeth 3–5 mm long, without mucro, 12–15 at each side . . . . 6. **A. sublata** Kom.

14. Calyx teeth incised-serrate or with long lateral teeth; leaves dense,
348 nearly imbricate, oblong-elliptic, regularly large-toothed . . . . 3. **A. tricuspidata** (Fisch.) A. DC.

\+ Calyx teeth entire or very small-toothed (teeth visible only under magnification) . . . . 15.

15. Leaves 7 cm long, 2 cm wide, not imbricate, ovate or broadly lanceolate, dentate. Flowers 1.5–2.3 cm long; calyx teeth declinate . . . . 5. **A. lamarckii** Fisch.

\+ Leaves 10 cm long, 6 cm wide; robust plant; stem ca. 1 cm thick at base and 120 cm high. Flowers 2–5–3 cm long; calyx teeth spreading . . . . 4. **A. golubinzevaena** Reverd .

Section 1. MICRODISCUS Fed. sect. nov. in Addenda XXIII, p. 337. – Platyphyllae Borbás in Magyar Bot. Lapok, 6/7 (1904) 191, 192, ex p. (pro subsect. gen. Campanulae s. latiore). – Disk (nectary) small, short, ca. 1 mm in diameter and long. Calyx teeth ovate-lanceolate, lanceolate or linear, erect or declinate, much shorter than the campanulate corolla, sometimes with lateral teeth; flowers in raceme or many-flowered in panicle. Cauline leaves alternate, ovate, elliptic, oblong or broadly lanceolate, serrate or dentate, sessile or distinctly petioled.

Type species of section: A. liliifolia (L.) Bess.

Note. This section more or less corresponds with the group Alternifoliae Borbás, which Borbás placed in the subsection Platyphyllae of the section Thyrsanthe; the taxonomic rank of Alternifoliae Borbás is not indicated, although there is a brief description, and it is impossible to determine precisely this rank (in all probability it corresponds to the series); we therefore feel that the name Alternifoliae Borbás cannot be used for our section and we give it the name Microdiscus.

Series 1. **Liliifoliae** Fed. – Style distinctly exserted, at least at flowering; corolla campanulate or narrowly so; inflorescence many- or few-flowered. Cauline leaves sessile or the lower partly short-petioled.

1. A. liliifolia (L.) Bess. Enum. pl. Volh. (1822) 90; Szaf., Kulcz., Pawl. Rosl. Polskie, 646; Nekrasova in Fl. yugo-vost. VI, 291; Maevskii, Fl. ed. 8-e, 548. – A. liliifolia Fisch. apud Ldb. in Ind. sem. hort. Dorpat. (1822) 1; A. DC. Monogr. (1830) 358; idem in DC. Prodr. VII, 2, 492; Ldb. Fl. Ross. II, 2, 894; Korsh. in Mém. Acad. Sc. Pétersb. 7 sér. XVII, 2 (1899) 39; ej. Tent. Fl. ross. or. 275; Fedch. and Fler., Fl. Evrop. Ross. 940; Syreishch., Ill. fl. Mosk. gub. III, 229; Kryl., Fl. Zap. Sib. XI, 2646,
349 p. p. – A. lillifolia f. genuina Korsh. l. c. (1899) 40. – A. l. f. stylosa Korsh. l. c. (1899) 40. – A. l. var. spreta A. DC. l. c. (1830) 359. – A. l. var. infundibuliformis A. DC. l. c. (1830) 359. – A. communis Fisch. in Mém. Soc. Nat. Mosc. IV (1823) 168. – A. stylosa Fisch. l. c. – A. polymorpha Ldb. Fl. alt. I (1829) 246, p. p. – A. suaveolens C. A. M. in Beitr. Pflanzenk. d. Russ. Reichs, IX (1854) 20. – A. polymorpha var. denticulata stylo exserto Herd. in Tr. Peterb. bot. sada, I (1873) 309, non A. denticulata Fisch. – A. communis var. lamarckii stylo exserto Herd. op. cit., non A. lamarckii Fisch. – A. communis var. integerrima Trautv. in Bull. Soc. Nat. Mosc. IV (1866) 406. – Campanula liliifolia L. Sp. pl. (1753) 233; Roem. et Schult. Syst. V, 110; Borbás in Magyar Bot. Lapok, 6/7 (1904) 195. – C. suaveolens Gilib. Fl. lithuan. I (1781) 50. – C. intermedia Roem. et Schult. l. c. non Ldb. – C. fischeri Roem. et Schult. l. c. 116. – C. spreta Roem. et Schult. l. c. 123. – C. stylosa Lam. Encycl. Meth. I (1783) 580. – Ic.: Ker-Gawl. Bot. Reg. tab. 236; Rchb. Ic. Fl. Germ. XIX, tab. 1618; Waldst. et Kit. Pl. rar. Hung. tab. 247; A. DC. l. c. (1830) tab. I, f. A (analysis); Fedch. in Fler., op. cit. Fig. 940 (mediocr.); Syreishch., op. cit. 229 (bona). – Exs.: Tarachkov and Poganko, Gerb. Orlovsk. gub. No. 216; Dorfler, Herb. norm. No. 4896; Fl. exs. austro-hung. No. 2236; Hayek, Fl. stir. exs. No. 578; Woloszczak, Fl. pol. exs. No. 847.

Perennial; root thick, fusiform or radish-shaped, simple or branching, with sweetish dry flesh; stems 50—70—150 cm, erect, simple or branching, leafy, cylindrical, longitudinally striate, glabrous; leaves glabrous or sparingly fine-soft-hairy at margins and along nerves; radical leaves sometimes drying early, petiolate, blade cordate-orbicular, dentate; cauline leaves spreading, not overlapping, ovate-elliptic or broadly lanceolate-oblong, rather large-toothed or serrate-dentate, sessile or sometimes short-petioled; upper leaves always sessile. Flowers drooping, ca. 1.5 cm, sometimes pleasantly aromatic, with thin pedicels in many-flowered pyramidal panicle; calyx completely glabrous, oval, teeth declinate, triangular at base, narrowly lanceolate, acuminate, sometimes very finely serrate or entire at margins, nearly twice as long as tube and much shorter than corolla; corolla campanulate, blue-gray or bluish, shallowly divided into slightly declinate or erect broad lobes; disk (nectary) glabrous, small, short-cylindrical; style at flowering distinctly, sometimes even markedly exserted; seeds flattened, rufous. June—July.

Inundated and forest meadows, shrubby formations, broad-leaved forests with sparse herbs, especially in birch and oak forests, sometimes pine
350 groves and spruce forests mixed with broad-leaved species, sometimes in steppes. — European part: V.-Kama, U. V., U. Dnp., M. D., V.-Don, Transv., L. V. (N.), Bl., U. Dns., Balt. (S.), L. Don; W. Siberia: U. Tob., Irt., Alt.; Centr. Asia: Balkh., Dzu-Tarb. **Gen. distr.**: Centr. Eur., Bal.-As. Min. (NW), Med. (N. Italy). Described from "Tartary and Siberia." Type in London.

Note. The northern border of the distribution of this species in the USSR extends from the Baltic area through Belorussia to Kostroma and up to the Urals approaching 60°N. Then, approximately from Tyumen, it lies to the southeast in N. Kazakhstan and reaches Altai where it grows with A. lamarckii which is predominant in the Sayan Mountains and Transbaikalia and also encountered in N. Mongolia. From Altai the two species are distributed in the mountains of Saur, Tarbagatai and Dzungarian Ala-Tau, extending nearly up to Tien Shan which is the southern border of A. liliifolia in Asia. A. liliifolia is rare in Siberia, in the north it nearly extends to the Yenisei. Earlier reports on the extension of this species to the Far East, China and Japan are erroneous. Here, other species are found, and the geographical vicarious species of A. liliifolia are A. tricuspidata Fisch., partly A. sublata Kom. and others.

A. liliifolia's southern border in the European part of the USSR extends from Bessarabia to Stalingrad, never reaching the Crimea or Caucasus. The separate but very close, high mountain species A. taurica (Sukacz.) Juz. is found in the Crimea in complete isolation, and it is completely absent in the Caucasus in spite of older data on Adenophora.

Roemer and Schultes (l. c.) considered that the epithet "liliifolia" was a result of a slip of the pen (lapsu calami) since the leaves bear little resemblance to those of the lily, rather the similarity lies in the shape of the flowers. Therefore, it seems more likely that Linnaeus accidentally wrote "liliifolia" instead of the epithet "liliiflora." However, in his monograph on Adenophora, Fischer (l. c.) correctly noted that the flowers of Adenophora were not in the slightest similar to those of the lily but the leaves of this and many other species were reminiscent of those of the lily because of the whorls and frequent absence of radical rosettes, so that the epithel "liliifolia" might actually be applicable to all species of Adenophora.

Of the two names proposed simultaneously and independently for A. liliifolia by Ledebour and Besser, the latter is preferred because it is accompanied by a reference to a basonym which is the Linnaeus name Campanula liliifolia L.

351 **2. A. taurica** (Sukacz.) Juz. in Bot. mat. Gerb. Bot. inst. AN SSSR, XIII (1950) 301, descr. nostra in Addenda XXIII, 469. — A. liliifolia ssp. taurica Sukacz. in Poplavskaya, Spisok rast. Krymsk. zapov. (1931) 39, in nota; Sukach. in Bot. zhurn. SSSR, XXV, 4—5, 297, in obs. — A. liliifolia auct. non Bess. nec Ldb.: A. DC. Monogr. 894, p. p.; Fedch. and Fler., Fl. Evrop. Ross. 940, p. p. — Campanula liliifolia auct. non L.: M. B. Fl. taur.-cauc. III (1819) 139. — Ic.: Sukach., op. cit. fig. 2 (photo) et f. 3, sinistra.

Perennial; root thick, 2 cm across, radish-shaped, sometimes branching in lower part, dark brown, rugose; stems usually few, simple, rarely slightly branching, 0.3 cm thick, usually not more than 50 cm, longitudinally striate, densely leafy, few-flowered; leaves all cauline, densely covering the middle part of stem, overlapping, 2—3—4 cm long, 1—2 cm wide, elliptic, acute, short-petioled, irregularly large-toothed at margins, with acute contiguous teeth, paler beneath than above; upper leaves lanceolate, sessile, more or less dentate, the lower early drying. Inflorescence racemiform, 5—10-flowered, rarely with more; flowers usually 1.3—1.5 cm long; calyx obconical, glabrous, teeth lanceolate-triangular, acute, as long as tube and much shorter than corolla; corolla glabrous, tubular-infundibular or campanulate but not broad, divided for one-fifth into acute short lobes; style much exserted. July. (Plate XXI, Figure 1.)

Mountain meadows and near the upper timberline. — European part: Crim. (Yaila). Endemic. Described from top of Bol'shaya Chuchel' Mountain. Type and topotype in Leningrad.

[Note]. Sukachev (Poplavskaya, l. c.) first designated this species as ssp. taurica, restricted to the peaks of the Crimean mountains, in particular Bol'shaya and Malaya Chuchel', but even here it is rather rare. It differs from A. liliifolia by its lower stature, poor inflorescence and small approximate leaves. Sukachev grew this plant in Leningrad alongside the type A. liliifolia and established that A. taurica retained all its typical characters and sharply differed from A. liliifolia. Thus, A. taurica is completely isolated from the lowland species of the Crimea and is a race with constant hereditary characters.

The first description of this species (as a subspecies), which was made by Sukachev (see Poplavskaya, op. cit.), is extremely brief. In view of the shortcomings of that diagnosis we include in the Addenda (see) a detailed description in Latin.

352 Series 2. **Tricuspidatae** Fed. — Style as long as or shorter than corolla or barely exserted; corolla broadly campanulate; inflorescence usually multiflorous. Cauline leaves sessile. Calyx teeth sometimes with lateral dentiform lobes.

3. A. tricuspidata (Fisch. ex Roem. et Schult.) A. DC. Monogr. (1830) 355; idem in DC. Prodr. VII, 2, 492; Kitagawa, Lineam. Fl. Manshur. 418. — A. denticulata Fisch. in Mém. Soc. Nat. Mosc. VI (1823) 167; Turcz. El. baic.-dahur. II, 188; Ldb. Fl. Ross. II, 2, 894; Korsch. in Mém. Acad. Sc. Pétersb. 7 sér. XLII, 2, 37, p. p.; Kom., Fl. Man'chzh. III, 2, 561; Kom. and Alis., Opred. rast. Dal'nevost. kr. II, 994. — A. communis var. denticulata Trautv. in Tr. Peterb. bot. sada, VI, 1 (1879) 97, p. p. — A. polymorpha var. denticulata α genuina Herd. in Tr. Peterb. bot. sada, I, 2 (1872) 308. — A. turczaninowi Feer in Engler's Bot. Jahrb. XII (1890) 618. — A. dahurica Turcz. in schedis. — Campanula tricuspidata Fisch. ex Roem. et Schult. Syst. V (1819) 116; Borbás in Magyar Bot. Lapok, III (1904) 193. — C. pilosa Less. in Ldb. Fl. Ross. II, 2 (1845—1846) 894, pro syn. non Pall. ex Roem. et Schult. l. c. (1819). — C. denticulata Spreng. Syst. I (1825) 735. — C. monadelpha Pall. ex Fisch. l. c. (1823) 168, pro syn. p. p. — C. richteri Borbás l. c. I (1902) 253 et III (1904) 193; idem apud Freyn in Oesterr. Bot. Zeitschr. (1890) 77, nom nud.; — Ic.: Rchb. Hort. bot. tab. 2. — Exs.: GRF, No. 1230.

Perennial; root thick, sometimes branching, whitish, rugose; the whole plant glabrous or scarcely hairy; stems straight, simple in lower part, branching above, 70—150 cm; leaves alternate, pale beneath, glabrous or sparingly pubescent along margin, sessile, oval or oval-lanceolate, acute, largely and acutely serrate, 2—3.5(7) cm long, slightly involute, stiff; radical leaves suborbicular, but rarely developing. Inflorescence branching, paniculate; flowers generally drooping; calyx completely glabrous, ovate, teeth recurved, oval or triangular-lanceolate, acute, incised-serrate or dentate or entire, nearly twice as long as tube and much shorter than the blue campanulate corolla; stamens half as long as corolla; disk (nectary) small, glabrous, thick; style nearly as long as corolla; capsule subglobose, crowned by relics of calyx teeth and dried corolla. July.

Broad-leaved forests, predominantly oak forests, thickets, meadows. — E. Siberia: Ang.-Say. (only up to Krasnodarsk in the west), Dau.; Far East: Ze.-Bu., Uda, Uss. **Gen. distr.**: NE Ch. Especially abundant in Dauria. Described from Dauria (Nerchinskii zavod). Type in Leningrad.

Note. In outward appearance some forms of A. tricuspidata are easily confused with A. coronopifolia but they are distinguished from
353 this species by the shape of the disk (nectary), which in A. tricuspidata is short and small and in A. coronopifolia narrowly tubular.

Fischer (l. c.) described A. tricuspidata under two different names though intending one and the same species in both cases, having the same specimen (type). The first name, Campanula tricuspidata, with a short description was sent by him to Roemer and Schultes (l. c.). Later he renamed his plant A. denticulata; this is clearly obvious from the comparison of names in his work and also from his inscription on the label of the authentic specimen: "denticulata mihi, tricuspidata olim."

Feer's (l. c.) A. turczaninowi is apparently a poorly developed (1-flowered) specimen of A. tricuspidata, which the author found among other specimens collected by Basnin (Turchaninov herbarium) in Transbaikalia ("prope acidulam Pogromecensem"). The specimens in Leningrad bearing such labels vary in size and number of flowers (there are also few-flowered specimens) but unquestionably they all refer to A. tricuspidata, even those marked by Turchaninov himself as "A. dahurica."

One of the forms related to A. tricuspidata, collected near Nerchinsk in Siberia, was described by Borbás (l. c.) as the species Campanula richteri. Judging from the specimens of Adenophora in our possession, which were collected near Nerchinsk, not one of them could be regarded as a "new species"; it is obvious from the diagnosis of C. richteri Borbás that the plant described by Borbás differs from the typical A. tricuspidata only by having more dissected calyx lobes. We therefore put C. richteri Borbás in the synonymy of A. tricuspidata. It should also be recalled that Nerchinsk, from where the so-called C. richteri was described, is a classical locality (locus classicus) for A. tricuspidata and it is very unlikely that two species would originate in one and the same geographical area.

4. **A. golubinzevaeana** Reverd. in Sist. zam. gerb. Tomsk. univ. 3–4 (1935) 4; Kryl., Fl. Zap. Sib. XI, 2648. – A. golubinzevaeana var. pilosa Reverd. op. cit. – Ic.: Reverd., op. cit. tab. separata, sine numero.

Perennial; stems usually few, 120 cm, 1 cm thick in lower part; root thick, grumosely tuberculate, yellow at surface, ca. 5 cm in diameter; leaves sessile, alternate, stem nearly approximate in upper part; median leaves 10 cm long, 6 cm wide, all leaves with fine short hairs along nerves and at margin, large-toothed, teeth slightly contiguous. Flowers in short raceme, pediceled; bracteoles oval or lanceolate-oval, long-acuminate, short-ciliate
354 and with hard teeth; calyx glabrous or pubescent (var. pilosa), teeth acutely dentate, with thin cartilaginous teeth, nearly as long as tube, slightly shorter than corolla; corolla broadly campanulate, blue, large, reaching 2 and even 3 cm, shallowly divided into broad, somewhat acute lobes; style as long as or shorter than corolla. August.

Subalpine meadows near the upper timberline. – W. Siberia: Kuznetsk Ala-Tau. Endemic. Described from Sakhchakh Mountain (upper reaches of Askyz River, tributary of the Abakan River in Kuznetsk Ala-Tau). Type in Tomsk.

Note. Sharply differing from most of the other species and in particular from the taxonomically close A. liliifolia and A. tricuspidata by having corollas unusually large, sometimes up to 3 cm long and widely open, and very large leaves. Undoubtedly, it is most related to A. lamarkii which sometimes has flowers just as large but its leaves are always narrower and differently toothed.

Besides the holotype and paratype reported from Khakass by Reverdatto in his diagnosis there are other specimens of this species. Undoubtedly the plants collected by Il'in in 1914 in the former Marinsk County refer to this species as do those from the Turkevich herbarium which were found near Arak in the former Achinsk County.

5. **A. lamarckii** Fisch. in Mém. Soc. Nat. Mosc. VI (1829) 168; A. DC. Monogr. 357, excl. var.; idem in DC. Prodr. VII, 2, 492; Ldb. Fl. alt. I, 245. – A. polymorpha var. lamarkii Herd. in Tr. Peterb. bot sada, I, 2 (1872) 309, partim. – A. communis var. lamarkii Trautv. in Tr. Peterb. bot sada, VI (1879) 98, partim. – A. denticulata × liliifolia f. lamarkii Korsh. in Mém. Acad. Sc. Pétersb. 7 sér. XLII, 2 (1894) 38. – A. liliifolia var. lamarkii Kryl. Fl. Alt. III (1904) 782, excl. f.;

Kryl., Fl. Zap. Sib. XI, 2647, excl. f. et var. — Campanula liliifolia Lam. Encycl. Méth. I (1783) 581, non L. — C. lamarkii Borbás in Magyar Bot. Lapok, III (1904) 193, p. p. — C. lamarkii var. botryantha Borbás l. c. — Ic.: Gmel. Fl. Sibir. III, tab. 26 (mediocr.).

Perennial; root thick, rugose, whitish, with dry mealy flesh; stems
usually ca. 70 cm, sometimes 30–100, glabrous, smooth, leafy, simple at base
but usually forming inflorescence above with short and thin branches;
radical leaves deciduous; cauline leaves alternate, numerous, ovate-lanceo-
late, ca. 5–7 cm long, 2 cm wide, acutely serrate, paler beneath than above,
slightly curling and white-ciliate at margin, glabrous at surface. Flowers
drooping, in many-flowered raceme or panicle, terminal and axillary, short-
pediceled; calyx slightly ovoid or obconical, longitudinally striate, teeth
355 lanceolate, erect and entire, nearly half as long as tube and much shorter
than corolla; corolla blue, broadly open, infundibular-campanulate,
1.5–2(3) cm long, shallowly divided into ovate erect and rather acute lobes;
filaments dilated from base to middle, pubescent; anthers yellow, shorter
than filaments; disk (nectary) cylindrical, thick; style as long as corolla or
shorter.

Predominantly in broadleaved and open forest and at forest edges. Rarely found in mountainous taiga, often in foothills and plains. — W. Siberia: Alt.: E. Siberia: Ang.-Say. (east to Baikal); Centr. Asia: Balkh., Dzu-Tarb. **Gen. distr.:** Mong. (N.). Described from "Siberian Altai." Type in Leningrad.

Note. This species is intermediate in characters between A. tricuspidata and A. liliifolia which is why Korzhinskii (l. c.) treated it as one of the conjectured hybrids between the two species. This may be so, but in view of the distinct constancy of A. lamarckii and its well-defined distribution area, it would be more correct to regard it as a separate species and maintain its original name. It is clearly distinguished from A. liliifolia by its broadly open and rather large flowers without exserted style.

The specimens from the Sayans and Tuva have the largest corollas, in addition to having grayish incised-crenate leaves.

The species epithet most often used was "Lamarkii," but its author Fischer (l. c.) in the first diagnosis wrote "Lamarckii," and since this contains no error it should be used.

Series 3. **Sublatae** Fed. — Style considerably exserted; corolla infundibular-campanulate; inflorescence few-flowered. Cauline leaves sessile, broad, usually large-toothed. Calyx teeth linear-subulate, much recurved.

6. A. sublata Kom. in Bot. mat. Gerb. Glavn. bot sada, VI, 1 (1926) 13; Kom. and Alis., Opred. rast. Dal'nevost. kr. II, 994.

Perennial; rhizome densely covered above with rufous-brown scarious
acute scales; stems erect, longitudinally striate, slightly sulcate below,
3 mm thick at base, tapering toward apex, ca. 1 mm thick beneath inflores-
cence, 30–50 cm long, leafless and glabrous below, rather densely leafy and
356 sparsely short-hairy above; leaves subsessile, alternate, obovate or broadly
lanceolate, cuneate toward base, acute, acutely toothed or serrate-dentate
at margin, twice as long as internodes, sparsely and finely pubescent
beneath, paler beneath than above. Raceme simple or sometimes slightly

branching with axillary branches; pedicels 5–7 mm long, erect, or arcuate-curved, with lanceolate-linear entire marginal or denticulate bracts and bracteoles; calyx elongate, obconical, glabrous, teeth linear, acute, recurved, as long as or longer than tube but much shorter than corolla; corolla glabrous, campanulate, 1.3 cm long; disk (nectary) short-cylindrical; style long, markedly exserted, slightly thickened, sometimes arcuate under stigma. August.

Mostly in oak forests, often in birch, mixed and coniferous forests, but not in shady and mossy places. Up to the upper timberline. – Far East; Uss., Uda. Endemic. Described from the Botchi River valley in Sikhote-Alin Range.

Note. This species is sometimes confused in herbaria with A. pereskiifolia, but it is immediately distinguished by the nonwhorled leaves, hardly branching or simple inflorescence and dentate bracts. In many characters except for the shape of the leaves it resembles A. jacutica Fed. (see Note to this species); it is less reminiscent of A. liliifolia, but sometimes it is quite similar so that even Komarov himself assumed that his own A. sublata was A. liliifolia, which in fact is completely absent in the Far East.

7. **A. jacutica** Fed. sp. nov. in Addenda XXIII, p. 338.

Perennial; almost completely glabrous plant, with thick radish-shaped slightly transversely rugose grayish-brown root; stems erect, simple or barely branching at apex with thin branches, cylindrical, glabrous and smooth, 2–3 mm thick, leafy, with brown squamous relics of dry petioles at base; radical leaves none or deciduous; cauline leaves alternate, subsessile or very short-petioled, rhombic, cuneate at both ends, deeply large-toothed, teeth 1–1.5–2 cm long, 7–8 at each side, triangular, erect, abruptly tapering at tip above into curved callous tip. Flowers drooping, with short thin
359 pedicels, in few-flowered raceme or a poor panicle, unattractive; calyx oval, glabrous, teeth declinate or spreading, linear, mucronate, entire, as long as tube and much longer than the campanulate corolla; stigma branching; capsule oval, membranous, opening by small pores at base and crowned with reflexed or spreading calyx teeth. July.

River valleys, Yezo spruce forests, not at valley bottoms, but on adjacent mountain slopes. – E. Siberia: Lena-Kol. (SE). Endemic. Described from Uchersk District in Yakutsk (along Uchuru River near Muguskan). Type in Leningrad.

Note. A. jacutica is taxonomically close to A. sublata Kom., but differs from it by its large leaf teeth, 1–1.5–2 cm long, declinate, not contiguous, abruptly tapering into curved callous mucro. Besides, there are 7–8 teeth on each side of the leaves, whereas in A. sublata there are usually 12–15 and more. The depth of the notches between the teeth in A. jacutia is always one-quarter the width of the lamina and in A. sublata it is not even one-tenth and often less.

Series 4. **Remotiflorae** Fed. – Style as long as corolla or exserted; corolla campanulate; inflorescence few-flowered. Cauline leaves distinctly petiolate, broad, ovate, attenuate at apex. Calyx teeth oblong or lanceolate.

PLATE XXI. 1 – Adenophora taurica (Sukacz.) Juz., plant, inflorescence, pedicels (cross section); 2 – A. pereskiifolia (Fisch.) G. Don.; 3 – Brachycodon fastigiatus (Duf.) Fed., plant and flowers.

If A. remotiflora Miq. is accepted as an independent species, then it is a vicarious (Chinese) of the series.

8. A. trachelioides Maxim. Prim. Fl. Amur. (1859) 186, in nota; Kitagava, Lineam. Fl. Manshur. 418. — A. isabellae Hemsl. in Journ. of Bot. ((1876) 207. — A. remotiflora auct. non Miq.: Kom. Fl. Man'chzh. III, 2 (1907) 556; Kom. and Alis., Opred. rast. Dal'nevost. kr. 994. — A. remotiflora Kom. f. longifolia et f. cordata Kom. op. cit. (1907) 558.

Perennial; root thick; long; stems flexuous or erect or arcuate at base, 50—120 cm, with longitudinal protruding ribs, simple or slightly branching above; radical leaves present only in young specimens with undeveloped stems, with petioles 8—9 cm long, cordate-orbicular or cordate, acutely bidentate, sparsely crisp-hairy; cauline leaves alternate, petiolate, cordate or oval or oval-lanceolate, long-mucronate, largely and irregularly acutely serrate, glabrous or with sparingly short hairs, thin, coriaceous in the shade
360 and stiff and robust in the light, 15 cm long, 6 cm wide, petioles 3—5 cm long, blade decurrent at petioles. Flowers in loose terminal leafless raceme; bracts lanceolate-linear, mucronate, crenulate, ciliate at ends; pedicels uniflorous, short, pendulous; calyx teeth oblong or ovate-lanceolate, acuminate, entire, longer than ovary [tube?] and much shorter than the light blue narrowly campanulate corolla; disk (nectary) short-cylindrical; style pubescent, exserted in most of the specimens. July—August.

Shady mixed and coniferous forests, humus-rocky soil covered with moss carpet. — Far East: Uss. (west of Suifun River). **Gen. distr.**: Korea, China. Described from China (mountains west of Peking near Na Tse T'ai Hsi pagoda). Type in Leningrad.

Note. In appearance this species is very similar to Campanula trachelium. It grows only in the western part of Siberia. Kitagawa (l. c.) reports the authentic A. remotiflora Miq. from Manchuria, but in the USSR only A. trachelioides Maxim., which Komarov (l. c.) accepted as A. remotiflora, has been found thus far.

It is highly probable that the so-called A. periplocifolia which Lamarck described under the name Campanula periplocifolia (Lam. Encycl. méth. (1783) 580) is none other than A. trachelioides Maxim. A. periplocifolia still remains obscure because of the difficulty of examining the original specimen and because of the erroneous comparison with the completely unrelated species A. stylosa, i. e. A. liliifolia. Ledebourg (l. c.) wrote that he did not know ("mihi ignota") A. periplocifolia and repeated de Candolle's (l. c.) suggestion that the species was related to A. stylosa, i. e. to A. liliifolia. However, as noted, this suggestion scarcely holds true. Lamarck's (l. c.) description of this species is based on a cultivated specimen grown in Paris from seeds from Siberia. Unfortunately, the place from which the seeds were obtained is unknown. Neither Lamarck's diagnosis nor the more detailed one made by de Candolle, who saw the real specimen, affords the possibility of comparing A. periplocifolia with A. stylosa. Fischer (Adumbratio, 6) was of the opinion that Lamarck had described Campanula stylosa Lam. from the Urals ("Patria ut videtur montes Uralenses"). Lamarck noted the locality of A. periplocifolia simply as Siberia. Undoubtedly the plants from which

Lamarck established Campanula stylosa and C. periplocifolia came from different parts of Europe and Asia, from areas which were possibly very remote from each other. The characters of the so-called A. periplocifolia are actually quite different from those of A. stylosa, i.e. A. liliifolia. According to de Candolle (l. c.), A. periplocifolia
361 is characterized by petioles as long as the blades, the blade cordate, stems not straight but slightly ascending, inflorescence few-flowered. The plant is unusually small, the stems being 2–3 inches, the leaves only 5 lines in length and 4 lines in width. If the presence of calyx lobes with lateral teeth had not been reported, one could think with full assurance that Lamarck was describing under the name C. periplocifolia, a poor dwarf Adenophora trachelioides Maxim. from Siberia. Of all the Siberian species of Adenophora this is the only one with long petioled leaves and cordate blades and the only species whose specimens might be as small as Lamarck's. If our hypothesis is proven, then Adenophora trachelioides Maxim. will become a synonym of A. periplocifolia (Lam.) A. DC. (see also List of Dubious Species).

Section 2. PLATYPHYLLAE (Borbás) Fed. comb. nov. — Platyphyllae Borbás in Magyar Bot. Lapok, 6/7 (1904) 191, 192, s. restr. (pro subsect. gen. gen. Campanulae s. latiore). — Disk (nectary) tubular-cylindrical, long or abbreviated; calyx teeth linear-subulate or linear-lanceolate, declinate, much shorter than the narrowly or broadly campanulate corolla; flowers in whorls or paniculate inflorescence. Cauline leaves in whorls of 3–6, ovate, elliptic or broadly lanceolate, dentate.

Type species of section: A. tetraphylla (Thunb.) Fisch.

9. A. tetraphylla (Thunb.) Fisch. in Mém. Soc. Nat. Mosc. VI (1823) 167; Kitagawa, Lineam. Fl. Manshur. (1939) 418. — A. verticillata Fisch. l. c. 165; A. DC. Monogr. 356; idem in DC. Prodr. VII, 2, 492; Ldb. Fl. Ross. II, 892; Turcz. Fl. baic.-dahur. II, 187; Rgl. Tent. Fl. Ussur. 100; F. Schmidt in Mém. Acad. Sc. Pétersb. 7 sér. XII, 2, 155; Trautf. in Tr. Peterb. bot sada, VI, 1, 96, p. p.; Korsh. in Mém. Acad. Sc. Pétersb. 7 sér., XLII, 2, 33, p. p.; Kom. Fl. Man'chzh. III, 366; Miyabe, Fl. Kurile Isl., Mem. Boston. Soc. Nat. Hist. VI, 246; Kom. and Alis., Opred. rast. Dal'nevost. kr. 993, 994. — A. verticillata var. typica Trautv. l. c. — A. verticillata var. hirsuta F. Schmidt l. c. 155. — A. verticillata var. typica, var. angustifolia, var. subintegrifolia Rgl. l. c. 100–101. — A. verticillata var. genuina, var. media, var. denticulata, var. maritima et earum formae numerosae Korsh. l. c. 34, 35. — A. verticillata proles princeps Korsh. l. c. 35. — Campanula tetraphylla Thunb. Fl. Jap. (1784) 87. — C. verticillata Pall. Reise, III (1776) 320, 422, 436 u. Anh. 719, non Hill. (1765); Roem. et Schult. Syst. V, 116; Borbas in Magyar Bot. Lapok, III, 192. — Ic.: Pall. l. c. tab. G, f. 1; Kom., op. cit. tabl. 299, f. 3.

Perennial; rooth thick, radish-shaped, rugose; stem straight, simple but with thin branches in inflorescence, glabrous, smooth, orbicular at cross
362 section, longitudinally striate, 70–150 cm, leafy, mostly in middle part; leaves in whorls of 3–4–6, sometimes only alternate, dense, sessile or

short-petioled, elliptic or lanceolate, serrate at margin, acute or acuminate. Inflorescence many-flowered, paniculate, branched, more or less in whorls with distinct internodes between them; flowers drooping, 1—1.3 cm long, short-pediceled; bracts subulate; calyx small, blackish, obconical, glabrous, teeth spreading, narrowly linear-subulate, longer than or as long as tube but much shorter than corolla; corolla glabrous, tubular or narrowly tubular-campanulate, blue, shallowly divided into very short lobes; disk (nectary) 3 mm long, tubular; style usually distinctly exserted, thickening slightly at apex, bluish; capsule pendulous, ovoid; seeds blackish. July.

Broadleaved forests and shrubby formations. — E. Siberia: Dau.; Far East: Ze.-Bu., Uss., Uda, Sakh. (S.). **Gen. distr.**: China, Korea, Japan. Described from Japan. Type in Uppsala.

Note. This species is so very variable that Korzhinskii (l. c.) and later authors were able to describe a wide array of varieties and forms distinguished mostly by their leaves. It is sometimes confused with A. pereskiifolia which also has whorled leaves, but A. tetraphylla is easily distinguished by its tubular corolla, whorled inflorescences and small, narrowly lanceolate-subulate calyx teeth.

Fischer (l. c.) listed this species in his "Adumbratio generis Adenophorae" under the name A. verticillata which is Campanula verticillata. Although de Candolle (l. c.) established its identity with Campanula tetraphylla Thunb., the earlier name A. tetraphylla has never been used; in works pertaining to flora of the USSR it has been ignored.

10. A. pereskiifolia (Fisch. ex Roem. et Schult.) G. Don in London, Hort. Brit. (1830) 74; Rgl. Tent. Fl. Ussur. 101; Kitagawa, Lineam. Fl. Manshur. 416. — A. latifolia Fisch. in Mém. Soc. Nat. Mosc. VI (1823) 168; A. DC. Monogr. 356; idem in DC. Prodr. VII, 2, 492; Ldb. Fl. Ross. II, 2, 894; Turcz. Fl. baic.-dahur. II, 188; Maxim. Prim. Fl. Amur. 186 (et var. parvifolia Maxim. l. c.); Kom. Fl. Man'chzh. III, 564; Kom. and Alis., Opred. rast. Dal'nevost. kr. 993. — A. verticillata × liliifolia f. latifolia Korsh. in Mém. Acad. Sc. Pétersb. 7 sér. XVII, 2 (1894) 36. — A. communis var. latifolia Trautv. in Tr. Peterb. bot. sada, VI (1879) 99. — A. polymorpha var. latifolia Herd. ibid. I (1873) 306. — A. polymorpha var. verticillata Franch. et Sav. Enum.
363 pl. Jap. II (1876—1879) 422, non A. verticillata Fisch. — Campanula pereskiaefolia Fisch. ex Roem. et Schult. Syst. V (1819) 116; Borbás in Magyar Bot. Lapok, III, 192. — C. pereskia Fisch. l. c. pro syn. — C. peirescifolia Spreng. Syst. I (1825) 735. — C. rhomboidea Borbás, l. c. — Ic.: Kom. and Alis., op. cit. tabl. 299, f. 2. — Exs.: Karo, Pl. Amur. et. Zeaënsae, No. 233 (sub A. verticillata auct. non Fisch.).

Perennial; root thick, straight, fleshy, radish-shaped, white or brown, rugose; stem simple and straight below, branching in inflorescence, 70—100 cm, rounded or longitudinally striate, glabrous, leafy; leaves dense, whorled, 3—4(5) leaves per whorl, slightly shorter than internodes; radical leaves usually deciduous, suborbicular when developed; cauline leaves ovate- or sometimes lanceolate-oblong, 5—7—10 cm long, 1.5—2.3 cm wide, acuminate, serrate or largely serrate-dentate, glabrous, scabrous at margin with sparse hairs or small teeth (visible only under magnification); upper cauline leaves reduced, bracts subulate-acuminate. Flowers many, in paniculate or sometimes racemiform inflorescence, drooping, two's or three's, sometimes solitary; pedicels 0.5—1 cm long, or much shorter; calyx ovate, blackish, teeth linear-lanceolate, dark like tube, declinate or reflexed, as long as tube or often 2—4 times longer, many times shorter than corolla, sometimes with small marginal teeth; corolla broadly campanulate or infundibular, blue, glabrous, 2—2.5 cm long, shallowly divided into nearly erect ovate acuminate lobes; stamens with narrow subfiliform anthers; disk (nectary) thick, cylindrical; style as long as or hardly longer than corolla. July. (Plate XXI, Figure 2.)

Shrubby formations, broadleaved and pine forests, forest edges, dry meadows, stony slopes. — E. Siberia: Dau.; Far East: Ze.-Bu., Uda. **Gen. distr.:** Japan, China, Korea, Mongolia (E.). Described from Dauria (Nerchinskii zavod). Type in Leningrad.

Note. Korzhinskii regarded this species as a hybrid of A. tetraphylla × A. liliifolia, but Komarov (Fl. Man'chzh. 566) rightly noted that "if this plant is a hybrid, then it is a hybrid that has long since become an independent race developing normal flowers and adapted to exist where not one of its parental species is present." A. pereskiifolia differs from the Far Eastern A. tetraphylla by more oblong leaves, broader and larger corolla with style slightly exserted, calyx teeth long, reflexed, lanceolate, not subulate. It is sharply distinguished from other species by
364 its whorled leaves. In addition the disk (nectary) is shorter than in A. tetraphylla and is usually nearly quadrate in cross section.

Borbás (l. c.) described one of the herbarium specimens collected near Blagoveshchensk on the Amur as Campanula rhomboidea, though it was undoubtedly typical of C. pereskiifolia; he was impressed by the fact that in that specimen the inflorescence was disposed in whorls and the leaves were rhombic in shape. Otherwise there were no essential differences from the type.

The epithet "pereskiifolia" is not a good one. Fischer (l. c.) used the word "pereskia" in order to denote the resemblance of the whorled leaves to an umbrella (or, more accurately, an umbel). He thought that "pereskia" meant an umbrella ("pereskia id est umbrella, ob verticillos"). However, the motive of de Candolle (l. c.), Roemer and Schultes in changing the name to "pereskiaefolia" seems to stem from the generic name Pereskia. Sprengler (l. c.) was also motivated by the same idea in naming the plant "peirescifolia." At the same time de Candolle wrote that Fischer's plant did not resemble the genus Pereskia whatsoever, and questioned the association of the word "pereskia" with "umbrella." He thus considered the initial name ambiguous and Fischer's later name A. latifolia unsuitable. Taking all this under consideration the chosen Russian epithet for A. pereskiifolia is "latifolia."

Section 3. THYRSANTHAE (Borbás) Fed. comb. nov. — Thyrsanthe Borbás in Magyar Bot. Lapok, 6/7 (1904) 190, 191, s. restr. (pro sect. gen. Campanulae s. latiore). — Disk (nectary) tubular, usually long and narrow, being longer than wide, sometimes rather short. Calyx teeth linear or lanceolate, straight or curved, one-half to one-third as long as the tubular-campanulate corolla. Flowers in racemiform or paniculate inflorescence, with declinate branches. Cauline leaves alternate, oblong, linear or lanceolate, entire or dentate.

Type species of section: A. stenanthina (Ldb.) Kitagawa.

Series 1. **Stenanthinae** Fed. – Style markedly exserted; corolla narrow, tubular-campanulate or campanulate, constricted beneath lobes. Leaves linear, narrowly lanceolate and entire, or oblong, attenuate, acutely crisp-dentate at margins.

11. A. stenanthina (Ldb.) Kitagawa, Lineam. Fl. Manchur. (1939) 418; Kryl., Fl. Zap. Sib. XI, 2645. – A. marsupiiflora Fisch. apud Ldb. in Ind. sem. hort. Dorpat. (1822) 1; Fisch. in Mém. Soc. Nat. Mosc. VI,
365 165; Ldb. Fl. atl. I, 244; ej. Fl. Ross. II, 2, 893; Turcz. Fl. baic.-dahur. II, 191; Maxim. Prim. Fl. Amur. 185; Korsh. in Mém. Acad. Sc. Pétersb. 7 sér. XLII, 2, 29, quoad var. typicam et pilosam; Kom. Fl. Man'chzh, III, 558; Kryl., Fl. Alt. III, 781; Kom. and Alis., Opred. rast. Dal'nevost. kr. 994. – A. marsupiiflora var. jaluensis Kom. Fl. Man'chzh, III (1907) 560. – A. coronata A. DC. Monogr. (1830) 366; idem in DC. Prodr. VII, 2, 494. – A. montana Turcz. Fl. baic.-dahur. II (1848) 190. – A. verticillata var. marsupiiflora Trautv. in Tr. Peterb. bot. sada, VI, 1 (1879) 96, excl. syn. nonnul. – A. intermedia Ldb. Cat. hort. Dorpat (1824), non Fisch. – A. insolens Reverd. in Sist. zam Gerb. Tomsk. univ. 3–4 (1935) 3. – Campanula stenanthina Ldb. in Mém. Acad. Sc. Pétersb. V (1814) 525. – C. coronata Ker-Gawl. in Bot. Reg. II (1816) tab. 149; Borbás in Magyar Bot. Lapok, III, 191. – C. marsupiiflora Roem. et Schult. Syst. V (1819) 116. – C. sajanensis Pall. ex Roem. et Schult. l. c. 102, non Fisch. – C. gmelini Roem. et Schult. l. c. 103, non Spreng. – S. monadelpha Pall. ex A. DC. l. c. 563, pro syn. – C. syngenesiflora Pall. ex A. DC. l. c. 563, pro syn. – Floerkea marsupiiflora Spreng. Anleit. Ed. 2, II (1818) 523. – Ic.: Ker-Gawl. l. c. tab. 149; Rchb. Hort. Bot. tab. 15 (bona); A. DC. l. c. (1830) tab. 1, f. B (analysis).

Perennial; root thick, branching or simple, fusiform; stems 50–100 (150) cm, straight or ascending from an arcuate base, cylindrical, leafy, always simple below, branching at inflorescence, appearing glabrous, somewhat scabrous with minute tuberculate hairs; radical leaves drying early and deciduous or withering at flowering; lowermost leaves with orbicular to ovate blade and developed petioles; cauline leaves usually the only present, dense, numerous, alternate, straight or slightly arcuate, linear or lanceolate-linear, with convolute margins, somewhat scabrous or hardly puberulent along nerves at both sides, glaucescent, 4–10 cm long, 3–6–8 mm wide. Inflorescence a long raceme or panicle, with thin branches; flowers drooping; calyx glabrous, obconical, teeth linear, acuminate, dorsally thickened, much shorter than corolla; corolla violet-blue, tubular-campanulate, slightly attenuate at apex, 1–1.5 cm long, divided into short lobes; disk (nectary) long, tubular; style considerably exserted; capsule oval, smooth (not ribbed); seeds ovate, somewhat flattened, yellowish. July.

Sandy-gravelly riverbanks, dry riparian meadows, coastal bluffs, sporadically encountered. — W. Siberia: Alt.; E. Siberia: Ang.-Say., Dau.; Far
366 East: Ze.-Bu. **Gen. distr.:** NE China, Korea, N. Mongolia. Described from sandy banks of Yenisei River near Krasnoyarsk. Type in Leningrad.

Note. The original specimen (holotype) and all topotypes known to us have leaves which are very narrow, linear, and, like stem, sparingly pubescent; panicle spreading; corolla narrow,

We cannot regard A. insolens, described by Reverdatto (l. c.) as an independent species. The specimen on which the description is based does differ from the type of A. stenanthina (mainly by its low habit and comparatively broad corollas) but there are still not sufficient indications to rank it as a separate species because of the scanty material at hand. This form apparently has no distribution area and may possibly be an impoverished specimen of A. stenanthina.

Although we were unable to see the type A. montana described by Turchaninov (l. c.) and although this author was inclined to refer this form to A. coronopifolia rather than to A. stenanthina, we still believe this form is a typical A. stenanthina, since the locus classicus of A. montana and A. stenanthina is one and the same (Krasnoyarsk) and A. coronopifolia has never been found in that area.

12. A. crispata (Korsh.) Kitagawa, Lineam. Fl. Manshur. (1939) 415. — A, crispata Turcz. ex Ldb. Fl. Ross. II (1844—1846) 893, nom. nud. pro syn.; Turcz. Fl. baic.-dahur, II, 192, pro syn. — A. marsupiiflora var. γ Turcz. II, 192. — A. marsupiiflora var. γ Ldb. l. c. cum diagn. brev. — A. marsupiiflora × denticulata f. crispata Korsh. in Mém. Acad. Sc. Pétersb. 7 sér. XLII, 2 (1894) 32. — A. marsupiiflora var. crispata Kitagawa in Rep. First Sc. Exped. Manch. sect. 4, II (1935) 108. — A. stenanthina f. crispata Kitagawa in Rep. Inst. Sc. Res. Mangh. IV, 7 (1940) 97. — Ic.: Kitagawa, l. c. (1935) f. 14, 4.

Perennial; stem 70 cm, 3 mm thick, longitudinally thinly striate, slightly flexuous, simple or with thin branches at upper part; lower leaves falling; median cauline leaves rather dense, sessile, alternate (spirally), short-lanceolate or short-elliptic, 1.5—3 cm long, 0.7—1.3 cm wide, crisply and acutely toothed, with elongate or sometimes nearly subulate teeth, strongly caudiform-attenuate at apex; upper leaves few, smaller and narrower. Inflorescence slightly paniculate, ca. 40 cm long, with branches 5—10 cm long, thin; flowers rather numerous, 1.5—2 cm long; calyx oval, subglabrous, teeth lanceolate, erect, acute, one-third to one-half as long as corolla; corolla narrowly tubular-campanulate, glabrous, divided for one-fifth into short acute lobes; style distinctly exserted; lobes of stigma directed forward. Fl., probably in July.

367 The habitat according to Turchanin's label is "in campis"; we assume he meant the entire steppe area. — E. Siberia: Dau. **Gen. distr.:** NE China, Inner Mongolia (Shara-Murun), Outer Mongolia (Ulan Bator and Khentei Mountain). Described from "Nerchinsk Dauria." Type in Leningrad, isotype in Tokyo.

Note. The Japanese botanist Kitagawa (l. c. 1939) initially called this species a variety but later, quite rightly, raised it to a species restoring the name given by Turchaninov in schedis. Although Kitagawa represented

Turchaninov as the author of the species, the latter only supplied a nomen nudum. Hence the real author should be regarded as Kitagawa and the combined names should include Korzhinskii in brackets since he described the plant (1894) in Latin under the rank of a form. Ledebour was the first to supply a short diagnosis to this plant but did not give it a specific name, applying Turchaninov's name pro synonimo. Turchaninov himself did the same later (Fl. baic.-dahur.).

In his description of Adenophora collina from NE China, Kitagawa (l. c. 1940) repudiated his earlier consideration of A. crispata as an independent species, and ranked it as a form (of A. stenanthina). In doing so, he pointed out that his decision to raise A. crispata to a species had been based almost exclusively on the herbarium specimen from which he established A. collina. At that point we are in agreement with him. However, after studying herbarium material we feel that Kitagawa was wrong in his final conclusion to rank A. crispata as a form. It is a completely independent species, none of which specimens, growing in different and remote areas, differ from Turchaninov's plant, which must be regarded as the real specimen for A. crispata.

Apparently this is a very rare plant in the USSR but grows abundantly in NE China. Chaney collected typical A. crispata in the above-noted locality in Inner Mongolia.

Series 2. **Gmelinianae** Fed. — Style as long as or shorter than corolla; corolla campanulate, not constricted under lobes. Leaves linear or narrowly lanceolate, entire or lanceolate, acutely and partially runcinate-dentate.

13. A. gmelinii (Spreng.) Fisch. in Mém. Soc. Nat. Mosc. VI (1823) 167; A. DC. Monogr. (1830) 366; idem in DC. Prodr. VII, 2, 493; Ldb. Fl. Ross.
368 II, 2, 893; Turcz. Fl. baic.-dahur. II, 192; Kitagawa, Lineam. Fl. Manshur. 416; Kom. Fl. Man'chzh. III, 561, in obs. — A. gmelini var. stylosa A. DC. l. c. (1830) 366. — A. erysimoides Nakai ex Kitagawa in Rep. Inst. Sc. Res. Mangh. II, 7 (1938) 298. — A. marsupiiflora typica × denticulata f. gmelini Korsh. in Mém. Acad. Sc. Petérsb. 7 sér. XLII, 2 (1824) 31. — Campanula gmelini Spreng. Fl. Hal. Mant. I (1807) 56; Borbás in Magyar Bot. Lapok, III, 191 (et var. pomponiifolia Borbás, l. c. 192). — C. erysimoides Vest ex Roem. et Schult. Syst. V (1819) 102. — C. rabelaisiana Roem. et Schult. l. c. 158. — C. pomponiifolia Fisch. ex A. DC. l. c. (1830) 366, pro syn. — Ic.: Gmel. Fl. Sibir. III, tab. 33 (mediocr.).

Perennial; root thick, sometimes branching above, brown, rugose; stems straight, simple or few-branched in inflorescence, low, not more than 40—50(60) cm, glabrous, cylindrical, longitudinally striate, leafy mainly at the middle; radical leaves drying early (very rarely 1—2 leaves with orbicular dentate blade and thin petioles); cauline leaves linear, narrow, usually not more than 2(3) mm wide and 5—7 cm long, glabrous, smooth, stiff, entire or sparingly toothed with slightly convolute margins. Flowers 3—10, rarely more, usually in a secund raceme or a few-branched inflorescence, with thin filiform pedicels up to 2—2.5 cm long; calyx glabrous, oval, teeth erect, lanceolate, as long as or longer than tube, entire, much shorter than corolla; corolla broad, orbicular, blue, glabrous, shallowly divided into ovate acuminate lobes; disk (nectary) usually short, sometimes subannular, rarely acutely toothed, if so then 3 mm long; style not exserted. July—August.

Steppes and dry slopes. — E. Siberia: Dau. **Gen. distr.:** NE China. Sprengler (l. c.) based his description on Gmelin's plant, diagnosis and drawing (l. c.) from Transbaikalia. The type apparently has been lost.

Note. Reminiscent of A. stenanthina in outward appearance but readily distinguished by the large, campanulate corolla and the nonexserted style. The disk, as we mentioned above, is very short and thus very characteristic of this species.

14. A. coronopifolia Fisch. in Mém. Soc. Nat. Mosc. VI (1823) 157; A. DC. Monogr. (1830) 362; idem in DC. Prodr. VII, 2, 493; Ldb. Fl. Ross. II, 2, 894; Turcz. Fl. baic.-dahur. II, 189; Maxim. Prim. Fl. Amur. 185; Kom. Fl. Man'chzh. III, 562; Kom. and Alis., Opred. rast. Dal'nevost. kr. 994; Kitagawa, Lineam. Fl. Manshur, 415; Kryl., Fl. Zap. Sib. XI, 2645. — A. coronopifolia var. pubicalycina Reverd. in Sist. zam.
369 Gerb. Tomsk. univ. 3—4 (1935) 5. — A. coronopifolia var. puberula Reverd. op. cit. — A. polymorpha var. coronopifolia Trautv. ex Herd. in Tr. Peterb. bot. sada, I (1873) 309. — A. marsupiiflora typica × denticulata f. coronopifolia et f. subcoronopifolia Korsh. in Mém. Acad. Sc. Pétersb. 7 sér. XLII, 2 (1894) 31. — A. lamarkii angustifolia Fisch. ex A. DC. l. c. (1830) 362; pro syn., non A. lamarckii Fisch. (1823). — Campanula coronopifolia Fisch. ex Roem. et Schult. Syst. V (1819) 157; Borbás in Magyar Bot. Lapok, III, 192. — C. monadelpha Pall. ex Fisch. l. c. (1823) 158, pro syn. p. p. — C. salicifolia Juss. ex A. DC. l. c. (1830) 362, pro syn. — C. fischeriana Spreng. Syst. (1825) 77. — Ic.: Rchb. Hort. Bot. tab. 23.

Perennial; root thick, simple or branching, grayish, rugose; stems simple, or sometimes paniculately branching in inflorescence, glabrous, 50—70 cm, longitudinally striate, leafy mainly at the middle; radical leaves rarely present, ovate-orbicular, cordate, crenate-dentate, paler beneath than above, petiolate, with blade ca. 2 cm wide and twice as long petioles; cauline leaves alternate, linear-lanceolate, ca. 0.5—0.7 cm wide, 7 cm long, glabrous, smooth, remotely and rather largely runcinate-dentate, with antrorse, sometimes almost falciform or contiguous teeth, with convolute margins. Inflorescence usually 16-flowered, sometimes less, very rarely more-flowered, more or less racemiform or slightly paniculate; flowers drooping, 2 cm long, short-pediceled; calyx glabrous, ovate, teeth declinate, lanceolate, entire, as long as tube and much shorter than corolla; corolla broad, campanulate, blue, divided into short suberect lobes; disk (nectary) cylindrical, thick; style as long as or shorter than corolla. July—August.

Sandy flooded meadows, shrubby formations. — E. Siberia: Dau.; Far East: Ze.-Bu. **Gen. distr.:** Mongolia (E.) N. China. Described from Dauria. The type is a specimen from the Fischer herbarium which bears his inscription and was grown from seeds obtained from Leningrad where the specimen is now preserved.

Note. This species is taxonomically close to A. gmelinii from which it differs by broader dentate leaves and larger and more robust habit.

15. A. rupestris Reverd. in Sist. zam. Gerb. Tomsk. univ. 3–4 (1935) 2.

Perennial; root 2–4 cm thick, brownish, multicipital, bearing above relics of stems, densely covered with light brown scarious-fibrous scales; stems
370 usually few, sometimes up to 10 and more, 15–20(35) cm high, 2 mm across, distinctly longitudinally striate, scabrous (with very short papillae and apically rotund hairs), nearly smooth above, rapidly tapering toward apex, 0.5 mm thick beneath inflorescence; rosetted leaves absent; cauline leaves densely crowded at the middle, in lower part drying early or squamose, in upper part sparse and reduced; the developed leaves 2–3 cm long, 5–7 mm wide, narrowly lanceolate, usually arcuately curved, attenuate, acute, acutely toothed at margins, sometimes slightly incised-dentate, sessile, glaucous-green, sometimes with violet tinge (like stems). Inflorescence few-flowered, racemiform; flowers drooping, blue-violet; calyx obconical, subglabrous, sometimes darkish, teeth distinctly recurved, lanceolate, acute, half as long as tube and many times shorter than corolla; corolla broadly campanulate, glabrous, divided for one-fifth into broad-oblong acute lobes; disk (nectary) short-cylindrical, as long as wide (nearly square at cross section); style not exserted. July.

Dry bluffs and rubbly slopes covered with steppe vegetation.— E. Siberia: Ang.-Say. Everywhere west of Yenisei and only within the confines of Khakass. Endemic. Described from Tatytgyzyn Mountain in Khakass. Type in Tomsk, isotype in Leningrad.

N o t e. A very unique and little-known species, apparently endemic to Khakass. It has been found in Tatytgyzyn Mountain, and in addition, near Shira Lake, Bogradskii District near Uti'che Lake, the Uibat River valley, Sagyr-Gaya Mountain, Kamysht tributary of the Abakan River, near Bol'shoi Syr, near Forpost station on the Belyi Yus River, Krestovaya Mountain near Lake Shira, Oshkol' Lake near Chernyi Yus, Abakan steppe near Letnik and Kaly, and near Verkhne-Bidzhinskaya.

Section 4. PACHYDISCUS Fed. sect. nov. in Addenda XXIII, p. 338.— Disk (nectary) thick, cylindrical, ca. 4 mm long and wide, lanate-pubescent; calyx teeth large, erect, oblong-triangular, one-fourth as long as the large campanulate corolla; flowers solitary or few. Cauline leaves alternate, lanceolate, faintly toothed.

Type species of section: A. h i m a l a y a n a Feer.

16. A. himalayana Feer in Engler's Bot. Jahrb. XIV (1890) 618; Korsh. in Mém. Acad. Sc. Pétersb. 7 sér. XLII, 2, 26; O. and B. Fedch. in Tr.
371 SPb. obshch estestvoisp. XXXV, 3, 56; O. and B. Fedch., Perech. rast. Turk. 720; Pavlov, Fl. Tsentr. Kazakhst. III, 200, in obs.

Perennial; root thick; glabrous, glaucescent plant; stem 15–40 cm, 3–4 mm thick below, ascending, simple, rounded at cross section, somewhat scabrous; lower leaves ovate or ovate-lanceolate, remotely and obtusely toothed, with winged petioles as long as blade; upper leaves sessile, linear-lanceolate, slightly falcate; leaves in inflorescence very small. Flowers few, solitary in the low-growing specimens, in the developed specimens arranged in spreading secund racemes; calyx teeth erect, oblong-triangular, very acute, coriaceous, one-fourth as long as corolla; corolla 10–25 mm long, broad, tubular-campanulate, rounded at base, glabrous, divided for one-fourth into lobes; anthers linear-oblong, tapering at base; disk (nectary) thick, cylindrical, at margin entire, shortly lanate; style nearly as long as corolla or exserted, hairy and slightly thickened in upper part, lobed into short stigma. June–July.

Alpine and subalpine mountain zones. – Centr. Asia: Pam.-Al., T. Sh., Dzu-Tarb. **Gen. distr.:** Ind.-Him., Ch. Jap. Described from W. Himalayas (Kumaun near Napalcha-Bayans, glacial region). Type in Geneva, isotypes in the large herbaria of India and Europe.

Note. The main characteristics of this distinct and "good" species are the cylindrical, pubescent disk (nectary) and its length (4 mm).

Dubious Species

1. A. periplocifolia (Lam.) A. DC. Monogr. (1830) 361; idem in DC. Prodr. VII, 2, 493; Ldb. Fl. Ross. II, 2, 895. – A. communis var. lamarkii Trautv. in Tr. Peterb. bot. sada, VI (1879) 96, quoad syn., non Adenophora lamarckii Fisch. – Campanula periplocifolia Lam. Encycl. Méth. I (1783) 580; Roem. et Schult. Syst. V, 111; Spreng. Syst. I, 735.

"Stem ascending, glabrous, ca. 2–3 inches, simple, leafless below; leaves all petiolate, crowded at the middle part, glabrous, ovate, acuminate, semi-cordate, ca. 5 lines in length and 4 in width, crenate-dentate, paler along margin and beneath than above, petioles ca. half an inch long. Flowers drooping, few, pediceled, in some specimens flowers solitary, in others paired; calyx 2 lines long, with broad spherical tube and triangular acute teeth often with 1–2 lateral denticles; corolla ca. 5 lines in length, pale blue
372 (Lamarck, l. c.), appearing truncate above; stamens broad at base, ciliate; disk (nectary) broad, thick, length three-quarters of a line; style ca. 3 lines long, longer than corolla, filiform at base and thickened at apex; stigmas 3, short."

"Habitat: Siberia; grown (from seeds) in the Paris Botanical Garden (Lamarck, l. c.)."

Note. The above diagnosis of A. periplocifolia, was made by de Candolle who saw the original specimen which had been used by Lamarck in describing Campanula periplocifolia. The translation is presented here to show that it is rather impossible to locate Lamarck's plant among the species of the flora of Siberia. It is more probable that this is the same plant described much later by Maksimovich as Adenophora trachelioides Maxim. (see Note to A. trachelioides). We have no way of checking this since the diagnosis does not coincide with the description of A. trachelioides; Lamarck's drawing of the plant is lost, and we have never seen the original specimen which may still be preserved in Paris.

Genus 1435. **POPOVIOCODONIA** * FED., gen. nov.

Fed. in Addenda XXIII, p.338

Corolla dissected to base into broad lanceolate lobes; disk (nectary) absent. Calyx ribbed with prominent nerves, divided into linear recurved teeth shorter than tube. Filaments dilated, very finely pubescent. Style distinctly exserted; stigmas 3. Capsule dehiscing at apex. Leaves all cauline; rosettes never present. Perennial plants, similar in appearance to Adenophora. The genus occupies an intermediate position between Adenophora and Campanula.

Type species: P. uyemurae (Kudo) Fed.

Endemic, oligotypic genus, represented in the Far East by two species.

1. Rhizome thin, with few filiform underground shoots; corolla 1.3 cm long, infundibular . . . . 1. **P. uyemurae** (Kudo) Fed.

\+ Root thickened, fleshy; corolla 2 cm long and longer . . . . 2. **P. stenocarpa** (Trautv. et Mey.) Fed.

373 **1. P. uyemurae** (Kudo) Fed. comb. nov. – Adenophora uyemurae Kudo in Bull. Exp. Forest. Kyûshû Imp. Univ. I (1931) 88; Ishiyama in Rep. Saghal. Centr. Exp. Stat. No. 1, 87; Sugawara, Pl. Saghal. 293. – Campanula uyemurae (Kudo) Miyabe et Tatewaki, Contrib. Fl. North Japan, VI, 81; Sugawara, Ill. Fl. of Saghal. IV (1940) 1749; Tolmachev in Bot. Zhurn. SSSR, XXXV, 4, 351; M. Popov, ibid, XXXVI, 4, 377. – Ic.: Ishiyama, l. c. tab. VII, f. b (photo); Miyabe et Tatewaki, l. c. 82, f. 8 (flos); Sugawara, l. c. (1940) 1748, tab. 802.

Perennial; rhizome thin, with few underground filiform short shoots; stems simple or slightly branching, 10–15 cm long, arcuate at base, then straight, bearing terminal corymbiform or short-racemiform inflorescences, sometimes 1-flowered; rosetted leves absent; cauline leaves large, crowded at the middle, tapering at both ends, broadly elliptic or obovate, acute or obtuse, largely and irregularly rather acutely toothed, sometimes slightly incised-dentate, rarely obtusely and obsoletely large-toothed, cuneate-attenuate, 1.5–2(3) cm long, 0.7–1.3 cm wide; uppermost leaves reduced, lanceolate, few-toothed or entire, like the whole plant completely glabrous or with few hairs at margins. Flowers 1.3 cm long, pale blue-violet; calyx glabrous, somewhat ribbed, oblong, teeth linear-lanceolate, recurved, subulate but obtuse, as long as tube to twice as long; corolla dissected into broadly lanceolate acute lobes nearly three times as long as calyx teeth; stamens with dilated lanceolate filaments finely pubescent, anthers narrow, linear; style with 3-lobed stigma, very finely pubescent in upper part, prominently exserted. July–August. (Plate XXII, Figure 2.)

Bare mountains, rocky slopes ca. 1,150–1,500 m above sea level. – Far East: Sakh. (E. Sakhalin Range, Lopatin Mountain, Kashipo, Nakapodake and Khoroto mountains). Endemic. Described from the mountains of S. Sakhalin. Type in one of the herbaria of Japan.

* Named in memory of M.G. Popov, corresponding member of the Academy of Sciences of the Ukrainian SSR, who devoted many years of his life to the flora of Siberia and earlier investigated Central Asia, the Caucasus, Carpathians and other parts of the USSR.

Note. This species resembles Adenophora in general appearance and it is therefore not surprising that Kudo (l. c.) described the plant as Adenophora. Miyabe and Tatewaki (l. c.) referred it to Campanula on the basis of the absence of a disk (nectary) typical for Adenophora. The Japanese botanists note that the American C. aparinioides Pursh is related to P. uyemurae by its deeply cleft corolla. We think that it is more accurate to establish Popovicodonia for P. uyemurae and
374 Campanula stenocarpa Trautv. et Mey., which is actually what we have done.

2. **P. stenocarpa** (Trautv. et Mey.) Fed. comb. nov. – Campanula stenocarpa Trautv. et Mey. Fl. Ochot. (1856) 61; Kom. and Alis., Opred. rast. Dal'nevost. kr. II, 993 (sphalm. nom. auct. "Fr. et Mey.").

Perennial; stems 10–15(25) cm, arcuate at base, straight, simple or sometimes branching, angular, completely glabrous, indurate, ca. 1 mm thick; rosetted leaves absent; cauline leaves largest at middle of stem, becoming smaller above and beneath, all sessile, sometimes the radical leaves in young plants with long filiform petioles; blades obovate, elliptic, sometimes nearly ovate-lanceolate or rhombic, obtuse or acute, 2–3 cm long, 0.5–1 cm wide, remotely crenate-serrate at margin, with depressed nerves above and with prominent ones beneath, glabrous. Flowers in corymbiform or short-racemiform inflorescences; bracts lanceolate, usually entire, as long as pedicels, reaching 1–1.5 cm; calyx clavate-obconical, ribbed with prominent nerves, teeth linear-lanceolate, recurved, half as long as tube; capsule straight, obconical, dehiscing by valves at apex, trilocular, many-seeded, glabrous; seeds oblong, sparingly pubescent, shiny. July–August.

Far East: Okh. Endemic. Described from Ukurundu Mountains of Segnek Peninsula. Type in Leningrad.

Note. Trautfetter and Meyer (l. c.) noted the dissimilarity of this species vis-a-vis Campanula. They wrote: "a Campanulae speciebus hucusque descriptis omnibus, ni fallimur, manifeste diversa."

Genus 1436. **ASTROCODON** * FED., gen. nov.

Fed. in Addenda XXIII, p.339

Corolla 5-merous, parted or dissected into broadly lanceolate lobes, lobes stellately spreading. Disk (nectary) not developed. Calyx 5-merous, divided into linear-lanceolate recurved teeth. Ovary trilocular. Filaments
377 finely pubescent, dilated at base. Style with tripartite stigma. Capsule dehiscing at base. Leaves only cauline. Perennials, with thick fleshy root.

This oligotypic genus is similar to Popoviocodoni a but differs in its capsule which dehisces at base and not at apex. Its two species are found only in the Far East.

Type species: Astrocodon kruhseanus (Fisch. ex Rgl. et Til.) Fed.

Note. One of the species of this genus was earlier described under the name Campanula kruhseana Fisch. ex Rgl. et Tiling and then referred

* From the words aster – star and codon – bell, i.e. star-shaped bells.

PLATE XXII. 1 – *Astrocodon kruhseanus* (Fisch.) Fed.; 2 – *Popoviocodonia uyemurae* (Kudo) Fed., plant and flowers (cross section, without perianth), capsule.; 3 – *Legouzia falcata* (Ten.) Fritsch.

to the genus Wahlebergia, but the plant by no means resembles the latter genus. Included in the tribe Wahlenbergieae are those genera in which the capsule dehisces above by one aperture and the ovary is half-inferior; for this reason we separate Campanula kruhseana and C. expansa into the independent Astrocodon and refer it to the tribe Campanuleae. Astrocodon differs from Campanula in overall characters.

Astrocodon, especially the species A. kruhseanus, bears a remarkable resemblance to Popoviocodonia Fed. In habit both genera are close to Adenophora except for the very unique shape of the flowers with their deeply dissected, stellate corollas and absence of disks (nectaries). Astrocodon differs sharply from Popoviocodonia by the capsule which dehisces by basal pores, while the capsule of Popoviocodonia dehisces at the apex. The two genera could possibly be united into one genus with two sections within it (according to the opening of the capsule), but we find it artificial and therefore unacceptable.

1. Flowers solitary; stems thin; leaves linear-oblong or lanceolate, entire; corolla flatly ringent . . . . 2. **A. expansus** (Rud.) Fed.

\+ Inflorescence racemiform, usually few-flowered; stems robust; leaves rhombic-oblong, dentate; corolla broadly infundibular . . . . 1. **A. kruhseanus** (Fisch.) Fed.

1. A. kruhseanus (Fisch. ex Rgl. et Til.) Fed. comb. nov. — Campanula kruhseana Fisch. ex Rgl. et Tiling, Fl. Ajan. (1858) 108, excl. syn. Herd. in Tr. Peterb. bot sada, I, 2 (1872) 299. — Wahlenbergia kruhseana Fisch. ex A. DC. in DC. Prodr. VII, 2 (1838) 425, pro syn.

Perennial; root thick, fleshy, sometimes branching, whitish, transversely finely rugose; stems one or few, 1–2 mm thick, 10–20(25) cm long, arcuate
378 at base, then straight, indurate, longitudinally somewhat striate, with brown squamous relics of leaves at base, densely leafy, glabrous; cauline 2–3(4) cm long, 0.5–0.7(1.5) cm wide, sometimes even much larger, the largest at the middle, reduced above and beneath, slightly overlapping or somewhat remote, rhombic-oblong, acute, largely and sometimes acutely toothed mostly in upper part, paler beneath than above, with prominent nerves; upper leaves narrower, sometimes sublinear, remote. Inflorescence usually few-flowered, racemiform; flowers with rather long pedicels, drooping; calyx obconical, slightly ribbed with prominent nerves, teeth lanceolate, acute, becoming arcuate, much shorter than corolla; corolla 2–3 cm in diameter, broadly infundibular, pinkish, deeply divided into broad acute triangular-lanceolate lobes, 1 cm long and even longer; stamens with large anthers and short filaments; style slightly exserted, stigma tripartite with spirally twisted lobes; capsule slightly ribbed, obconical-oblong, dehiscing at base by small pores. June–August. (Plate XXII, Figure 1.)

Gravelly soil of the dry coastal tundra. — Far East: Okh., Anad. Endemic. Described from the Sea of Okhotsk near Ishige. Type in Leningrad.

Fischer apparently suspected that this plant was identified with the earlier described Campanula hemallanthina Ldb. (described by Rudolph under the name C. expansa). In a letter to de Candolle he informed of his tentative name Wahlenbergia kruhsiana being synonymous with Wahlenbergia homallanthina (Ldb.) A. DC. (l. c.).

This problem was settled by Regel and Tiling who observed that the capsule opened below, thus it was no longer possible to refer this species to Wahlenbergia. The plant does not refer to Campanula either, being in many respects similar to Adenophora. Consequently we think that it would be more correct to establish Astrocodon, based on the type C. kruhseana.

2. A. expansus (Rud.) Fed. comb. nov. — Campanula expansa Rud. in Mèm. Acad. Sc. Pètersb. IV (1809) 340; Trautv. et Mey. Fl. Ocht. 60; Trautf. in Tr. Peterb. bot sada, IV, 1, 80. — C. homallanthina Ldb. in Mèm. Acad. Sc. Pètersb. V (1811) 524; Cham. in Linnaea, IV (1829) 41. — C. redowskiana Cham. l. c. pro syn. non C. redowskyi Fisch. — Platycodon homallanthinum A. DC. Monogr. (1830) 126. — Wahlen-
379 bergia homallanthina A. DC. Prodr. VII, 2 (1838) 425, p. p.; Ldb. Fl. Ross. II, 2, 871, p. p. — W. expansa (Rud.) Soczava in sched. (1930). — Ic.: Rud. l. c. tab. 2, f. 1—3 (mediocr.).

Perennial; plant 10—15(20) cm, with thin, simple or slightly branching rhizome, sometimes producing leafy shoots; leaves only cauline, crowded at the middle, sessile, linear, oblong or broadly lanceolate, prominently nerved, reduced, petiolate. Flowers solitary, bluish; calyx obconical, slightly ribbed, teeth linear-lanceolate, acute, curved, much shorter than corolla; corolla stellately open, flat, divided nearly to base into broadly lanceolate acute lobes; stamens with filaments dilated below and narrow yellow anthers, at first contiguous, later diverging; style with tripartite stigma; capsule oblong-obconical, dehiscing at base by several rather large pores.

Far East: Uda (Shantar Islands). Endemic. Described from "East Siberia." Type lost but Ledebour's specimen intact.

Note. Despite the assertion by Trautfetter and Meyer that the capsule dehisces near the base, this still would have to be verified since the specimens at hand have been poorly preserved and new plants have not been collected. It should also be noted that de Candolle provisionally referred this plant to Platycodon and Wahlenbergia, not having the fruits. The plant has nothing in common with the types of these genera. The species epithet "expansa" is used because Rudolph selected this name due to the nearly rotate corolla, with its elongate lobes, characters which specified the plant he was describing. A later homonym was Campanula expansa Friv. (1836) which was subsequently renamed. Reports on the discovery of this species in Kamchatka are erroneous.

Tribe 2. **PERACARPEAE** Fed. tr. nov. in Addenda XXIII, p. 339. — Corolla regularly obconical, deeply dissected into erect long-acuminate lobes, 5-merous (like calyx); calyx without appendages between teeth, obconical; stamens free, filaments and anthers linear; style with 2—3 curved stigmas; ovary trilocular; capsule thinly membranous, nearly transparent, indehiscent or dehiscing by 3 small valves near base, pyriform, pendulous. Perennials with thin branching stems; leaves alternate, petiolate and only cauline; flowers axillary. Rhizome creeping, thin, with shoots.

Type genus: Peracarpa Hook. f. et Thoms.

380 Genus 1437. **PERACARPA** * HOOK. f. et Thoms.

Hook. f. and Thoms. in Journ. Linn. Soc. II (1858) 26; Benth. et Hook. Gen. pl. II (1862–1883) 558.

Calyx-tube obconical, lobes triangular, straight, acute. Corolla infundibular-campanulate, deeply 5-lobed, lobes equal, straight, linear, acuminate. Filaments free. Anthers narrowly lanceolate or linear. Style elongate, with 2 or 3 filiform curved stigmas. Ovary trilocular. Capsule oblong, pendulous, thinly membranous, constricted at apex, few-seeded, indehiscent or dehiscing at base by 3 pores. Seeds large, short-fusiform or oblong, smooth, with coriaceous coating. Weak, delicate herbs, thin or slightly fleshy, prostrate or creeping, nearly completely glabrous; leaves petiolate, broadly ovate or suborbicular; pedicels thin, and flowers small, inconspicuous.

Type: P. carnosa (Wall.) Hook. f. et Thoms.

An oligotypic genus distributed in SW China, Himalayas, Japan, Sakhalin, Kurile Islands and Kamchatka.

A genus taxonomically unique in all characters.

P. circaeoides (Fr. Schmidt) Feer in Engler's Bot. Jahrb. XII (1890) 621 (Feer scripsit – "Perocarpa"); Kudo, Veget. of Yezo in Jap. Journ. of Bot. II, 4, 279; Kom. Fl. Kamch. III, 117; Miyabe et Miyake, Fl. Saghal. 293; Kom. and Alis., Opred. rast. Dal'nevost. kr. II, 992; Sugawara, III, Fl. of Saghal. IV, 1755; M. Popov in Bot. zhurn. SSSR, XXXVI, 4, 378. – P. carnosa Matsumura, Ind. pl. Jap. II, 2 (1912) 617, non Hook. f. et Thoms. nec Campanula carnosa Wall. – P. carnosa Hult. Fl. of Kamtch. IV (1930) 152, non Hook. f. et Thoms., nec Campanula carnosa Wall. – C. circaeoides Fr. Schmidt in Mém. Acad. Sc. Pétersb. 7 sér. XII, 2 (1862) 154; Trautf. in Tr. Peterb. bot. sada, VI, 1, 78. – Ic.: Fr. Schmidt, l. c. tab. III, f. 14–19 (bona); Feer, l. c. tab. VII, f. B (habitus); Sugawara, l. c. tab. 804. – Exs.: Rein, No. 5 (sec. Feer).

Perennial; small, very thin and delicate plant, with whitish creeping underground shoots; stems straight, up to 7 cm, weak, glabrous; leaves membranous, serrate, sparsely fine-hary above, glabrous beneath; lower leaves small, alternate, orbicular-obovate, attenuate into petioles; upper
381 leaves crowded as if in whorls, larger, orbicular, or broadly ovate, slightly cordate at base, then abruptly tapering into petioles. Flowers 3–4 mm long, solitary, with straight filiform pedicels, later curved and pendulous; calyx obconical, teeth triangular, acute, green, narrowly membranous at margin, half as long as corolla; corolla whitish, infundibular-campanulate, parted nearly to middle into linear-lanceolate straight acute lobes; stamens 5, with very short filaments dilated at base; style not exserted, pubescent near apex, with 2 stigmas; ovary obovate, tapering at base; capsule pendulous, membranous, obovoid, dehiscing at base by 3 apertures, transparent when ripe so the seeds are visible; seeds attenuate at both ends, oblong-fusiform, brown, smooth. July–August. (Plate XXIII, Figure 2.)

Mountain slopes, coniferous taiga, alder thickets, moist mossy places. – Far East: Sakh., Kamch. **Gen. distr.:** N. Japan, China (Taiwan). Also reported for continental China, but another species – P. carnosa – grows there. Described from the slopes of Niburipo Mountain in Sakhalin, between Manyu and Mogunkotan. Type in Leningrad.

* From the Greek pera – bag, and karpos – fruit.

Note. Feer (l. c.) pointed out that the spelling Perocarpa is more accurate than Peracarpa, but in view of Peracarpa as an orthographic variation, the name should be retained.

The rather cumbersome Russian species epithet "dvulepestnikovidnaya" has been derived from the Russian generic name "dvulepestnik," i. e. Circaea alpina, to which Peracarpa circaeoides bears a remarkable resemblance and which is the reason for its Latin name.

Tribe 3. **OSTROWSKIEAE** Fed. in Addenda XXIII, p. 339. – Corolla campanulate, shallowly divided into short triangular lobes, 7- rarely 5–9-merous; calyx short-cylindrical or flattened-spherical, without appendages, dissected below teeth into numerous large oblong slits, divided into 7, rarely 5–9, long broadly linear acute teeth. Ovary 7- rarely 5–9-locular; capsule chartaceous, turbinate, dehiscing by 14, rarely 10–18 large apertures; style thick, cylindrical, usually with arcuate stigmas; stamens free, filaments short, anthers linear, later spirally twisted. Perennial plants, with whorled leaves and tuberiform thickened root.

Type: Ostrowskia Rgl.

382 Genus 1438. **OSTROWSKIA** * RGL.

Rgl. in Tr. Peterb. bot. sada, VIII, 3 (1884) 686.

Calyx-tube short-cylindrical or flattened-sperhical, broadly conical at base, tube dissected by large oblong longitudinal parallel slits as if forming a belt around the calyx, teeth lanceolate, usually 7, rarely 5–9, 5–7 cm long, 4 times as long as tube, acute, three-fourths as long as corolla; corolla campanulate, 8 cm long, 5–9-lobed, usually 7-lobed, lobes short, ovate, acuminate, one-fifth to one-fourth as long as tube. Stamens usually 7, rarely 5–9, hidden and half as long as corolla; filaments free at base, short, ovate, broadening at the lower part and with carinate folds above; anthers linear, 2-locular, short-mucronate, three times as long as filaments, straight, later spirally twisted. Ovary inferior, often 7-locular, rarely 5–9, cells with numerous ovules; style thick, cylindrical, barely longer than stamens; stigma 7-lobed, lobes straight, later arcuately curved below, twisting at tip. Capsule dry, chartaceous, trubinate up to the middle, constricted at apex and furnished with numerous oblong holes. Seeds ovate-oblong, flattened, narrowly winged.

Monotypic, endemic genus whose single species is found in the mountains of Central Asia.

1. O. magnifica Rgl. in Tr. Peterb. bot. sada, VIII, 3 (1884) 686; O. and B. Fedch., Perech. rast. Turk. III (1909) 353; B. Fedch. in Rastit. bogatstv. SSSR., 3, 60; B. Fedch. in Izv. Tadzh. bazy AN SSSR, I, 1, 79, II (1936) 199 (distr. geogr.). – Ic.: Regel', op. cit. tabl. I; Fedtsch. a. Vvedensky, Sp. of wild flowers of the USSR, tab. XXVII (color).

* Named after M.N. Ostrovskii, a field botanist with E. Regel.

Perennial; glabrous plant, 1 m and taller, with thick tuberiform root; stems sturdy, straight, thick, 1.5 cm in diameter, fistular; leaves in remote whorls, ovate, oblong or ovate-oblong, largely incised-dentate with mucronate or glandular teeth; the lower leaves much like the median; uppermost leaves reduced, one-third to one-half as long as median cauline leaves. Flowers in spreading pyramidal inflorescence; pedicels 5—10 cm long, thickened at apex; calyx teeth narrowly linear-lanceolate, straight, nearly
383 coriaceous, ca. 5 cm long, three-fourths as long as corolla; corolla 8 cm long, glabrous, lilac or nearly white, with shallow acute lobes; capsule enclosed in calyx. June. (Plate XXIII, Figure 3.)

Central mountain belt. — Centr. Asia: Pam.-Al. (Darvaza, Kulyab, Baldzhuan), T. Sh. (Ugamskie Mountains). Endemic. Described from Darvaza Mountains. Type in Leningrad.

Note. Cultivated for a long time because of the beauty of its flowers and unique habit.

Fedchenko (op. cit.) noted the occurrence of yellow flowers, apparently having in mind Ostrowskia found by the Baranov expedition in 1921 in the Ugamskii Range of W. Tien Shan and not in Pamir-Alai. Unfortunately, there is no evidence from the Baranov specimens and the plants later collected by Lepeshkin, which are in the Tashkent herbarium, that Ostrowskia really grows in Tien Shan. Vvedenskii (letter communication) finds it impossible to settle this problem, positively or negatively, until the plants of Ugamskii are investigated on the spot.

Tribe 4. **MICHAUXIEAE** Fed. tr. nov. in Addenda XXIII, p. 340. — Corolla dissected nearly to base into narrow linear lobes, like calyx 8—10-merous; calyx with recurved triangular appendages between teeth; stamens free, filaments considerably dilated at base, anthers short-mucronate; style cylindrical, pubescent, with stellate stigma; ovary 8—10-locular; capsule pendulous, dehiscing at base by valves, ribbed-sulcate. Monocarpic plants, with alternate cauline leaves and rosetted radical ones.

Type: Michauxia L'Hér.

Genus 1439. **MICHAUXIA** * L'HÉR.

L'Hér. Dissert. monogr. Lam. Illustr. 2 (1788) tab.295; A. DC. Monogr. (1830) 211.— Mindium Juss. Gen. pl. (1789) 164.

Calyx 8—10-partite, with recurved appendages in notches between teeth and broadly turbinate or semispherical tube; corolla 8—10-partite, lobes narrow, spreading or recurved. Stamens 8—10, free, filaments dilated at base, anthers short-mucronate. Style thick, pubescent; stigma 8—10-partite
384 into long lobes. Ovary inferior, 8—10-locular; placentas many-ovuled. Capsule cylindrical, pendulous, dehiscing by 8 valves. Large, tall, monocarpic herbs, with thick stem, lyrate radical leaves and interrupted spicate or paniculate inflorescence.

* After the French botanist Michaux.

PLATE XXIII. 1 – Michauxia laevigata Vent.; 2 – Peracarpa circaeoides (Fr. Schmidt) Feer, plant, flowers, capsule; 3 – Ostrowskia magnifica Rgl. upper part of stem with flowers, whorl of cauline leaves, inside flower (without corolla).

Type: M. campanuloides L'Hér.

A Near Asian genus of 6 species, only one of which is found in the USSR.

According to Grossgeim (Rastitel'nye bogatstva Kavkaza, ed. 2-e (1952)), some species of this genus contain latex. The leaves of M. laevigata contain 1.5–2%, sometimes up to 3%. There is also a small quantity of latex in the stems. The largest yields are obtained at flowering and fruiting.

1. **M. laevigata** Vent. Hort. Cels. (1800) 81; A. DC. Monogr. 212; idem in DC. Prodr. VII, 457; Ldb. Fl. Ross. II, 875; Boiss. Fl. or. III, 891; Trautf. in Tr. Peterb. bot. sada, VI, 55; Fomin in Mater. Fl. Kavk. IV, 6, 128; Grossg. Fl. Kavk. IV, 78; Takht. and Fed., Fl. Erevana, 291. – Ic.: Vent. l. c. tab. 81; Bot. Reg. tab. 1451; Bot. Mag. tab. 3128.

Biennial; root thick, vertical, fusiform, sometimes branching, often 3 cm in diameter at apex; stems 1 m and even taller, 1–2 cm thick, straight, cylindrical, smooth, glabrous, glaucescent-white, simple, leafy, with abundant milky juice inside; leaves greenish-glaucous, scabrous-hirsute with short bristly hairs, brittle, thick; radical leaves petiolate, ovate-oblong, irregularly bidentate, often with developed lower lobes, slightly lyrate, 15–20 cm long, 5–7 cm wide, with prominent bristly nerves beneath, blade winged-decurrent along petiole; cauline leaves sessile, gradually tapering toward stem apex, oblong-lanceolate, often with auriculate lobes at base of blade compactly attached to stem, serrate-dentate. Flowers solitary, rarely together, along
387 a spicate raceme, spreading, with short pedicels; calyx obconical, almost glabrous, teeth hirsute, spreading, ovate-lanceolate, one-third as long as corolla, appendages recurved, lanceolate, bristly, completely covering tube; corolla 2–3 cm long, 8-partite, white, glabrous outside, sparingly pubescent inside; stamens 8, with very short filaments ciliate-glandular at margin, anthers filiform, yellow, three times as long as filaments; style straight, stiff-hairy, as long as corolla; stigma 8-partite; capsule coriaceous, obpyramidal, 8-ribbed, deeply sulcate, 8-locular, when mature enclosed by the persistent acuminate incurved calyx teeth; seeds many, ovate, smooth, brown. July–August. (Plate XXIII, Figure 1.)

Dry pebbly mountain slopes. – Caucasus: S. Transc. (Armenia, Nakhichevan ASSR), Tal. (Zuvand). **Gen. distr.:** N. Iran. Described from N. Iran (Elburz Range). Type in Paris.

Note. A characteristic plant in the low-shrub formation ("phrygana"), growing on dry slopes covered with the rubble of wind-blown shales or tufaceous sandstone in S. Armenia, Nakhichevan ASSR and the peaks of Talysh Range (Zuvand or Diabar basin). It never grows in thickets and is found mostly in much exposed and sunny habitats.

Tribe 5. **PHYTEUMATEAE** Fed. tr. nov. in Addenda XXIII, p. 340. – Corolla dissected nearly to base into narrow, linear or lanceolate, free or basally connate lobes, rarely rotate with broader lobes, 5- rarely 4-merous; calyx variable in shape; stamens free or imbricate, as if connate at base; style elongate or short, with 2–3 stigmas; ovary 2–3-locular; capsule prism-shaped, obconoidal, cylindrical or subglobose, usually constricted at apex, dehiscing between walls at apex, base or sides by rounded or slitlike apertures. Perennial or annual herbs, with alternate leaves, sometimes with rosettes of radical leaves, with roots or rhizomes sometimes markedly thickening.

Type: Phyteuma L.

Note. The tribe includes many genera that appear to have little in common. Nevertheless, they are all distinctly related and there is a trend in change of characters from one genus to the other. Thus, the genus Legouzia is most different from the tribe type, the genus Phyteuma because of its prismatic capsule and nearly rotate corolla. The genus
388 Cylindrocarpa, however, with its long, cylindrical capsule dehiscing at apex by slits, very much like the capsule of Legouzia, represents a phylogenetic bridge between the dissimilar genera Legouzia and Phyteuma. Certain characters of Cylindrocarpa bring it close to Phyteuma, Sergia, Cryptocodon and also Asyneuma.

Genus 1440. **PHYTEUMA*** L.

L. Sp. pl. Ed. 1 (1753) 170; A. DC. Monogr. (1830) 180, quoad sect. 2; R. Schultz, Monogr. Bearbeit. d. Gatt. Phyteuma (1904) 20; Schönland in Pflanzenfam. IV, 5 (1894) 53, p.p.– Hedranthum G. Don, Gen. hist. dichlam. pl. VII (1831–1838) 746, pro sect.

Flowers with 2- or 3-merous gynoecium, with 5-merous (in one species 4-merous) perianth. Calyx and corolla deeply 5- or (in one species) 4-partite, corolla lobes connate in upper part throughout flowering, forming a tube around the style, arcuate and parted in the lower part, later usually separated and strongly declining to the side. Filaments linear, dilated at base; anthers narrow, linear. Style elongate, with 2–3 filiform stigmas. Ovary inferior, 2–3-locular, with free central placentas and anatropous ovules. Capsule 2–3-locular, short-obconical or subglobose, many-seeded, dehiscing at apex by two pores. Perennial herbs, usually with thick fleshy root, alternate leaves and subsessile small flowers in rather short spicate or subglobular-capitate inflorescence.

Lectotype: P. spicatum L.

When separating Asyneuma (Podanthum), Cylindrocarpa, Petromarula, Synotoma and others as independent genera, Phyteuma s. str. should be singled out as being a natural taxon. It is a typical Central European genus and is also found in parts of the Mediterranean and the more western and northern regions of Europe. Many of its species are concentrated in the Alps, Pyrenees and Carpathians where they most probably have been derived from the same ancestor of P. spicatum L., broadly distributed at one time. In his monograph on Phyteuma s. str., Schultz (l. c.) enumerated 29 species and 15 hybrid forms in this genus, in addition to the many varieties and subspecies among which many good species are probably
389 hidden, if a species is defined as a geographical race. There are 14 such races and hence it can be assumed that there are altogether 33 [43?] species in Phyteuma, 4 of which grow in the USSR (the introduced P. nigrum is not regarded as a member of Soviet flora).

* The name appears in the works of Pliny and Dioscorides. Pliny's name for the genus can be translated as: plant adding strength, due to his conception of its aphrodisiac properties.

Schultz successfully grouped the species of Phyteuma into series. We have used part of those when classifying those species growing in the USSR. Less successful is his division into sections which cannot be applied without alterations. On the basis of a 4-merous perianth and androecium he established a monotypic series for P. tetramerum. We think that this series should be promoted to a new monotypic section in contrast to sect. Spicata R. Schulz, whose species have 5-merous flowers and a 2-locular ovary. There is little justification to split the section Capitata R. Schulz. Section Hedranthum G. Don, which chronologically was the first proposed, cannot be used because it fully corresponds with Phyteuma as it is accepted nowadays.

1. Corolla and calyx 4-merous, ovary 2-locular; flowers bluish-lilac, arcuately curved before flowering; the lower leaves cordate or obtuse at base, the cauline usually cuneate, in inflorescence linear, all leaves serrate . . . . 5. **P. tetramerum** Schur.
\+ Corolla and calyx 5-merous, ovary 2-locular . . . . 2.
2. Leaves in inflorescence ovate or sometimes broadly ovate, acuminate; inflorescence globular, capitate; flowers dark blue; lower leaves lanceolate or ovate-lanceolate . . . . 4. **P. orbiculare** L.
\+ Leaves in inflorescence linear or lanceolate; inflorescence short-ovoid, ovoid or cylindrical . . . . 3.
3. Corolla white, rarely bluish, sometimes yellow but pale; inflorescence cylindrical, spicate; leaves biserrate, the lower with blades as long as wide . . . . 1. **P. spicatum** L.
\+ Corolla dark-, sometimes blackish-blue-violet, rarely pale; inflorescence ovoid or short-ovoid, sometimes capitate and globose . . . . . . 4.
4. Blades of lower leaves nearly equal in length and width, deeply cordate, often obtuse [?] crenate [?] . . . . 2. **P. vagnerii** A. Kerner.
\+ Blades of lower leaves two times longer than wide, oblong-ovate, acute . . . . 3. **P. nigrum** Schmidt.

390 Section 1. SPICATA R. Schultz, Monogr. Bearbeit. d. Gatt. Phyteuma (1904) 57. — Corolla, calyx and androecium 5-merous, ovary 2-locular; flowers sessile or subsessile, usually in dense spicate inflorescences.
Type: P. spicatum L.

Series 1. **Cordifoliae** R. Schultz, Monogr. Bearbeit. d. Gatt. Phyteuma (1904) 64. — Lower leaves with blades nearly as long as wide, deeply cordate; cauline leaves cordate, truncate or reniform at base. This series includes, besides those species found in the USSR, the West European P. halleri All. and P. pyrenaicum R. Schultz.

1. P. spicatum L. Sp. pl. (1753) 242; A. DC. Monogr. 197; idem in DC. Prodr. VII, 2, 453; Ldb. Fl. Ross. II, 2, 872; Trautf. in Tr. Peterb. bot sada, VI, 1 (1879) 49, excl. var. secund.; Shmal'g., Fl. II, 172; R. Schultz, Monogr. Bearbeit. d. Gatt. Phyteuma, 64; Fedch. and Fler., Fl. Evrop. Ross. 941; Maevskii, Fl. ed. 8-e, 549; M. Popov, Rast. in fl. Karpat. 259; Szaf., Kulcz.,

Pawl. Rosl. Polskie, 649. — *P. spicatum* var. *typica* Trautv. op. cit. — *Rapunculus spicatus* Gilib. Fl. lithuan. III (1781) 59. — Ic.: Fedch. and Fler., op. cit. fig. 941; Curtis Bot. Mag. tab. 2347; Gilib. Exerc. phytol. I, 75; Szaf. et al. l. c. 640. — Exs.: GRF, No. 180; Woloszczak, Fl. pol. exs. No. 750; Bge. Fl. exs. Liv.-Ehst. u. Kurlands, VII, No. 468.

Perennial; root thick, fleshy, cylindrical or terete-fusiform, usually constricted and throat-shaped beneath stem, brownish, hardly branching, ca. 5—10 cm long, 1—1.5 cm wide; stems straight, simple, glabrous, smooth, slightly angular and faceted or longitudinally striate and slightly sulcate, 50—60 cm high, usually less, leafy, sometimes up to 0.3—0.4 cm thick; leaves glabrous, the radical long-petioled, cordate, acute or obtuse, with blades 5—7 cm long and nearly as wide or narrower, biserrate-dentate; lower cauline leaves with much shorter petioles, not longer than blade, less deeply cordate; upper leaves sessile, linear-lanceolate, spreading, irregularly dentate. Inflorescence dense, spicate, oblong-ovate or cylindrical, 5—7(8) cm long, elongating after flowering, obtuse as if truncate at apex; bracts linear, mucronate, subentire; flowers white or somewhat yellowish, sometimes bluish but not dark blue; calyx ovate, 10-nerved, teeth spreading, linear-lanceolate, entire, as long as tube and much shorter than corolla; corolla with linear lobes, free below but connate above, later becoming almost free;
391 filaments dilated, membranous, triangular, pubescent; anthers narrow, filiform; style exserted from corolla, pubescent from middle to apex, with 2 or 3 stigmas; capsule globular, yellowish, 10-nerved, rather short, dehiscing at sides by two pores, with dry spreading relics of calyx teeth approximately up to middle; seeds small, ovate, flattened, straw-colored. June—July. (Plate XXIV, Figure 1.)

Broadleaved and mixed forests, mountains and lowlands. — European part: U. Dns., Bes., Balt., U. Dnp., M. D. **Gen. distr.**: Bal. (N.). Described from W. Europe (Switzerland, Italy, England, France). Type in London.

Note. This predominantly C. European and mountain plant grows in the Baltic area of the USSR up to Estonia and Ostrovskii District and is apparently associated with broadleaved forests. It does not grow beyond the western half of Belorussia in the east (?).

2. P. vagnerii A. Kerner in Vágner Máramarosmegye Nov. Ism. (1875) 192; Simonkai, Enum. fl. Transs. 378; R. Schultz, Monogr. Bearbeit. d. Gatt. *Phyteuma*, 76; M. Popov, Rast. i fl. Karpat, 259. — *P. spiciforme* Roch. Bot. Reis. Banat (1835) 69, nomen: Szaf., Kulcz., Pawl. Rosl. Polskie (1953) 639. — Exs.: Fl. exs. austro-hung. No. 964.

Perennial; root fleshy, thick, fusiform, usually slightly constricted in upper part below stem, brown, longitudinally rugose when dry; stems straight, simple, not densely leafy, 12—50 cm high, thickish, glabrous, smooth; lower leaves rather long-petioled, obtuse, deeply cordate, usually reniform, with blades as long as wide or often wider than long; cauline leaves oblong, cordate or truncate at base; all leaves crenate-dentate, rarely largely bidentate. Inflorescence short, ovoid- or capitate-spicate; bracts linear or linear-lanceolate; corolla black-blue or black-violet, very rarely whitish, arcuately curved before flowering; ovary 2-locular; style with 2 stigmas. July.

Subalpine and alpine meadows and slopes of shale, gneiss, trachyte and limestone, sometimes matgrass and edges of alder thickets. European part: U. Dns. (Carpathians). **Gen. distr.:** Centr. Eur. (Carpathians and Transylvanian Alps). Described from Marmarosh Mountains in the Carpathians. Type in Budapest or Vienna.

Note. This species has been collected in the USSR on the mountains of Velikii Verkh, Cherna Gora, Goverla, Petrosh, Pikui, Pop-Ivan, Bliznitsa (or Blizhnitsa), and along the Evir River near Yasinya.

392 Popov (op. cit.) regards this species as the hybrid P. spicatum × P. orbiculare.

Series 2. **Lanceolatae** R. Schultz, Monogr. Bearbeit. d. Gatt. Phyteuma (1904) 83. — Lower leaves with blades nearly twice as long as wide; cauline leaves cordate. P. gallicum, the second species of the series, is found in volcanic highlands in the central part of France.

3. P. nigrum Schmidt, Fl. boem. I (1793—1794) 87; Roem. et Schult. Syst. V, 81; A. DC. Monogr. 199; idem in DC. Prodr. VII, 2, 453; Petunn. in Tr. Peterb. obshch. estestvoisp. XXX, 3, 65; R. Schultz, Monogr. Bearbeit. d. Gatt. Phyteuma, 84; Maevskii, Fl. ed. 8-e, 549. — P. spicatum var. nigrum Goroshankin in Byull. Mosk. obshch. ispyt. prir 2 (1888) 3. — P. spicatum var. ovatum Syreissch. Ill. fl. Mosk. gub. III (1910) 219, videtur non P. ovatum Schmidt. — P. spicatum Kaufm. Mosk. fl. ed. 2-e (1889) 319, non L.; Petunn., Rastit. Mosk. gub. 239. — Ic.: Syreishch,, op. cit. 218; Rchb. Ic. Fl. Germ. et Helv. tab. 19, f. 1586 (1).

Perennial; root thick, fleshy; stems straight, 40—75 cm, simple, glabrous, longitudinally striate, leafy; leaves glabrous, pale green, paler beneath; radical leaves cordate or orbicular-ovate, usually obtuse, ca. 4 cm long (without petioles) obtusely crenate but not bidentate, with long narrowly winged petioles; lower cauline leaves short-petioled, ovate, acute, narrower than the radical and less deeply cordate, serrate to nearly entire; upper leaves sessile, becoming narrower and smaller toward stem apex, linear-lanceolate, obtusely serrate, very remote from each other. Inflorescence subcapitate or ovoid, sometimes elongate especially after flowering; bracts recurved, linear-lanceolate, mucronate, subentire; calyx glabrous, subglobular, teeth linear-lanceolate, as long as tube but much shorter than corolla; corolla gray-blue or blue, not white, deeply divided into linear lobes connate in upper part, later becoming free and spreading; filaments dilated and pubescent at base; style dark-colored, as long as corolla lobes, with 2, sometimes 3, stigmas. June.

Broadleaved forests in the USSR. — European part: U. V. (introduced into Moscow area). **Gen. distr.:** Centr. Eur. Described from Karlovy Vary (Carlsbad), Czechoslovakia. Type possibly in Prague.

Note. This West European plant was found near Moscow (Petrovsko-Razumovskoe), now the site of the K. A. Timiryazev Agricultural Academy,
393 by Schroeder in 1884. According to Academician S. G. Navashin, to whom both Kauffmann (op. cit.) and Syreishchikov (op. cit.) refer, the plant appeared every year on the banks of a pond, apparently being introduced into the park. On this basis Syreishchikov and other authors concluded that it might be

included in the plants of the USSR. In fact, we do not know if it still grows near Moscow nor do we know how accurate are the earlier reports since we have never seen the herbarium specimens.

Series 3. **Orbiculatae** R. Schultz, Monogr. Bearbeit. d. Gatt. Phyteuma (1904) 110. – Lower leaves with blades usually much longer than wide, rounded, cordate or truncate at base; cauline leaves narrowly cuneate at base; leaves in inflorescence ovate, acuminate. This series includes besides P. orbiculare L. of the USSR, the West European P. tenerum R. Schultz and P. hispanicum R. Schultz.

4. P. orbiculare L. Sp. pl. (1753) 242, excl. var. nonnull.; A. DC. Monogr. 187; idem in DC. Prodr. VII, 2, 451; Ldb. Fl. Ross. II, 2, 872; Boiss. Fl. or. III, 958; Trautf. in Tr. Peterb. bot. sada, VI, 1, 48; Shmal'g., Fl. II, 172; R. Schultz, Monogr. Bearbeit. d. Gatt. Phyteuma, 110; Fedch. and Fler., Fl. Evrop. Ross. 941; M. Popov, Rast. i fl. Karpat. 259; Szaf., Kulcz., Pawl. Rosl. Polskie, 640. – P. orbiculare var. lanceolatum Trautv. l. c. videtur non Pers. – Rapunculus orbicularis Gilib. Fl. lithuan. III (1781) 59. – Ic.: Sims in Bot. Mag. tab. 1466.

Perennial; root thick, cylindrical, 7–10 cm long, 1–1.5 cm thick, sometimes with many reduced shoots above; stems straight or ascending, 30–50 cm, glabrous, smooth, cylindrical, not densely leafy; leaves glabrous or slightly pubescent; radical leaves many, broadly lanceolate or ovate, cordate or somewhat so, acute, crenate, 3–5 cm long, 1–2 cm wide (without petioles), petioles varying in length, sometimes long and thin; cauline leaves alternate, spreading, sessile, linear or narrowly ovate-lanceolate, rather small, crenate. Inflorescence bluish, spherical, ovoid in fruit; flowers bracteate, slightly curved; bracts wide, amplexicaul, ovate or ovate-lanceolate, acute, entire, usually ciliate; calyx somewhat ovoid, glabrous, teeth straight, linear, acute, subglabrous, shorter than corolla; corolla dark blue, paler at base, glabrous, with linear lobes connate in upper part to form a tube
394 around the conspicuously exserted style, somctimes lobes separate and bend to one side; filaments white ciliate, dilated at base, anthers long but shorter than filaments; pollen violet; style exserted from connate corolla lobes but as long as the latter at flowering; capsule globose and crowned by the dry calyx teeth and corolla lobes; seeds ovate, shiny. May–July.

Subalpine meadows, slopes, forest glades in mountains, very rarely at lowland meadows. – European part: U. Dns. (Carpathians), U. Dnp. (Chernigov), M. D. (W.). **Gen. distr.:** Centr. Eur., Atl. Eur. (except for England), Med. (Pyrenees), Bal. (N.). Described from W. Europe. Type in London.

Note. The Carpathian plants are distinguished from those collected in the lowlands (for example, in Chernigov and Zhitomir districts) by their narrow leaves, cuneate at base, and the rather thin roots. In view of the lack of material these and other specimens are regarded as belonging to the same species.

Section 2. **Tetramera** (R. Schultz) Fed. comb. nov. – Series Tetramera R. Schultz, Monogr. Bearbeit. d. Gatt. Phyteuma (1904) 90. – Corolla, calyx and androecium 4-merous, ovary 2-locular; flowers sessile or subsessile, in long spicate cylindrical inflorescence. Monotypic section.

Type: P. tetramerum Schur.

5. **P. tetramerum** Schur in Verh. Siebenburg. Ver. Naturw. X (1859) 105 (err. "tetrameriucum" et "tetramerium"), III (1852) 88 et IV (1853) 47, nomina nuda; Fuss. Fl. Transs. exc. (1866) 414; Simonkai, Enum. fl. Transs. 378; R. Schultz, Monogr. Bearbeit. d. Gatt. Phyteuma, 90; M. Popov, Rast. i fl. Karpat, 259; Szaf., Kulcz., Pawl. Rosl. Polskie, 639.

Perennial; root 1 cm thick, fleshy, sometimes producing in addition to stem, sterile shoots with reniform leaves; stem 40–80 cm, straight, simple, nearly leafless in upper third; lower leaves cordate, long-petioled, blades nearly twice as long as wide; lower cauline leaves ovate or lanceolate, cordate or truncate at base; median leaves lanceolate, obtuse or cuneately tapering, very short-petioled, sometimes sessile; upper leaves lanceolate or linear-lanceolate, sessile; all leaves crenate or dentate at margin, glabrous. Inflorescence cylindrical or ovoid, spicate; bracts linear, usually as long as or shorter than width of inflorescences; calyx, corolla and androecium 4-merous; corolla blue [lobes?] arcuate before flowering; ovary 2-locular; style with 2 stigmas. June.

395 Sandstone, igneous rocks at the upper timberline, mountainous meadows 1,500 m above sea level. – European part: U. Dns. (Carpathians: Cherna Gora, Goverla; Bogdan Rakhovskii District). **Gen. distr.:** Centr. Eur. (adjacent to the Soviet part of E. Carpathians). Described from E. Carpathians. Type in Vienna.

Dubious species

P. angustifolium Ldb. Fl. Ross. II, 2 (1845–1846) 874. – P. canescens auct. non Waldst. et Kit.: Trautf. in Tr. Peterb. bot. sada, VI, 1 (1879) 50, p. p.

"A completely glabrous plant with branching, angular stem. Leaves sessile, lanceolate-linear, remotely glandular-serrate; upper leaves entire, bracteate. Spikes apical at stem and branches, interrupted beneath; flowers in 3's or solitary, subsessile; calyx glabrous; capsule turbinate-obovate (2-locular?) . . . Style bipartite at apex."

Found in the lowlands of Kherson District in the south European part of the USSR ("in campestribus"). Described from a lone specimen. Type in Leningrad.

Note. The original specimen (type) of P. angustifolium Ldb. has been preserved. After scrutiny we have come to the conclusion that Trautfetter was wrong in referring it to P. canescens (i. e. Asyneuma), for it has nothing in common with it. The plant unquestionably belongs to Phyteuma and not Asyneuma, whose type is A. canescens. However, Ledebour's specimen cannot be identified as any of the known species of Phyteuma, including those found in the southwestern part of the USSR. The presence of lateral, undeveloped and very thin branches on the stem, the uncommonly small inflorescences and the general appearance of the plant leads us to assume that P. angustifolium is a defective and extremely narrow-leaved form of P. spicatum L., which occurs in the southwestern part of the USSR. In support of our assumption is the fact that nothing resembling P. angustifolium has ever been found anywhere in the south European part of the USSR, particularly near the locus classicus of Ledebour's plant.

P. sibiricum Vest ex Roem. et Schult. Syst. V (1819) 77; Ldb. Fl. Ross. II, 2, 872; Trautf. in Tr. Peterb. bot sada, VI, 1, 54.— P. pauciflorum Roem. et Schult. l. c. 872, pro syn.

"Head many-flowered; flowers all bracteate; bracts oblong, stiff, acute; all leaves cuneate-linear, dentate; stem straight, leafy."

396 According to Roemer and Schultes (l. c.) the plant is found in Siberia.

Note. In addition to the above-quoted diagnosis (in translation) Roemer and Schultes noted that in the original specimen, the "stem is 2 inches (?) high, leaves 3–4 lines long, nearly straight, flat, slightly dentate, nearly lanceolate. Head (inflorescence) spherical, nearly like in Globularia."

Ledebour pointed out in his "Flora rossica," that the plant is unknown to him ("mihi plane ignotum"). Trautfetter (l. c.) submitted that the plant described by Roemer and Schultes is of outside origin (not from Siberia) since none of the Russian travelers in Siberia had ever found anything resembling P. sibiricum ("species a peregrinatoribus rossicis in Sibiria nusquam observata et verosimiliter alienigena"). Later explorations in Siberia corroborated this and there is no doubt now that neither Phyteuma nor Asyneuma, as characterized by Roemer and Schultes, grow in Siberia. It is quite possible that another plant, even a Siberian specimen, but of another family, might have been described under the name Phyteuma sibiricum. We present the translated diagnosis in order to bring this situation to the attention of botanists.

Genus 1441. **ASYNEUMA** * GRISEB. et SCHENK

Griseb. et Schenk in Wiegm. Archiv, XVIII, 1 (1852) 335.— Sect. Podanthum G. Don, Gen. hist. dichlam. pl. III (1831–1838) 748.— Podanthum Boiss. Fl. or. III (1875) 945.

Calyx quinquepartite. Corolla divided nearly to base into 5 lobes, lobes linear-liguliform, always free, at first elongate, later recurved and much longer than calyx teeth, bluish or nearly white. Anthers free, elongate. Style pubescent, with tripartite stigma. Capsule 3-locular, dehiscing at apex by pores or sometimes at the middle of the membranous lateral walls (between ribs), short-cylindrical or short-obconical, angular at cross section, not constricted at apex, rarely short-urceolate. Perennial or monocarpous herbs, with solitary or few, more or less leafy stems, thin rhizome or fusiform root. Leaves alternate, subsessile, clustered or solitary; radical leaves sometimes very approximate, even forming a more or less dense rosette, and then differing noticeably in shape and size from the cauline
397 leaves. Flowers in long spicate-racemiform, often interrupted inflorescence.

* The Greek word Asyneuma could be translated as "not together," meaning, in this case, the character typical of the genus: separation of the corolla lobes, as opposed to Phyteuma in which the lobes are connate at the apex in the shape of a cylinder.

Type: A. canescens (Waldst. et Kit.) Griseb. et Schenk.

Asyneuma is typical for the Mediterranean area, Caucasus and Transcaucasia. One of its species, A. japonicum, is found in the Soviet Far East and eastern Asia.

Grossgeim writes that the root of some species (A. salignum, A. pulchellum) contains rubber (Rastitel'nye bogatstva Kavkaza, ed. 2-e (1952)). Microchemical analysis reveals that the rubber content is large, but the plants have never been thoroughly investigated.

1. Radical leaves not forming a rosette or absent; all leaves except bracts similar in shape and size; stems more or less densely and regularly leafy; rhizome rather thin, usually oblique, rarely straight . . . . 2.
\+ Radical leaves very approximate or forming more or less dense rosettes, distinctly differing in shape and size from the narrower and much smaller cauline leaves; stems slightly leafy, sometimes nearly leafless; root fusiform . . . . 15.
2. Leaves (1)2–3(4) cm wide, distinctly and rather largely acutely toothed, rounded, obtuse or cordate at base; inflorescences dense especially in upper part, densely spicate-racemiform, rarely subcapitate . . . . 3.
\+ Leaves not wider than 1 cm, small-toothed, crenate, sometimes nearly entire, cuneate or attenuate at base and apex; inflorescences elongate, narrow, usually interrupted . . . . 10.
3. Median cauline leaves with amplexicaul cordate base, serrate or biserrate, ovate-oblong; calyx teeth hamately curved. Plant 50–70 cm high, with rather thick stem and dense spicate-racemiform inflorescence . . . . 7. **A. amplexicaule** (Willd.) Hand.-Mazz.
\+ Median cauline leaves not amplexicaul, rounded, obtuse or cuneate at base . . . . 4.
4. Plants glabrous . . . . 5.
\+ Plants more or less pubescent . . . . 6.
5. Plant with spicate, sometimes paniculate inflorescence; root thick, somewhat radish-shaped, fleshy stem 70–100 cm; leaves broad ovate-rhombic, dentate . . . . 6. **A. japonicum** (Miq.) Briquet.
398 + Plant with capitate main inflorescence and loosely flowered poor lateral inflorescences; stem 25–40 cm . . . . . 3. **A. ramosum** Pavl.
6. Plant very short-pubescent; rhizome oblique . . . . 7.
\+ Long-hairy plant, hairs spreading, sometimes unequal; root straight . . . . 8.
7. Flowers 0.5–0.6 cm long; corolla overtopping calyx at aestivation . . . . 8. **A. talyschense** Fed.
\+ Flowers ca. 1 cm long; corolla not constricted . . . . 5. **A. campanuloides** (M. B.) Bornm.
8. Stem thin, not thicker than 1 mm, weak, long; leaves 3–5 cm long, 2–4 cm wide, ovate . . . . 4. **A. debile** Fed.
\+ Stem usually 2–3–7 mm thick; leaves elongate, 5–8 cm long, (1)2–3(4) cm wide, oblong or ovate-oblong . . . . 9.
9. Leaves 8 cm long, 4 cm wide; inflorescence branching; stem thick, 0.5–0.7 cm . . . . 2. **A. baldshuanicum** (B. Fedtsch.) Fed.

\+ Leaves 5–7 cm long, 1–2 cm wide; inflorescence often simple, sometimes slightly branching . . . . 1. **A. argutum** (Rgl.) Bornm.

10\. Leaves nearly entire, narrowly lanceolate, sometimes nearly linear-lanceolate, long-attenuate at both ends, especially at apex . . . . . . 11.

\+ Leaves serrate, finely crenate or finely acute-serrate . . . . 12.

11\. Stem 35–70 cm high; leaves 5 cm long, 5–8 mm wide, lanceolate . . . . 13. **A. attenuatum** (Franch.) Bornm.

\+ Stem 10–15(20) cm high; leaves lanceolate, at lower part of stem reduced, squamose . . . . . . 14. **A. trautvetterii** (B. Fedtsch.) Bornm.

12\. Leaves finely crenate, canescent, scabrous, lanceolate, cuneate-ovate at base. Inflorescence slightly branching; stem sparsely leafy, 30 cm high; root fibrous . . . . 12. **A. canescens** (Waldst. et Kit.) Griseb. et Schenk.

\+ Leaves finely and acutely serrate or with cartilaginous teeth . . . 13.

13\. Leaves finely and acutely serrate, with approximate teeth; stem 20–40 cm high, glabrous, leafy, with simple or branching interrupted spicate-racemiform inflorescence . . . . 9. **A. salignum** (Waldst. et Kit.) Fed.

\+ Leaves serrate, with remote or approximate cartilaginous teeth . . . 14.

14\. Leaf teeth ± remote; calyx and corolla subglabrous . . . . 10. **A. lanceolatum** (Willd.) Hand.-Mazz.

399 + Leaf teeth approximate; calyx and corolla densely pubescent . . . . 11. **A. rigidum** (Willd.) Grossh.

15\. Flowers sessile . . . . 16.

\+ Flowers pediceled . . . . 18.

16\. Calyx teeth oval-lanceolate, obtuse, one-fifth as long as corolla; bracts small, acute; radical leaves oblong-elliptic, obtusely serrate . . . . 17. **A. leianthum** (Trautv.) Bornm.

\+ Calyx teeth broadly lanceolate or linear . . . . 17.

17\. Radical leaves oblong-lanceolate, acuminate, dentate; bracts reniform-orbicular, with mitriform elongate teeth; calyx teeth broadly lanceolate . . . . 16. **A. pulchellum** (Fisch. et Mey.) Bornm.

\+ Radical leaves oblong or oblong-lanceolate, obtuse, irregularly dentate, undulate at margin; bracts oval, acuminate, nearly flat; calyx teeth linear or linear-lanceolate . . . . 15. **A. cichoriforme** (Boiss.) Bornm.

18\. Calyx urceolate, rounded beneath, elongate-narrowed at apex below teeth; capsule short, also slightly urceolate . . . . 18. **A. urceolatum** (Fom.) Fed.

\+ Calyx rounded beneath but not narrowed-elongate below teeth . . . 19.

19\. Calyx pubescent; entire plant finely villous-pubescent; stem thin, simple, scapiform, slightly leafy; radical leaves rosetted; inflorescence spicate-racemiform, with scattered flowers . . . . 20. **A. woronovii** (Fom.) Bornm.

\+ Calyx glabrous or scabrous and sparsely puberulent . . . . 20.

20\. Radical leaves oblong-lanceolate or narrowly lanceolate, tapering at both ends, obscurely toothed at margin; pedicels as long as or longer than calyx tube; calyx teeth linear-lanceolate, acute . . . . 19. **A. lobelioides** (Willd.) Hand.-Mazz.

\+ Radical leaves oblong-lanceolate, attenuate into distinct petioles, crenate at margin but nearly edentate; pedicels as long as calyx tube; calyx teeth thick, triangular-lanceolate, obtuse . . . . 21. **A. otites** (Boiss.) Bornm.

Section 1. EUPODANTHUM (Boiss.) Fed. emend. in Addenda XXIII, p. 340. — Sect. Eupodanthum Boiss. Fl. or. III (1875) 945, p. p. — Radical leaves not forming a rosette or absent; all leaves similar in shape except for bracts; stems more or less densely and regularly leafy. Inflorescences
400 spicate-racemiform. Perennial herbs, with thin usually oblique rhizome, rarely with a straight root.

Type: Asyneuma canescens (Waldst. et Kit.) Griseb. et Schenk.

Series 1. **Argutae** Fed. — Leaves (1)2–3(4) cm wide, distinctly and acutely toothed; inflorescence densely spicate-racemiform especially in upper part.

1. A. argutum (Rgl.) Bornm. in Beih. z. Bot. Zentralbl. XXXVIII, 2 (1921) 339. — Phyteuma argutum Rgl. in Bull. Soc. Nat. Mosc. III (1867) 183; Trautf. in Tr. Peterb. bot. sada, VI, 1, 52; Fedch., Rast. Turk. 721, p. p.; Grigor'ev, Opred. rast. Stalinabada 259. — Podanthum argutum O. et B. Fedtsch. in Tr. Peterb. obshch. estestvoisp. XXXV, 3 (1906) 53, excl. var.; O. and B. Fedch., Perech. rast. Turk. IV, 354.

Perennial; root straight, 0.5–0.7 cm thick, pale brown; plant covered with short and long hairs, usually subglabrous in upper part; stems usually rather tall, 50–70 cm and more, simple or somewhat branching, rather densely leafy; leaves ovate-oblong-lanceolate or linear-lanceolate, rounded or cuneate at base, acutely toothed margin. Inflorescence spicate-racemiform, sometimes branching, dense above, interrupted below; bracts lanceolate, usually crenate; flowers 0.5–0.7 cm long, bluish; calyx teeth subulate; capsule ovoid-globose, trihedral, umbilicate at both tips, 3-locular, 3-ribbed and 2-furrowed along edges, dehiscing at sides between ribs by small slits; seeds oblong-elliptic, flattened, obtuse at both ends, brown, smooth. July–August.

Stony places with herbs, juniper stands, mountain slopes. — Centr. Asia: T. Sh., Pam.-Al. Endemic. Described from W. T. Sh. Type in Leningrad.

Note. The most widespread species in the mountains of Central Asia.

2. A. baldshuanicum (O. Fedtsch.) Fed. comb. nov. — Podanthum argutum var. baldshuanicum O. Fedtsch. in Tr. Peterb. obshch. estestvoisp. XXXV (1906) 53; O. and B. Fedch., Perech. rast. Turk. IV, 355. — P. argutum var. foliosum Fedtsch. op. cit. 53 et al. op. p. p. — Phyteuma argutum B. Fedtsch. Rast. Turk. (1915) 721, p. min. p., non Rgl. — P. baldshuanicum O. et B. Fedtsch. in sched. — P. warsobicum B. Fedtsch. in schedis.

Perennial; root nearly straight, pale brown, ca. 0.5–0.7 cm thick, rugose; stems 1 m, usually rather thick, 0.5–0.6 cm, scabrous with few, rather short
401 hairs, striate, densely leafy, much branching in upper part, rarely simple; leaves ca. 7–8 cm long, 3–4 cm wide, very short-petioled to subsessile, rounded or slightly cordate at base, ovate-oblong, acute or obtuse, irregularly serrate-dentate, sparingly pubescent or subglabrous, upper leaves reduced. Inflorescence paniculate, branching into interrupted spikes; bracts lanceolate, usually dentate; flowers 0.6 mm long, pale, densely crowded in upper part of branches; calyx orbicular, with linear teeth; corolla dissected to base into linear lobes; capsule short, trihedral, prominently ribbed, dehiscing at the middle between ribs by pores. June–July.

Mountainous meadows, juniper stands, river valleys, stony places but not on bluffs. — Centr. Asia: Pam.-Al. (from Gissar Range nearly to Pamir), T. Sh. (Fergana Range). Endemic. Described as a variety of A. argutum from Baldzhuan, but not collected there. The lectotype is a specimen from the Varzov River valley near Pugus, No. 191, labeled by Fedchenko as "Phyteuma baldshuanicum O. et B. Fedtsch. (P. warsobicum B. Fedtsch.)." It is preserved in Leningrad.

Note. A. baldshuanicum is closely related to A. argutum, the main main differences being its large habit, branching inflorescence, and broad leaves rounded at base. It is connected to A. argutum by transitional forms. We ranked A. baldshuanicum as a species in view of the presence of other races (A. trautvetteri and A. attenuatum) within the section Eupodanthum interrelated by transitional forms though distinguished from the first described A. argutum more than A. baldshuanicum. Thus, the series Argutae and Canescentes are undoubtedly young groups of closely related species without definite borders.

3. A. ramosum Pavl. in Vestn. AN Kazakhsk. SSR. 8, (1954) 135. — Phyteuma ramosum Pavl. olim in schedis. — Podanthum argutum var. foliosum Fedtsch. in O. and B. Fedch., Perech. rast. Turk. III (1909) 354, p. p.

Perennial; root thick, short, woody; stems 25—45 cm, cylindrical, longitudinally striate, glabrous, much branching from base with branches arcuately ascending, as long as or half as long as main stem, bearing poor inflorescences; leaves small, sessile, ovate or ovate-lanceolate, rounded or slightly cordate at base, acuminate, 1—2 cm long, 0.5 —1.3 cm wide, glabrous irregularly serrate-dentate to subentire, prominently nerved beneath.
402 Inflorescence terminal, subcapitate, with very approximate many-flowered whorls on the main stem, poorly and sparsely flowered on branches; bracts linear or linear-lanceolate, longer than pedicels; flowers with pedicels 1—2 mm long; calyx campanulate, glabrous, 5—6 mm long, with narrowly linear teeth, 2—3 mm long, 0.3—0.4 mm wide, corolla narrowly campanulate, 8—10 mm long, 3—4 mm wide, dissected nearly to base into 5 linear narrow obtuse slightly scabrous lobes; stamens as long as corolla; capsule unknown. July—August.

On rubble, bluffs, subalpine meadows, stony slopes. — Centr. Asia: T. Sh. (S. Kazakhstan, Ugamskii Range). Endemic. Described from Bostandyn District in E. Kazakhstan near the source of the Ispaisai River. Type in Alma-Ata, isotypes in Moscow and Tomsk.

Note. Pavlov (op. cit.) noted that A. ramosum "is related to A. argutum but distinguished by the stems strongly branching from base (and not simple), leaves sessile (not petioled), ovate (not lanceolate), glabrous (not scabrous-pubescent) and by the shorter and more capitate inflorescence (not interruptedly whorled)."

4. A. debile Fed. sp. nov. in Addenda XXIII, p. 340.

Perennial; plant glaucescent, sparsely covered with white and somewhat spreading hairs; stems thin, 70 cm long, weak, leafy, with more or less thin branches; leaves short, ovate, rounded or slightly cordate at base, short-petioled or subsessile, acuminate or obtuse, acutely and irregularly dentate,

PLATE XXIV. 1 – *Phyteuma spicatum* L.,plant and flowers; 2 – *Asyneuma pulchellum* (Fisch. et Mey.). Bornm.; 3 – *A. campanuloides* (M.B.) Bornm.

teeth antrorse, blades 3–5 cm long, 2–4 cm wide, reduced in upper leaves. Flowers small, pediceled, bluish, in small corymbiform or somewhat paniculate raceme; calyx subglabrous but scabrous, teeth linear, three times as long as tube and half as long as corolla; corolla glabrous, divided into linear lobes; anthers linear, yellow, much longer than filaments and nearly as long as corolla; style filiform, blue, not exserted, with tripartite stigma; capsule (unripe) with 3 lateral pores at the middle by which dehiscing later. July.

Habitat apparently the subalpine mountain zone. — Centr. Asia: Pam.-Al. (S. Tadzhikistan). Endemic. Described from Zevistan Mountain. Type in Leningrad.

405 Note. Related to A. argutum but differing by short and ovate leaves, thin weak stems and branches, and also by the pubescence comprising slightly spreading, white and scabrous hairs.

5. A. campanuloides (M. B.) Bornm. in Beih. z. Bot. Zentralbl. XXXVIII, 2 (1921) 340. — A. campanuloides (M. B.) D. Sosn. in Grossg., Opred. rast. Kavk. (1949) 426; Kharadze in Fl. Gruzii, VIII, 195. — Phyteuma campanuloides M. B. Fl. taur.-cauc. I (1808) 156, III (1819) 148; A. DC. Monogr. 206, p. p.; idem in DC. Prodr. VII, 2, 455; C. A. M. Verzeichn. 83; Ldb. Fl. Ross. 2, 874; Trautf. in Tr. Peterb. bot sada, VI, 1, 51; Shmal'g., Fl. II, 172. — A. campanuloides var. gracile Mand. et. Kuthath. in Zam. po sist. i geogr. rast. Tbil. bot. inst. 18 (1955) 64. — Podanthum campanuloides Rupr. in Bull. Acad. Sci. Pétersb. XI (1867) 172. — P. campanuloides Boiss. Fl. or. III (1875) 949; Somm. et Lev. Enum. pl. Cauc. 327; Fomin in Mater. Fl. Kavk. IV, 6, 133; Medv. in Tr. Tifl. bot sada, XVIII, 328; Grossg., Fl. Kavk. IV, 80. — Exs.: Fl. Caus. exs. No. 249.

Perennial; rhizome oblique, whitish, rather thin; plant scabrous-puberulent to subglabrous; stems simple, straight, 50–70 cm, sometimes 0.5 cm across, usually thinner, rounded or very slightly angular at cross section, usually densely leafy at the lower part, less densely so near apex; leaves crenate or rarely serrate-dentate, often obscurely crenate, 5–8 cm long, 2–4 cm wide, lower leaves ovate, obtuse, short-petioled, the rest sessile, cordate or cordate-semiamplexicaul at base, acute, ovate-oblong, becoming lanceolate at stem apex. Inflorescence spicate, rather short, interrupted at base, consisting of subsessile clusters of small flowers; flowers bluish-violet, 1 cm long; calyx glabrous, narrowly obconical, grooved, teeth straight, lanceolate, acute, nearly twice as long as tube and one-third as long as corolla; corolla deeply cleft into linear recurved lobes; stamens with membranous, ovate, slightly dilated filaments and long anthers; style as long as corolla, with 3 stigmas; capsule oblong, dehiscing at apex by small valves. July–August. (Plate XXIV, Figure 3.)

Subalpine zone, meadows, mountainous forest belt. — Caucasus: everywhere except Talysh. **Gen. distr.:** Iran (?). Described from near Kislovodsk and slopes of Besh-Tau Mountain. Type in Leningrad.

Note. Mandenova and Kutateladze (op. cit.) have described var. gracile Mand. et Kuthath. Judging from the specimen sent to us, this variety could be accepted as an independent species if its distribution area would
406 be established. Its characters are as follows: stems thin, 15–30 cm; leaves

4–4.5 cm long, 2–2.5 cm wide, cuneately tapering at base; flowers small, 10 mm long, in few-flowered clusters. Classical locality: Migaria Mountain in W. Georgia.

6. **A. japonicum** (Miq.) Briquet in Candollea IV (1931) 335. – Phyteuma japonicum Miq. in Ann. Bot. Lugd.-Bat. II (1866) 192; Gerd. in Tr. Peterb. bot. sada, I, 2, 305; Kom., Fl. Man'chzh. III, 569; Kom. and Alis., Opred. rast. Dal'nevost. kr. II, 1001; Kitagawa, Lineam. Fl. Manshur. 420.

Perennial; root thick, fleshy, radish-shaped or terete-cylindrical, usually branching below, brownish, becoming longitudinally rugose when dry, 7–10 cm long, ca. 1 cm thick; stems 1–3, sometimes more, straight, simple or rarely somewhat branching in upper part, 50–100 cm high, glabrous, thinly longitudinally sulcate, sometimes slightly reddish, leafy; leaves alternate, the lower and median rather short-petioled, the upper subsessile, all ovate-rhombic, tapering at both ends, acute, nearly crenate, sometimes acutely bidentate, glabrous, thin. Inflorescence racemiform-spicate, sometimes slightly paniculate; bracts lanceolate-linear, the upper nearly subulate; flowers small, pale blue; calyx nearly ovate, persistent, teeth subulate-linear, three times as long as tube and nearly one-third as long as corolla; corolla divided into narrowly linear, completely free lobes, markedly divergent after flowering, elongating; filaments pubescent, dilated at base; style long, exserted from corolla; capsule short, ribbed, covered with recurved subulate calyx teeth, opening above middle by small pores. Fl. June–July. Fr. September.

Forest edges, glades, dense shrubby formations. – Far. East: Uda, USS., Ze.-Bu. Gen. distr.: Jap., NE China, Korea. Described from Japan. Type in London.

7. **A. amplexicaule** (Willd.) Hand.-Mazz. in Ann. Hofmus. Wien, XXVII (1913) 431; Born. in Beih. z. Bot. Zentralbl. XXXVIII, 2, 338. – A. amplexicaule (Willd.) Sosn. in Grossh., Opred. rast. Kavk. (1949) 426; Kharadze in Fl. Gruzii, VIII, 194. – Phyteuma amplexicaule Willd. Sp. pl. I (1798) 925; A. DC. Monogr. 207, p. p.; idem in DC. Prodr. VIII, 2, 455, p. p.– Podanthum amplexicaule Boiss. Fl. or. III (1875) 948, p. p.; Fomin in Mater. Fl. Kavk. IV, 6, 132, p. p.; Medv. in Tr. Tifl. bot. sada, XVIII, 328; Grossg., Fl. Kavk. I V, 79.

Perennial; plant glabrous or leaves sparingly hairy beneath; stems 50–70 cm, 4 mm across, leafy, straight, simple, slightly angular-sulcate,
407 with simple or slightly branching more or less long inflorescence interrupted at base and consisting of numerous small flowers; leaves strongly serrate or biserrate with acute teeth, lower leaves ovate-oblong, short-petioled, the other sessile, cordate-amplexicaul, acute, lanceolate at stem apex. Flowers 7–8 mm long, short-pediceled, in clusters forming rather dense spikes; calyx glabrous, ovate-cylindrical, teeth shortly triangular-lanceolate, obtuse, usually hamately curved, slightly longer than tube and much shorter than corolla; corolla violet, glabrous, divided to base into linear lobes; capsule ovoid-globose, dehiscing above middle by pores. July.

Open forests along river valleys in arid areas. – Caucasus: S. and E. Transc. (Armenia, Azerbaidzhan). **Gen. distr.:** As. Min., Iran. Described from Asia Minor. Type in Paris.

8. A. talyschense Fed. sp. nov. in Addenda XXIII, p. 341.

Perennial; plant apparently with thin rhizome; stems straight, simple, thickish below, otherwise thin, 50–70 cm, densely leafy at base, becoming less so toward apex, longitudinally striate, glabrous; all leaves cauline, slightly scabrous with short whitish hairs beneath, the lower short-petioled, the rest sessile, truncate or hardly cordate at base, oblong-ovate, acute, 2–5 cm long, 1–2.5 cm wide, prominently nerved beneath, distinctly dentate or biserrate, uppermost leaves reduced. Flowers ca. 0.5–0.6 cm long, short-pediceled, violet-blue, clustered at apex in rather dense spifate interrupted below inflorescence; calyx obconical, glabrous, furrowed, spicate teeth lanceolate-linear, recurved, nearlyas long as or longer than tube, four times longer than corolla; corolla elongate, glabrous, overtopping calyx at aestivation, dissected nearly to base into linear liguliform lobes; anthers elongate, filiform; style filiform, stigmas 3. July.

Upper mountain belt, rocky places in broad-leaved forests. – Caucasus: Tal. Endemic. Described from Lenkoran District, Azerbaidzhan, from the wooded slopes of the Tarakech Mountains south of Alar. Type and paratype in Leningrad.

Note. This species is distinguished from the related A. amplexicaule (Willd.) Hand.-Mazz. by the leaves being truncate or barely cordate
408 and not amplexicaul at base, the inflorescences being shorter, their leaves smaller and recurved, the calyx teeth not being hamate.

Series 2. **Canescentes** Fed. – Leaves up to 1 cm wide, small-toothed, crenate, sometimes subentire. Inflorescences elongate, narrow, usually interrupted.

9. A. salignum (Waldst. et Kit. ex Besser) Fed. comb. nov. – A. salicifolium (A. DC.) Flerov, Spisok rast. Sev. Kavk. i Dag. (1938) 543 (err. "salicifolia"). – A. salicifolium (A. DC.) D. Sosn. in Grossg., Opred. rast. Kavk. (1949) 426; Kharadze in Fl. Gruzii, VIII, 195. – Phyteuma salicifolium Kit. ex Schult. Oesterr. Flora, I, 2, Aufl. (1814) 400, in obs.; Roem. et Schult. Syst. V, 86, in obs. – P. salicifolium Waldst. et Kit. ex A. DC. Monogr. (1830) 205; idem in DC. Prodr. VII, 2, 455, cum nom. auct. Bess. – P. salignum Waldst. et Kit. ex Bess. Prim. Fl. Galiciae, I (1809) 368; in obs.; Ldb. Fl. Ross. II, 2, 873. – P. canescens Waldst. et Kit. s. str. – Podanthum salicifolium Rupr. in Bull. Acad. Sc. Pétersb. XI (1867) 203; Grossg. Fl. Kavk. IV, 79. – Podanthum canescens Boiss. Fl. or. III (1875) 950, p. p.; Lipskii, Fl. Kavk. 380. – P. canescens var. salicifolium Fomin in Mater Fl. Kavk. IV, b (1906) 136.

Perennial; plant more or less glaucescent-green; stems straight, 20–40 cm and even longer, simple or slightly branching in upper part, glabrous or hairy at margin, paler beneath and prominently nerved, oblong-ovate-lanceolate to lanceolate, obtuse or acute, 2–5 cm long, acutely serrate, lower leaves petiolate, the upper narrower, short, gradually becoming linear subentire bracts. Inflorescence more or less dense above, interrupted in lower part, spicate, sometimes slightly paniculate; flowers 1 cm long, subsessile, or pediceled in clusters; calyx glabrous, ovate, longitudinally striate, teeth linear-lanceolate, straight, acute, entire or with lateral teeth, one-fourth to one-third as long as corolla; corolla deeply divided into linear lobes. June—July.

Dry subalpine meadows. — Caucasus: Dag., S. and E. Transc.; European part: U. Dnp., M. D.? **Gen. distr.:** Centr. Eur. (Danube countries), Bal. (W.). A plant from Hungary was apparently implied in the description but Besser marked his specimen, which must be acknowledged as the type Phyteuma salignum, to be from "Podolia."

409 Note. The synonymy of this species is utterly confused and full of errors. It was unintelligibly described by Besser in "Primitiae Florae Galiciae." He indicated Waldstein and Kitaibel as having established this species, therefore in transferring it to Asyneuma the new, above-proposed combination is required. De Candolle is reported as author of most of the earlier combinations, but in his monograph (l. c.) de Candolle included two different nomina nuda (P. salicifolium and P. salignum) and referred to the work of Besser to whom he attributed Phyteuma salicifolium in his "Prodromus." Here he made a mistake because Besser's epithet is "salignum," and not "salicifolium." It is impossible to ascribe the published name "Phyteuma salicifolium" to de Candolle, as Schultes published earlier this name with a short diagnosis (in observat.) in "Oesterreichische Flora," with a reference to Kitaibel. Therefore, the combination of names made by Flerov (op. cit.), Sosnovskii (op. cit.) and Grossheim is inaccurate. It is unjustified to attribute the published, valid name of this plant to Besser since he did refer to Waldstein and Kitaibel. The combination proposed by Ruprecht is also ruled out, since on transferring this species to Podanthum, he repeated de Candolle's error and used Besser's name for the epithet "salicifolium" never used by Besser.

A. salignum might be viewed as a variety of A. canescens (described earlier); both forms grow together in the European part of the USSR and the Danube countries. A. salignum alone is found in the Caucasus, a fact reported long ago by Fomin (op. cit.). These two related species share a common habitat only in the west.

10. A. lanceolatum (Willd.) Hand.-Mazz. in Ann. Hofmus. Wien, XXVII (1913) 431; Bornm. in Beih. z. Bot. Zentralbl. XXXVIII, 2, 341, p. p. — Phyteuma lanceolatum Willd. Sp. pl. I (1797) 924, non Desf.; A. DC. Monogr. 204, p. p.; idem in DC. Prodr. VII, 2, 454, p. p. — Podanthum lanceolatum Boiss. Fl. or. III (1875) 951, p. p.; Fomin in Mater. Fl. Kavk. IV, 37, excl. var.; Grossg., Fl. Kavk. IV, 79, excl. var. — Campanula tauricola Boiss. et. Bal. in Boiss. Diagn. ser. II, 3 (1856) 116.

Perennial; root thick, whitish, simple; plant finely scabrous-puberulent, pale green; stems thin, virgate, robust and slightly woody below, straight, but ascending at base, simple or slightly branching, leafy below up to the middle, terminating in long inflorescence clusters of small flowers; leaves
410 alternate, lanceolate, with more or less remote cartilaginous teeth at margin, attenuate at base, lower leaves short-petioled, soon deciduous, gradually becoming the sessile reduced upper leaves. Inflorescence interruptedly spicate; bracts linear, subentire; flowers small, 1—3 in subsessile clusters, approximate at stem apex, sparse below; calyx scabrous, teeth lanceolate, nearly straight, slightly longer than elongate obconical tube and three times as long as corolla; corolla scabrous, dissected nearly to base into narrowly lanceolate

declinate or recurved whitish or pinkish lobes; style slightly shorter than corolla; capsule oblong-elliptic, dehiscing by 3 pores at apex. July (?).

Apparently growing in dry and stony places. — Caucasus: possibly occurring in the extreme west of Armenia and at the Georgian-Turkish border. **Gen. distr.:** As. Min. Described from Armenia (at its present border with Turkey). Type in Paris.

Note. A. lanceolatum was recorded for the Caucasus by Fomin and Grossheim, authors of "Flora Kavkaza," as a species which might grow also within the USSR. Since Boissier (l. c.) and recently Bornmüller (l. c.) have reported this plant from Trabzon and other parts of Turkey which are geographically close to Transcaucasia, we have to agree with Fomin and Grossheim and include this species in the flora of the USSR. This is even necessary due to the fact that A. rigidum, which is related to A. lanceolatum and has already been found in the USSR, may be regarded only as a variety.

11. A. rigidum (Willd.) Grossh. Opred. rast. Kavk. (1949) 426. — Phyteuma rigidum Willd. Sp. pl. I (1797) 925; A. DC. Monogr. 204, p. p.; idem in DC. Prodr. VII, 2, 545, p. p. — P. canescens Trautv. in Tr. Peterb. bot. sada, II, 2 (1873) 561, VI, 1 (1897) 51, p. p. — P. grossheimii Karjag. in Grossg., Opred. rast. Kavk. (1949) 425, nom. et diagn. brev. ross. in clave. — Podanthum lanceolatum var. rigidum Boiss. Fl. or. III (1875) 952; Fomin in Mater. Fl. Kavk. IV, 6, 137; Bornm. in Beih. Bot. Centralbl. XXXVIII, 2, 343 (subvar. Asyneumatis); Grossg., Fl. Kavk. IV, 79. — Campanula dracunculifolia Boiss. l. c. pro syn. (Boiss. in herb. Kotschyan.).

Perennial; subglabrous plant; stems straight, simple, leafy, slightly virgate, indurate, 40—50(70) cm, ending with long narrow interrupted spicate inflorescence; leaves stiff and thick, narrowly lanceolate to linear-lanceolate, with sectors of cartilaginous acute teeth at margin, lower leaves short-petioled, the upper sessile, smaller and narrower. Flowers small, 8 mm long, sessile or subsessile in clusters of 1—3; bracts short, linear; calyx
411 rather densely pubescent, teeth lanceolate, slightly longer than elongate obconical tube and nearly one-third as long as corolla; corolla pubescent, cylindrical at aestivation, later with spreading narrowly linear lobes; style slightly exserted from corolla; capsule oblong, dehiscing at apex by pores. June.

Dry hills covered with highland xerophytes. — Caucasus: S. Transc. (Nakhichevan ASSR). **Gen. distr.:** As. Min. Described "from the East" ("ex Oriente"). Type in Berlin or Paris.

Note. According to Grossheim (op. cit.) the plant grows in the Nakhichevan ASSR, but unfortunately we have never seen it. Fomin (op. cit.) noted that it differs from A. lanceolatum, yet undiscovered in the USSR, "by the narrower, subglabrous and indurate leaves, beset at margin with sectors of cartilaginous teeth, and the more intensive pubescence of the calyx and corolla." Apparently A. grossheimii Karjag. ex Grossh., which Grossheim included in his key in "Opredelitel' rastenii Kavkaza," should be regarded as synonymous to A. rigidum. Karyagin, to whom Grossheim attributed the establishment of this species, had never described it nor did he accept its existence in nature (in a letter communicated to us). It is impossible to identify this "species" in herbaria; such type has never been collected. Its only difference from A. grossheimii is the narrowness and faint crenation of the leaves, and hence it is quite impossible to accept the species independence of A. grossheimii.

12. A. canescens (Waldst. et Kit.) Griseb. et Schenk in Wiegmann's Archiv, XVIII, 1 (1852) 335; Fuss, Fl, Transs. exc. 415; Born. in Beih. z. Bot. Zentralbl. XXXVIII, 2, 340; Visyulina in Vizn. rosl. URSR, 511; Szaf., Kulcz., Pavl. Rosl. Polskie, 639. – Phyteuma canescens Waldst. et Kit. Pl. rar. Hung. I (1802) 12; Roem. et Schult; Syst. V, 86; M. B. Fl. taur.-caus. III, 149; A. DC. Monogr. 205; idem in DC. Prodr. VII, 2, 455; Ldb. Fl. Ross. II, 2, 873; Trautf. in Tr. Peterb. bot. sada, VI, 1, 50, p. p.; Shmal'g., Fl. II, 182; Fedch. and Fler., Fl. Evrop. Ross. 941; Maevskii, Fl. ed. 8-e, 538. – Podanthum canescens Boiss. Fl. or. III (1875) 950, p. p.; Fomin in Mater. Fl. Kavk. IV, 6, 136, excl. var. – Ic.: Waldst. et Kit. l. c. tab. 14; Geid., Opred. rast. Moldavsk. SSR, 258, fig. 31.

Perennial; root fibrous, pale brown, branching, producing above white wintering tufts of leaves of the next year; stems straight, from 30 cm to 1 m, cylindrical, longitudinally striate, more or less scabrous, simple, leafy; leaves scabrous-canescent, paler beneath, crenate, rather remote along stem, alternate, lower leaves sessile, ovate-lanceolate, ca. 2.5 cm long, 1.3 cm wide; median cauline leaves narrower and smaller, the uppermost
412 gradually becoming the linear-lanceolate subentire bracts. Inflorescence long, spicate, sometimes with lateral many-flowered branches thus appearing paniculate; flowers 1–3 in axils of bracts, erect before and after flowering; calyx more or less scabrous, sometimes canescent, ovate, angular at cross section, obsoletely striate; teeth straight, linear, acute, as long as tube; corolla blue violet, cylindrical in aestivation, acute and tapering above, erect, glabrous or pubescent, three times as long as calyx teeth, broadly open at flowers, incised nearly to base into spreading or recurved narrow liguliform lobes; stamens with membranous, scabrous filaments broadening at margin, anthers rather long; style hardly longer than corolla, stigmas 3; capsule ovate, flattened, dehiscing in upper part by 3 pores; seeds oblong, flattened, marginate, brown, black at ends. July–August.

Stony places in the forest-steppe belt. – European part: Crim., Bes., Bl., U. Dnp. (S.), M. D. **Gen. distr.:** Centr. Eur., Bal. (N.). Described from the environs of Budapest ("prope Budam"). Type in Vienna.

13. A. attenuatum (Franch.) Bornm. in Beih. z. Bot. Zentralbl. XXXVIII, 2 (1921) 340. – Phyteuma attenuatum Franch. Pl. du Turkest. Mission Capus (1883) 114; Fedch., Rastit. Turk. 721. – Podanthum attenuatum O. et B. Fedtsch. in Tr. Peterb. obshch. estestvoisp. 3, XXXV (1906) 53; O. and B. Fedch., Perech. rast. Turk. IV (1911) 355. – P. artutum var. elegans O. Fedtsch. in Tr. Peterb. obshch. estestvoisp. XXXV (1906) 53; O. and B. Fedch. op. cit. (1911) 355.

Perennial; plant scabrous-puberulent; stems straight, 30–70 cm, simple, or many-branched at base; leaves sessile, lanceolate, attenuate at both ends, lower and upper leaves reduced, the median 5 cm long, 5–8 mm wide or wider, subentire or denticulate, acuminate. Raceme simple, sometimes half as long as stem, interrupted at base, dense above; bracts small; pedicels half as long as calyx; calyx glabrous, teeth linear, one-third as long as corolla; corolla pale blue, with linear lobes; capsule unknown. June.

Mountain river gorges. – Centr. Asia: Pam.-Al. (Zeravshan River valley). Endemic. Described for Chyullyuk in the Zeravshan River valley. Type in Paris.

Note. No specimens of Franchet collections exist in the USSR. Fedchenko, who saw the original specimen in Paris, claims that Franchet was
413 wrong in considering this species as something between A. argutum and Cylindrocarpa sewerzowii. Fedchenko reports that A. attenuatum is reminiscent of var. elegans O. Fedtsch which was described as a variety of Podanthum argutum (Rgl.) O. et B. Fedtsch. (O. and B. Fedchenko, op. cit. (1911) 355). This variety cannot be referred to Asyneuma argutum (Rgl.) Bornm. as is now accepted and it has to be placed as another synonym of A. attenuatum (Franch.) Bornm.

14. A. trautvetteri (B. Fedtsch.) Bornm. in Beih. Bot. Zentralbl. XXXVIII, 2 (1921) 350. – Phyteuma sp. Trautv. in Tr. Peterb. bot. sada, VI, 1 (1879) 54, absque nom. diagn. tantum. – P. trautvetteri B. Fedtsch. Rastit. Turk. (1915) 721. – Podanthum trautvetteri B. Fedtsch. in Tr. Peterb. obshch. estestvoisp. XXXV, 3 (1906) 53; O. and B. Fedch., Perech. rast. Turk. IV, 355, also in Beih. Bot. Zentralbl. XL, 2 (1923) 213.

Perennial; plant subglabrous or hispid; root fusiform, ca. 0.5–0.6 cm in diameter; stems straight, densely leafy, 10–15(20) cm, simple or slightly branching with thin branches; lower leaves reduced, squamate, scarious, small; cauline leaves oblong-lanceolate, acute, narrowly cuneate at base, nearly entire or slightly denticulate, noticeably smaller in lower and upper part of stem, the lower short-petioled, the upper sessile, median leaves 4 cm long, 12 mm wide, leaves on branches the same shape as those on main stem but smaller, uppermost leaves very small, linear-acuminate. Flowers with thin, sometimes rather long pedicels, spreading, forming loose spicate-racemiform or slightly paniculate (branching) inflorescence; bracts minute, linear-acuminate; calyx glabrous, teeth linear, mucronate, nearly twice as long as the angular and grooved tube but shorter than corolla; corolla glabrous, dissected nearly to base into linear lobes. July.

Stony or dry herbaceous places, sometimes juniper stands in the central mountain zone. – Centr. Asia: Pam.-Al. (Zeravshan Range, Gissar Range, Alai Range, Angren River valley). Endemic. Described from "Kokan and also near Zeravshan River." Type in Leningrad.

Note. Trautfetter (l. c.) in his diagnosis described the species without naming it. Fedchenko gave the name but based his description on the specimen collected by O. A. Fedchenko atop Naubid Mountain in the Zerav-
414 shan valley. Nevertheless, the specimen described by Trautfetter ("Kokania, Isfairam inter Langar et Tengis-bai, leg. O. Fedtschenko") should be regarded as the authentic one. It should be noted that both specimens are in poor condition, the main stem was torn or cut off by cattle. The best specimens are the autotypes from the Mur mountain pass and Angren, according to which we established the description given here.

Section 2. SCEPTRUM Fed. in Addenda XXIII, p. 342. – Sect. Eupodanthum Boiss. Fl. or. III (1875) 945, p. p. – Radical leaves forming more or less dense rosettes or generally approximate, differing in shape and size from cauline leaves; stem slightly leafy, sometimes nearly leafless. Inflorescence long, interruptedly spicate-racemiform. Mostly monocarpic herbs, rarely perennials, with fusiform root.

Type: Asyneuma pulchellum (Fisch. et Mey.) Bornm.

Series 1. **Pulchellae** Fed. – Flowers sessile. Stems coarse, robust. Radical leaves in dense rosettes.

15. A. cichoriforme (Boiss.) Bornm. in Beih. z. Bot. Zentralbl. XXXVIII, 2 (1921) 340. – Phyteuma cichoriforme Boiss. Diagn. ser. 1, 4 (1844) 37. – Podanthum cichoriforme Boiss. Fl. or. III (1875) 946; Fomin in Mater. Fl. Kavk. IV, 6, 131; Grossg. Fl. Kavk. VI, 79. – Exs.: Pinard, Pl. exs., isotypus!

Biennial; plant covered with very fine and short white hairs; root ca. 0.8 cm in diameter, fusiform, brown; stems thick, short [!?], 70 cm and even taller, somewhat sulcate, angular at cross section, more or less leafy below, nearly leafless at middle and apex, terminating in paniculate spreading inflorescence; radical leaves undulate at margin but not curly, oblong-lanceolate, obtuse, attenuate into rather long petioles, together with petioles 25 cm long and longer, blades 2 cm wide and more; cauline leaves few, small, linear. Inflorescence composed of rather thin, long and strict, interruptedly spicate branches; bracts ovate, acuminate but not mitriform; flowers 0.8 cm long, 3–5 in sessile clusters; calyx teeth glabrous, linear, half as long as corolla and nearly twice as long as oblong tube covered with whitish dots or very short papilliform hairs; corolla dissected into linear lobes; style not exserted. June.

415 Clayey bluffs. – Caucasus: possibly near the Turkish border. **Gen. distr.**: As. Min., Iran. Described from Karia Mountain (As. Min.) and Kadma near Denizli. Type in Geneva, isotypes in exsiccatae.

Note. A rare plant, collected only near Kars, Kagyzmen and Ol'ta. Very probably it grows also at the Soviet border with Turkey. It differs from the very similar A. pulchellum by the much larger and longer leaves and the shape of the bracts (not mitriform).

16. A. pulchellum (Fisch. et Mey.) Bornm. in Beih. Bot. Centralbl. XXXVIII, 2 (1921) 348. – A. pulchellum (Fisch. et Mey.) D. Sosn. in Grossg., Opred. rast. Kavk. (1949) 425. – Phyteuma pulchellum Fisch. et Mey. Ind. sem. hort. Petrop. I (1835) 35; A. DC. in DC. Prodr. VII, 2, 455; Hohenack. in Bull. Soc. Nat. Mosc. VI, 292; Ldb. Fl. Ross. II, 2, 873; Trautf. in Tr. Peterb. bot. sada, VI, 1 (1877) 50. – Podanthum pulchellum Boiss. Fl. or. III (1875) 947; Fomin in Mater. Fl. Kavk. IV, 6, 131; Medv. in Tr. Tifl. bot. sada, XVIII, 2, 328; Sosnovsk. in Tr. Azerbaidzh. otd. Zakavk. fil. AN SSSR, sect. bot. 1, 46; Grossg., Fl. Kavk. IV, 79. – P. limonifolium Hohenack. Enum. Talysch. (1838) 62; C. A. M. Verzeichn. 83, non Sibth. – Campanula pulchella Boiss. in Nouv. Mem. Soc. Nat. Mosc. XII (1860) 144.

Biennial; plant finely short-pubescent; root thick, 1 cm across, fusiform, simple, pale brown; stems 1 cm across below, leafy, coarse, robust, 70 cm, paniculately branching above into long upright branches; radical leaves oblong-lanceolate, acuminate, attenuate into short petioles, slightly crisp-dentate at margin, 7 cm long, 1–1.5 cm wide; cauline leaves sessile, gradually reduced toward stem apex, narrower and more remote above. Inflorescence spicate-paniculate, consisting of 2–5 clusters of sessile flowers; bracts slightly mitriform, dentate, rounded-reniform below, semi-amplexicaul; calyx subglabrous or sparingly pubescent, teeth broadly lanceolate, twice as long as the short turbinate tube and half as long as corolla; corolla 6–8(10) mm long, dissected nearly to base into linear lobes; capsule ovoid, rounded at base, dehiscing at apex by rather large pores; seeds brownish, shiny. June–July. (Plate XXIV, Figure 2.)

Stony slopes covered with stiff dwarf shrubs in the upper and central mountain zones. — Caucasus: S. Transc. (Armenia, Daralagez, Nakhichevan
416 ASSR, Karabakh), Tal. **Gen. distr.:** N. Iran. Described from Zuvand (treeless clearings atop Talysh Range). Type in Leningrad.

17. A. leianthum (Trautv.) Bornm. in Beih. z. Bot. Zentralbl. XXXVIII, 2 (1921) 343. — Phyteuma leianthum Trautv. in Tr. Peterb. bot. sada, III, 2 (1874) 274, VI, 1 (1877) 49, 50, in obs. — Podanthum leianthum Boiss. Fl. or. Suppl. I (1888) 334; Fomin in Mater. Fl. Kavk. IV, 6, 132; Grossg. Fl. Kavk. IV, 78.

Biennial or perennial? Root rather thick, fusiform, simple; stems coarse, rather thick in lower part, 50 cm and even taller; virgate, slightly leafy or leafless, antrorsely branching in upper part, bearing loose inflorescence of long interruptedly spicate branches; leaves obtusely but distinctly large-serrate-dentate or biserrate, the lower 5–7 cm long, 1.5–2 cm wide, obovate-oblong, slightly spatulate, obtuse, long-cuneate, gradually attenuate into obsolete petioles; other leaves probably smaller (absent in specimens at hand). Flowers whitish (?), sessile, in clusters of 3; bracts glabrous, very small, acute ovate; calyx glabrous, teeth ovate-lanceolate, obtuse, one-fifth as long as corolla, persistent and erect in fruit; corolla dissected nearly to base into linear lobes; style as long as corolla, with 3 rather short, slightly curved stigmas; capsule oblong, markedly elongating under the persistent calyx, dehiscing with apical pores. July.

Apparently in dry stony places. — Caucasus: might possibly be found near the Turkish border. **Gen. distr.:** As. Min. (Olta). Described from the banks of the Olta-Chaya. Type in Leningrad, isotype in Tbilisi (Georgian Museum).

Note. The only existing specimens are those collected by Radde in 1874 in the above-noted locality.

Series 2. **Urceolatae** Fed. — Flowers pediceled. Stems thin-virgate; radical leaves approximate, sometimes rosetted.

18. A. urceolatum (Fom.) Fed. comb. nov. — Podanthum lobelioides var. urceolatum Fom. in Mater. Fl. Kavk. IV, 6 (1906) 138; Grossg. Fl. Kavk. IV, 79. — P. lobelioides Lipsky, Fl. Kavk. (1899) 380, p. p. non Boiss. — P. lobelioides Hausskn, in Fomin, l. c. pro syn. — P. urceolatum B. Schischk. in Grossh. et Schischk. Schedae ad herb. "Pl. orient. exs." I–VIII (1924) 38. — Exs.: Pl. or. exs. No. 149.

Perennial; root fusiform, pale brown, rather thick, branching in lower part; stems straight or slightly flexuous, rather thin, 30–40 cm, glabrous,
417 smooth, often branching with flexuous thin nearly leafless antrorse branches; radical leaves glabrous (like the whole plant), long-petioled, blade oblong-ovate or lanceolate, obtuse or acute, nearly entire or very obscurely dentate; cauline leaves reduced, sessile, lanceolate-linear, the uppermost linear-subulate in inflorescence. Flowers small, in paniculate or rarely spicate-racemiform inflorescence, with short pedicels elongating in fruit; calyx glabrous or slightly scabrous, teeth linear, nearly twice as long as the urceolate elongate tube and one-third to one-half as long as corolla; corolla dissected nearly to base into linear lobes; style as long as corolla; capsule short, broadly cylindrical to urceolate, slightly transversely rugose, pale brown, crowned by the persistent, antrorse and elongated calyx teeth. May–July.

Dry stony slopes, bluffs in the central mountain zone. – Caucasus: possibly near Turkish border. **Gen. distr.**: As. Min. (Kars, Olta, Kagyzman). Described from the upper reaches of Akh-Chai River near Kagyzman. Type in Leningrad, isotype in Tbilisi.

Note. Distinguished from the related A. lobelioides (Willd.) Hand.-Mazz. by its urceolate calyx, globular below and elongating above, and the short, slightly urceolate capsule.

19. A. lobelioides (Willd.) Hand.-Mazz. in Ann. Hofmus. Wien, XXVII (1913) 431; Bornm. in Beih. z. Bot. Zentralbl. XXXVIII, 2, 345. – Phyteuma lobelioides Willd. Phytogr. I (1794) 20; Roem. et Schult. Syst. V, 84; A. DC. Monogr. 203; idem in DC. Prodr. VII, 2, 454. – P. lanceolatum Desf. Choix de pl. du Coroll. de Tournef. (1808) 34, non Willd. – P. stylidioides Boiss. Diagn. ser. 1, 11 (1849) 73. – P. pestalozzae Boiss. l. c. 74. – P. gracile Boiss. l. c. 74. – Campanula willdenowii Boiss. l. c. ser. II, 3 (1856) 117. – Podanthum lobelioides Boiss. Fl. or. III (1875) 953; Fomin in Mater. Fl. Kavk. IV, 6, 137, excl. var. Grossg. Fl. Kavk. IV, 79, excl. var. – Ic.: Willd. l. c. tab. 4, f. 2.

Perennial; root rather thick; plant glabrous or short-pubescent; stems thin-virgate, slightly leafy, 20–60 cm, simple or branching, terminating in loose racemiform-spicate inflorescence; radical leaves petiolate, narrowly lanceolate, rarely oblong, attenuate at both ends, acute, crenate-dentate with glandular teeth, usually undulate at margins; cauline leaves few, very narrowly linear. Flowers pediceled, solitary in axils of bracts or in
418 clusters of 2–3; bracts subulate mucronate; calyx glabrous or scabrous, teeth linear-lanceolate, slightly shorter than tube and nearly one-third as long as corolla; tube oblong, constricted below; corolla violet, dissected into linear lobes; style as long as corolla, with 3 thick stigmas; capsule ovoid-oblong, dehiscing at apex by pores. May–June.

Dry herbaceous places and bluffs. – Caucasus: possibly near Turkish border. **Gen. descr.**: As. Min. Described from Turkish Armenia. Type in Paris.

Note. We cannot understand why Boissier (Fl. or l. c.) found de Candolle's diagnosis ("Campanulaceae" (l. c.) and "Prodromus" (l. c.)) as not corresponding with the type collected by Willdenow (according to Tournefort) inasmuch as the description completely fits the diagnosis made by Boissier himself. It is possible that de Candolle's specimen (from the Lamarck herbarium) appeared to Boissier to have little resemblance to Willdenow's type. Fomin (op. cit.) claims that this species was collected by König in Kars near Olta and by Mikhailovskii on Karadash Range near Kagyzman. We have not seen these specimens, which are in Tbilisi. The specimens in Leningrad refer to A. urceolatum (Fom.). We respect Fomin's authority and therefore include A. lobelioides in the flora of the USSR, since its discovery in the USSR is possible; it is probable, however, that the specimens from the herbaria of König and Mikhailovskii may prove to be A. urceolatum and not typical A. lobelioides.

20. A. woronovii (Fom.) Bornm. in Beih. z. Bot. Zentralbl. XXXVIII, 2 (1921) 351. — Podanthum woronovii Fom. in Vestn. Tifl. bot. sada, X (1908) 35.

Perennial; plant shortly villous-puberulent (hairs visible only under magnification); stems thin, slightly leafy, simple, scapiform, with dry leaves around base; radical leaves rosetted, oval or oblong, remotely toothed at margin, attenuate into petioles several times longer than blade; lower cauline leaves linear-lanceolate, acuminate, crenate-dentate, the upper sessile, gradually reduced. Flowers solitary, small, pediceled, in sparsely spicate raceme; bracts small, ciliate at margin; calyx pubescent, teeth half as long as tube and one-third as long as the glabrous, violet corolla. June—July (probably).

Forest-edges in arid regions. — Caucasus: possibly in S. Transc. **Gen.**
419 **distr.:** As. Min. (Artvin District). Described from Artvin District (between Ardanuch and Akhisar). Type in Tbilisi.

Note. We are adding this species to the list of those which may one day be found in Transcaucasia. It is important to include it in "Flora of the USSR" since it has never appeared in the "Flora" series and if found might be considered as a new species. It was mentioned only twice in the literature and has been almost completely forgotten by the botanists. Grossheim does not list it in his "Flora Kavkaza" among species not yet discovered in the Caucasus. Fomin (op. cit.) reports that it is close to A. lobelioides. We have never seen the original specimen.

21. A. otites (boiss.) Bornm. in Beih. z. Bot. Zentralbl. XXXVIII, 2 (1921) 347. — Podanthum otites Boiss. Fl. or. III (1875) 952; Fomin in Mater. Fl. Kavk. IV, 6, 137; Grossg. Fl. Kavk. IV, 79. — Phyteuma otites Trautv. in Tr. Peterb. bot. sada, VI, 1 (1897) 49, quoad pl. Ledebourianam. — P. limonifolium Ldb. Fl. Ross. II, 2 (1846) 873, excl. syn. plur. non Sibth.

Perennial; finely pubescent plant; stems 70 cm, thin and virgate, usually simple but sometimes with thin, somewhat spreading branches in upper part, somewhat leafy in lower part, terminating in very long and thin inflorescence; radical leaves narrowly oblong-lanceolate, attenuate into short petioles, obtuse, obscurely crenate-dentate at margin; cauline leaves very narrow, linear-lanceolate, sessile. Inflorescence interruptedly spicate; flowers very small, short-pediceled, in clusters of 2—3 at axils of short acuminate bracts; calyx glabrous, teeth thick, short-triangular-lanceolate, obtuse, often ob-conical, two-fifths as long as corolla; corolla small glabrous, dissected to base into linear lobes; capsule small, ovate, strongly furrowed, ribbed, transversely rugose, dehiscing at apex by pores. June (?).

Probably growing in dry places in the central or lower mountain zones. – Caucasus: possibly in Adzharistan or somewhere in Dzhavakhetiya. **Gen. descr.**: As. Min. Described from the Taurus Mountains in Cilicia near the villate of Gülek Boğaz.

Note. This species was collected by Norman near Rize at the Turkish
border; this only existing specimen (at Ledebour herbarium) serves as the
basis for assuming the possibility of discovering it within the USSR. It
differs from all the species of Asyneuma by its exceptionally small
420 flowers (3–4 mm long) and minute capsules which Boissier (l. c.) described
as "magnitudine seminis Cannabeos." It is also distinguished by its nearly
branchless stem, long and narrow, interruptedly spicate inflorescences, and
the sparse leaves.

Genus 1442. **SERGIA*** FED., gen. nov.

Fed. in Addenda XXIII, p.342

Calyx semi-spherical, dissected nearly to base into 5 lanceolate acute teeth. Corolla dissected nearly to base into 5 lanceolate lobes nearly as long as calyx teeth or elongating after flowering. Stamens 4, with very short thick filaments and elongate nearly sessile anthers, sometimes sterile and not completely developed. Style cylindrical, blue, covered with small papillae; stigmas 3. Capsule spherical, ribbed due to the protruding septa, tapering at apex, membranous, villous or hispidulous, dehiscing below between ribs by pores covered with small valves, crowned by persistent calyx teeth. Seeds smooth, globular, small, straw-yellow. Perennial plants, with numerous stems, indurate distally thickened grumous root and narrowly lanceolate or ovate leaves. Type Sergia sewerzowii (Rgl.) Fed.

Oligotypic genus endemic to W. Tien Shan and Pamir-Alai. It is distinguished from other genera of Campanulaceae by its spherical, angular-ribbed, tapering and not broadening above capsules, fusiform subsessile anthers, very short thickened filaments, by shape of its flowers and other parts, and by its general appearance.

1. Calyx and corolla densely and finely hispid with short thick hairs outside; leaves ovate-lanceolate, dentate . . . . . 2. **S. regelii** (Trautv.) Fed.
\+ Calyx with fine papilliform very short hairs; corolla glabrous; leaves linear-lanceolate, nearly entire . . . . 1. **S. sewerzowii** (Rgl.) Fed.

1. S. sewerzowii (Rgl.) Fed. comb. nov. – Campanula sewerzowii
Rgl. in Bull. Soc. Nat. Mosc. III (1867) 188; Trautf. in Tr. Peterb. bot.
421 sada, VI, 1, 82 ("Sewerzowii"; sphalm. nom. auct. "Rgl. et Herd."); B. Fedch.,
Rast. Turk. 720. – Ic.: Rgl. l. c. tab. V.

Perennial; root thick, woody, brown, broadening in upper part; stems sometimes more than 100, almost rounded or with irregular grumose

* Named after Professor S.V. Yuzepchuk, who worked on the classification of bellflowers.

thickenings, ca. 10–15 cm long, 4–5 cm [mm] across, branching, often geniculate and many times curved or flexuous, leafy, glabrous, bearing paniculate inflorescence; leaves only cauline, subequal in shape, diminishing toward stem apex, linear-lanceolate, acute, with fine contiguous teeth at margin, short-petioled or sessile, 2–4 cm long, ca. 3 mm wide, the uppermost much narrower and shorter. Flowers small, 0.7 cm long, blue, with short thin pedicels; calyx short, subspherical, teeth lanceolate, acute, prominently nerved, nearly as long as corolla, appendages absent; corolla cup-shaped-campanulate, glabrous, divided deeply or nearly to base into oblong, obtuse or acute lobes; stamens with very short filaments and subsessile fusiform-narrowly-lanceolate anthers, very often stamens sterile, with undeveloped anthers; style blue, fusiform at apex, otherwise cylindrical, with 3 arcuate stigmas, covered with small papillae, exserted after flowering; capsule spherical, ribbed, dehiscing by valves at the lower part between septa; seeds small, globular, yellowish. June–July. (Plate XXV, Figure 1.)

Rock-crevices in dry places. – Centr. Asia: T. Sh. (Talass Ala-Tau, Kara-Tau). Endemic. Described from Boroldai in the mountains of Kara-Tau. Type in Leningrad.

Note. Regel based his description on poorly dried, fruitless specimens which he referred to Campanula. Trautfetter (op. cit.), also lacking fruits, could not find a systematic place for this species ("Fructus ignoti ideoque speciei in systemate locus mihi omnino dubius"). The new collections have ripe capsules which are dissimilar in shape to those of Campanula; this, together with other characters, makes it possible to separate Campanula sewerzowii and Phyteuma regelii as an independent genus, as we have done.

2. S. regelii (Trautv.) Fed. comb. nov. – Phyteuma regelii Trautv. in Tr. Peterb. bot. sada, VI, 1 (1879) 53; B. Fedch., Rast. Turk. 721. – P. multicaule Franch. Pl. du Turkest. Mission Capus (1883) 115. – Asyneuma regelii Bornm. in Beih. z. Bot. Zentralbl. XXXVIII (1921) 348; Popov et Vvedensky in Sched. ad. Herb. Fl. As. Med. XVIII (1928)
422 No. 449; V. Nikitin in Fl. Turkm. VI, 375. – Podanthum regelii O. et B. Fedtsch. in Tr. Peterb. obshch. estestvoisp. XXXV, 3 (1906) 53; O. and B. Fedch., Perech. rast. Turk. IV, 355. – Ic.: Franch. l. c. tab. 15, f. B. – Exs.: HFAM, No. 449.

Perennial; root thick, indurate, grumose, producing stems and sterile shoots; stems many, 10–15–25 cm long, thin, scabrous, with small reduced scarious-squamous brown leaves at base, rather densely leafy above, straight or slightly arcuate at base, with thin branches in upper part forming somewhat paniculate inflorescence; leaves of sterile shoots ca. 2 cm long, gradually attenuate into long petioles, obovate-spatulate; cauline leaves 2–3 cm long, ovate or oblong-lanceolate, acuminate, tapering at base, acutely dentate with teeth sometimes aristate, short-petioled or subsessile; upper leaves sessile, usually subentire, reduced; all leaves slightly scabrous-hispidulous, glaucescent or glaucous-green. Flowers rather long-pediceled, 0.6–0.7 cm long, bluish; bracts acuminate. entire; calyx subspherical, verrucose-dotted, prominently nerved, teeth linear, acuminate, straight, scabrous, accrescent after flowering, without appendages; corolla finely verrucose-dotted outside, divided nearly to base into linear lobes, as long

PLATE XXV. 1 – Sergia sewerzowii (Rgl.) Fed., plant and capsule; 2 – Cryptocodon monocephalus (Trautv.) Fed.; 3 – Cylindrocarpa sewerzowii Rgl., plant and capsule.

as calyx teeth at first, later hardly as long; capsule trilocular, subglobose, scabrous and finely verrucose-hairy, dehiscing at base by small pores, covered with dried calyx teeth. June–July.

Crevices in mostly limestone bluffs, – Centr. Asia: Pam.-Al. (Zeravshan, Gissar and Turkestan ranges, Kugitang). Endemic. Described from the Zeravshan River valley (Iora mountain pass). Type in Leningrad.

Genus 1443. **CRYPTOCODON** * FED., gen. nov.

Fed. in Addenda XXIII, p.342

Flowers 5-parted. Ovary trilocular. Stigmas 2-parted. Style not exerted. Corolla tubular, deeply 5-lobed, lobes straight, lanceolate. Calyx 5-toothed, teeth nearly as long as corolla and nearly three times longer
425 than tube, with small downcurved appendages in between. Capsule elliptic, opening at middle by 3 pores. Perennial low plants, with very thick and indurate grumous root and densely leafy low stems. Leaves [more or less] oblong, denticulate. Flowers in dense capitate inflorescence, surrounded by involucre of dense terminal leaves longer than flowers.

Type: C. monocephalus (Trautv.) Fed.

Taxonomically, Cryptocodon is closely related to Asyneuma but in appearance it does not resemble the latter in any characters. It differs by the straight (not curved) lanceolate corolla lobes as long as calyx teeth, the tubular corolla, style with bipartite stigma, not exserted, by the small, but conspicuous appendages opposite calyx incisions, the dense capitate inflorescence surrounded by involucral leaves, and by the thick, woody, markedly grumous root. Such characters have not been observed in any of the species of Asyneuma.

A monotypic genus endemic to mountains in the Zeravshan River valley.

1. C. monocephalus (Trautv.) Fed. comb. nov. – Campanula monocephala Trautv. in Tr. Peterb. bot. sada, VI, 1 (1879) 64; O. and B. Fedch., Perech. rast. Turk. III, 356; B. Fedch., Rast. Turk. 720. – Phyteuma monocephalum (Trautv.) Pavl. in sched. – P. occultans M. Pop. et Vved. in sched.

Perennial; low plant, 8–13 cm; root very thick, woody, grumous, short-branching in upper part, covered above with short, squamous remnants of the dead petioles; stems many, straight, indurate, densely leafy, scabrous; lower leaves ovate-elliptic or oblong, 7 cm long, 1.5 cm wide, obtuse or rounded at apex, gradually tapering at base, denticulate, short-petioled; cauline leaves hispidulous, elongate, oblong, tapering at both ends, acute, acutely toothed, sessile, gradually becoming the involucral leaves, involucral leaves shorter, ovate, acute, rounded, amplexicaul, dentate, somewhat overlapping. Flowers axillary, erect, crowded in small head, subsessile; bracts oblong-spatulate, dentate, as long as flowers; calyx elliptic, teeth two to three times as long as tube, oblong-lanceolate, acute, entire, scabrous at margin, as long as corolla; corolla blue, 1.5–2 cm long, tubular-campanulate, glabrous, deeply quinquepartite; style nearly not exserted, stigmas 2; capsule
426 elliptic, pentagonal at cross section, dehiscing at the middle by 3 pores, scabrous, erect. June. (Plate XXV, Figure 2.)

* From the Greek cryptos – hidden and codon – bell, i.e. hidden bell.

Mountainous steppes of couch grass and wormwood. — Centr. Asia: Pam.-Al. Endemic. Described from mountains in Zeravshan River valley near Dasht-i-Kaza. Type in Leningrad.

Genus 1444. **CYLINDROCARPA** * RGL.

Rgl. in Tr. Peterb. bot sada, VI, 1 (1877) 258.

Calyx tube cylindrical, deeply quinquepartite. Corolla with very short tube and deeply quinquepartite into linear-oblong lobes, at first closed together, later free and spreading. Stamens 5, filaments very short, ovate-triangular below, tapering beneath anthers; anthers oblong-linear, much longer than filaments, free, straight. Style with bipartite or tripartite stigma. Capsule narrowly cylindrical, attenuate, 2- or 3-locular, dehiscing at apex by slitlike holes. Root woody, perennial; stems leafy, numerous, branching; leaves narrowly lanceolate.

Type: Cylindrocarpa sewerzowii Rgl.

Monotypic genus, endemic to the Kara-Tau Mountains in W. Tien Shan, in appearance resembling Campanula, corolla like Asyneuma, and capsule like Legouzia (Specularia).

1. C. sewerzowii Rgl. in Tr. bot. sada, V, 1 (1877) 258; O. and B. Fedch., Perech. rast. Turk. III, 356. — Phyteuma sewerzowi Rgl. et Herd. in Bull. Soc. Nat. Mosc. III (1876) 184; Trautf. in Tr. Peterb. bot sada, VI, 1 (1879) 54 ("Ssewerzowii"); B. Fedch., Rast. Turk. 721.

Perennial; completely glabrous plant; root thick, woody, with upper part large, grumous, 7 cm in diameter, bulbous nodes at the upper part bearing short brown branches in addition to rudiments of stems sometimes numbering several hundred; stems 20—40 cm long, growing in very dense obconical or subspherical tuft, leafy, erect or ascending, rather thin, strict, becoming somewhat virgate, slightly flexuous, glabrous in lower part, yellow-brown, with obliquely ascending branches; leaves linear-lanceolate, spreadingly dentate, attenuate at base, acute, (1)2—4(5) cm long, 3—4 mm wide, bright green, subsessile, the lower deciduous. Flowers solitary; calyx with cylindrical tube, teeth narrowly lanceolate, acute, one-third as long as corolla;
427 corolla blue, 1—1.5 cm long, dissected nearly to base into lanceolate or narrowly oblong, acute, spreading lobes; stamens barely as long as calyx lobes, with long anthers and short filaments dilated below; style filiform, with 2—3 filiform stigmas; capsule terete or narrowly cylindrical, urceolate-tapering in upper part where open by longitudinal, parallel, narrow slits, crowned by dry calyx teeth. June—July. (Plate XXV, Figure 3.)

Dry rocky places and pebbly mountain slopes. — Centr. Asia: T. Sh. (Kara-Tau Mountains). Endemic. Described from Kara-Tau Mountains. Type in Leningrad.

Note. E. Regel (op. cit.) erroneously reported that this plant had been collected by A. Regel in the mountains of "Alatau, ad fl. Tschotkal, Kara-Tschok." There is no specimen bearing such a label, and the plant could not have been collected in Chatkal because it has never been encountered there.

* From the Greek cylindros — cylinder and carpos — fruit, i.e. cylindrical fruit.

Genus 1445. **LEGOUZIA** * DURAND

Durand Fl. Bourg. II (1782) 26.– Specularia A. DC. Monogr, (1830) 344.– Prismatocarpus L'Hér. Sert. Angl. (1788) 2, p.p.– Apenula Neck. Elem. bot. I (1790) 234.

Calyx quinquepartite, with elongate prismatic or long obconical tube. Corolla rotate or broadly open, somewhat campanulate or infundibular, 5-lobed, lobes nearly as long as calyx teeth or noticeably shorter. Stamens 5, free, half as long as corolla, with very short filaments and long subfiliform anthers. Style not exserted, pubescent with fine hairs in 10 longitudinal rows. Stigmas 3, rarely 2, filiform, usually curved, rather short. Capsule long-prismatic, trilocular or unilocular as a result of abortive septa, dehiscing at apex or a little above middle by slits or small valves. Seeds ovoid or lentiform, slightly flattened, shiny. Low annual herbs, without rosettes, with small alternate leaves and terminal or axillary, bluish, pink or whitish, rather small flowers; widespread in the Mediterranean area.

Type: L. speculum-veneris (L.) Fisch.

One of the synonyms of the now approved generic name Legouzia is Specularia, proposed by Heister (Heister, Syst. pl. gen. ex fructific. etc., 1748, 8) who till now has been designated as the author. However, though
428 the legitimate synonym is still Specularia, the author's name should be de Candolle, since Heister had established his genus before the nomenclatural date of 1753.

All the species of Legouzia found in the USSR refer only to section Gerontogeae A. DC. As a result no description is given here. Specularia perfoliata (L.) A. DC., a representative of section Americana A. DC., which was reported by Shmalhauzen (Fl. 180) as a weed in the Crimea, has apparently disappeared. For this reason we are not including this species in "Flora of the USSR."

1.

1. Corolla as long as or slightly longer than calyx teeth; stems with spreading branches . . . . 2.

\+ Corolla nearly half as long as calyx teeth; stem usually simple or branching in lower part . . . . 3.

2. Capsule not constricted at apex; calyx teeth half as long as tube at flowering . . . . 2. **L. pentagonia** (L.) Thell.

\+ Capsule conspicuously constricted at apex; calyx teeth as long as or almost as long as tube . . . . 1. **L. speculum-veneris** (L.) Fisch.

3. Calyx teeth antrorse, much shorter than tube . . . . 3. **L. hybrida** (L.) Delabr.

\+ Calyx teeth curved, later falcately reflexed, hardly as long as tube . . . . 4. **L. falcata** (Ten.) Fritsch.

1. L. speculum-veneris (L.) Fisch. ex A. DC. Monogr. (1830) 347 pro syn. ("L. speculum"); Grossg., Opred. rast. Kavk. 425; Kharadze in Fl, Gruzii, VIII, 191; Szaf., Kulcz., Pawl. Rosl. Polskie, 647. – Campanula speculum-veneris L. Sp. pl. (1753) 238; Trautf. in Tr. Peterb. bot. sada, VI, 1, 91. – C. hirta Roem. et Schult. Syst. V. (1819) 153. – Prismatocarpus speculum L'Hér. Sert. Angl. (1788) 2. – Specularia speculum A. DC. l. c. 346; idem in DC. Prodr. VII, 2, 490; Ldb. Fl.

* Named after the 17th century French traveler Le Gouz.

Ross. II, 2, 892; Boiss. Fl. or. III, 959; Fomin in Mater. Fl. Kavk. IV, 6, 127; Grossg. Fl. Kavk. IV, 77; Neimark in Sorn. rast. SSSR, IV, 180. — Ic.: Rchb. Ic. Fl. Germ. et Helv. tab. 255.

Annual; root thin, whitish, fibrous; stems 15—30 cm, usually strongly spreading-branching, straight, glabrous or sparingly scabrous-puberulent; leaves alternate, finely crenate, glabrous or slightly scabrous; lower leaves obovate, obtuse, attenuate into short petiole, median leaves ovate, acute,
429 sessile, upper leaves lanceolate to narrowly lanceolate, subentire, all leaves 1—2 cm long. Flowers erect, terminal or axillary, sessile, numerous; calyx glabrous or slightly scabrous, cylindrical, teeth linear-lanceolate, erect at aestivation, then spreading and recurved, usually slightly shorter than corolla; corolla gaping, nearly rotate, whitish-green at the center, intensively violet at margins, shallowly divided into ovate acute-mucronate slightly pubescent lobes; stamens with very short filaments, dilated and scalelike at base, anthers narrow, filiform; style longer than stamens but shorter than corolla, cylindrical, with 3 ascending later curved stigmas; capsule erect, cylindrical, broadening at the middle, tapering at both ends, ending in the recurved calyx teeth, 10-nerved, 3-furrowed, dehiscing at apex (opposite calyx teeth) by 3 valves; seeds ovoid, shiny.

Fields in the lower and central zones. — Caucasus: Cisc. (C. Caucasus), E. Transc. (Georgia). **Gen. distr.**: Med., Centr. Eur., Bal.-As. Min. Described from S. Europe. Type in London.

Note. Although this species is listed among the weeds of the USSR (Neimark, op. cit.) the report that this plant grows in the Caucasus is highly doubtful. All authors who record this species for the flora of the Caucasus base it on Lagovskii's specimen of *L. speculum-veneris* which he collected "at the upper reaches of the Ardon" and on Fischer's specimen "ex Iberia" (or, more accurately, according to Ledebour (l. c.), not Fischer's but Wilhelm's). It is not known where this specimen is at present. As to Lagovskii's specimen, it is apparent that there has been a mix-up in the labels, as is the case in all his collections from the Caucasus. Trautfetter noted "bei Kars" on Lagovskii's specimens of *L. speculum-veneris*; it is possible that they were really found near Kars in Turkey, but the specimen from "the upper reaches of the Ardon" is not likely from there. By the nature of the collection, its drying and phase of development it gives the impression of actually being a fragment of the Kars specimens (it is typical that no date is given). The species has never been found again in the C. Caucasus or Georgia, nonetheless we are leaving it in the flora of the USSR.

2. L. pentagonia (L.) Thell. in Vierterjahrsschr. Gen. Zürich, XLVI, (1908) 465; Grossg. Opred . rast. Kavk. 425; Kharadze in Fl. Gruzii, VIII, 192. — *L. pentagonia* (L.) Druce, List Brit. Pl. (1908) 46. — *Campanula pentagonia* L. Sp. pl. (1753) 169; Trautf. in Tr. Peterb. bot sada, VI, 1, 91. — *Prismatocarpus pentagonia* L'Hér. Sert. Angl. (1788) 2. — *Specularia pentagonia* A. DC. Monogr. (1830) 344; idem
430 in DC. Prodr. VII, 2, 489; Vatke in Linnaea, XXXVIII, 713; Ldb. Fl. Ross. II, 892; Boiss. Fl. or. III, 959; Fomin in Mater. Fl. Kavk. I V, 127; Grossg. Fl. Kavk. IV, 77. — Ic.: Rchb. Ic. Fl. Germ. tab. 256.

Annual; root simple, thin, filiform; stems straight, 30 cm, branching, glabrous or scabrous-puberulent; leaves alternate, spreading, glabrous, green, paler beneath, subentire or obscurely crenate, slightly overlapping at margin; lower leaves obovate, attenuate into petioles, obtuse, median cauline leaves lanceolate-obovate, the upper linear-lanceolate. Flowers terminal, rarely axillary, always erect; calyx long, prismatic, at base with bracteoles little differing from pedicels, 10-nerved, glabrous or with spreading white hairs; teeth linear-lanceolate, spreading, entire, more or less ciliate; corolla pentagonal at cross section before aestivation, ovoid, later infundibular and rotate, wide, as long as or longer than calyx teeth, glabrous but hairy at the very apex and even villous at base, white below, blue at middle, violet at apex, lobes ovate, acute mucronulate, one-third as long as corolla, 3–5-nerved; stamens half as long as corolla, with membranous dilated slightly ciliate filaments and filiform anthers longer than filaments; style pubescent, with 3 thick stigmas emerging laterally after flowering; capsule very long, up to nearly 3 cm, prismatic, erect, not tapering at apex, many-seeded; seeds globular, shiny, light brown. May.

Dry and grassy places. – Caucasus: W. Transc. (Batumi). **Gen. distr.**: Bal.-As. Min., Med. (east and south of France). Described from Thrace (Greece). Type in London.

Note. There is only one specimen of this species in the USSR. It was collected near Batumi by Smirnov in 1885. Another specimen from a neighboring place but beyond the Turkish border was found by Nordmann.

3. L. hybrida (L.) Delabr. Fl. Auv. ed. II (1800) 47; Grossg. Opred. rast. Kavk. 425; Visyulina in Viznachn. rosl. URSR, 511; Kharadze in Fl. Gruzii, VIII, 193. – Campanula hybrida L. Sp. pl. (1753) 239; M. B. Fl. taur.-cauc. I (1808) 156; Trautf. in Tr. Peterb. bot. sada, VI, 92. – C. spuria Pall. ex Roem. et Schult. Syst. V (1819) 154. – Prismatocarpus hybridus L'Hér. Sert. Angl. (1788) 2. – P. confertus Moench, Meth. pl. (1794) 496. – Specularia hybrida A. DC. Monogr. (1830) 348; idem in DC. Prodr. VII, 2, 490; Ldb. Fl. Ross. II, 2, 892; Stev.
431 in Bull. Soc. Nat. Mosc. IV, 417; Boiss. Fl. or. III, 960; Shmal'g., Fl. II, 180; Fomin in Mater. Fl. Kavk. IV, 6, 128; Grossg., Fl. Kavk. IV, 78; Neimark in Sorn. rast. SSSR, IV, 180; Szaf., Kulcz., Pawl. Rosl. Polskie, 647. – Ic.: Rchb. Ic. Fl. Germ. tab. 255.

Annual; root whitish, fibrous, thin, usually not branching; stems straight, simple, 10–15 cm, leafy, longitudinally striate, more or less scabrous-puberulent; leaves alternate, crenate or obscurely so, slightly undulate at margin, sparingly pubescent; lower leaves obovate, 1–1.5 cm long, 0.5 cm wide, obtuse, attenuate into rather short petioles, median leaves ovate, slightly acuminate, sessile, almost the same size as lower leaves, upper leaves small, ovate-lanceolate. Flowers few, crowded above, erect, sessile; calyx long, prismatic, scabrous, later very accrescent and slightly expanding at the middle; teeth straight and spreading, ovate-lanceolate, rather wide, ciliate at margin, half as long as corolla; corolla pink or bluish-pink, paler outside, infundibular-rotate, often abortive, slightly hairy, lobes ovate, acute; stamens shorter than corolla, with filaments very short, at base, and anthers glabrous, narrow, subfiliform; style shorter than corolla, cylindrical, pubescent in upper half, with 3 stigmas; capsule prismatic, tapering at both ends, crowned by the straight calyx teeth, scabrous, 10-nerved and 3-furrowed dehiscing at apex by 3 valves; seeds ovoid, shiny. May.

Crops, roadsides and similar habitats. — European part: Bl., Crim.; Caucasus: W. Transc., Tal. **Gen. distr.:** Med., Atl. and Centr. Eur., Bal. Described from England and France. Type in London.

4. L. falcata (Ten.) Fritsch in Mitteil. Naturwiss. Ver. Wien, V (1907) 100; Grossg. Opred. rast. Kavk. 425 (sphalm. comb. nom. auct.: (DC.) Fritsch); Kharadze in Fl. Gruzii, VIII, 192. — Prismatocarpus falcatus Ten. Prodr. (1812) 16. — Campanula falcata Roem. et Schult. Syst. V (1819) 154; Trautf. in Tr. Peterb. bot. sada, VI, 1 (1879) 91. — C. syriaca Willd. ex Roem. et Schult. l. c. 133. — Specularia falcata A. DC. Monogr. (1830) 345; idem in DC. Prodr. VII, 2, 489; Boiss. Fl. or. III, 960; Fomin in Mater. Fl. Kavk. IV, 6, 127; Grossg. Fl. Kavk. IV, 78. — S. hybrida auct. non A. DC.; Linch., Rast. resursy Turkm. SSR, 1 (1935) 74. — Ic.: Ten. Ic. Fl. Nap. tab. 20; Rchb. Ic. Fl. Germ. tab. 225.

Annual; root thin, subfiliform, fibrous, whitish; stems straight and slightly branching below, branches arcuate below but straight above, cylindrical, glabrous or barely scabrous, leafy, 20—30 cm high; leaves alternate, glabrous or rarely finely hairy (mostly the lower leaves), green, paler
432 beneath, slightly crenate, hardly reflexed at margin; lower leaves orbicular-obovate, attenuate into rather short petioles, median leaves ovate, acute, sessile, upper leaves ovate-lanceolate, spreading. Flowers always axillary, sessile, solitary, erect, approximate in upper part of stem, otherwise spreading and forming long interrupted spicate inflorescence; calyx long, prismatic, 10-nerved, scabrous at angles, teeth linear-lanceolate, long-acuminate, with reflexed margins, sometimes scabrous, straight before flowering, later falcately curved, twice as long as infundibular-rotate corolla; stamens with ovate-broadening membranous and glabrous filaments, anthers narrow, filiform; style shorter than corolla, with 3, rarely 2, stigmas; capsule prismatic, 3-furrowed, not tapering toward apex, dehiscing by 3 valves; seeds ovoid or slightly lentiform, shiny. May. (Plate XXII, Figure 3.)

Dry places at field edges, crops, also coastal sands. — Caucasus: Cisc. (C. Caucasus), W. Transc. (Abkhaziya), S. Transc. (S. Karabakh); Centr. Asia: Mtn. Turkm. **Gen. distr.:** Med., As. Min., Iran., Canary Islands. Described from Italy. Type in Rome.

Note. According to reports, this plant was collected for the first time in the Caucasus, near Alagir, by Lagovskii. This report might be doubtful since all the labels in the Lagovskii Herbarium referring to the Caucasus are in disorder. The new, reliable reports on Legouzia falcata being found on Pitsunda Cape and near Gagry in Abkhaziya and also in Azerbaijan make Trautfetter's description, which was based on the Lagovskii Herbarium, more credible. This predominantly W. Mediterranean species of Legouzia grows as a weed in the Caucasus and also Turkmenia (Linchevskii, op. cit.).

The epithet "falcata" is due to the nearly falcate shape of the calyx teeth after blossoming.

Tribe 6. **WAHLENBERGIEAE** Endl. Gen. I (1874) 513; A. DC. in DC. Prodr. VII, 2, 413, s. restr. — Corolla rotate, 5-merous (like calyx); style usually with tripartite stigma; ovary usually trilocular, half inferior; capsule dehiscing at apex by valves in one slit. Generally perennial herbs, leaves alternate, opposite or whorled, or in fascicles of 3—4 leaves.

Type: Wahlenbergia Schrad.

Note. We comprehend the tribe Wahlenbergieae in the narrow sense (sensu restricto) in comparison with its treatment by Endlicher, A. de Can-
433 dolle, Schoenland and others. The main diagnostic characters of the genera included in this tribe are the half inferior ovary and the capsule dehiscing by one terminal opening with the deflexion of the valves. In the diagnosis of the tribe we exclude here characters not common in the representatives of this tribe in the flora of the USSR.

Genus 1446. **CODONOPSIS** * WALL.

Wall. in Roxb. Fl. Ind. ed. Carey, II (1824) 103; Kom. in Tr. Peterb. bot. sada, XXIX, 1 (1908) 102; Chipp in Journ. Linn. Soc. XXXVIII (1908) 374.— Glossocomia D. Don, Prodr. Fl. Nepal. (1825) 158.

Calyx quinquepartite, tube glabrous or pubescent, striate, teeth acuminate, glabrous or pubescent with obtuse or acute incisions in between; corolla campanulate, infundibular or tubular-campanulate, green, whitish, purple or bluish, glabrous or sparsely pubescent, lobes ovate or triangular, acute or obtuse separated by shallow incisions. Filaments dilated at base, glabrous or partly ciliate or densely pubescent from base; anthers attached at base, straight, oblong, bilocular, keeled or flat outside, obtuse, with pectinate bristles along keel. Style dilated at base, erect, rounded at middle, glabrous or pubescent; stigma 3- to 5-lobed, scabrous outside, glabrous inside, obtuse or attenuate. Ovary half inferior, inferior or superior, obconical or subglobose. Placentas fleshy, with numerous ovules. Capsule 3- or 5-locular. Seeds numerous, winged or wingless, elliptic, oblong or ovoid, brownish or whitish, sometimes reticular, with erect embryo and abundant fleshy albumin. Perennial herbs, with globular-tuberiform or oblong thick indurate roots. Stems branching, thin or thick, climbing, creeping or rarely straight, glabrous or pubescent. Leaves opposite, alternate or in tufts, petiolate or sometimes sessile, simple, glabrous or pubescent. Flowers terminal, axillary or between leaf axils, with glabrous or pubescent pedicels.

Type: C. viridis Wall. (Himalayas).

Codonopis (Wall., l. c.) comprised at first only three species, but in the nearly simultaneously published monographs by Komarov (op. cit.) and
434 Chipp (l. c.) there were already 22—23. At present there are 40 species which are mostly found in E. and C. Asia. In the USSR this genus is encountered only in the Far East and the mountains of Central Asia.

All the species growing in the USSR refer to the subgenus Eucodonopsis Kom., characterized by the broadly campanulate or tubular-campanulate corollas, the half inferior ovary, and the oval-subglobose capsule. Pseudocodonopsis Kom., a second subgenus common to China, includes 2 species and is remarkable for its rotate, deeply dissected corolla. Some species such as C. lanceolata, C. ovata, C. rotundifolia, C. ussuriensis,

* From the Greek codon — bell (shape of campanulate corolla, resembling the genus Campanula).

C. tangshen and C. viridiflora are sometimes cultivated in gardens. The root of C. tangshen is used in Chinese folk medicine as a tonic and an aphrodisiac.

1. Leaves in fascicles of 3–4 at apex of lateral branches . . . . 2.
+ Leaves opposite or alternate . . . . 3.
2. Seeds winged; root radish-shaped . . . . 2. **C. lanceolata** (Sieb. et Zucc.) Benth. et Hook.
+ Seeds wingless; root subglobular-tuberiform . . . . 3. **C. ussuriensis** (Rupr. et Maxim.) Hemsl.
3. Stems climbing; corolla greenish with violet spots . . . . 1. **C. pilosula** (Franch.) Nannfeldt.
+ Stems straight or flexuous, branching from base; corolla bluish with dark blue nerves, without spots . . . . 4. **C. clematidea** (Schrenk.) C. B. Clarke.

Series 1. **Volubiles** Kom. – Stems climbing; leaves alternate, opposite or in fascicles of 3–4.

The Chinese C. henryi Oliver and the Himalayan C. rotundifolia (A. DC.) Benth. are closely related to the Russian species C. pilosula.

1. C. pilosula (Franch.) Nannfeldt in Acta Hort. Goth. V (1929) 29; Kitagawa, Lineam. Fl. Manshur. 419. – C. silvestris Kom. in Tr. Peterb. bot. sada, XVIII (1901) 425; idem, Fl. Man'chzh. III, 573; idem in Tr. Peterb. bot. sada, XXIX, 1 (1908) 105; Kom. and Alis., Opred. rast. Dal'nevost. kr. II, 1001. – C. rotundifolia Chipp in Journ. Linn. Soc. XXXVIII, 382, p. p. non Royle. – Campanumoea pilosula Franch. Pl. David. I (1884) 192. – Ic.: Kom. op. cit. (1908) tabl. IX (mediocr.).

Perennial; root thick, radish-shaped, indurate, transversely rugose and slightly grumous, canescent, ca. 1.5 cm in diameter; stems climbing, finely
435 hirsute at base, glabrous and smooth above, very long, branching; leaves petiolate, ovate or broadly ovate, acute, glaucous beneath, thin, entire or nearly so, slightly undulate at margin, hispid above, later glabrous, with appressed fine hairs beneath. Flowers solitary, yellowish with violet tinge and darker violet spots, thin-pediceled; calyx teeth glabrous or sparingly pubescent, oblong; corolla glabrous, broadly campanulate, parted to middle into oblong-lanceolate, acute, straight (not recurved) lobes; capsule 3-locular, dehiscing by valves at the flattened apex; seeds wingless, dark, shiny. August–September.

Dense shrubby thickets, shade of trees at forest edges, banks of rivers and streams, often in quantities. – Far. East: Uss. **Gen. distr.**: Mong., Korea, NW China. Described from near Peking. Type in Paris.

Note. Chipp (l. c.) lumped for no reason C. silvestris (namely C. pilosula) with the earlier described Himalayan C. rotundifolia, although the leaves of the latter are more distinctly crenate and the flowers are bluish. Komarov (op. cit., 1908) thought that his species C. silvestris could be synonympus with the earlier described Campanumoea pilosula Franchet (l. c.), as proved later.

Economic importance. Used in medicine in China.

2. C. lanceolata (Sieb. et Zucc.) Benth. et Hook. Gen. pl. II (1876) 557; Trautf. in Tr. Peterburg. bot. sada, VI, 1, 46, excl. var.; Kom., op. cit. XXIX, 1, 109; idem, Fl. Man'chzh. III, 571; Chipp in Journ. Linn. Soc. XXXVIII, 379; Kom. and Alis., Opred. rast. Dal'nevost. kr. II, 1001; Naimark. in Sorn. rast. SSSR, IV, 181. — Campanumoea lanceolata Sieb. et Zucc. Fl. Jap. I (1835) 174. — Glossocomia lanceolata Maxim. in Mél. Biol. VI (1881) 487. — G. lanceolata Rgl. in Bull. Phys. Math. Acad. Pétersb. XV (1857) 223; ej. Tent. Fl. Ussur. 100, p. p. — Ic.: Sieb. et Zucc. l. c. tab. 91; Planch. in Fl. des Serres, tab. 927; Kom. and Alis., op. cit. tab. 300.

Perennial; root radish-shaped, thick, long; stems glabrous or sparsely pubescent, climbing, branching; leaves broadly lanceolate or rhombic, glaucous-gray, in fascicles of 4 at ends of short lateral branches, rarely solitary, short-petioled, acute or obtuse, entire or sometimes dentate, glabrous, finely ciliate or rarely spreading-hairy, 3.5—7.5 cm long, 2—3.5 cm wide. Flowers terminal, with 1—9 cm long glabrous pedicels; calyx subglobular, glabrous, teeth lanceolate or oblong, acute, glabrous; corolla campanulate, glaucescent-green or yellowish, with purple margin and purple
436 spots, glabrous, 3 cm long, lobes recurved, triangular, acute; filaments glabrous, dilated below; style glabrous, with tripartite stigma; ovary trihedral, half inferior, 3-locular; capsule obconical, glaucescent, dehiscing at apex by valves; seeds winged, smooth, dull, 2—3 mm long. July—August.

Shrubby thickets along banks of rivers and streams in valleys. — Far East: Uss., Ze.-Bu. **Gen. distr.**: Korea, N. Ch., Jap. Described from a specimen cultivated in Japan. Type, possibly, in Leningrad.

Note. Sometimes a weed among crops, like C. ussuriensis.

3. C. ussuriensis (Rupr. et Maxim.) Hemsl. in Journ. Linn. Soc. XXVI (1889) 6; Kom. in Tr. Peterb. bot. sada, XXIX, 1 (1908) 111; idem, Fl. Man'chzh. III, 570; Chipp in Journ. Linn. Soc. XXXVIII, 380; Kom. and Alis., Opred. rast. Dal'nevost. kr. II, 1001; Neimark in Sorn. rast. SSSR, IV, 181. — C. lanceolata var. ussuriensis Trautv. in Tr. Peterb. bot. sada, VI, 1 (1879) 47. — Glossocomia ussuriensis Rupr. et Maxim. in Bull. Phys. Math. Acad. Pétersb. XV (1857) 209; Maxim. Prim. Fl. Amur. 184 et in Mél. Biol. XII, 487; Korzhinsk. in Tr. Peterb. bot. sada, XII, 364. — G. lanceolata Rgl. Tent. Fl. Ussur. (1861) 100, p. p., non Campanumoea lanceolata Sieb. et Zucc. — G. lanceolata β obtusa Rgl. in Bull. Phys. Math. Acad. Pétersb. XV (1857) 223. — G. lanceolata var. ussuriensis Rgl. in Ind. sem. hort. Petrop. (1866) 92.

Perennial; root tuberiform, fleshy, half-globular or slightly oblong, 1—2.5 cm in diameter; stems glabrous or sparsely pubescent at internodes, thin, climbing, greenish or whitish or dark violet; leaves in fascicles or in pseudowhorls of 3—5 at ends of short thin axillary branches, appearing like one palmatipartite leaf, short-petiolate, lanceolate or elliptic, entire, acute or obtuse, attenuate at base, glabrous or few hairs, finely ciliate, 3—5 cm long, 1.5—2.5 cm wide, petioles glabrous or pubescent, 2—5 cm long. Flowers terminal or axillary, with short glabrous pedicels 1—2 cm long; calyx subglobular, glabrous, teeth lanceolate, acute, glabrous, 1—2 cm long, 6—8 mm wide; corolla campanulate, with linear lobes, dark violet or dirty purple, inside with darker lines and blackish spots; filaments glabrous; style glabrous; stigma tripartite, lobes ascending; ovary 3-locular, half inferior; capsule conical, flattened above, dehiscing at apex by 3 valves forming one opening; seeds dark brown, shiny, smooth. July–August. (Plate XXVI, Figure 3.)

437 Valleys and inundated meadows, often sandy soils. — Far East: Uss., Uda, Ze.-Bu. **Gen. distr.**: NE China, Korea, Jap. Described from estuary of Ussuri River near Turm. Type in Leningrad.

Note. Komarov (op. cit.) writes that this species is "particularly characterized by its tuberiform roots, small leaves, rather short internodes and meadow habitat." The first description of the species should be attributed to Ruprecht and Maksimovich together and not only to Ruprecht as Chipp (l. c.) and others assumed. Maksimovich (Prim. Fl. Amur., l. c.) mentioned both authors for Glossocomia ussuriensis. In the first description the name Maksimovich is at the end of the diagnosis and Ruprecht at the end of the note. The labels of the original specimens are written in the handwriting of Maksimovich, who discovered the plant.

Series 2. **Erectae** Kom. — Stem straight, scapiform or leafy or somewhat creeping but not climbing; leaves opposite or alternate, solitary.

One species occurs in the USSR, the others of this series, including C. ovata Benth., are related to the one described below and are found in the Himalayas and China.

4. C. clematidea (Schrenk) C. B. Clarke in Hook. Fl. Birt. Ind. III (1872–1883) 433; Kom. in Tr. Peterb. bot. sada, XXIX, 1, 115; Grigor'ev, Opred. rast. Stalinabada, 259. — C. ovata auct. non Benth.; Trautf. in Tr. Peterb. bot. sada, VI, 1, 47; O. and B. Fedch. in Tr. Peterb. obshch. estestvoisp. XXXV, 3, 52; O. and B. Fedch., Perech. rast. Turk. IV, 353; B. Fedch., Rast. Turk. 721. — C. ovata var. cuspidata et var. obtusa Chipp in Journ. Linn. Soc. XXXVIII (1908) 385. — Wahlenbergia clamatidea Schrenk, Enum. pl. nov. Songar. 1 (1841) 38; Ldb. Fl. Ross. II, 2, 871. — Glossocomia clematidea Fisch. in Rgl. Gartenfl. (1856) 226. — Ic.: Fisch. in Rgl. l. c. tab. 267, f. 2.

Perennial; root thick, fusiform; stems straight or flexuous, 1 m, glabrous or hirsuta, obtusely angular; branches thin, spreading, sometimes subfiliform, densely leafy; leaves petiolate, opposite, alternate at ends of branches, ovate or ovate-oblong or broadly lanceolate, sometimes slightly cordate, acute, entire, short-hairy. Flowers terminal, pediceled; pedicels finely white-hirsute; calyx glabrous, glaucous, teeth glabrous or ciliate, oblong or ovate-lanceolate, markedly accrescent and spreading after flowering; corolla broadly campanulate, twice as long as calyx, whitish or bluish, with darker blue nerves; filaments dilated below; anthers longer than filaments; style short; stigma tripartite; capsule compressed, obconical or ovoid,
438 dehiscing at apex by 3 valves; seeds narrowly elliptic, obtuse at both ends, slightly flattened, smooth, pale brown, dull. June–August. (Plate XXVI, Figure 2.)

Banks of mountain streams, springs and other similar places in the central and subalpine zones. — Centr. Asia: T. Sh., Pam.-Al. Dzu-Tarb. (Dzungarian Ala-Tau). **Gen. descr.**: Dzu.-Kash., Ind.-Him. (Kashmir, Gilgit), Iran. (Afghanistan). Described from the mountains of "Ala-Tau and Dzhilkaragai." Type in Leningrad.

Note. Many authors have erroneously identified this plant with the Himalayan and Chinese C. ovata Benth., but Clark (in Hook., l. c.) established C. clematidea and Komarov (op. cit.) indicated that the real C. ovata "is an alpine or subalpine plant, low, with small, somewhat different, leaves . . . stem subscapiform," etc.

Genus 1447. **PLATYCODON** * A. DC.

A. DC. Monogr. (1830) 125, excl. sp. alt.

Calyx quinquepartite. Corolla 5-lobed, large, broadly infundibular. Stamens 5, free, filaments dilated at base and hairy. Style with 5 stellately divergent stigmas. Ovary 5-locular. Capsule dehiscing by 5 valves between septa, valves opposite calyx lobes, alternating with loculi and stamens. Seeds ovate, large, shiny, flattened. Perennial plants, usually glaucescent, with alternate or nearly opposite sessile leaves and fleshy thick root.

Alphonse de Candolle described the genus as being oligotypic (2 species). The nomenclatural type should be P. grandiflorus (Jacq.) A. DC. whereas the second species accepted by de Candolle should be included in the genus Astrocodon (see); Platycodon is thus considered as a monotypic genus.

1. P. grandiflorus (Jacq.) A. DC. Monogr. (1830) 125; idem in DC. Prodr. VII, 2, 422; Ldb. Fl. Ross. II, 2, 870; Turcz. Fl. baic.-dahur. II, 474;
441 Korzhinskii in Tr. Peterb. bot. sada, XII, 2, 364; Kom., Fl. Man'chzh. III, 2, 574; Karo in Oesterr. bot. Zeitschr. 10, 399; Kom. and Alis., Opred. rast. Dal'nevost. kr. II, 1001; Kitagawa, Lineam. Fl. Manshur. 420. — P. chinense Lindl. et Paxton, Fl. Gard. II, 121. — P. autumnale Dcne. in Rev. hort. ser. 3, II (1846) 561. — P. glaucum Nakai in Tokyo Bot. Mag. XXXVIII (1924) 301. — P. grandiflorum var. glaucum Sieb. et Zucc. in Abh. Acad. Wiss. München. IV, 3 (1846) 179. — Campanula grandiflora Jacq. Hort. Vinob. III (1776) 4; Roem. et Schult. Syst. V, 111. — C. glauca Thunb. Fl. Jap. (1784) 88. — C. gentianoides Lam. Encycl. Méth. I (1789) 781. — Wahlenbergia grandiflora Schrad. Cat. hort. Götting. (ann. 1814). — Ic.: Jacq. l. c. tab. 2 (fig. optima); Gmel. Fl. Sibir. III, tab. 28 (fig. bona); Curtis Bot. Mag. tab. 252; Lindl. et Paxton, l. c. tab, 61; Kom. and Alis., op. cit. Plate 301, fig. 1 and 4–6 (mediocr.). — Exs.: Karo, Pl. Amur. et Zeaënsae, No. 209.

Perennial; plant wholly glabrous, slightly glaucescent; root perennial, white, fleshy, radish-shaped, finger-thick, with abundant milky juice; stems ascending from base or straight, simple, ca. 40–50 cm, herbaceous, glabrous or smooth, longitudinally striate in lower part; radical leaves not distinguished in shape, all leaves alternate or sometimes partly nearly opposite, arranged along the lower half of stem or even higher, ovate-lanceolate, sessile, tapering at base, 2.5–3.4 cm long, 2–3 cm wide, rather large-toothed, pale beneath, glaucescent, upper leaves reduced. Flowers usually 1, sometimes 2, large, terminal, broadly open; pedicels erect; calyx glaucescent,

* From the Greek platos – broad and codon – bell.

(439)

PLATE XXVI. 1 – Platycodon grandiflorus (Jacq.) A. DC., plant and capsule; 2 – Codonopsis clematidea (Schrenk) C. B. Clarke, plant and flowers; 3 – C. ussuriensis (Rupr. et Maxim.) Hemsl., part of plant and lower part of stem with tuberiform root.

obconical, broadening above, teeth straight or slightly recurved but parallel to corolla, triangular-acuminate, entire, much shorter than corolla; corolla 4–5 cm across, broadly infundibular, blue, divided for one-third into ovate, acute, recurved lobes; stamens half as long as corolla, with membranous ciliate filaments triangularly dilated at base, anthers longer than filaments; style slightly longer than stamens, thickened at base, cylindrical and hairy above, hairs obscurely disposed in 5 longitudinal rows, stigmas 5, thick, straight or slightly curved, glabrous, white; capsule erect, ovoid; seeds blackish, shiny. August–October (Plate XXVI, Figure 1.)

Dry meadows, stony and pebbly places on slopes, shrubby formations. — E. Siberia: Dau.; Far East: Ze.-Bu., Uda, Uss. **Gen. distr.**: Jap., Korea, NE Chi. Described from a garden specimen, seeds from "Siberia and Tatary." Type in Vienna.

Economic importance. The root is used in the Far East in folk medicine for which it is collected in great quantity.

442 Tribe 7. **EDRAJANTHEAE** Fed. in Addenda XXIII, p. 343. — Corolla tubular, divided into short lobes, 5-merous (like calyx); calyx without appendages; stamens free, dilated, anthers narrow-linear; style with 2 stigmas; capsule splitting from above downwards by unequal valves like a funnel. Perennial herbs, stems numerous, leaves alternate, sometimes imbricate, forming more or less dense cushions. Root thick, woody. Inflorescence densely capitate, bracteate, sometimes flowers solitary, terminal.

Type: Edrajanthus A. DC.

Genus 1448. **EDRAJANTHUS*** A. DC.

DC. Prodr. VII (1839) 449.— Sect. Edrajantha A. DC. Monogr. (1830) 129.— Hedraianthus Griseb. Spicil. fl. Rumel. (1844) 292; Wettst. Monogr. d. Gatt. Hedraeanthus (1887) 1.— Hedranthus Rupr. in Bull. Acad. Sc. Pétersb. XI (1867) 203.— Muehlbergella Feer in Engler's Bot. Jahrb. XII (1890) 615.

Calyx quinquepartite. Corolla quinquepartite for one-third to one-half, tubular rarely nearly campanulate or infundibular. Stamens 5, free, filaments dilated at base. Style not exserted, cylindrical, Stigma bipartite, rarely tripartite, with linear later spreading lobes. Capsule 2- to 3-locular subglobose, ovoid or prismatic, angular, flattened above, dehiscing by unequal deciduous fragments or valves, appearing like a deep funnel after dispersing seeds, long persistent (in calyx). Seeds numerous, ovate, flattened, large. Perennial woody plants; leaves alternate usually linear; bracts and calyx teeth ciliate at margin. Flowers sessile, solitary or in heads, with blue or violet corolla.

Only the following species, one of the 11 species growing in the Mediterranean countries, is found in the USSR.

Note. The names given to this genus by different authors are not orthographically alike (Edrajanthus, Hedraianthus, Hedraeanthus, Hedranthus). Since de Candolle's name Edrajanthus is not
443 an orthographical mistake it should be maintained, and with his name as the author.

* From the Greek edrajos — sessile and anthos — flower.

1. **E. owerinianus** Rupr. in Bull. Acad. Sc. Pétersb. XI (1867) 203 (sub "Hedrantho"); Boiss. Fl. or. III, 886; Wettst. Monogr. d. Gatt. Hedraeanthus (1887) 6; Trautf. in Tr. Peterb. bot. sada, VI, 1, 45; Fomin in Mater. Fl. Kavk. IV, 6, 140 (sub "Hedraeantho"); Grossg. Fl. Kavk. IV, 80 (sub "Hedrantho" autem); idem, Opred. rast. Kavk. 426. — Muehlbergella oweriniana (Rupr.) Feer in Engler's Bot. Jahrb. XII (1890) 616. — Ic.: Feer, l. c. tab. VIII, B (bona).

Perennial; rhizome indurate but not woody, producing numerous short slightly columelliform stems and shoots, densely covered with imbricate dry relics of old leaves and crowned by small rosettes of young leaves forming dense tufts; leaves small, short, linear, obtuse, shiny, flat and smooth above, furrowed beneath, glabrous, glaucescent, involute and with retrorse cilia at margin. Flowers small, solitary, sessile, terminal, ca. 1 cm long, wrinkled post anthesis and long persistent; calyx glabrous, angular, teeth recurved, almost appressed to calyx after flowering, lanceolate-linear, acute, later as long as the oblong-obconical tube and more than half as long as corolla, appendages short, acute; corolla tubular, glabrous; stamens with very short filaments; capsule obconical or prismatic, rounded, ca. 0.7 cm long, 0.2 cm wide, obtuse; seeds elliptic, brown, blackish above. May. (Plate XXVII, Figure 2.)

Limestone in the central mountain zone. — Caucasus: Dag. Endemic. Described from N. Dagestan (Koisubu beneath Erpeli Mountain). Type in Leningrad.

Note. Besides the indicated locality, the plant has so far been found on the southern slopes of Salatau near Chirkaty, on the left bank of the Koisu River below Gimri and on the slopes of Mukhita Mountain.

Note. Feer (l. c.) attributed much taxonomic importance to the dehiscence and shape of the capsule of E. owerinianus among the Campanulaceae; consequently, he separated it into the genus Muehlenbergella and described it in great detail. The description was accompanied by excellent drawings (analysis) of the plant. Nevertheless, we are not inclined to accept Muehlenbergella, because there are many species of Edrajanthus s.l. that resemble E. owerinianus, particularly E. pumilio. Moreover, there are different types of dehiscing capsules among the other genera and species, even in the genus Campanula, which are very variable in shape. The other diagnostic characters pointed out by Feer to justify his division are insignificant. For this reason we feel that it would be feasible just to establish a special section within Edrajanthus, under the name Muehlenbergella, for its specific representative in the
444 Caucasus. Neither Grossheim nor Fomin (op. cit.), in compiling his monographic study of the Campanulaceae of the Caucasus, knew of Feer's work.

Tribe 8. **JASIONEAE** Fed. in Addenda XXIII, p. 343. — Corolla dissected nearly to base into narrow lobes, 5-merous (like calyx); calyx without appendages; stamens connate at base of anthers; style with 2, rarely 3 stigmas; ovary 2-locular; capsule globose or ovoid, flattened, dehiscing at apex by 2 broad but short valves. Perennial and monocarpic herbs, with alternate leaves. Flowers in dense globose heads.

Type: Jasione L.

Genus 1449. **JASIONE*** L.

Lp. Sl. pl. (1753) 1317; Schönland in Pflanzenfam. IV, 5 (1889) 59.— Ovilla Adans, Fam. pl. (1763) 134; Rupr. Fl. Ingr. I (1860) 652.

Calyx quinquepartite. Corolla 5-sect nearly to base, into linear-lanceolate lobes. Stamens 5, connate at base of anthers, filaments thin. Pollen grains bluish or purple. Style with hairs in 10 longitudinal rows from the middle upwards. Stigmas 2, very rarely 3, short, thick. Capsule 2-locular, globose or ovoid, flattened above, dehiscing at apex by short valves but with broad opening. Seeds many, very small, ovoid, shiny. Perennial or monocarpic herbs, outwardly reminiscent of Scabiosa, with alternate narrow leaves and flowers disposed in compact terminal head, bracteate.

Type: J. montana L.

An oligotypic genus distributed in Europe and in the Mediterranean area to N. Africa.

1. Heads of inflorescence 1—2.5(2) cm across; stem pubescent below, glabrous above; bracts ovate, entire or crenate, long-acuminate . . . . 1. **J. montana** L.

447 + Heads of inflorescence not more than 1 cm across; stem completely scabrous-puberulent with stiff short hairs; bracts lobate-dentate, with acicular mucro . . . . 2. **J. heldreichii** Boiss. et Orph.

1. **J. montana** L. Sp. pl. (1753) 1317; A. DC. Monogr. 101, excl. var. nonnul.; idem in DC. Prodr. VII, 2, 415, excl. var. nonul.; Ldb. Fl. Ross. II, 2, 870, p.p.; Boiss. Fl. or. III, 885, excl. var.; Trautf. in Tr. Peterb. bot. sada, VI, 44, excl. var. dentata; Shmal'g., Fl. II, 171; Fomin in Mater. Fl. Kavk. IV, 6, 141, excl. var.; Syreishch., Ill. Fl. Mosk. gub. III, 217; Maevskii, Fl., ed. 8-e, 549; Szaf., Kulcz., Pawl., Rosl. Polskie, 638.— J. umbellata Gilib. Fl. lithuan. I (1781) 58.— J. montana var. tylica Trautv. l. c.— J. montana var. umbellata (Gilib.) Trautv. l. c.— J. montana var. glabra et var. hispida Syreishch. l. c. (in obs.), non Peterm. nec G. Beck.— Globularia vulgaris Gorter, Fl. Ingr. (1761) 20.— Ovilla globulariaeflora Rupr. Fl. Ingr. I (1860) 653.— Ic.: Rchb. Ic. Fl. Germ. tab. 217; Syreishch., op. cit. 218; Maevskii, op. cit. 207; Szaf. et al. l. c. 637.— Exs.: GRF, No. 324.

Biennial; root thin, fibrous, white, without milky juice; stems straight or ascending, 20—40 cm, simple or slightly branching, pubescent below glabrous above, leafy except beneath inflorescence, terminating in dense inflorescence; leaves alternate, linear-lanceolate, acute, sessile, undulate at margins, membranous and slightly crisp, 1—1.5 cm long, 2—3 mm wide, more or less pubescent; Inflorescence spherical-capitate, 1—1.5(2) cm across; flowers ca. 5 mm long, blue, short-pediceled, densely crowded; bracts imbricate, ovate, acute, entire or crenate, 2—4 mm long, thickened at margin; calyx usually glabrous, ovate, small, teeth linear, acuminate, as long as tube but shorter than corolla; corolla blue or sometimes white, divided nearly to base into narrow liguliform lobes; filaments not dilated, white, thin;

* From the Greek jasion, the name given to the plant by Theophrastes. Sometimes this name is interpreted as derived from the Greek word jasis, meaning medical treatment (indicating the purgative properties of the plant).

(445)

PLATE XXVII. 1 – Jasione montana L., plant and flowers; 2 – Edrajanthus owerinianus Rupr., plant and flowers (at end of blossoming), with perianth (partly in cross section); 3 – Sphenoclea zeylanica Gaertn., plant, flowers (section) and capsule with open cap.

anthers connate below, straight; style cylindrical, blue, not exceeding stamens when young but markedly elongating after flowering and exceeding corolla, white-hairy in upper part, thickening at apex and dividing into 2 short toothlike green stigmas; capsule erect, ovoid; seeds ovoid, slightly flattened. June—August. (Plate XXVII, Figure 1.)

Dry meadows, light pine groves, sands.— European part: nearly everywhere except for the North. **Gen. distr.**: Centr. and W. Eur., Scand. (S.). Described from W. Europe. Type in London.

Note. This species was reported from the Caucasus and the Mediterranean countries; actually it is a vicarious species of the southern races
448 of which only the following species might be found in the USSR.

2. **J. heldreichii** Boiss. et Orph. in Boiss. Diagn. sér. II, 6 (1859) 120; Grossg. Fl. Kavk. IV, 80.— J. montana var. dentata A. DC. in DC. Prodr. VII, 2 (1839) 415; Boiss. Fl. or. III, 885; Trautf. in Tr. Peterb. bot. sada, VI, 1, 44; Fomin in Mater. Fl. Kavk. IV, 6, 142.— J. montana auct. Fl. Cauc. non L.: Lipskii, Fl. Kavk. (1899) 376.— Exs.: Baenitz, Herb. Europ. sine numero.

Annual; plant covered with very short fine blunt but stiff hairs; stems arcuately ascending, 10—15 cm long, simple or with few short branches in upper part, densely leafy below, leafless in upper part; leaves linear, incised-dentate, crisp-hairy at margins, ca. 1 cm long. Bracts lanceolate, nearly as long as capitate inflorescence, pinnatipartite or crenate-dentate, with bipartite or tripartite lateral lobes thickening at margin, prominently nerved below, mucronate at apex; pedicels slightly longer than calyx; calyx teeth subulate-acuminate, one-third as long as tube and more than half as long as corolla. June.

Dry places in the central mountain zone.— Caucasus: possibly near the Turkish border. **Gen. distr.**: Bal.-As. Min. Described from Greek Macedonia. Type in Geneva.

Note. Since the times of Ledebour and Trautfetter only one very poor and incomplete specimen of this species has been preserved. It was collected by Nordmann near Rize, Turkey, at the border of Transcaucasia. The probability of finding J. heldreichii within the USSR is rather slight, nevertheless it should not be ruled out (taking for granted that the collector did not confuse the geographical data, his label is correct). This species differs from J. montana mainly by the oblong or lanceolate bracts acute-acicular at apex and the rather small inflorescence only 1 cm across.

Subfamily 2. **SPHENOCLEOIDEAE** (Mart.) Fed. comb. nova.— Stamens adnate to tube above the base of the petals, between lobes; calyx lobes 5, orbicular, concave, open at flowering but later accrescent and closing above ovary; corolla 5-lobed, lobes slightly auricled at base; stamens 5, with very short filaments; anthers subsessile, orbicular, 2-locular, opening by longitudinal slits; style very short, stigma bifid, smooth, not bearded; ovary 2-locular; fruit a capsule dehiscing by a cap, membranous, 2-locular, many-
449 seeded, short conical below, flattened above. Placentas broadening like a mushroom, at each side of septa. Seeds very small, many, with 3 teeth, very short cotyledons and very little endosperm.

Monotypic subfamily with one genus and the species Sphenoclea distributed in the tropics and subtropics of the Old World.

Note. We have established a special subfamily for the monotypic, tropical genus Sphenoclea; for a truer representation of the phylogenetic relationships within the order Campanulatae, it would be more accurate to separate it as an independent monotypic family, as Martius (l. c.) did. We have ranked this group as a subfamily only because it was not designated as a separate family in the earlier published key of families of the Sympetalae (see Flora of the USSR, Vol. 18), there being only a remark that it belongs to the family Campanulaceae. At any rate, treating the Lobeliaceae as a family separate from the Campanulaceae should also be applied to the Sphenocleaceae. The separation of the genus Sphenoclea into a special family is based upon certain characters which are completely alien to the typical Campanulaceae. The stamens in Sphenoclea are adnate to the walls of the corolla tube and not to its base or disk as is the case in the Campanulaceae. At the same time they are not opposite the corolla lobes but alternating them. The stigma of Sphenoclea is capitate, on a very short style. The capsule dehisces by cap splitting along the periphery. In habitus Sphenoclea resembles none of the genera of Campanulaceae, rather more does it resemble Phytolacca.

Genus 1450. **SPHENOCLEA** GAERTN.

Gaertn. Fruct. et sem. pl. I (1788) 113; Boiss. Fl. or. III, 963; Schönland in Pflanzenfam. IV, 5 (1894) 6.– Pongati Adans., Hist. nat. Senegal (1759) 152, nom. rejic.– Pongatium Juss. Gen. pl. (1787) 423, nom. rejic.– Gaertnera Retz. Obs. bot. VI (1791) 24, non Lam. et al.– Rapinia Lour, Fl. Cochinchin. I (1793) 156.

Characteristics of the genus are given in the description of the subfamily Sphenocleoideae.

1. S. zeylanica Gaertn. Fruct. et sem. pl. I (1788) 113; Clarke in Hook. Fl. Brit. Ind. III (1882) 438; O. and B. Fedch. in Tr. Bot. muz. AN, X, 122 and in Izv. AN, VII, 4, 218; O. and B. Fedch., Perech. rast. Turk. V and Dopol. (1913) 199; B. Fedch., Rast. Turk. 721; Vvedensky in Sched. ad Herb. Fl. As. Med. XVIII, November (1928) No. 450; Neimark in Sorn. rast. SSSR, IV, 182; Vasil'chenko in Sorn. rast. Tadzhikistana, I, 433; Yu. Grigor'ev,
450 Opred. rast. Stalinabada, 259.– S. pongatium DC. Prodr. VII, 2, 548; Boiss. Fl. or. III, 963.– Pongatium indicum Lam. Tabl. Encycl. I (1791) No. 1991.– Schrebera pongati Retz. Obs. bot. VI (1791) 24.– Rapinia herbacea Lour, Fl. Cochinchin. I (1793) 156.– Ic.: Gaertn. l. c. tab. 24; f. 5 (analysis); Rheede, Hort. Malabar. II (1692) tab. 24 (n. v.); Schönland in Pflanzenfam. IV, 5 (1894) f. f. 23 J. (flos), 27 J. (capsula), 38 (habitus); Naimark, op. cit. f. 442 (semen).– Exs.: HFAM, No. 450.

Annual; glabrous plant, with short fibrous root; stems straight, 70 cm, 0.7 cm across, sulcate, simple or slightly branching, branches antrorse, brown, fibrous, basal; leaves all cauline, alternate, 7 cm long, 1.5 cm wide, exstipulate, lanceolate, acute, entire, attenuate into short petioles. Flowers small, sessile, with 3 bracts to each flower, disposed in very dense cylindrical or attenuate spikes at ends of the lateral branches, the median spike longer than the lateral; seeds finely dotted, smooth, cylindrical. July–September. (Plate XXVII, Figure 3.)

Rice inundated fields, ditches. — Centr. Asia: Pam.-Al. (S. Tadzhikistan, S. Uzbekistan). **Gen. distr.:** Ind.-Him. (India, Ceylon, Indochina), Iran., tropical Am. (introduced). Description based on a specimen grown in the Leiden Botanical Garden from seeds from Ceylon. Type probably in Leiden.

Note. This interesting plant was first found in the USSR in 1878 by Nevesskii. Since then it has only been found in Central Asia, from Surkhan-Darya to Kurgan-Tyube.

Family CLIX. **LOBELIACEAE** * JUSS.

Flowers hermaphrodite; calyx 5-lobed, adnate to ovary; corolla zygomorphous or actinomorphous, sympetalous, persistent; stamens 5, filaments free or more or less connate; anthers connate, 2-locular, dehiscing longitudinally; ovary inferior or half inferior, 2—3-locular, rarely 1-locular, with 1 style and 2-lobed, rarely simple, stigma; ovules many, fleshy; fruit dry or fleshy; seeds many, with straight small embryo and abundant endosperm.

About 26 genera and more than 300 species distributed in the tropical and subtropical regions, rarely in the moderate zones.

451 Genus 1451. **LOBELIA** ** L.

L. Sp. pl. (1753) 929; Wimmer in Pflzr. 107 Heft (1953) 408.— Dortmannia Neck.— Elem. I (1790) 132.

Calyx deeply 5-lobed, tube obconical, ovate or subglobular; corolla 2-lipped, the tube straight-cylindrical or infundibular, the upper lip with 2 narrow lobes, erect, the lower lip with 3 broad lobes, rarely 3-toothed, spreading; stamens connate at upper parts of filaments and anthers, two anthers or rarely all barbate at apex; style filiform; stigma 2-lobed; ovary 2—3-locular, rarely 1-locular; fruit a capsule dehiscing above by 2 valves.

Herbaceous (in the USSR) plants containing milky juice; leaves spirally arranged; flowers in racemes.

Lobelia includes about 250 species distributed in all parts of the world with the exception of W. Asia. They are beautiful ornamental plants. The species L. erinus L., from S. Africa, with its blue flowers, and L. fulgens L., with its red flowers, are widespread in the USSR as hedge plants.

1. Aquatic plant; radical leaves linear, in rosettes; scapes with sparse small linear leaves and few solitary flowers at upper part on drooping 3—4 mm long pedicels emerging from water; calyx nearly half as long as corolla tube; corolla blue or white; capsule oval . . . . 1. **L. dortmanna** L.

\+ Terrestrial plant; radical leaves absent; stems densely leafy; lower leaves oblong, obtuse, the rest oval-lanceolate or lanceolate; flowers in terminal raceme, pedicels 1—1.2 cm long; calyx as long as corolla tube; corolla dark blue; capsule globose . . . . 2. **L. sessilifolia** Lamb.

* Treatment by S.G. Gorshkov.

** After the Dutch botanist M. Lobelia (1538—1616).

1. **L. dortmanna** L. Sp. pl. (1753) 929; DC. Prodr. VII, 376; Ldb. Fl. Ross. II, 868; Shmal'g., Fl. II, 170; Maevskii, Fl., ed. 8-e, 549; Wimmer in Pflzr. 107 Heft, 444. — Ic.: Rchb. Ic. Fl. Germ. XIX, tab. 1618; Fedch and Fler., Fl. Evrop. Ross. fig. 943; Hegi, Ill, Fl. VI, I, tab. 255, f. 4—4d. — Exs.: Herb. Fl. Ingr. No. 382; Pl. Finl. exs. Nos. 968, 969 and 970; Fl. pol. exs. No. 57.

Perennial; aquatic plant, with numerous thin white fibrous roots; radical leaves rosetted, linear-terete, (2)3—7 cm long, 1.5—3.5(4) mm wide, below with 2 lobes, smooth, obtuse, recurved above. Scapes 30—70 cm high, simple,
452 straight, green or yellowish-green (var. decolor Lindb. fil.), with 3—6 mm long, 5 mm wide, sparse, linear, obtuse leaves; flowers few, on 3—4 mm long drooping pedicels, emerging from water; calyx 3—5 mm long, nearly half as long as corolla tube, smooth, tube obconical, lobes lanceolate-linear 1—1.8 mm long, 0.2 mm wide, acute, straight, entire; corolla blue or white (var. decolor Lindb. fil.), 1.2—1.5 cm long, 3—4 times as long as calyx, smooth, 2-lipped, the tube cylindrical, 5—7 mm long, somewhat tapering at upper part, spreading-hairy inside, lobes unequal, ciliate, the upper two linear, 3—5 mm long, 0.5 mm wide, erect, acute, the lower slightly longer, 4—7 mm long, 1—3 mm wide, drooping, ovate-lanceolate, obtuse, all pubescent inside at base; stamens hidden in corolla, filaments free below, connate in upper part like anthers, anthers gray-violet, long-white-hairy at apex, the lower two pubescent; style glabrous, with 2-lobed stigma; ovary adnate to calyx, oval, 2-locular; capsule oval, 7 mm long, 3.5 mm wide, glabrous, greenish, finely ribbed, scarious; seeds oblong or oblong oval, 0.7 mm long, 0.3 mm wide, brown, inconspicuously and finely longitudinally tuberculate, shiny. July—August.

Sandy bottoms of lakes, rivers and creeks. — European part: Kar.-Lap., Dv.-Pech. (Arkhangel area, Priozernyi district near Sez-Lakhty River, Karpozero Lake), Balt., Lad.-Ilm., U. Dnp. (Minsk area, Sintez Lake near Novogrudok, Slutsk district). **Gen. distr.:** Scand., Centr. and Atl. Eur., N. Am. Described from W. Europe. Type in London.

2. **L. sessilifolia** Lamb. in Trans. Linn. Soc. X (1811) 260; DC. Prodr. VII, 380; Turcz. Fl. baic.-dahur. II, 178; Ldb. Fl. Ross. II, 869; Kom. and Alis., Opred. rast. Dal'nevost. kr. II, 1001; Wimmer in Pflzr. 107 Heft, 647. — L. saligna Fisch. in Mém. Soc. Nat. Mosc. III (1812) 65. — L. salicifolia Fischer ex Trautv. in Tr. Peterb. bot. sada, VIII, 2 (1882) 557. — L. kamtschatica Pall. ex Ldb. Fl. Ross. II (1846) 869. — Rapuntium kamtschaticum Presl. Prodr. Lobel. (1836) 24. — Ic.: Lamb. l. c. tab. VI, f. 2; Fisch. l. c. tab. IX; Gartenfl. XII, tab. 392.

Perennial; plant 25—100 cm high, glabrous; rhizome 3—4 cm long, oblique, with numerous thin white fibrous roots; stems solitary, terete, straight, simple, sometimes branching at the middle, finely sulcate, densely leafy at the middle and above; radical leaves absent; lower leaves oblong, 1.5—2(4) cm long, 0.4—0.8 cm wide, obtuse, the rest oval-lanceolate or lanceolate, (2.5)4—7 cm long, 0.5—1 cm wide, acute, sometimes mucronate,
453 all sessile, cordate at base or short-cuneate, more or less 3-nerved, acutely denticulate at margin. Flowers sessile or with 1—1.2 cm long, more or less flexuous antrorse pedicels, arranged in loose, erect, 6.5—16 cm long, 3.5—5 cm wide, terminal, nearly secund raceme; calyx 0.8—1.1 cm long, as long as

corolla tube, tube subglobular or pyriform, angular, 3–5 mm long, lobes lanceolate, 5–6 mm long, 1 mm wide, mucronate, entire; corolla 2-lipped, dark blue, 2.5–3 cm long, nearly three times longer than calyx, glabrous, the tube semicylindrical, 1.2–1.4 cm long, 3 mm wide, more or less tapering above, pubescent inside, the two upper lobes narrowly linear, 1.3–1.5 cm long, 0.5–1.5(2) mm wide, oblique, twisted, ascending, the 3 lower oblong or lanceolate, 1.3–1.5 cm long, 2–3(4) mm wide, all acute, densely white-ciliate; stamens with filaments free at base, connate above, anthers gray, adnate above, forming angular-curved cylinder, oblique above, the two lower anthers white-bearded at tips; style filiform, glabrous; stigma 2-lobed; ovary globose; capsule more or less globose, inflated, glabrous, finely ribbed, 0.9–1 cm long, 0.7–1 mm wide, 2-locular, many-seeded; seeds ovoid, 1.6–1.8 mm long, 1 mm wide, flattened, angular, dark brown, shiny, smooth. July–August.

Moss, reed, sedge and sphagnum marshes, lake banks, hot springs, flooded marshy meadows. – E. Siberia: Lena-Kol. (Tungir River basin), Dau.; Far. East: Kamch., Okhot., Ze.-Bu., Uda, Uss., Sakh. **Gen. distr.:** Jap. (N.), Ch. (Manchuria), Korea. Described from Kamchatka. Type in London.

Economic importance: A weed among crops and in rice fields.

ADDENDA XXIII *

DIAGNOSES PLANTARUM NOVARUM IN TOMO XXIV FLORAE URSS COMMEMORATARUM

(DIAGNOSES OF NEW SPECIES MENTIONED IN VOLUME XXIV)

Februari 1957

457 SCABIOSA L.

1. **Prismakena** Bobr. sect. n. — Involucellum fere tetraedrum, foveolis sulcisque deficientibus, quadricostatum, costis intermediis debilibus.

Typus sectionis: *S. comosa* Fisch.

2. **S. austro-altaica** Bobr. sp. n. (sect. *Asterocephalus*, ser. *Isetenses* Bobr.).

Planta suffrutescens, radix crassa lignosa pluriceps; caules 1—4, erecti, paulo foliati, 20—60 cm alt. non dense breviterque praesertim in parte superiore pilosi, capitula 1—3 emittentes; folia imprimis radicalia ambitu elliptica 5—10 cm long., petioli eorum 1—2 cm long., caulina minora opposita 2—3-paria; lamina utrinque breviter pilosa, axibus plus-minusve setosis, omnia pinnatisecta lobis lanceolatis 5—20 mm long. 2—3 mm lat. interdum vix fissa. Capitula floralia 2—3 cm in diametro, fructifera globosa 1.5—2 cm diam.; involucri phylla elongato-ovata superne angustata breviter denseque subtomentoso pilosa, floribus marginalibus duplo-triplo breviora; paleae in triente superiore carinatae anguste lanceolatae tenuiter denseque pilosae, inferne filiformes; involucellum 3—4 mm long. basi longe hirsutum inter foveolas costatum; corona membranacea 2—3 cm lat., calycis aristae 4—5 mm long., apice magis atratae; corolla pallide violaceo-cyanea, extus appresse pilosa; flores mediani 6—8 mm long., marginales ad 15 mm long. subbilobati. Fl. VI; fr. VIII.

Habitat: in schistosis regionis stepposae montium Altaj australis (Kazachstania orientalis et Dzhungaria chinensis).

Affinitas: a *S. isetensi* L. primo intuitu habitu peculiari differt: caulibus parum foliatis, foliis imprimis radicalibus, lobis lanceolatis 5—20 mm long., 2—3 mm lat., floribus violaceo coeruleis.

Typus: Kazachstania orientalis, distr. Zaisan in valle fl. Kaldzhir, inter Ajna-bulak et Terekty, leg.. B. A. Keller, initio Aug. 1908; typus in Leninopoli asservatur.

CEPHALARIA SCHRAD.

3. **C. nachiczevanica** Bobr. sp. n. (subgen. *Denticarpus* Szabo, sect. *Atrocephalae* Szabo, ser. *Procerae* Bobr.).

Perennis, caules ca. 0.8 m alti, sulcati, superne subglabri, inferne sensim densius retrorsum setosi, pedunculis paulo pilosis; folia infima

* [This appendix has been reproduced photographically from the Russian original.]

458 longepetiolata subintegra, media et superiora ambitu lanceolata, 10—15 cm long., 2—3 cm lat. segmentis 2—4 lateralibus minutis, lamina utrinque pilosa, margine vix dentata vel integra, segmenta foliorum superiorum sublinearia. Capitula globosa ca. 2 cm diam. involucri phylla subcoriacea ovato-lanceolata tantum in apice fusca, breviter longeque appresse pilosa, paleae coriaceae lanceolatae ad 10 mm long. breviter acuminatae dorso margineque pilosae apice fuscae. Achenium 6—8 mm long. tetragonum pilosum, dentibus angularibus ad 3 mm long., dentibus intermediis duplo brevioribus in fructificatione calycem superantibus; calyx margine inaequaliter setosus; flores sulphurei, extus pilosi, interni ca. 10 mm long. exteriores majores.

Habitat: Transcaucasia, respublica autonoma Nachiczevan, in subalpinis: in monte Ljakatach, 11 VII 1933 leg. Gadzhiev et Gurvicz — typus!; ibidem, inter p. p. Ljakatach et Bajachmed 28 VIII 1933 iidem legerunt; prope Bitschenach ca. 1800 m. leg. Heideman et Aliev 18 VIII 1932. Typus in Leninopoli.

Affinitas: differt a *C. hirsuta* Stapf foliis minus pilosis (non hirsutis) segmentis 2—4 (non 6—8) minoribus, involucri phyllis ovato-lanceolatis (non rotundatis), paleis lanceolatis (non elongato-cuneatis).

BRYONIA L.

4. **B. transoxana** Vass. sp. n.

Perennis, dioica. Caules graciles glabri vel subglabri. Folia ambitu hastato-triangularia 5-lobata (vel 5-fida), lobis lanceolato-triangularibus vel oblongo-ovatis, terminali multo longiore acuminato, lateralibus minoribus obtusis. Lamina foliorum undulata interdum cum dentibus singularibus et lobulis parvulis instructa, superne glabra vel subglabra, subtus calloso aspera basi profunde emarginata. Flores feminei in racemis laxissimis elongatis, partim basi pedunculi sessiles; pedunculo communi (2) 3—7 cm lg. (cum rhachide). Fructus ruber indehiscens, seminibus sordide flavescentibus breviter emarginatis 4.5—4 mm lg., 3.5—4 mm lt. Plantas masculas non vidi.

Habitat: ad ripas et margines lacuum in fruticetis.

Area geographica: Pamiro-Alai occidentalis.

Affinitas: a *B. dioica* Jacq. racemis foemineis laxissimis elongatis et lobo terminali foliorum valde elongato differt.

Typus: Schahrissjabs, in loco dicto „Tschopuk" 12 VI 1896 fr., Lipsky; in Herb. Inst. Bot. nomine V. L. Komarovii Ac. Sc. URSS conservatur.

5. **B. lappifolia** Vass. sp. n.

Perennis, dioica. Caules glabri vel sparse pilosi. Folia integra margine ± undulata, basi profunde et anguste emarginata, 9—13 cm lg.,

459 6—9 cm lat., supra glabra subtus et margine calloso-aspera, formam foliorum Lappae aemulans, petiolus 3.5—4.5 cm lg. setulis brevibus patulis obsitus. Flores ignoti. Fructus rubri globosi sessiles vel brevissime pedicellati in axillis foliorum aggregati, seminibus sordide-flavescentibus emarginatis ca. 5 mm lg., 4 mm lt.

Habitat: ad rupes in juniperetis.

Area geographica: Pamiro-Alai meridionalis.

Affinitas: a *B. dioica* var. *subsessili* Boiss. foliis lappiformibus et seminibus brevioribus differt.

Typus: Tadzhikistania meridionalis, in jugo Hardany-Uschty declivibus orientalibus in juniperetis ad rupes 2200 m s. m. 24 VI 1939, leg. M. I. Prjachin; in Herb. Inst. Bot. nom. V. L. Komarovii Ac. Sc. URSS conservatur.

CAMPANULA L.

Sect. *Medium* A. DC., subsect. *Triloculares* Boiss.

6. **C. Charkeviczii** Fed. sp. n.

Perennis, multicaulis, radice indurata crassa verticali vel obliqua, 1.5—2 cm in diam., ad collum residuis elongatis griseo-fuscis petiolorum emortuorum instructa. Caules 0.5—0.7 cm crassi robusti, 20—40 cm longi, e basi arcuata adscendentes dein erecti crebre foliosi ob pilos breves densos simplices scabridi, striatelli, saepe purpurascentes. Lamina foliorum rosularium magna, 10—12 cm longa, 3—4 cm lata, obovato-spathulata vel elliptica, in petiolos alatos laminam subaequantes sensim et cuneatim decurrentia, ad margines conspicue vel obsolete et inaequaliter crenata, apice obtusiuscula, supra subtusque pilis scabridis obsita vel fere setulosa, irterdum purpureo tincta; folia caulina inferiora radicalibus similia, sed deminuta; superiora angustiora subsessilia. Flores sat magni, ca. 3 cm longi, speciosi in racemos densos parum ramosos aggregati, coeruleo violacei, erecti vel nutantes, breviter pedicellati. Calycis obconici laciniae triangulari-lanceolatae acutae setoso-ciliatae corolla tubuloso-campanulata glabra vel vix papillosa in lobos ovatos acutiusculos ad trientem partem partita duplo vel fere triplo breviores, appendiculatae, appendicibus reflexis lanceolatis setoso ciliatis tubum calycinum obtegentibus; stylo incluso. VI.

Planta luxurians, ob caules robustos crassos, racemos densos, et ob pubescentiam densam scabram, foliorum magnitudinem, florum numerositatem inter species fere omnes subsectionis *Sibiricae* proeminens.

Crescit in rimis rupium verticalium in regione subalpina.

Typus: Tauria. In monte Baidarskaja jaila dicto, ad rupes. Rarissima planta. 14 VI 1952 leg. S. S. Charkevicz. Conservatur in Herb. Inst. Bot. Acad Sci. in Leningrad; isotypus — in Herb. Horti Bot. Kioviensis.

Nomen datur in honorem amiciss. S. S. Charkevicz. qui Campanulas tauricas ac caucasicas accurate investigavit et legit.

7. **C. elatior** (Fomin) Grossh. sp. n. (Descripsit Fed.).

Biennis; radix fusiformis duriuscula pallida fusca; caulis solitarius, sat crassus, simplex vel in parte superiore parum ramosus, sulcato angulatus, us lue ad 60 cm altus et 4, rarius 5 mm crassus, pilis brevibus et plus minus setulosis scabridis praecipue ad basin sat dense tectus. Folia radicalia et caulina inferiora elliptico lanceolata obtusiuscula, in petiolum late alatum interdum obsolete lobulatum vel denticulatum sensin attenuata, margine irregulariter crenata et interdum bicrenata, cum petiolo 8—10 cm longa et 2 cm lata; media et superiora — angustiora et minora, anguste oblongo-lanceolata ver fere linearia, sessilia et semiamplexicaulia, saepe etiam obtusiuscula. Inflorescentia angusta racemiformis vel saepius fere paniculata. Flores majusculi, subpatentes, coeruleo-violacei, usque ad 2.5—3 cm longi, pedicellati, pedicellis brevibus. Calycis brevissimi laciniae anguste lanceolatae acutae setuloso ciliatae corolla angusta tubuloso-infundibuliformi glabra vel vix puberula in lobos acutos subpatentes ad $^1/_4$ longitudinis partita duplo vel triplo breviores, appendiculatae, appendicibus obtuse-lanceolatis laciniis calycinis manifesto brevioribus; stylo incluso.

Affinis est *C. tauricae* Juz. et *C. sibiricae* L., sed a prima differt caulibus solitariis, inflorescentiis elongatis, dentibus calycinis longioribus corolla duplo brevioribus; a secunda — calycis dentibus non reticulato venosis et pubescentia sparsiora.

Habitat: in pratis, locis herbidis ad margines agrorum silvarumque, in planitiebus ad pedes montium et in montibus adjacentibus.

Area geographica: Ciscaucasia et regio stepposa partis Europaeae URSS.

Typus: in regione Kubanensi Caucasi septentrionalis, prope pag. Kavkasskaja. Leg. V. Lipsky, 18 V 1892. Conservatur in Herb. Inst. Bot. Acad. Sci. in Leningrad.

8. **C Schelkownikowii** Grossh. sp. n. (Descripsit Fed.).

Biennis, radice anguste fusiformi, fusca duriuscula; caules numerosi, nunc erecti, nunc prostrati vel arcuatim adscendentes, sat tenues, usque ad 25—30 cm longi, 1—2 mm crassi, superne parum ramulosi, foliosi, pilis brevibus albidis scabridis vestiti. Folia radicalia et caulina inferiora oblanceolata, obtusa vel acutiuscula, in petiolum anguste alatum sensim abeuntia, obsolete crenato-denticulata, scabrida, usque ad 5 cm longa et 0.8 cm lata; superiora et media — deminuta, sessilia, radicalibus conformia vel lanceolata et basi rotundata. Flores majusculi, 2.5—3 cm longi, pallidi, violaceo-coerulei, nutantes, pedicellati, pedicellis tenuibus filiformibus. Calycis brevissimi dentes lineari-lanceolati acuti interdum subulato lineares ciliati corolla late campanulata extus glabra in lobos

461 ovatos acutiusculos usque ad $^1/_4$ longitudinis partita multo breviores, appendiculati, appendicibus angustis, obtusiusculis tubum calycinum aequantibus, sed dentibus duplo vel triplo brevioribus, ciliatis; stylo non exserto.

Inter omnes species ex affinitate *C. sibiricae* s. l. corollis late campanulatis, dentibus calycinis angustissimis fere subulatis et pedicellis ramisque filiformibus proeminens.

Habitat: videtur in locis siccis regionis montanae mediae.

Typus: Azerbaidzhania, in districtu Karabagh. Ad pedes montis Ziarat, prope Chascham-czai, ca. 2000 m s. m. 27 VI 1909 leg. A. Schelkownikow. Conservatur in Herb. Inst. Bot. Acad. Sci. in Leningrad.

A cl. auctore speciei nomen datum in memoriam beati A. B. Schelkownikow, investigatoris florae Caucasi laboriosissimi ac praeclari.

9. Subsectio **Phasidianthe** Fed. subsect. n.

Calycis sinus appendicibus obtecti. Flores axillares inconspicui. Corolla parvula calycem tantum duplo superans, coerulea, tota glabra. Folia rosularia videtur nulla; caulina — breviter petiolata vel sessilia ambitu fere rhombea. Caules quasi dichotome ramosi ex apice rhizomatis in fissuris rupium dispositis emittentes, debiles, flexuosi. Subsectio monotypica.

Typus subsectionis: *C. imeretina* Rupr.

Subsectio caucasica inter subsectiones *Triloculares Annuasque* quasi media et inter eas collocanda.

Nomen subsectionis per conjunctionem verborum graecorum „Phasis“ (i. e. Rhion — flumen in Megrelia) et „anthos“ (flos) constitutum.

10. Subsectio **Tulipella** Fed. subsect. n.

Calycis sinus appendiculati, appendicibus ovatis acutis. Tubus, appendices dentesque calicini pilis albidis patentibus obsiti. Corolla magna cylindraceo-campanulata in parte media vix inflata tulipiformis in lobos erectos purpureo-punctatos partita. Flores nutantes sat longe pedicellati. Capsula nutans basi poris tribus dehiscens. Caules ramosi foliosi, foliis cordato-ovatis.

Typus subsectionis: *C. punctata* Lam.

Nomen subsectionis ob formam corollae datum.

11. Subsectio **Dasystigma**[1] Fed. subsect. n.

Calycis sinus appendiculati, appendicibus minutis lanuginosis. Tubus calycinus brevis obconicus glaber; dentes longe acuminati lanati tubum multo superantes. Corolla campanulata basi rotunda in lobos acuminatos partita. Stylus ad apicem pilosus. Flores longepedicellati in racemum

[1] i.e. stigma pilosum.

462 ramosum pauci- vel multiflorum quasi paniculatim aggregati. Folia lineari — lanceolata. Radix crassa fusiformis, pulpa spongiosa.

Typus subsectionis: *C. alpina* Jacq.

Subsect. **Involucratae** (Fom.) Fed.

12. **C. cephalotes** Nakai (descriptio nova).

Perennis, tota planta pilis brevibus incanis densis pubescens, scabridula vel interdum fere velutina, rarius glabriuscula, rhizomate brevi radicibus numerosis atrofuscis fibrosis donato. Caulis altus saepe fere metralis, erectus, firmus, striatellus, foliosus, viridis vel rubescens, simplex vel parum ramosus, ramis tenuibus axillaribus floriferis. Folia radicalia longissime petiolata, oblongo ovata, acuminata, basi cordata, ad margines minute et irregulariter serrato-denticulata, supra viridia, subtus ob pilos plus minus incana, in petiolos graciles lamina triplo longiores subito angustata; lamina foliorum 10—15 cm longa, 3—7 cm lata; folia caulina angustiora, ad basin caulis plus-minus longe petiolata, in parte media et ad apicem — sessilia, semiamplexicaulia; omnia late lanceolata vel oblongo-ovata, longe acuminata, serrato denticulata, dentibus inaequalibus obtusiusculis. Inflorescentia elongata interrupta vel interdum ramosa et fere paniculata, e fasciculis florum subsessilium distanter dispositis et ad apicem caulis in capitulum densum congestis formata, plus minus foliosa. Flores mediocres violacei ca. 2 cm longi; calycis obconici scabridi dentes lanceolati acuminati virides inappendiculati pubescentes corolla tubuloso-campanulata extus pubescens intus fere glabra in lobos ovatos acuminatos usque ad trientem parten fissa triplo breviores; stylo corollam non excedente.

Inter species affines seriei *Glomeratae* ob foliorum formam petiolorumque longitudinem et florum pubescentiam eminens. Proxima *C. farinosae*, sed ad eo corollis pubescentibus. foliis radicalibus longissime petiolatis et habitu magis robusto distinguenda.

Habitat: in pratis, nemoribus dumetisque in montibus et in planitiebus adjacentibus.

Area geographica: Dahuria et Oriens extremus.

Obs.: Descriptio nostra ad plantas numerosas in Oriente extremo et in Dahuria crescentes constituta.

13. **C. Maleevii** Fed. sp. n.

Ob pilos setaceos hyalinos patentes hirsutissima, perennis, rhizomate obliquo radicibus numerosis nigrescentibus praedito; caulis erectus crassiusculus, firmus, striatus, viridis vel rufescens, ca. 50 cm altus; folia radicalia ovata acuminata basi subtruncata vel cuneata in petiolos lamina subduplo breviores abrupte contracta, irregulariter crenato-denticulata; caulina radicalibus similia breviter petiolata vel sessilia, semiamplexicaulia, 5—7 cm longa, 1.5—2.5 (3) cm lata; flores in capitulum densum

463 ad apicem caulis congesti foliiis superioribus involucratis obtecti, coerulei; calycis obconici glabriusculi dentes lanceolati ad margines albo setulosi corolla tubuloso campanulata extus glabra intus vix barbellata in lobos ovatos acuminatos usque ad trientem partem fissa duplo breviores, virides, erecti, stylo pubescente corollam non superante. VI—VII.

Habitat: in pratulis inter sylvas frondosas ad pedes montium.

Typus: ad viam inter Tuapse et Dzhubga in parte occidentali Caucasi Magni. 3 VII 1952. Leg. Al Theodorov, An. Fedorov et M. Kirpicznikov. Paratypus: in monte Michailovskaja prope urbem Gelendzhik versus Tuapse, in pratulis 850 m s. m. 28 VI 51, leg. I. V. Vassiljev. Conservatur in Herb. Inst. Bot. Acad. Sci. URSS in Leningrad.

Ab affinibus speciebus seriei *Glomeratae* pilis valde setaceis patentibus hirsutis, foliis brevioribus omnibusque conformibus, superioribus capitulum circumcirca obtegentibus corollis autem glabris et toto habitu distinguenda.

In memoriam beati V. P. Maleev, qui antea Campanulam speciem detexit et descripsit.

14. **C. Trautvetteri** Grossh. sp. n. (Descripsit Fed.).

Perennis, plus minus pubescens vel glabriuscula, rhizomate brevi obliquo fusco radicibus fibrosis instructo. Caulis saepe humilis interdum elongatus, 10—30 cm altus, erectus, simplex, viridis vel rubescens, striatellus, parce foliosus, in capitulum solitarium pauciflorum abiens. Folia radicalia, petiolata, ovato oblonga, acuminata, ad margines minute serrata, in petiolum gracilem limbi aequilongum e basi truncata cuneatave abrupte anguastata; caulina — sensim deminuentia radicalibus conformia, in parte inferiore caulis breviter petiolata, caetera — sessilia et interdum semiamplexicaulia; superiora — involucrum obconicum formantia capitulum non superantia acuminata, interdum subulato acuminata basi dilatata membranacea ad margines inciso dentata vel integerrima. Flores speciosi sat magni vel majusculi, 2—4 cm longi, lilacini. Calycis obconici dentes lanceolati subulato acuminati ciliati corolla tubuloso-campanulata extus vix pubescens intus fere glabra in lobos oblongos acuminatos usque ad dimidium fissa triplo breviores; stylo corollam non superante. VI—VII.

Ob florum magnitudinem, habitum alpinum, humilitatem caulium et foliorum superiorem formam insignis. Campanulae Panjutini e montibus Abchasiae plus minus accedere, tamen ab ea optime differe videtur (folia involucri membranacea, subito et interdum subulato acuminata breviora, pubescentia sparsa etc.).

Habitat: in pratis alpinis rupibusque montium Caucasi Orientalis et Minoris.

Lectotypus (post mortem cl. auctoris speciei selectus): Armenia, Geseldara, 2 VII 1871, leg. G. Radde, n° 319. Conservatur in Herb. Inst. Bot. Acad. Sci. in Leningrad.

15. Subsectio **Dictyocalyx** Fed. subsect. n.

Calycis sinus appendiculati, appendicibus post anthesin valde accrescentibus et quasi inflato-ampliatis tubum obtegentibus, reticulatis. Flores parvuli, sessiles, in spicam interruptam vel in racemum laxum dispositi. Folia rigida ovato oblonga. Plantae ob pilos setaceos basi tuberculatos scabridae.

Typus subsectionis: *C. stricta* L.

Subsectio ab omnibus subsectionibus ex affinitate sectionis *Medium* A. DC. generis Campanulae indumento e setis basi tuberculatis constante distinctissima. An prope subsectionem *Triloculares* Boiss. locanda?

Subsectionis nomen constitutum per conjunctionem graecorum verborum „dictyon" (reticulum) et „calyx".

16. Subsectio **Latilimbus** Fed. subsect. n.

Calycis sinus appendicibus minutis inconspicuis vel interdum filiformibus instructi vel fere exappendiculati; dentes calycini post anthesin non accrescentes saepe erecti. Folia radicalia in petiolum subito contracta, basi cordata vel truncata vel etiam late cuneata, oblongo ovata, denticulata vel crenata. Caules simplices vel plus minus ramosi. Flores in racemum obliquum compositi vel ad ramos paniculatim dispositi, jam ante evolutionem nutantes. Capsula nutans.

Typus subsectionis: *C. collina* M. B.

Nomen subsectionis per conjunctionem verborum „latus" et „limbus" (i. e. lamina foliorum) constitutum.

17. **C. Siegizmundi** Fed. sp. n.

Perennis, fere tota glabra vel vix puberula, rhizomate incrassato apice ramoso surculos apice residuis numerosis setiformibus petiolorum emarcidorum tectos emittente. Caules adscendentes vel erecti, simplices, graciles, glabri, 20—40 cm alti, parce foliosi, in racemos elongatos abeuntes. Folia radicalia longissime et graciliter petiolata, cum petiolo 10—15 cm longa, 1.5—2 cm lata, oblongo-ovata vel triangulari ovata, apice attenuata et acuta, basi oblique cordata vel truncata, ad margines duplicato et acuti-serrata utrinque vel supra tantum glabra, subtus interdum puberula; petioli laminam triplo superantes, filiformes glabri; folia caulina inferiora petiolata, caetera sessilia, parva, lanceolata. Flores pedicellati, pedicellis calycem subaequantibus. Calycis laciniae lanceolatae acutae acuminatae erectae glabrescentes vel vix puberulae corolla glabra duplo breviores, appendiculatae, appendicibus brevibus minutis subulatis petentibus. Capsula turbinata vel orbiculari-obconica ob nervos prominulos costata, pilis brevibus plus-minus pubescens, basi tribus valvulis parvulis dehiscens.

Ab omnibus speciebus seriei Sarmaticae Fed. surculis hypogaeis

465 rhizomatis e rupium fissuris emittentibus, petiolis filiformibus gracilibus longissimis, foliorum cauliumque glabritiae et tote habitu distinctissima.

H a b i t a t: in rupium rimis in regione silvatica montana.

T y p u s: Caucasus centralis, Kabarda, in districtu Sovjetsky, prope pag. Babugent, ad lac. „Goluboje osero" (lacus coeruleus), in fissuris rupium 3 VIII 1949, leg. S. Charkevicz. Conservatur in Herb. Inst. Bot. Acad. Sci. in Leningrad; isotypus — in Herb. Horti Kioviensis.

Speciei nomen dedicatum in honorem amiciss. Siegizmundi Charkevicz, qui plantam hoc loco descriptam et nominatam detexit et legit.

18. Subsectio **Trigonophyllon**[1] Fed. subsect. n.

Calycis sinus fere exappendiculati vel appendicibus minutis instructi, dentes calycini post anthesin valde accrescentes tubum conspicue vel multiplo superantes. Folia radicalia et surculorum sterilium longe longissimeve et saepius graciliter petiolata, ambitu trigona basi cordata vel truncata cuneataque. Flores jam ante evolutionem nutantes. Capsula nutans. Caules uniflori et simplices vel parum ramosi, ramulis gracilibus floriferis. Rhizoma ramosum surculos steriles emittens vel eos destitutum.

T y p u s s u b s e c t i o n i s: *C. dzyschrica* Kolak.

19. Subsectio **Oreocodon** Fed. subsect. n.

Calycis sinus appendiculati vel exappendiculati vel etiam plicis minutis inter dentes dispositis donati. Flores in paniculas corymbiformes vel in racemos dispositi, nutantes, pedicellati, interdum cleistogami. Folia caulina omnia breviter petiolata vel subsessilia ambitu rotunda vel late ovata; inferiora manifeste petiolata et nunquam rosulam formantia. Indumentum densum lanatum vel velutinum, incanum vel albo tomentosum, interdum scabridum. Caules plures fragiles adscendentes vel arcuati interdum debiles et fere filiformes ex ramulis rhizomatis incrassati scrobiculati emittentes.

T y p u s s u b s e c t i o n i s: *C. incanescens* Boiss.

Subsectionis nomen per conjunctionem verborum graecorum „oros" (i. e. mons) et „codon" (i. e. campana vel campanula) constitutum.

20. Subsectio **Rupestres** (Boiss.) Fed. comb. n.

Calycis sinus appendiculati vel exappendiculati vel etiam plicis minutis inter dentes dispositis instructi; dentes calycini lanceolati vel lineari-lanceolati Flores pedicellati, ante evolutionem erecti, post anthesin nutantes. Caules pauciflori vel rarius uniflori. Folia ad basin caulis et ad apices surculorum sterilium rosulas saepe densas formantia, rosularia et caulina — spathulata, interdum lanceolata vel linearia, petiolata vel sessilia, rarissime cordata vel ovata, radicalia rosularibus conspicue latiora et longiora. Rhizoma

[1] i. e. folium triangulare.

466 saepe crassum, pluriceps, non raro columniforme et reliquiis petiolorum emarcidorum donatum, caespitosum, interdum fere repens. Herbae pubescentes vel glabriusculae.

Typus subsectionis: *C. petrophila* Rupr.

21. **C. Eugeniae** Fed. sp. n.

Perennis, humilis, glabriuscula, rhizomate caespitoso plurucipiti ramoso, surculis numerosis abbreviatis griseis donato; surculi reliquiis brevibus squamiformibus petiolorum emarcidorum instructi, rosulas laxas steriles formantes vel caules floriferos emittentes; caules simplices filiformes graciles folia radicalia et surculorum sterilium paulo superantes foliigeri uniflori vel biflori, 10—15 cm alti; folia basalia rosularia, (5) 8—10 cm longa, longissime et graciliter petiolata, petiolis filiformibus; lamina foliorum radicalium anguste lanceolata vel lineari-lanceolata utrinque attenuata acuta petiolo fere duplo breviora, glabra, tenera margine subintegerrima vel distanter et obsolete denticulata; folia caulina lineari-lanceolata vel linearia, alterna, sessilia. Flores erecti vel nutantes solitarii, 1.5 cm longi; calycis obconici glabriusculi dentes angusti-lineares corolla angusti campanulata glabra coerulea in lobos triangulares acutos partita subbreviores, tubum calycinum multiplo superantes, exappendiculati. Stylus inclusus. Capsula (immatura) obconica brevis demum fere globosa, poris ad basin sitis dehiscens. VIII.

Plus minus affinis *C. Lehmannianae* Bge. et *C. Capusii* (Franch.) Fed., sed ab utraque caulibus filiformibus petiolisque longissimis gracilibus et omnino habitu, a primo autem, praeter notas indicatas, floribus solitariis optime dignoscitur.

Habitat: in rimis rupium marmorearum, in regione subalpina ac media montana.

Typus: Asia Media, in districtu Dzhelalabad, prope pagum Kirov (olim Arslanbob). Ad latera montis Baubaschata jugi Ferganici, in rupibus marmoreis. 1945 VIII 19. Leg. An. Fedorov et E. Iljina. In Herb. Inst. Bot. Acad. Sci. URSS in Leningrad conservatur.

Nomen datum in honorem Eugeniae Iljinae, quae mecum plantam hoc loco descriptam legit.

22. **C. aldanensis** Fed. et Karav. sp. n.

Perennis, rhizomate brevi ramoso rosulam laxam formante surculoso surculis griseis subtilibus ad apicem reliquiis membranaceis pellucidis foliorum maceratorum instructis. Caules debiles filiformes a basi arcuatim adscendentes tandem erecti foliorum rosulas duplo superantes uniflori pilosiusculi parce foliosi, 10—20 cm longi. Folia radicalia et surculorum sterilium lanceolata 5—20 cm longa, 0.3—0.5 cm lata, in petiolos filiformes laminam subaequantes sensim attenuata apice obtusiuscula ad margines inaequaliter et obsolete denticulata, denticulis obtusiusculis

467 inter se distantibus sursum orientibus, utrinque glabra laete viridia; caulina linearia subsessilia deminuentia. Flores parvi, 0.5 cm longi, videtur pallide coerulei. Calycis hirsuti dentes lanceolati apice arcuatim incurvati hirsutiusculi corolla infundibuliformi extus vix puberula in lobos crispule pubescentes ad tertiam partem divisa subduplo breviores, appendiculati, appendicibus brevibus reflexis triangularibus hirsutiusculis; stylo non exserto. VII—VIII.

Ob caules petiolosque filiformes atque calycinos dentes arcuatos, foliorum formam et distributionem geographicam inter omnes species seriei *Dasyanthae* insignis. A *C. dasyantha* M. B., quae plusminusve proxima, differt foliis glabris et omnibus partibus tenuioribus ac gracilioribus.

Crescit ad rupes calcareas secus ripas fluviorum.

Typus: Sibiria, in districtu Aldanensi, in ripa fluminis Aldan supra ostium confl. Mili, in rimis rupium calcarearum. n° 882. Leg. 6 VIII 1928, V. Korzhevin. In Leningrad conservatur.

23. Subsectio **Hypopolion** Fed. sect. n.

Calycis sinus appendiculati, appendicibus triangularibus reflexis tubum calycinum brevem hemisphaericum aequantibus. Dentes calycini lanceolati vel fere lineares tubum multo superantes subpatentes arcuati. Corolla campanulata extus glabra calycem fere triplo superans. Flores in corymbos laxos dispositi, ante evolutionem erecti dein nutantes. Capsula suberecta. Caules flexuosi, debiles, ramosi ex ramulis superioribus tenuibus rhizomatis emittentes, foliosi. Folia omnia caulina, linearia in parte superiore obsolete denticulata, subtus ob pilos tomentosos canescentia.

Typus subsectionis: *C. hypopolia* Trautv.

Nomen subsectionis per mutationem nominis speciei typicae constitutum.

24. Subsectio **Rotula** Fed. subsect. n.

Calycis sinus appendicibus non obtecti; dentes calycini e basi triangulata lanceolati acuti integri, non raro minute denticulati, post anthesin patentes. Corola late infundibuliformis et fere rotata. Capsula erecta cylindraceo-ovoidea, ob 10 nervos longitudinaliter striata poris tribus ad apicem dehiscens. Caules ramosi foliosi. Folia rosularia nulla; caulina inferiora longe petiolata ovato-cordata irregulariter crenato-dentata in petiolum plus-minus longum gracilem abrupte attenuata, superiora — subsessilia angustiora. Plantae perennes.

Typus subsectionis: *C. carpatica* Jacq.

Nomen subsectionis ob formam corollae rotatam datum.

25. Subsectio **Melanocalyx** Fed. subsect. n.

Calycis sinus exappendiculati; dentes calycini corollam subaequantes vel ea vix breviores, erecti. Calycis tubus in florendi tempore nec non

468 post anthesin apice inflatus et prope originem loborum constrictus basin versus sensim angustatus, nigricans, puberulus, ob nervos prominulos striatus. Corolla in lobos lanceolatos obtusiusculos suberectos fere usque ad basin fissa. Capsula apice inflata subclavata lobis calycinis persistentibus erectiusculis coronata, valvulis tribus prope apicem sitis dehiscens. Caules uniflori plus-minus foliosi ex ramulis partim procumbentibus rhizomatis emittentes. Folia praecipue ad basin caulis congesta et tunc quasi rosulata, lanceolata vel interdum fere spathulata, obsolete crenulata vel subintegra. Plantae praeter calyces pilosos glabrae.

Typus subsectionis: *C. uniflora* L.

Subsectionis nomen per conjunctionem verborum graecorum „melas" (i. e. obscurus) et „calyx" constitutum.

26. Subsectio **Odontocalyx** Fed. subsect. n.

Calycis sinus exappendiculati; dentes caiycini e basi dilatata lineari-lanceolati ad utraque latera laciniato-denticulati. Corolla glabra, tubus calycinus villosus. Capsula erecta cylindrica poris tribus ad apicem sitis dehiscens. Folia rosularia caulinaque oblanceolata vel lanceolato-spathulata ad margines acuti vel fere setoso-denticulata. Rhizoma caules uniflores et surculos steriles emittens, laxe caespitosum.

Typus subsectionis: *C. lasiocarpa* Cham.

Nomen subsectionis ob calycis lobos laciniato-dentatos datum.

SYMPHYANDRA A. DC.

27. Sectio **Parageranion** Fed. sect. n.

Folia palmatipartita ambitu pentagona lobata, lobis incisodentatis; inferiora — longipetiolata. Corolla late campanulata fere rotata. Antherae in tubum connatae.

Typus sectionis: *S. zangezura* Lipsky.

Sectionis nomen per conjunctionem verborum graecorum „para" et „geranion" (ob nom. genericum Geranium) constitutum.

28. Sectio **Petrocodonia** Fed. sect. n.

Folia integra oblongo-lanceolata basi cuneata, breviter petiolata vel sessilia. Corolla late campanulata fere rotata. Antherae in tubum connatae.

Typus sectionis: *S. lezgina* Alexeenko.

Sectionis nomen per conjunctionem verborum graecorum „petros" (i. e. rupes) et „codon" (i. e. campana vel campanula) constitutum.

29. **BRACHYCODON**[1] FED. gen. n.

Corolla minima obconica, ad $^2/_3$ longitudinis suae 5-loba, calyce aequilonga dein dimidio brevior. Stylus corollae dimidio brevior, staminibus autem longior, tripartitus. Stigmata tria abbreviata. Ovarium trilo-

[1] i.e. campana brevis.

469 culare. Calyx fere ad basin laciniatus, laciniis exappendiculatis lineari-lanceolatis strictis corollam fere obtegentibus. Capsula obconica pentagona, 5-costata, valvulis prope apicem sitis dehiscens. Flores axillares brevissime pedicellati inconspicui. Plantae annuae vel biennes humiles radice tenui. Caulis ramosus quasi dichotomus, ramis divergentibus fastigiatis. Folia alterna vel subopposita parvula, rosularia — nulla.

Typus generis: *Brachycodon fastigiatus* (Desf.) Fed.

Genus monotypicum, in regione arida demissa praecipue mediterranea a Hispania et Africa boreali per totam Asiam minorem et Transcaucasiam usque ad Asiam mediam sporadice distributum. Inter Campanulas et Legouzias quasi medium; ob capsulas pentagonas et 5-costatas, sed triloculares, corollas breves e calyce non exsertas atque habitum, ramificationem et staturam inter omnia genera Campanulacearum insigne.

ADENOPHORA FISCH.

30. Sectio **Microdiscus**[1] Fed. sect. n.

Discus (nectarium) brevis parvus, 1 mm circiter longitudine et in diametro. Dentes calycini ovato-lanceolati, lanceoiati vel lineares, erecti vel patentes, interdum utroque latere laciniato denticulati corolla campanulata multiplo breviores. Flores in racemos vel in paniculas aggregati. Folia caulina alterna, ovata, elliptica, oblonga vel late-lanceolata, serrata vel dentata, sessilia vel manifesto petiolata.

Typus sectionis: *A. liliifolia* (L.) Bess.

31. **A. taurica** (Sukacz.) Juz. (Descripsit Fed.).

Perennis, radice crassa, ca. 2 cm in diam., rapiformi, inferne interdum ramosa, atrofusca, facie rugosa; caules innumerosi, simplices, rarius parum ramosi, crassiusculi, usque ad 0.3 cm in diam., longitudinaliter striati, crebre foliosi, saepius ca. 50 cm alti, oliganthi; folia omnia caulina, praecipue in parte media caulis congesta et tunc subimbricata, sat brevia, 2—3—4 cm longa, 1—2 cm lata, elliptica, acutiuscula, breviter petiolata, ad margines grosse et irregulariter dentata, dentibus acutiusculis decurrentibus, subtus quam supra pallidiora; superiora — lanceolata, sessilia, plus-minus dentata, inferiora — cito marcescentia et decidua; inflorescentia racemosa, e 5—10-floribus constans, rarius sat multiflora; flores saepe ca. 1.3—1.5 cm longi; calycis glabri obconici dentes trinagulari-lanceolati acuti tubum aequantes corolla tubuloso campanulata vel angusti-campanulata usque ad $^1/_5$ longitudinis in lobos acutiusculos breves partita multiplo breviores; stylus longe exsertus. VII.

Ab affini *A. liliifolia* (L.) Bess. caulibus humilioribus crebre foliosis, foliis subimbricatis in parte media caulis congestis et inflorescentiis oliganthis sat differt.

[1] i.e. nectarium parvum.

 Habitat: in pratis montanis ad margines superiores silvarum.

Typus: Tauria, in cacumine montis Czuczel majoris, n° 832, 18 VII 1925, leg. H. Poplawska, conservatur in Herb. Inst. Bot. Acad. Sci. URSS in Leningrad.

32. **A. jacutica** Fed. sp. n.

Perennis, fere omnino glabra, radice crassa napiformi transverse vix rugulosa griseo fusca, caulis erectus, simplex vel apice vix ramosus, ramulis tenuibus, cylindricus, glaber et laevis, 2—3 mm crassus, foliosus, ad basin reliquiis squamaeformibus fuscis petiollorum emortuorum instructus, ca. 70 cm altus; folia radicalia nulla vel cito emarcida; caulina alterna, subsessilia vel breviter petiolata, ambitu late rhomboidea, utrinque cuneatim angustata, grosse et profunde dentata, dentibus magnis usque ad 1—1.5—2 cm longis, utrinque latere 7—8, triangularibus, erectis, apice in mucronem incurvatum callosum subito angustatis; flores nutantes, breviter et tenuiter pedicellati, in racemum pauciflorum vel in paniculam depauperatam aggregati, inconspicui; calycis ovoidei glabri dentes patentes lineares subulati integri tubum subaequantes corolla campanulata (?) multiplo breviores; stylo exserto; capsula ovoidea membranacea poris ad basin sitis dehiscens et dentibus calycinis reflexis vel patentibus terminata.

Habitat: in vallibus fluviorum inter silvas e Picea ajanensi constantes, tamen ad declivitates montium adjacentium.

Typus: Respublica autonoma Jacutensis, in districtu Uczur, ad ripam sinistram fluminis ejusdem nomine, in loco Muguskan infer. dicto, 25 VIII 1954, n° 604, leg. L. N. Tjulina, conservatur in Herb. Inst. Bot. Acad. Sci. URSS in Leningrad.

Valde affinis *Adenophorae sublatae* Kom., sed primo aspectu differt dentibus foliorum magnis erectis apice incurvato mucronulatis et foliis latioribus ambitu fere rhomboideis.

33. Sectio **Pachydiscus**[1] Fed. sect. n.

Discus (nectarium) cylindricus crassus majusculus, 4 mm circiter longitudine et in diametro, lanatus. Calycis dentes majusculi, erecti, oblongo-triangulares corolla majuscula campanulata quadruplo breviores. Flores solitarii vel pauci. Folia caulina alterna, lanceolata, vix dentata.

Typus sectionis: *A. himalayana* Feer.

34. **POPOVIOCODONIA** gen. n.

Corolla 5-mera in lobos late lanceolatos usque ad basin fissa, disco (nectario) destituto. Calyx ob nervos prominulos costatus in dentes lineares tubo breviores reflexos partitus. Ovarium trimerum. Filamenta

[1] nectarium crassum.

471 dilatata pilosiuscula. Stylus conspicue exsertus, stygmata tria. Capsula ad apicem dehiscens. Folia omnia caulina, nunquam rosularia. Plantae perennes habitu Adenophorae.

Genus oligotypicum inter Adenophoram Campanulamque quasi medium et tantum in Oriente extremo distributum.

Typus generis: *P. Uyemurae* (Kudo) Fed.

Nomen genericum in memoriam beati botanici illustrissimi ingenio sublimi M. G. Popov, nuperrime investigatoris optimi florae Sibiriae, antea Asiae Mediae, Caucasi, montium Carpatorum aliorumque regionum dedicatum.

35. **ASTROCODON** Fed. gen. n.

Corolla 5-mera, in lobos late lanceolatos infra medium partita, lobis stellatim patentibus, disco (nectario) non evoluto. Calyx 5-merus, in dentes lineari-lanceolatos patentes vel reflexos partitus. Ovarium triloculare. Filamenta basi dilatata pilosiuscula. Stylus in stigmata tria divisus. Capsula valvalis ad basin sitis dehiscens. Folia tantum caulina. Herbae perennes, radice incrassata.

Genus oligotypicum, habitu Popoviocodoniae, sed ob capsulas ad basin dehiscentes sximie ab ea distinctum.

Area: Oriens extremus.

Typus generis: *A. Kruhseanus* (Fisch. ex Rgl. et Tiling) Fed.

Nomen genericum ob corollam stellatam datum.

36. Tribus **Peracarpeae** Fed. trib. n.

Corolla exacte obconica in lobos 5 erectos anguste acuminatos profunde incisa. Calyx exappendiculatus 5-merus obconicus. Stamina libera. Filamenta et antherae lineares. Ovarium triloculare. Stylus apice 2—3-partitus, stigmata revoluta. Campsula tenuiter membranacea et fere pellucida, clausa vel valvulis minutis prope apicem pedicelli sitis dehiscens, piriformis, pendula. Plantae perennes, caulibus ramosis debilibus, foliis alternis petiolatis tantum caulinis, floribus axillaribus. Rhizoma repens tenue surculos emittens.

Typus tribus: *Peracarpa* Hook. f. et Thoms.

37. Tribus **Ostrowskieae** Fed. trib. n.

Corolla campanulata, 7, rarius 5—9-mera, in lobos breves trigonos partita. Calyx exappendiculatus breviter cylindricus vel depresso-globosus prope originem loborum poris magnis numerosis oblongis perforatus et in dentes 7, rarius 5—9, longos late-lineares acutos incisus. Ovarium 7, rarius 5—9-loculare. Capsula turbinata pergamacea, poris magnis numerosis, 14, rarius 10—18, dehiscens. Stylus crassus cylindricus, stigmata saepius 7, arcuata. Stamina libera. Filamenta brevia. Antherae lineares

472 demum spiraliter tortae et revolutae. Plantae perennes, foliis verticillatis, radice tuberoso-incrassata.

Typus tribus: *Ostrowskia* Rgl.

38. Tribus **Michauxieae** Fed. trib. n.

Corolla in lobos lineares usque ad basin fissa, 8—10-mera. Calyx appendicibus trigonis reflexis instructus, 8—10-merus. Ovarium 8—10-loculare. Stamina libera. Filamenta basi latissima, antherae breviter cuspidatae. Stylus cylindricus pilosus, stigmata stellatim disposita. Capsula nutans valvulis ad basin sitis dehiscens, sulcato-costata. Plantae monocarpicae foliis caulinis alternis et radicalibus rosulatis.

Typus tribus: *Michauxia* L'Her.

39. Tribus **Phyteumateae** Fed. trib. n.

Corolla in lobos anguste lineares vel lanceolatos liberos vel apicibus coalitos usque ad basin fissa, rarius fere rotata, lobis latioribus, 5-, rarius 4-mera. Calycis forma maxime varians. Ovarium 2—3-loculare. Stylus elongatus vel brevis; stigmata 2—3. Stamina libera vel apicibus coalita. Antherae saepius elongatae, filamenta autem brevia. Capsula prismatica, cylindrica vel fere globosa, saepe ad apicem constricta, poris vel fissuris ab basin apicemque vel in media parte sitis dehiscens. Plantae perennes vel annuae, foliis alternis, interdum rosularibus, radice vel rhizomate interdum valde incrassato.

Typus tribus: *Phyteuma* L.

ASYNEUMA GRISEB. et SCHENK

40. Sectio **Eupodanthum** (Boiss.) Fed. descr. emend. — Sect. *Eupodanthum* Boiss. Fl or. III (1875) 945, p. p.

Folia radicalia non rosularia vel deficientia; praeter apicalia omnia conformia. Caules plus-minus dense et regulariter foliosi, inflorescentiis racemoso-spiciformibus terminati. Herbae perennes, rhizomate tenui obliquo, rarius radice erecta.

Typus sectionis: *Asyneuma canescens* (Waldst. et Kit.) Griseb. et Schenk.

41. **A debile** Fed. sp. n.

Perenne, pilis albidis subpatentibus parce puberulum, vix glaucescens. Caules tenues elongati, ca. 70 cm longi, debiles, foliosi, plus minusve ramulosi, ramis gracilibus; folia brevia, ovata, basi rotundata vel subcordata, breviter petiolata vel subsessilia, acuminata vel subobtusa, argute et irregulariter denticulata, dentibus inaequalibus sursum orientibus, ca. 3—5 cm longa, 2—4 cm lata, versus apicem caulis deminuentia. Flores parvuli, padicellati, coerulescentes in racemum parvum corymbi-

473 formem vel subpaniculatum aggregati. Calycis subglabri tricostati dentes lineares tubo triplo longiores corolla glabra in lobos lineares usque ad basin fissa duplo breviores. Antherae lineares filamenta multo superantes, corollam subaequantes, luteae. Stylus filiformis non exsertus, stigmate bifido terminatus, coerulescens. Capsula (immatura) poris lateralibus medio sitis dehiscens.

Ex affinitate *Asyneumatis arguti* (Rgl.) Bornm., tamen foliis brevioribus ovatis, caulibus ramisque tenuioribus debilibus, inflorescentiis quasi corymbosis et indumento e pilis albidis subpatentibus (nec scabridulis) constante diversum.

Habitat: Tadzhikistania australis. In cacumine montis Zevistan, ca. 2000 m s. m. 10 VII 1883 leg. A. Regel. Conservatur in Herb. Inst. Bot. Acad. Sci. URSS in Leningrad.

42. **A. talyschense** Fed. sp. nova.

Perenne, ob pilos breves albidos ad nervos paginae inferioris foliorum dispositos scabridulum, rhizomate videtur tenui. Caulis erectus simplex inferne crassiusculus caeterum sat tenuis et gracilis, 50—70 cm altus, ad basin crebre, sed apicem versus distanter foliosus, striatellus, glaber. Folia omnia caulina, inferiora subpetiolata, caetera — sessilia, a basi truncata vel vix cordata oblongo ovata acuminata, 2—5 cm longa, 1—2.5 cm lata, subtus prominenter nervosa, ad margines manifesto denticulata vel biserrata; suprema valge deminuentia. Flores parvuli breviter pedicellati, ca. 0.5—0.6 cm longi, violaceo-coerulei in spicam brevem apice densiusculam inferiore interruptam quasi fasciculatim aggregati. Calycis obconici glabri sulcati dentes lanceolato lineares tubum subaequantes vel eo sublongiores patentes vel reflexi corolla elongata glabra in alabastra supra calycem strangulata in lobos lineares et liguliformes usque ad basin fissa quadruplo multiplove breviores; stylo filiformi, stigmata tria, antheris elongatis filiformibus.

Ex affinitate *Asyneumatis amplexicauli* (Willd.) D. Sosn., sed discrepat foliis basi truncatis vel vix cordatis non amplexicaulibus, spicis brevioribus floribusque minoribus et laciniis calycinis reflexis vel patentibus nec circinato revolutis.

Crescit in regione superiori montium Talyschensium ad rupes inter silvas frondosas.

Typus: Azerbaidzhania, in districtu Lenkoran, meridiem versus a pag. Alara, ad latera silvosa montis Tarakecz. N° 584. Leg. 15 VII 1931, E. Matveeva.

Paratypus: Azerbaidzhania. In valle prope montem Tarakecz, ad limites silvarum quercetarum, in abruptis et rupibus, ca. 2300 m s. m. in LeninN° 857. Leg. 16 VII 1931. N. Schipczinsky.

Specimina hoc loco indicata in Herb. Inst. Bot. Acad. Sci. URSS in Leningrad conservatur.

474 43. Sectio **Sceptrum** Fed. sect. n.—Sect. *Eupodanthum* Boiss. Fl. or. III (1875) 945, p. p.

Folia radicalia rosulas plus-minus densas formantia vel valde approximata; rosularia a caulina magnitudine et forma distinctissima. Caules parce foliosi, interdum subaphylli, inflorescentiis elongatis interruptis racemoso-spiciformibus terminati. Herbae praecipue monocarpicae, rarius perennes, radice fusiformi.

Typus sectionis: *Asyneuma pulchellum* (Fisch. et Mey.) Bornm.

44. **SERGIA** FED. gen. n.

Calyx haemisphaericus in dentes 5 lanceolatos acutos fere usque ad basin partitus. Corolla in lobos lanceolatos calycem subaequantes vel superantes usque ad basin fissa. Stylus cylindricus coeruleus papillosus; stigmata tria. Stamina 5, filamentis brevibus ampliatis et incrassatis antherisque elongatis lanceolatis fere sessilibus interdum sterilibus et obsolete evolutis. Capsula globosa ob septa prominula costata apice constricta, membranacea, papillosa vel breviter setulosa, poris valvulis parvis instructis ad basin inter costas dispositis dehiscens et lobis calycinis marcescentibus coronata. Semina laevia rotunda parvula straminea. Herbae perennes multicaules, radice dura apice valde incrassata scrobiculata, foliis anguste lanceolatis ovatisque.

Typus generis: *S. Severzovii* (Rgl.) Fed.

Genus oligotypicum, inter omnia genera Campanulacearum capsulis globosis angulato-costulatis apice constrictis, antheris foliformibus subsessilibus, filamentis incrassatis brevissimis, florum et aliorum partium forma atque omnino habitu proeminens et incomparabile.

Area geographica: Montes Tjanschanici occidentales et promontorium Karatau atque in montibus Pamiroalaicis.

Nomen generis datum in honorem botanici clarissimi, systematici ingeniosissimi Sergii Juzepczuk, qui Campanulas novas in territorio Patriae nostrae crescentes optime ac venuste descripsit.

45. **CRYPTOCODON** FED. gen. n.

Flores ovario trimero excepto pentameri. Stigma autem bifidum. Stylus inclusus. Corolla tubulosa profunde 5-loba, lobis lanceolatis erectis. Calyx 5-dentatus, dentibus corollam subaequantibus tubo calycino duplotriplove longioribus, sinubus minute appendiculatis. Capsula ellipsoidea angulata medio poris tribus dehiscens. Herba perennis humilis, foliis oblongis denticulatis. Flores in fasciculum capituliformem densum compactum aggregati inter folia involucralia flores superantia dispositi.

Typus generis: *C. monocephalus* (Trautv.) Fed.

475 Affine *Asyneumati*, differt corollae lobis erectis calycem subaequantibus, stylo incluso, sinubus calycinis appendiculatis, floribus in capitulum terminalem congestis, foliis floralibus inflorescentiam occultantibus et allis notis numerosis.

Area geographica: Montes Zeravschanici Asiae Mediae.

Nomen generis per conjunctionem verborum graecorum „cryptos“ (i. e. obscurus) et „codon“ (i. e. campana vel campanula) constitutum.

46. Tribus **Edrajantheae** Fed. trib. n.

Corolla tubulosa in lobos breves partita, 5-mera. Calyx exappendiculatus. Stylus stigmatibus duobus terminatus. Stamina libera, filamentis dilatatis antheris autem angustioribus linearibus. Capsula maturitate valvis inaequalibus longitudinaliter dehiscens et post disseminationem quasi infundibuliformis. Plantae perennes, caulibus numerosis, foliis alternis interdum imbricatis non raro pulvinos plus minus densos formantibus. Radix crassa, lignoso-indurata. Inflorescentia capitata foliis floralibus praedita; interdum flores solitarii terminales.

Typus tribus: *Edrajanthus* A. DC.

47. Tribus **Jasioneae** Fed. trib. n.

Corolla usque ad basin in lobos angustos fissa, 5-mera. Calyx exappendiculatus. Ovarium biloculare. Stylus in stigmata dua, rarius tria partitus. Stamina partibus basalibus antherarum coalita. Capsula globosa vel ovata apice depressa et tunc valvis duabus latiusculis et brevibus dehiscens. Plantae perennes vel monocarpicae foliis alternis. Flores in capitulum globosum densum congesti.

Typus tribus: *Jasione* L.

Scabiosa Ulugbekii Zak. sp. n. (sect. *Asterocephalus* Coult. ser. *Isetenses* Bobr.)

A ceteris „*Isetensibus*“ differt caulibus parum foliatis breviter pubescentibus, longe denseque albopilosis foliis pinnatisectis albopilosis, lobis linearibus subfiliformibus.

Asia Media, Pamiroalaj; in vicinitate urbis Samarkand, prope Tubkhana non procul ab Urgut, 14 VI 1935 leg. Pimenov; sp. auth. in Leninopoli conservatur.

INDEX ALPHABETICUS*

nominum specierum atque synonymorum plantarum in tomo XXIV Florae URSS commemoratarum

* [This index has been reproduced photographically from the Russian original.]

** [Russian page numbers appear in the left-hand margin of the text.]

VEGETATION REGIONS OF THE USSR

Abbreviated name	Full name
I. Arctic	
1. Arc. Eur	Arctic (European part)
2. Nov. Z.	Novaya Zemlya
3. Arc. Sib.	Arctic (Siberia)
4. Chuk	Chukchi
5. An	Anadyr
II. European part	
6. Kar.-Lap.	Karelia-Lapland
7. Dv.-Pech.	Dvina-Pechora
8. Balt.	Baltic States
9. Lad.-Ilm.	Ladoga-Il'men
10. U. V.	Upper Volga
11. V.-Kama	Volga-Kama
12. U. Dnp.	Upper Dnieper
13. M. Dnp.	Middle Dnieper
14. V.-Don.	Volga-Don
15. Transv.	Transvolga area
16. U. Dns.	Upper Dniester
17. Bes.	Bessarabia
18. Bl.	Black Sea area
19. Crim.	Crimea
20. L. Don	Lower Don
21. L. V.	Lower Volga
III. Caucasus	
22. Cisc.	Ciscaucasia
23. Dag.	Dagestan
24. W. Transc.	Western Transcaucasia
25. E. Transc.	Eastern Transcaucasia
26. S. Transc.	Southern Transcaucasia
27. Tal.	Talysh
IV. West Siberia	
28. Ob	Ob region (from the eastern slopes of the Urals to the Yenisei River)
29. U. Tob.	Upper Tobol
30. Irt.	Irtysh
31. Alt.	Altai

V. East Siberia

32. Yenis	Yenisei
33. Lena-Kol.	Lena-Kolyma
34. Ang.-Say.	Angara River-Sayans
35. Dau	Dauria

VI. Far East

36. Kamch	Kamchatka
37. Okh.	Okhotsk
38. Ze.-Bu.	Zeya-Bureya
39. Uda	Udar River area
40. Uss.	Ussuri
41. Sakh.	Sakhalin

VII. Soviet Central Asia

42. Ar.-Casp.	Aral-Caspian
43. Balkh.	Lake Balkhash area
44. Dzu-Tarb.	Dzungaria-Tarbagatai
45. Kyz. K.	Kyzyl-Kum
46. Kara K.	Kara-Kum
47. Mtn. Turkm.	Mountainous part of Turkmenistan
48. Amu D.	Amu Darya
49. Syr D.	Syr Darya
50. Pam.-Al.	Pamir-Alai
51. T. Sh.	Tien Shan

Accepted Regions for Indication of General Distribution of Species Appearing in "Flora of the U. S. S. R."

I. Arc.	Arctic (Spitsbergen, Greenland and farther)
II. Scand.	Scandinavia (Norway, Denmark, Sweden, Finland)
III. Centr. Eur.	Central Europe (Germany, Poland, Czechoslovakia, Hungary, Austria, Switzerland)
IV. Atl. Eur.	Atlantic Europe (Netherlands, Belgium, Britain, France, Portugal)
V. Med.	Mediterranean (including North Africa)
VI. Bal.-As. Min.	Balkan Peninsula and Asia Minor
VII. Arm.-Kurd.	Lesser Armenia and Kurdistan
VIII. Iran	Iran and Afghanistan
XI. Ind.-Him.	India and Himalayas
X. Dzu.-Kash	[Dzungaria-Kashgar area] Eastern or Chinese Turkestan (Sinkiang)
XI. Mong.	Mongolia
XII. Jap.-Ch.	Japan and China
XIII. Ber.	North American coast of the Bering Sea
XIV. N. Am.	North America (U. S. A. and Canada)
XV. Tib.	Tibet

Other Geographical Abbreviations

Afr.	Africa
Aust.	Australia
Centr.	Central
E.	East(ern)
Gr.	Great, Greater
I.	Island
Is.	Islands
Mt.	Mount
Mts.	Mountains
N.	North(ern)
R.	River
S.	South(ern)
W.	West(ern)

TRANSLATOR'S NOTE

1. The Russian term "Srednyaya Aziya" is, in English, Central Asia (or Soviet Central Asia). Therefore the term Middle Asia has been used for Russian "Tsentral'naya Aziya," which is non-Soviet inner Asia, comprising western China (Sinkiang and Tibet) and Mongolia.

2. According to Russian usage, the European part of the USSR is "eastern Europe." Therefore "western Europe" includes the whole of Europe outside the USSR.

EXPLANATORY LIST OF ABBREVIATIONS OF RUSSIAN INSTITUTIONS AND PERIODICALS APPEARING IN THIS TEXT

Abbreviation	Full name (transliterated)	Translation
Bot.-Geogr. issled. v Turkest	Botaniko-geograficheskie issledovaniya v Turkestane	Botanical and Geographical Investigations in Turkestan
Bot. Mat. Gerb. Bot. inst. AN SSR	Botanicheskie Materialy Gerbariya Botanicheskogo instituta AN SSSR	Botanical Materials of the Herbarium of the Botanical Institute of the Academy of Sciences of the USSR
Bot. Mat. Gerb. Gl. Bot. Sada	Botanicheskie Materialy Gerbariya Glavnogo Botanicheskogo Sada	Botanical Materials of the Herbarium of the Main Botanical Gardens
Bot. zap. SPb. univ.	Botanicheskie zapiski Sankt-Peterburgskogo universiteta	Botanical Notes of St. Petersburg University
Bot. zhurn. SSSR	Botanicheskii zhurnal SSSR	Botanical Journal of the USSR
Byull. Glavn. Bot. Sada	Byulleten' Glavnogo Botanicheskogo Sada	Bulletin of the Main Botanical Gardens
Byull. Obshch. lyubit. estestvozn., antrop. i etnogr.	Byulleten' Obshchestva lyubitelei estestvoznaniya, antropologii i etnografii	Bulletin of the Naturalists', Anthropologists' and Ethnographers' Society
Byull. Voronezh. obshch. estestv.	Byulleten' Voronezhskogo obshchestva estestvoispytatelei	Bulletin of the Voronezh Society of Naturalists
Dendr.	Dendrarii	Arboretum
Der. i kust	Derev'ya i kustarniki	Trees and Shrubs
Der. i kust. Kavk.	Derev'ya i kustarniki Kavkaza	Trees and Shrubs of the Caucasus
Dikie polezn. i tekhnich. rasteniya SSSR	Dikie poleznye i tekhnicheskie rasteniya SSSR	Useful Wild Plants and Industrial Crops of the USSR
Dikorastushchie r. Kavkaza, ikh rasprostranenie, svoistva i primenenie	Dikorastushchie rasteniya Kavkaza, ikh rasprostranenie, svoistva i primenenie	Wild Plants of the Caucasus, Their Distribution, Properties and Uses
Dokl. AN Azerb. SSR	Doklady Akademii Nauk Azerbaidzhanskoi SSR	Reports of the Academy of Sciences of the Azerbaijan SSR

Fl.	Flora	Flora
Fl. Abkh.	Flora Abkhazii	Abkhasian Flora
Fl. Almat. Zapovedn.	Flora Alma-Atinskogo Zapovednika	Flora of the Alma-Ata Reserve
Fl. Alt.	Flora Altaya	Altai Flora
Fl. Alt. i Tomsk. gub.	Flora Altaiskoi i Tomskoi gubernii	Flora of Altai and Tomsk Provinces
Fl. Az. Ross.	Flora Aziatskoi Rossii	Flora of Asiatic Russia
Fl. Evrop. Rossii	Flora Evropeiskoi Rossii	Flora of European Russia
Fl. Gruzii	Flora Gruzii	Georgian Flora
Fl. Kamch.	Flora Kamchatki	Kamchatkan Flora
Fl. Kavk.	Flora Kavkaza	Caucasian Flora
Fl. Man'chzh.	Flora Man'chzhurii	Manchurian Flora
Fl. Mosk. gub.	Flora Moskovskoi gubernii	Flora of Moscow Province
Fl. Poles'ya	Flora Poles'ya	Flora of Polesie
Fl. Sev. Kraya	Flora Severnogo Kraya	Flora of the Northern Territory
Fl. Sakh.	Flora Sakhalina	Flora of Sakhalin
Fl. Sib.	Flora Sibiri	Siberian Flora
Fl. Sib. i Dal'n. Vost.	Flora Sibiri i Dal'nego Vostoka	Flora of Siberia and the Far East
Fl. Sr. i Yuzhn. Ross.	Flora Srednei i Yuzhnoi Rossii	Flora of Central and Southern Russia
Fl. Sr. Ross.	Flora Srednei Rossii	Flora of Central Russia
Fl. Tadzhik.	Flora Tadzhikistana	Flora of Tadzhikistan
Fl. Talysh.	Flora Talysha	Talysh Flora
Fl. Tsentr. Kazakhst.	Flora Tsentral'nogo Kazakhstana	Flora of Central Kazakhstan
Fl. Vost. Evr. Ross.	Flora Vostochnoi Evropeiskoi Rossii	Flora of East European Russia
Fl. Yugo-Vost.	Flora Yugo-Vostoka	Flora of Southeast
Fl. Yugo-zap. Ross.	Flora Yugo-zapadnoi Rossii	Flora of Southwest Russia
Fl. Yur. Bot. sada	Flora Yur'evskogo botanicheskogo sada	Flora of Yur'ev Botanical Garden
Fl. Zap. Sib.	Flora Zapadnoi Sibiri	Flora of West Siberia
Gerb. donsk. fl.	Gerbarii donskoi flory	Herbarium of Don Flora
Gerb. Orlovsk. gub.	Gerbarii Orlovskoi gubernii	Herbarium of Orel Province
Gerb. Ukr. fl.	Gerbarii Ukrainskoi flory	Herbarium of Ukrainian Flora
GRF	Gerbarii Russkoi Flory	Herbarium of Russian Flora
Ill. Fl. Mosk. gub.	Illyustrirovannaya Flora Moskovskoi gubernii	Illustrated Flora of Moscow Province
Izv. AN SSSR	Izvestiya AN SSSR	Bulletin of the Academy of Sciences of the USSR
Izv. Bot. Sada	Izvestiya Botanicheskogo Sada	Bulletin of the Botanical Gardens
Izv. Bot. Sada Petra Vel.	Izvestiya Botanicheskogo Sada Petra Velikogo	Bulletin of Peter the Great Botanical Gardens

Izv. Gl. Bot. Sada	Izvestiya Glavnogo Botanicheskogo Sada	Bulletin of the Main Botanical Gardens
Izv. Kavk. Muzeya	Izvestiya Kavkazskogo Muzeya	Bulletin of the Caucasian Museum
Izv. Kazakhst. fil. AN SSSR	Izvestiya Kazakhstanskogo Filiala Akademii Nauk SSSR	Bulletin of the Kazakhstan Branch of the Academy of Sciences of the USSR
Izv. Kievsk. Bot. Sada	Izvestiya Kievskogo Botanicheskogo Sada	Bulletin of the Kiev Botanical Gardens
Izv. Obshch. lyubit. estestvozn., antrop. i etnogr.	Izvestiya Obshchestva lyubitelei estestvoznaniya, antropologii i etnografii	Bulletin of the Naturalists', Anthropologists' and Ethnographers' Society
Izv. Obshch. Sadov.	Izvestiya Obshchestva Sadovodov	Bulletin of the Horticulturists' Society
Izv. Tadzhik. Bazy AN SSSR	Izvestiya Tadzhikskoi Bazy Akademii Nauk SSSR	Bulletin of the Tadzhikistan Base of the Academy of Sciences of the USSR
Konsp. rast. okr. Khar'kova	Konspekt rastenii okruga Khar'kova	Compendium of Plants of Kharkov District
Korm. rast. Estestv. senokosov i pastb. SSSR	Kormovye rasteniya estestvennykh senokosov i pastbishch SSSR	Fodder Plants of Natural Hay Meadows and Pastures of the USSR
Mat. (dlya) Fl. Kavk.	Materialy dlya Flory Kavkaza	Material on Caucasian Flora
Mat. (dlya) fl. Sredn. Azii	Materialy dlya flory Srednei Azii	Materials on Soviet Central Asian Flora
Mat. (dlya) Fl. stepei Khersonsk. Gub.	Materialy dlya Flory stepei Khersonskoi Gubernii	Materials on the Flora of Kherson Province Steppes
Nov. obozr.	Novoe obozrenie	New Review
Ob. rast. Kievsk. uch. okr.	Obzor rastitel'nosti Kievskogo uchebnogo okruga	Survey of Vegetation in the Kiev Educational District
Obz. Krym.-Kavk. *Medicago*	Obzor Krymsko-Kavkazskogo *Medicago*	Survey of Crimean-Caucasian *Medicago*
Och. obozr. i fl. Karpat	Ocherki rastitel'nosti i flory Karpat	Survey of Carpathian Vegetation and Flora
Ocherk. Tifl. fl.	Ocherki Tiflisskoi flory	Survey of Tiflis [Tbilisi] Flora
Opis. Amur. obl.	Opisanie Amurskoi oblasti	Description of the Amur Region
Opis. ist. razv. fl. vost. Tyan'-Shanya	Opisanie istorii razvitiya flory vostochnogo Tyan'-Shanya	Description of the History of the Development of Flora of Eastern Tien Shan
Opis. nov. rast. Turk.	Opisanie novykh rastenii Turkestana	Description of New Plants of Turkestan
Opis. nov. vidov	Opisanie novykh vidov	Description of New Species
Opred. der. i kust.	Opredelitel' derev'ev i kustarnikov	Key to Trees and Shrubs

Opred. rast. Dal'nevost. kr.	Opredelitel' rastenii Dal'nevostochnogo kraya	Key to Plants of the Far Eastern Territory
Opred. rast. Kavk.	Opredelitel' rastenii Kavkaza	Key to Caucasian Plants
Opred. vyssh.	Opredelitel' vysshikh rastenii	Key to Higher Plants
Opred. (vyssh.) rasten. Evrop. chasti SSSR	Opredelitel' (vysshikh) rastenii Evropeiskoi chasti SSSR	Key to Higher Plants of the European USSR
Opyt Russko-Kavk. Fl.	Opyt Russko-Kavkazskoi Flory	Attempted Russian-Caucasian Flora
Perech. rast. Turk.	Perechen' rastenii Turkmenii	List of Turkmenian Plants
Pochv. eksped. v bass. r. Syr-Dar'i i Amu-Dar'i	Pochvennaya ekspeditsiya v basseiny rek Syr-Dar'i i Amu-Dar'i	Soil Science Expedition to the Syr-Darya and Amu-Darya River Basins
Priroda	Priroda	Nature
Protok. Zased. Kievsk. Obshch. Estestv.	Protokol Zasedaniya Kievskogo Obshchestva Estestvoispytatelei	Protocol of a Conference of Kiev Naturalists' Society
Putesh.	Puteshestviya	Travels
Rast. i fl. Karp.	Rasteniya i flora Karpat	Plants and Flora of the Carpathians
Rast. letn. pastb. Gandzh.	Rasteniya letnikh pastbishch Gandzhi	Vegetation of Gandzha [now Kirovabad] Summer Pastures
Rast. res. Turkm.	Rastitel'nye resursy Turkmenii	Plant Resources of Turkmenia
Rast. resursy Kavkaza	Rastitel'nye resursy Kavkaza	Plant Resources of the Caucasus
Rast. Sib.	Rastitel'nost' Sibiri	Vegetation of Siberia
Rast. Sr. Az.	Rastitel'nost' Srednei Azii	Vegetation of Soviet Central Asia
Rast. Turkest.	Rastitel'nost' Turkestana	Vegetation of Turkestan
Rast. Zakasp. obl.	Rastitel'nost' Zakaspiiskoi oblasti	Vegetation of the Transcaspian Region
Rastit. Kavk.	Rastitel'nost' Kavkaza	Vegetation of the Caucasus
Rastit. pokrov. vost. Pamira	Rastitel'nyi pokrov vostochnogo Pamira	Plant Cover of the Eastern Pamirs
Rastit. syr'e Kazakhst.	Rastitel'noe syr'e Kazakhstana	Plant Resources of Kazakhstan
Rastit. zapovedn. Guralash i Zaaminsk. lesn. ugodii	Rastitel'nost' zapovednika Guralash i Zaaminskikh lesnykh ugodii	Vegetation of Guralash Reserve and Zaamin Forest Lands
Rezul't dvukh puteshevstv. na Kavk.	Rezul'taty dvukh puteshestvii na Kavkaz	Results of Two Travels to the Caucasus

Russk. Fl.	Russkaya Flora	Russian Flora
Russk. lek. rast.	Russkie lekarstvennye rasteniya	Russian Medicinal Plants
Sbor, sushka i raz. lek. rast.	Sbor, sushka i razvitie lekarstvennykh rastenii	Gathering, Drying and Development of Medicinal Plants
Sorn. rast. SSSR	Sornye rasteniya SSSR	Weed Plants of the USSR
Sots. Rastenie-vodstvo	Sotsialisticheskoe Rastenievodstvo	Socialist Plant Growing
Sov. Bot.	Sovetskaya Botanika	Soviet Botany
Sov. Farmats.	Sovetskaya Farmatsevtika	Soviet Pharmaceutics
Spis. rast.	Spisok rastenii	List of Plants
Spis. Rast. Krymsk. Zapovedn.	Spisok Rastenii Krymskogo Zapoved-nika	List of Plants of the Crimean Reserve
Tr. Bot. inst. AN SSSR	Trudy Botanicheskogo instituta AN SSSR	Transactions of the Botanical Institute of the Academy of Sciences of the USSR
Tr. Bot. Inst. Azerb. Filiala Akad. Nauk.	Trudy Botanicheskogo Instituta Azerbaid-zhanskogo Filiala Akademii Nauk	Transactions of the Botanical Institute of Azerbaijan Branch of the Academy of Sciences
Tr. Bot. Sada	Trudy Botanicheskogo Sada	Transactions of the Botanical Gardens
Tr. Bot. Sada Yur'evsk. Univ.	Trudy Botanicheskogo Sada Yur'evskogo Universiteta	Transactions of the Botanical Gardens of Yur'ev [now Tartu] University
Tr. Byuro prikl. Bot.	Trudy Byuro po prikladnoi botanike	Transactions of the Bureau of Applied Botany
Tr. Dal'nevost. bazy AN SSSR	Trudy Dal'nevostochnoi bazy AN SSSR	Transactions of the Far Eastern Base of the Academy of Sciences of the USSR
Tr. Inst. nov. lub. syr'ya	Trudy Instituta novogo lubyanogo syr'ya	Transactions of the Institute of New Fiber Raw Materials
Tr. Nauk.-Doslid. Inst. Bot. Khar. Derzh. Univ.	Trudy naukovo-doslid-noho instytutu botaniky Kharkivs'koho Derzhav-noho Universytetu	Transactions of the Botanical Research Institute of the Kharkov State University
Tr. Obshch. isp. prir. Khark'k. univ.	Trudy Obshchestva ispytatelei prirody Khar'kovskogo universiteta	Transactions of the Natural-ists' Society of Kharkov University
Tr. Obshch. sadov. v. Odesse	Trudy Obshchestva sadovodov v Odesse	Transactions of the Odessa Horticulturists' Society
Tr. odessk. obshch. sadov.	Trudy odesskogo obshchestva sadovodov	Transactions of Odessa Horti-culturists' Society
Tr. Peterb. obshch. estest-voisp.	Trudy Peterburgskogo obshchestva estest-voispytatelei	Transactions of St. Peters-burg Naturalists' Soviety

Tr. pochv.-bot. eksp. Peresl. upr.	Trudy pochvennobotanicheskoi ekspeditsii Pereslavskogo upravleniya	Transactions of the Soil-Botanical Expedition of Pereslavl Administration
Tr. po geobot. obsled. pastb. Azerb.	Trudy po geobotanicheskim obsledovaniyam pastbishch Azerbaidzhana	Transactions of Geobotanical Investigations of Azerbaijan SSR Pastures
Tr. Odessk. otd. R. obshch. sadov.	Trudy Odesskogo otdeleniya Rossiiskogo obshchestva sadovodov	Transactions of Odessa Branch of the Russian Horticulturists' Society
Tr. prikl. bot. (gen. i sel.)	Trudy po prikladnoi botanike, genetike i selektsii	Transactions of Applied Botany, Genetics and Selection
Tr. Ross. Obshch. sadov.	Trudy Rossiiskogo obshchestva sadovodov	Transactions of the Russian Horticulturists' Society
Tr. SAGU	Trudy Sredneaziatskogo Gosudarstvennogo Universiteta	Transactions of the Soviet Central Asian State University
Tr. Sarat. obshch. estestvoisp.	Trudy Saratovskogo obshchestva estestvoispytatelei	Transactions of the Saratov Naturalists' Society
Tr. Sil'skogospod. komit. bot.	Trudy sil'skohospodar'skoho komiteta botaniky	Transactions of the Botanical Agricultural Committee
Tr. SPb. obshch. estestv.	Trudy Sankt-Peterburgskogo obshchestva estestvoispytatelei	Transactions of the St. Petersburg Naturalists' Society
Tr. Tadzh. bazy AN SSSR	Trudy Tadzhikskoi bazy AN SSSR	Transactions of the Tadzhikistan Base of the Academy of Sciences of the USSR
Tr. Tbil. bot. inst.	Trudy Tbilisskogo botanicheskogo instituta	Transactions of Tbilisi Botanical Institute
Tr. Tbil. (or Tifl.) bot. sada	Trudy Tbilisskogo (Tiflisskogo) botanicheskogo sada	Transactions of the Tbilisi (Tiflis) Botanical Garden
Tr. Turkmensk. bot. sada	Trudy Turkmenskogo botanicheskogo sada	Transactions of the Turkmenian Botanical Garden
Tr. Turk. nauchn. obshch.	Trudy Turkmenskogo nauchnogo obshchestva	Transactions of the Turkmenian Scientific Society
Uchen. Zapiski Kazansk. Gos. Univ.	Uchenye Zapiski Kazanskogo Gosudarstvennogo Universiteta	Scientific Reports of the Kazan State University
Vest. Akad. Nauk. (or AN) Kazakhsk. SSR	Vestnik Akademii Nauk Kazakhskoi SSR	Bulletin of the Academy of Sciences of the Kazakh SSR
Vestn. estestv. nauk	Vestnik estestvennykh nauk	Bulletin of Natural Sciences
Vestn. Ross. Obshch. sadov	Vestnik Rossiiskogo Obshchestva sadovodov	Bulletin of the Russian Horticulturists' Society
Vest. Tifl. bot. sada	Vestnik Tiflisskogo botanicheskogo sada	Bulletin of Tiflis Botanical Garden

Visn. Kyyivsk. bot. sadu	Visnyk Kyyivs'koho botanichnoho sadu	Bulletin of the Kiev Botanical Garden
Vizn. (or Vznachn.) rosl. URSR	Viznachnyk roslyn URSR	Key to Plants of the Ukrainian SSR
V obl. polupustyni	V oblasti polupustyni	(In the) Semidesert Region
Yadov. rast. lugov i pastb.	Yadovitye rasteniya lugov i pastbishch	Poisonous Plants of Meadows and Pastures
Yubil. sbornik V. L. Komarovu	Yubileinyi Sbornik Posvyashchennyi V. L. Komarovu	Jubilee Collection Dedicated to V. L. Komarov
Zam. po sist. i geogr. rast. Tbil. bot. inst.	Zametki po sistematike i geografii rastenii Tbilisskogo botanicheskogo instituta	Notes on Taxonomy and Geography of Plants of the Tbilisi Botanical Institute
Zam. o Rast. Russk. Flory	Zametki o Rasteniyakh Russkoi Flory	Notes on Plants of the Russian Flora
Zam. po fl. EL'T	Zametki po flore El'tona	Notes on the Flora of Elton
Zap. Kievsk. Obshch. Estestv.	Zapiski Kievskogo Obshchestva Estestvoispytatelei	Reports of the Kiev Society of Naturalists
Zap. Kyyivsk. Inst. Nar. Osv.	Zapysky Kyyivs'koho Instytutu Narodnoho Osvichennya	Reports of the Kiev Institute of Public Education
Zap. Nauchno-Prikl. Otd. Tifl. Sada	Zapiski Nauchno-Prikladnogo Otdeleniya Tiflisskogo Sada	Reports of the Applied Sciences Section of the Tiflis [Tbilisi] Botanical Garden
Zap. NOVO-ROSS. obshch. Estestv.	Zapiski Novorossiiskogo obshchestva Estestvoispytatelei	Reports of the Novorossiisk Society of Naturalists
Zap. Russk. geogr. obshch.	Zapiski Russkogo geograficheskogo obshchestva	Reports of the Russian Geographical Society
Zhurn. Bot. obshch.	Zhurnal Botanicheskogo obshchestva	Journal of the Botanical Society
Zhurn. opytn. agron. Yugo-Vost.	Zhurnal opytnoi agronomii Yugo-Vostoka	Journal of Experimental Agronomy of the Southeast